# The Ecological Web

# The Ecological Web

## More on the Distribution and Abundance of Animals

H. G. Andrewartha

and

L. C. Birch

The University of Chicago Press
Chicago and London

H. G. ANDREWARTHA is professor emeritus of zoology at the University of Adelaide and honorary visiting research fellow in the Department of Entomology at the Waite Agricultural Research Institute. L. C. BIRCH is the Challis Professor of Biology in the School of Biological Sciences, the University of Sydney.

The University of Chicago Press, Chicago 60637
The University of Chicago Press, Ltd., London

© 1984 by the University of Chicago
All rights reserved. Published 1984
Printed in the United States of America

93 92 91 90 89 88 87 86 85 84      5 4 3 2 1

LIBRARY OF CONGRESS CATALOGING IN PUBLICATION DATA

Andrewartha, H. G. (Herbert George), 1907–
    The ecological web.

    Bibliography: p.
    Includes index.
    1. Animal ecology. 2. Zoogeography. I. Birch,
L. Charles, 1918-    . II. Title
QH541.A524   1984     591.5          84-70
ISBN 0-226-02033-9

# Contents

# Preface

This book covers much the same ground as *The Distribution and Abundance of Animals*. During the thirty years that have elapsed since the publication of that volume there have been some noteworthy advances in the theory of population ecology. For our part, we have learned to think more rigorously about this difficult subject. We hope this book will be accepted as a new contribution to population ecology—not merely a revision of our previous work.

In *The Distribution and Abundance of Animals* the chapter on general theory follows the chapter that summarizes fourteen empirical studies of populations. In this book the order is reversed: theory precedes practice. To put theory last would imply that theory is little more than a summary of empirical knowledge, which is to approach perilously close to the philosophy of induction by simple enumeration (sec. 9.02). We did not intend to give this impression in 1954 and we wish to avoid it now, because we hold that good theory depends not only on knowledge but also on subjective judgment, supported by insight and imagination.

We have tried to use technical terms sparingly. Nevertheless, there are a number of words whose technical meanings must be strictly observed for the sake of a consistent argument. Most of the technical words are defined in chapters 1 and 2, and there are a few more definitions in later chapters. Two important words, "habitat" and "niche" have been used widely and loosely in the ecological literature, so let us make it clear from the beginning that we follow Charles Elton's original usage. A "habitat" is a place that might be habitable for the animal whose ecology is being studied. The boundaries of the habitat and the qualities that determine the boundaries are fixed arbitrarily by the ecologist (Elton 1949; Elton and Miller 1954). A "niche" is a quality of the animal which determines how the animal responds to its environment—for example, the ability to live on a particular sort of food or to coexist with a particular sort of predator. For every component of environment there is a corresponding niche (Elton 1927, 63). In a nutshell, the habitat makes the environment; the niches of the animal are its response to the environment.

Insofar as we have achieved rigor in our analysis of environment, a large share of the credit belongs to Dr. B. S. Niven. Fortunately for us, while this book was still in the formative stages, Dr. Niven saw an opportunity to apply the rules of formal logic to our proposed analysis of environment. After much research on her part, after many hours' discussion with her and with the benefit of her advice on a number of specific problems, there emerged the attempt at a rational analysis of environment which forms Part 1. We are grateful to Dr. Niven for her help and for writing the Appendix and allowing us to publish it here. The subject may be followed up in two papers (Niven 1980, 1982).

All references to Andrewartha and Birch (1954) in this book are to the 1954 complete edition. However, chapters 3, 6, 12, 13, and 14 of *The Distribution and Abundance of Animals* are reprinted in Andrewartha and Birch (1982), *Selections from the Distribution and Abundance of Animals.*

We acknowledge help from a number of our colleagues. Professor P. R. Ehrlich read all the chapters. Professor T. O. Browning read chapters 1 and 8. Dr. A. J. Underwood and his students provided the information on intertidal animals given in chapter 2. Professor R. T. Paine and Dr. G. Fitt read chapter 5. Dr. R. O. Peterson provided much of the information on the moose and its predator the timber wolf cited in chapter 5. Dr. A. C. Hodson read chapter 9. Dr. P. J. den Boer provided us with unpublished data relevant to chapter 9. Dr. C. E. Taylor and Dr. P. Wellings assisted greatly with chapter 11. Dr. K. Myers and Dr. B. Cooke placed unpublished data at our disposal for chapter 12. The discussion on the fruit flies of the genus *Dacus* in chapter 14 owes much to Dr. M. A. Bateman and Dr. B. S. Fletcher. And we have benefited from discussions with many other students and colleagues during the formative stages of the book.

We are indebted to Mrs. J. Jeffery, Miss S. Warren, and Mrs. S. Suter for the painstaking preparation of the manuscript. Mrs. Jeffery made all the line drawings. Miss Warren typed and checked the text and the bibliography through many transitions. Mrs. Suter transcribed the final typescript onto word-processor diskettes, complete with coding for computerized typesetting.

# PART 1

# *The Theory of Environment*

# 1

# The Theory of Environment

## 1.0   The Scope of Ecology

In this book "ecology" means the study of the distribution and abundance of
particular species of animals or, in certain contexts, the knowledge that might come
from such studies. We build on the theory of environment that was outlined in
Andrewartha and Birch (1954); that is, we seek to explain the distribution and
abundance of species by studying the environments of individuals in natural populations.

The theory of environment depends on a definition (an axiom) that says that *the
environment of an animal consists of everything that might influence its chance to
survive and reproduce*. The expectation of life and the fecundity of the individual
are reflected in the birth rate and death rate of the population, which is reflected in
the distribution and abundance of the species. The theory is outlined by three
propositions given in section 1.1. A little detail is supplied in section 1.3; and the
detail is completed, with examples, in chapters 2–7.

## 1.1   Three Propositions That Outline the Theory of Environment

*The first proposition:* Any environment is made up of a *centrum* of directly acting
components and a *web* of indirectly acting components (fig. 1.01).

Any animal's chance to survive and reproduce depends on the probability that
certain "things" will impinge on its body, influencing its physiological condition
and thereby its chance to survive and reproduce. Changes in physiological condition might be reflected as changes in behavior, growth, or expectation of life.
"Things" is used broadly to imply any accumulation or configuration of matter or
energy. "Impinge" is also used broadly: thus a fox (a predator) may impinge on
a rabbit by eating it; leaves of a grass (food) may impinge on the rabbit by providing
nourishment. The fox and the grass are said to be "directly acting" components in
the environment because each is the immediate cause of a characteristic change in
the rabbit's condition. That is, no intermediate link can be discerned between the
environment (fox) and the animal (rabbit). So to say that the fox is a directly acting
component in the environment of the rabbit means that the fox is likely to be the
proximate cause of the rabbit's death. When a component of environment makes
its influence felt through an intermediate link or links it is said to be acting
indirectly, so its place is in the web. Note that physical contact, even aggressive

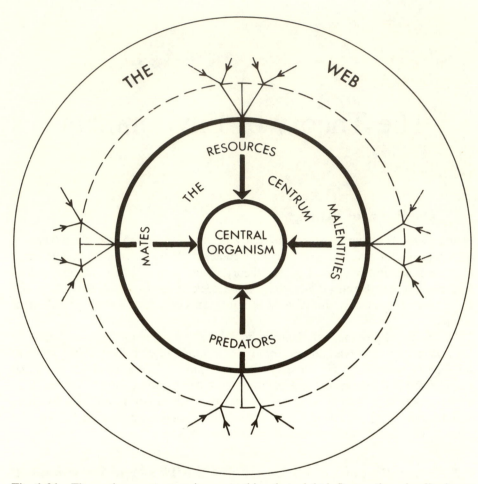

**Fig. 1.01**   The environment comprises everything that might influence the animal's chance to survive and reproduce. Only those "things" that are the proximate causes of changes in the physiology or behavior of the animal are placed in the centrum and recognized as "directly acting" components of environment. Everything else acts indirectly, that is, through an intermediary or a chain of intermediaries that ultimately influences the activity of one or other of the components in the centrum. All these indirectly acting components are placed in the web.

physical contact, is not a sufficient criterion for "directly acting". If by overt physical aggression one animal preempts the food that was coveted by another (of the same or different species), shortage of food (not aggression) will be recognized as the proximate cause of the second animal's disability. The aggressive animal will be said to have exerted its influence indirectly through food, and it will properly be placed in the web of the environment of the victim.

The probability that a rabbit will be eaten by a fox may depend on the number of foxes that are hunting rabbits in that place and on the number of obstacles that may impede the hunt (e.g., a fox may be more successful in digging out the nestlings in a friable sandy soil than in a hard rocky soil sec. 12.432). To express this idea, not only with respect to predators but quite generally for all components of environment, we speak of the "activity" of the component. In the present instance the "activity" of predation might be represented by a factor for the abundance of foxes multiplied by a factor for their success in hunting. So it becomes necessary to study the ecology of the fox in order to fulfill our study of the ecology of the rabbit. Let us imagine that the fox supports itself by eating a certain sort of bird during the season when rabbits are hard to get; and the bird eats certain insects that eat certain grasses and other herbaceous plants that are likely to be more widespread and abundant after copious rain has fallen at the right season. The rain, the plants, the insects, and the birds all have their place in the centrum of one or other of the organisms in this chain, but insofar as they influence the rabbit's chance to survive and reproduce they do so indirectly, through their influence on the distribution and abundance of foxes. The fox belongs in the centrum of the rabbit's environment, but all the other organisms and the rain belong in the web. The web is made up of many such chains (chap 2).

*The second proposition:* the centrum of the environment of any animal comprises four divisions (or "compartments"); each division houses a characteristic set of components that we call resources, mates, malentities, and predators.

This proposition is developed in section 1.31. Briefly the partitioning of the centrum is based on the function of the component in the environment. The "function" of the component is defined by (1) the influence of the component on the animal's chance to survive and reproduce and (2) the influence that the density of the parental population has on the activity of the same component in the environments of the animals of the filial population.

*The third proposition:* The web comprises a number of systems of branching chains (figs. 2.01–2.06). A link in the chain may be a living organism (or its artifact or residue) or inorganic matter or energy (chap. 2).

Any living organism that is in the web or contributes an artifact or residue to the web will have an environment of its own. Consequently, to investigate its contribution to the activity in the web requires only that it be put into its own centrum and that the standard ecological inquiry be broadened accordingly. For a link that is not a living organism or does not have such an origin, the inquiry might lead into almost any branch of science.

The web in our theory of environment is essentially the same idea that Darwin (1859, 73) developed when he wrote about "the web of complex relations". The food chains that are familiar to students of communities may be counted in Darwin's "web" and in the web of our theory too. But our theory differs from Darwin's web because we emphasize the difference between directly acting and indirectly acting components of environment, and we recognize nonliving components of environment in both centrum and web.

We conceive of the web as a number of branching chains. In practice they are uncovered as each one is traced away from its particular component in the centrum. But the action flows in the opposite direction, inward toward the centrum and through the centrum to the primary animal whose ecology is being studied. Because the web includes everything that might influence the activity of the centrum, it is convenient to have one word that can be applied quite generally to any component of the web. We use "modifier", in the sense that a component in the web might modify another component in the web or a component in the centrum.

It is the purpose of population ecology to focus the ecological action on one particular species at a time. This is what distinguishes population ecology from community ecology, which is more concerned to identify the ecological action in the community at large. We believe it is this difference that makes population ecology the sharper tool when it comes to pursuing such practical goals as pest control, wildlife management and conservation.

Because population ecology focuses on one species at a time, the questions that arise in population ecology are different from those of community ecology. Population ecology is not primarily concerned, for example, with species diversity, community stability, energy pathways through different trophic levels, ecological succession, and "trophic complexity".

## 1.2   The Origin and Growth of the Theory of Environment

Despite the fundamental differences between the goals and methods of ecology and evolution, both sciences depend on the concept of environment, or the struggle for existence, as developed by Darwin. When we first put forward our theory of environment (Andrewartha and Birch 1954) we leaned heavily on Darwin. The three brief passages from *The Origin of Species* that we quote below show that "food", "weather" and "other animals" were seen by Darwin as parts of the "web of complex relations" that shaped the struggle for existence; we borrowed these three components of environment from Darwin and added a fourth of our own contriving, which we called "a place in which to live". Darwin wrote:

I use the term Struggle for Existence in a large and metaphorical sense, including dependence of one being on another, and including (which is more important) not only the life of the individual, but success in leaving progeny. Two canine animals, in time of dearth, may truly be said to struggle with each other which shall get food and live. But a plant on the edge of a desert is said to struggle for life against drought, though more properly it may be said to be dependent on the moisture. (Darwin 1859, 62).

A struggle for existence inevitably follows from the high rate at which all organic beings tend to increase. Every being, which during its natural lifetime produces several eggs or seeds, must suffer destruction during some period of its life, and during some season or occasional year, otherwise, on the principle of geometrical increase, its numbers would quickly become so inordinately great that no country could support the product. Hence, as more individuals are produced than can possibly survive, there must in every case be a struggle for existence, either one individual with another of the same species, or with individuals of distinct species, or with the physical conditions of life. (Darwin 1859, 63).

In looking at Nature it is most necessary to keep the foregoing considerations always in mind—never to forget that every single organic being around us may be said to be striving to the utmost to increase its numbers; that each lives by a struggle at some period of its life; that heavy destruction inevitably falls either on the young or the old during each generation or at recurrent intervals. (Darwin 1859, 66).

From these passages it is clear that Darwin saw the *individual* "struggling" against what we now call its "environment" in an effort to survive and to leave progeny. In Andrewartha and Birch (1954), again following Darwin's lead, we defined the environment of an animal as everything that might influence the animal's chance to survive and reproduce. We use the same definition in this book. We think of the mean of the environments that are experienced by the individuals that constitute the population. So we speak of "the animal" in the singular, meaning the "average" animal that has experienced the average environment. In this way we avoid the error of circularity when analyzing a whole class of important reactions associated with overcrowding.

This definition relates to a hypothetical animal that is said to represent the population because the values for such statistics as "expectation of life" or "fecundity" that are attributed to it are, by definition, the same that would be gotten by sampling the population. Expectation of life and "expected fecundity" for the "average" individual are calculated from "age-specific death-rate" and "age-specific birth-rate" as estimated by sampling the population. In theory, we sum these statistics, expressed as probabilities, when we speak of the animal's "chance to survive and reproduce". We take for granted their relevance to the corresponding statistics for the population.

We frequently have occasion to refer to this hypothetical "average" animal which represents the population. When no ambiguity threatens we call it "the animal"; when there is risk of ambiguity we call it the "primary animal" to distinguish it from all those others that are in the web.

In 1954 we reviewed the current concepts of environment and found them anthropocentric; we criticized this approach as too "descriptive". By contrast we were seeking an analysis that would depend more on how the environment influenced the animal's chance to survive and reproduce and less on what the environment looked like in the eyes of a man. We said we were seeking a "functional" analysis of environment. This is still our aim; we hope to have approached it more closely in this book.

The first step toward a more functional concept of environment was taken by Browning (1962, 1963, 90) who defined two new components—"resources" and "hazards"—allowing him to discard two of the original ones: "food" because it was too narrow and "a place in which to live" because it was too diffuse. We have extended the scope of hazards a little and now call this component "malentities" but the idea is essentially the same. Maelzer (1965) drew attention to the important distinction between the directly acting and indirectly acting components of environment that we now call the centrum and the web. Niven (1980) in a seminal paper in which she used the concepts of mathematical logic to define the basic principles of the theory, sharpened the boundaries between the four divisions of the centrum. And in doing so she greatly clarified our thinking about all components of environment, but especially those that popularly constitute "the weather".

Insofar as "progress" contains an original idea, it stems, as all originality must, from the fundamental epistemological process of deciding what shall be called "alike" and what "unlike". There is the story from the history of mathematics, perhaps apocryphal but nevertheless instructive, that Dantzig (1947, 101) told about Pythagoras.

Pythagoras believed that a number was a number only if it could be represented by a heap of stones or by other discrete objects that could be counted. He set a religious value on *rational* numbers because he knew many sets of such numbers that defined geometric or numerical "perfections", so perfect that it seemed they must have been designed by a god. For example, he knew that a perfect right angle was defined by any set of three numbers such that $a^2 + b^2 = c^2$. Then it was shown that for the sides and diagonal of a square the ratios were 1, 1, $\sqrt{2}$. Now the square root of 2 cannot be represented by a heap of stones and cannot be counted. The very idea of a "number" that could not be counted was a sacrilege to Pythagoras, and he tried to suppress all knowledge of and thought about this terrible idea, calling it unutterable. Dantzig (1947, 101) quotes Proclos:

It is told that those who first brought out the irrationals from concealment into the open perished in shipwreck, to a man. For the unutterable and the formless must need to be concealed. And those who uncovered and touched this image of life were instantly destroyed and shall remain forever exposed to the play of the eternal waves.

The first protagonists of irrational numbers may have been destroyed as Proclos said, but their idea lived on. Within a hundred years there came to be general agreement among scholars that irrationals (like $\sqrt{2}$) should be recognized as numbers because they can be multiplied and divided like the rationals even though they cannot be represented by a heap of stones and counted. This consensus marked a big step forward in the power of mathematics. The whole of our intellectual evolution is punctuated by such milestones, usually established at the cost of great anguish for conservatives like Pythagoras. At several places in this book we draw attention to occasions when ecological progress may have been held up because an established theory was too strongly entrenched (secs. 3.131, 9.34, 9.4).

## 1.3   The Theory of Environment in Greater Detail

In this section we try to explain the theory of environment in greater detail, beginning with the idea of the centrum. Chapters 2–7 are also largely devoted to this purpose. We are seeking a general but realistic theory of environment, remembering that the ultimate aim is to have a workable framework of theory within which to ask questions that will help explain the distribution and abundance of animals in natural populations. The sort of question may reflect the theory that is currently accepted (chaps. 2, 10; sec. 9.0).

### 1.31   The Centrum

There are two questions to be asked about any directly acting component of environment: (1) Does the animal's chance to survive and reproduce increase or decrease as the environmental component increases in abundance (or "activity")? (2) Does the abundance (or "activity") of the environmental component increase or decrease as the population increases? The answer to the first question permits only two alternatives. The reaction must be either positive or negative because, by definition, only those "things" that influence the animal's chance to survive and reproduce are counted in its environment (fig. 1.02). The answer to the second

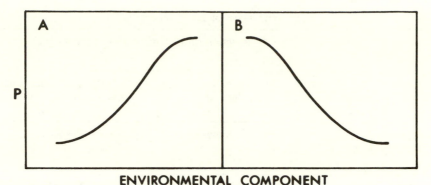

**ENVIRONMENTAL COMPONENT**

**Fig. 1.02** The influence of the environmental component on the probability that the animal will survive and reproduce (P) is positive in A and negative in B. In each cell of the diagram the activity or abundance of the environmental component increases to the right.

question permits three alternatives because there are important components of environment that do not react to the density of the population (fig. 1.03B). For example, the amount of fruit produced by a fruit tree is independent of the number of fruit flies that ate the fruit in the previous generation. Or the supply of cow dung in one generation is independent of the number of dung beetles that fed on the dung in the previous generation. The reaction is said to be zero when the component does not react to the density of the population. It is said to be positive when the curve slopes upwards from left to right (figs. 1.02A, 1.03A) and negative when the curve is downward from left to right (figs. 1.02B, 1.03C).

Six interactions can be described by different combinations of the two reactions in figure 1.02 with the three reactions in figure 1.03. If the zero reaction as shown in figure 1.03B is confounded with the negative reaction as shown in figure 1.03C we get a 2 by 2 interaction table as shown in table 1.01. The 2 by 2 table is more convenient than the 2 by 3 table, and very little information is lost—but see section 6.2. Each cell of table 1.01 defines a directly acting component of environment. The centrum of any environment comprises these four components and only these four components—no other situation is logical.

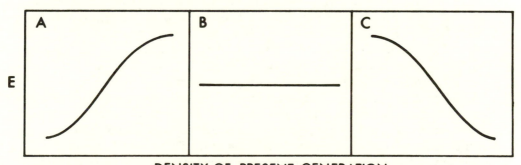

**DENSITY OF PRESENT GENERATION**

**Fig. 1.03** The influence of the density of the present generation on the activity of the environmental component (E), as it is experienced by future generations, is positive in A, zero in B, and negative in C. Components that conform to curves A and C are called interactive; those that conform to B are called noninteractive.

**Table 1.01**   Directly Acting Components of Environment

| Response of the Animal to the Component in the Centrum (fig. 1.02) | Reaction of the Component to the Animals (fig. 1.03) | |
| --- | --- | --- |
| | Activity of Component Increased (+) | Activity of Component Decreased or No Change (− or 0) |
| Expectation of life or fecundity increased (+) | Mates (chap. 4) | Resources (chap. 3) |
| Expectation of life or fecundity decreased (−) | Predators (chap. 5) | Malentities (chap. 6) |

*Note:* The directly acting components of environment are classified into the four divisions of the centrum according to the response of the animal to the component and the consequent reaction of the component to the animals. In the first instance it is the *immediate* response of the animal that is important (left side of table); in the second instance it is the *delayed* response of the component of environment that counts (right side of table). The signs +, 0, and − refer to the slope of the curves in figs 1.02 and 1.03. See text for further explanation.

To illustrate the procedure, consider the fox which eats rabbits. It is already clear that the fox is a directly acting component in the environment of the rabbit because there is no intermediary standing between rabbit and fox (sec. 1.1). So it is proper to proceed toward table 1.01. From figure 1.02 the reaction is negative, because the rabbit's chance to survive and reproduce will decline as the activity of the fox increases. From figure 1.03 the reaction is positive, because when rabbits are numerous in the present generation the numbers of foxes will build up and be likely to prey more heavily on the next generation of rabbits. Now go to table 1.01. A negative reaction in figure 1.02 and a positive reaction in figure 1.03 bring the component of environment into the bottom left cell of table 1.01, which houses "predators". If the procedure is reversed and the fox is designated the primary animal, the rabbit finds its way to the top right cell and therefore is called a resource in the environment of the fox.

But why use all this formality merely to arrive at a conclusion that was already obvious? True of the fox and the rabbit! But not so true of all the interactions that might turn up. For some of the more subtle ones the formal analysis may give conclusions that are surprisingly different from the conventional ones. We stand by the formal analysis because it gives a model that is consistently functional and therefore useful for opening up an investigation in ecology.

The four cells of table 1.01 represent the four divisions (or "compartments") of the centrum that were mentioned in section 1.1. The general names (resources, mates, malentities, and predators) are given to these major groupings of the directly acting components of environment. There can be no directly acting component of environment that does not fall into one of these categories. We devote a separate chapter to each category, so here we shall be content with a few general comments about them. The names "resources" and "mates" have well-understood colloquial meanings. Browning (1962) was the first to recognize the component of environment that we call malentities. He called it "hazards". Andrewartha (1970, 130) renamed it malentities. Malentities are to be distinguished from predators as described in section 1.42.

Colloquial English distinguishes between carnivore and herbivore, predator and parasite, and so on, but it has no word to denote generally any organism whose food includes any other sort of living organism. Yet this general meaning is important in ecology. Any organism—animal, plant, or microbe—whose food includes any other living organism, be it animal, plant, or microbe, belongs in the bottom left cell of table 1.01. We have put such organisms there and called them "predators", thereby greatly expanding the colloquial meaning of this word to include herbivores, carnivores, parasites, and pathogens.

We say that the fox is a *predator* in the environment of the rabbit; but to be precise we should say that *predation* by the fox reduces the rabbit's chance to survive and reproduce. For convenience we shall always use the concrete noun to name the component of environment, with the understanding that the concrete noun implies the *action* that would be more properly named by an abstract noun—thus predator (fox) implying predation, food (grass) implying the nutrition that comes from eating grass, heat (absorbed from a hot medium or hot body in the medium) implying the heating of the animal, with the consequent change in temperature leading to changes in physiology or behavior that influence the animal's chance to survive and reproduce.

This convention makes for easier reading by avoiding a surfeit of abstract nouns. It also gets us off the horns of another dilemma that arises when one natural entity is or gives rise to more than one component in the environment of the same animal (sec. 2.32 figs. 2.01–2.06). For example, the sheep tick *Ixodes ricinus* lives near the surface of the ground under the matted vegetation that is grazed by sheep (see summary of Milne's work in Andrewartha and Birch 1954, sec. 13.32). Once a year during its life of three years the tick may crawl up a grass stem and climb onto a passing sheep. For several days the tick clings to the sheep, engorging with blood, while the sheep wanders, perhaps a long way from where the tick was picked up. In due course the tick drops off and seeks its usual shelter under the vegetation. The sheep is both food and an agent for dispersal in the environment of the tick. The context in which we use the concrete noun "sheep" will tell whether we are implying the action of nutrition or the action of dispersal.

Another example. In inland Australia immature scale insects, *Saissetia oleae*, living on orange trees, thrive when heat absorbed from a warm medium raises their body temperature moderately, say to 20–25°C; they die when heat from a hot medium raises their body temperature extremely, say to 35–40°C. Heat absorbed from a warm medium is classified into the top right cell of table 1.01 and called a resource; heat absorbed from a hot medium is classified into the bottom right cell of table 1.01 and called a malentity. Again, there need be no ambiguity because the context will tell whether we are dealing with heating that maintains the body within the temperature range that favors healthy metabolism or raises the body temperature to a fatally high range (sec. 2.32).

Many of the "things" that we shall have to consider and allot to their proper places in the centrum bear names that are familiar to us humans as components in our own environment—food, weather, and disease to mention only a few. This is not surprising because, ecologically at least, the human animal is much more like other animals than many philosophers, theologians, or politicians would have us believe (Birch and Cobb 1981, chap. 4). But remember that in our theory of environment these words have special meanings that are different from the anthropocentric meanings of colloquial English. For example, contemplating oxygen in our own environment and realizing the fatal consequences of an acute shortage, we might be tempted to say that oxygen is very important to us. Yet in the general run of ecological texts and papers oxygen scarcely gets a mention: it is of little interest to ecologists because the supply of it is usually so reliable in the places where aerobic animals live. For many species the same is true of food and water. In general it can be said that beneficial components (fig. 1.02A) whose activity is not likely to fall below the optimum and harmful components (fig. 1.02B) whose activity is not likely to rise above a negligible threshold are unimportant. It is the highly variable ones that attract all our attention, especially those whose activity is likely to exceed such limits, or those that might be artificially manipulated so that their activity exceeds a critical limit.

## 1.32   The Web

The concept of the web is necessary because the activities of the components in the centrum have to be explained. In other words, we must study the ecologies of the organisms that are seen to be important in the centrum; and we must also seek to understand the "events" that determine the activity of the important nonliving components of the environment. This program must not be planned indiscriminately. On the contrary, we must follow our hunches in the search for "key" components, not only in the centrum but also in the web. The web and centrum are equally essential to the model. Indeed, the limelight is more often than not focused on some part of the web, as it is in the ecology of *Thrips imaginis*.

According to Davidson and Andrewartha (1948a,b) and Andrewartha (1970, 27), *T. imaginis* lives and breeds exclusively in certain flowers; a diet of pollen is essential for the growth of nymphs and the production of eggs by adults (sec. 10.12). Also, no doubt as an adaptation to a life-style based on exploiting short-lived flowers, the thrips are highly dispersive; they readily exploit the turbulence of surface breezes to go seeking new flowers as the old ones senesce. This dispersiveness pays big dividends when suitable flowers are abundant and densely distributed, but it entails a severe risk of premature death when flowers are few and sparsely distributed. Armed with this sort of knowledge of the physiology and behavior of thrips, Davidson and Andrewartha postulated that fluctuations in the population of *T. imaginis* were caused by fluctuations in their supply of food, which in turn were related to certain aspects of the weather that were thought to be important in the ecology of the plants that provided food for the thrips. They formulated an equation of the form:

$$y = a + b_1x_1 + b_2x_2 + b_3x_3 + b_4x_4,$$

where $y$ stands for the number of thrips; $x_1,...x_4$ are quantities calculated from standard meteorological records of temperature, rainfall, and evaporation and $b_1...b_4$ are constants, calculated from the empirical data; they measure the independent association between $x_1...x_4$ and $y$.

Two of the independent variates $x_2$ and $x_3$, might have had a direct influence on both plants and thrips, but the other two might have had a direct influence on the plants alone.

The equation accounted for 78% of the variance in the thrips' populations. So there seemed good reason for accepting the hypothesis that the thrips' chance to survive and reproduce depended largely on fluctuations in their supply of food, which in turn depended largely on fluctuations in certain aspects of the weather.

In this example certain aspects of weather (temperature, rainfall, and evaporation) influencing heat and water in the environments of certain plants turned out to be important components in the environment of *T. imaginis*. A prior knowledge of the natural history, physiology, and behavior of *T. imaginis* and some knowledge of the biology of the plants led Davidson and Andrewartha to recognize the importance of these components; and empirical testing confirmed the hypothesis. When the theory of environment, with its concepts of centrum and web, is used in this way we think it may often lead accurately and economically to the solution of ecological problems.

In subsequent chapters, especially 2, 5, 6, and 7, we unfold many complexities of the web. This is not merely an intellectual pastime. The purpose is to establish guidelines that might help to direct an ecological inquiry (secs. 2.0, 2.1).

## 1.4 The Sorts of Components That Might Be Found in Any Environment

The web is discussed in chapter 2. The four components of the centrum are discussed in chapters 36. Some introductory comments on resources, predators, and malentities follow in sections 1.41 and 1.42.

### 1.41 Resources

A resource is defined by its position in the top right cell of the matrix diagram that is table 1.01. Food is the most familiar resource. The others are water, oxygen, heat, and tokens. Food and water contribute the chemicals from which the body is built; and food, with oxygen, provides energy. Heat, radiant or ambient, that is absorbed by the animal may help keep the body at a temperature that favors a healthy metabolism. What we call tokens are not usually recognized in colloquial speech as resources, but they clearly are such by our definition. We speak of a token when a caterpillar or a bird, having measured the length of day, changes its body chemistry to suit the season, preparing for diapause or migration as the case may be. In this instance the length of day is the token. It is in the centrum because it impinges directly on the animal (through the appropriate photoreceptors) and causes the animal to change its body chemistry. It is a resource because, by virtue of the animal's positive response to the token (fig. 1.02), and by virtue of the token's neutral response to the animal (fig. 1.03), the token fits into the top right cell of table 1.01.

The dictionary gives token the meaning "symbol" or "sign". We use "token" in the context that the token signals to the animal that it is time to prepare for the next season or day, or the next stage in the life cycle, the reproductive cycle, the cycle of diurnal behavior, and so on. Light, especially the relative length of day and night, makes the most spectacular and most important tokens. In Andrewartha and Birch (1954, chap. 8) we discussed how light might synchronize life cycles with the seasons and synchronize the life cycles of individuals with each other, but we now see tokens a little more broadly than that.

## 1.42   Predators and Malentities

Predators and malentities between them comprise everything that conforms to the negative reaction illustrated in figure 1.02B. But table 1.01 shows that the predator's reaction to animals is positive, whereas the malentity's reaction to animals is negative or zero (sec. 6.2). A malentity may kill the primary animal. If the malentity is an organism it may even eat the animal. But such behavior is not selective; the animal may be eaten incidentally along with some other sort of food that is sought. The primary animal does not feature as food and rarely, if at all, features in any other capacity in the environment of the malentity. Another characteristic of a predator that distinguishes it from a malentity is that it usually shows some adaptation that equips it for feeding on the primary organism.

Malentities were originally conceived as "unfortunate accidents". The typical malentity was considered to be an inanimate object, perhaps an artifact left behind by an animal. Browning (1963, 91) mentioned deep hoofprints left by a bullock walking across a marsh. The hoofprints fill with water and drown small insects (Collembola) that are trapped in them. There is no way that the death of the insects might benefit the bullock. "Malentities" is still used in the context of "unfortunate accident" but there may be a number of variations on this theme. Also, it is well to remember that the same natural object may occupy more than one place in the environment of the same animal (secs. 1.31, 2.32). For example, in section 6.2 we tell how the ant *Myrmica scabrinodis* is likely to aggressively evict *M. rubra* from any of its nests that are in open, sunny, well-drained sites. The aggression is direct and overt: the ants may bite legs off workers of *M. rubra* and eat soldiers. Members of the evicted species build nests in less-favored situations where they seem to be safe from attack. By their violence the aggressors, *M. scabrinodis*, are directly responsible for the death of their victims, and the more aggressors there are, the more victims there will be. Hence the reaction of *M. rubra* (the primary animal) to the aggressor (component of environment) is negative. On the other hand, the abundance of the aggressors does not increase as the numbers of its victims increase except insofar as they may have gained some food by eating a few soldiers. There is no suggestion that this reaction is described by the curve in figure 1.03A. The reaction is illustrated by figure 1.03B or possibly 1.03C; that is, it is zero or negative. It would be negative if a dense population of victims exhausted the aggression of the aggressor to the aggressor's disadvantage. But in no circumstance does the presence of *M. rubra* in either low or high numbers aid *M. scabrinodis*. *M. scabrinodis* unambiguously falls into the lower right cell of table 1.01 as a malentity.

In the environment of *M. rubra*, *M. scabrinodis* has a place in the centrum as a malentity by virtue of its direct attacks on the bodies of its victims. But its aggressive activities also drive the surviving *M. rubra* out of their favored sites. In less-favored sites the chance that *M. rubra* will survive and reproduce is less.

Hence the aggressor has a place in the web of *M. rubra* as a modifier of all those components of environment that differ in the two sorts of sites. *M. scabrinodis* profits from its activities as a modifier because it gains favored sites that might otherwise be occupied by *M. rubra*. It is important not to confuse the roles of the aggressor as malentity and as modifier, since they are two quite different actions.

## 1.5   The Importance of Beginning with the Physiology and Behavior of the Animal

In Andrewartha and Birch (1954, 10, 557, 558) we said that an ecological study should begin with a general appreciation of the natural history, environmental physiology, and behavior of the animal so far as it is known or can be observed. Such knowledge, together with a naturalist's appreciation of the sorts of places where the animal normally lives, is sufficient foundation on which to build, tentatively, a hypothetical environment that can be tested step by step. A critical appreciation of the theory and practice of sampling and a working acquaintance with statistical methods and statistical inference are essential parts of ecological talent.

We would not change this advice today, except perhaps to be a little more particular in emphasizing the importance of beginning with a knowledge of the natural history, environmental physiology, and behavior of the animal chosen for study. The better the knowledge, the more surely it leads to key components in the centrum and, through them, to the web. If the theory of environment provides a sure guide to the important components in the animal's environment, the theory will have served its purpose. If here and there we seem to have been too preoccupied with the attempt to place a component of environment in its proper category, we would not want to give the impression that classification is an end in itself. The model should work better if we can be consistent in identifying the components of environment, but the chief aim remains to present a general model of how environment works, hoping that it might point to the most effective questions to be asked at each stage of an investigation (chap. 2) and perhaps to the most effective explanation for the empirical results (chap. 9).

## 1.6   The Natural Population

We have made it clear in sections 1.0 and 1.1 and elsewhere that the theory of environment concerns the distribution and abundance of animals in natural populations. Because population, in the ecological context, has only an elusive meaning, we shall try at this point to establish a meaning for "natural population" as we use this term (see also chap. 8). We begin by saying what a natural population is not: it is neither artificial nor abstract. "Artificial" implies that the animals have been physically deprived of some part of their natural experiences, as in an experiment. For example, consider the laboratory experiments summarized in Andrewartha and Birch (1954, 421), that Gause did with microbes or those that Park did with flour beetles or Birch with grain beetles, or the field experiments that Myers did with rabbits in enclosures (sec. 12.2). On the other hand, an abstract population may exist only in the mind, perhaps having been deduced as a mathematical model from an imaginary premise. For example, consider the models that were put forward by Volterra and Lotka, and summarized in Andrewartha and Birch (1954, sec. 10.1). Or, more fruitfully, an abstract population may be conceived when a real population is contemplated but certain of its qualities are arbitrarily abstracted from it so that they can be ignored. Both the experiment and the abstraction make analysis simpler. (That is why we practice them). But simplicity may be dearly bought because the price is measured in departure from reality. This debt must be repaid before a realistic knowledge of the natural population can be claimed (chap. 9).

We approach a realistic meaning for "natural population" through an abstraction which we call a "local population". In southern Australia at any time during late winter or spring, a carcass of, say, a sheep lying in the open is likely to be occupied by large numbers of maggots of the blowfly *Lucilia sericata* and many other species of scavengers. When the maggots are fully grown they pupate either under the carcass or in some other sheltered place. In due course they emerge as adult flies. By this time the carcass may have been fully consumed, but in any case it is no longer any good for *Lucilia*, which requires a fresh carcass (Fuller 1934). An adult will fly away from a spent carcass and seek a fresh one to lay eggs on. Also dispersing from spent carcasses and joining in the search may be other *Lucilia*, several sorts of predators of *Lucilia*, and a number of species of scavengers that would share the carcass with *Lucilia*. Thinking in probabilities, as we must, it is clear that all the spent carcasses from which these animals might come and all the fresh carcasses that any one of these animals might find are linked, and none can be ignored without abstracting something from nature. In other words, our meaning for "natural population" must allow for the probability of interaction between these "local populations" centered on single carcasses. In other examples the local population may be less of an abstraction, as, for example, the black pineleaf scale *Nuculaspis californica*, where the population on a single pine tree constitutes a local population (sec. 11.1) or the "interaction groups" of carabid beetles studied by den Boer (sec. 8.41). With other species, especially those that disperse great distances, such as the bush fly *Musca vetustissima*, the local populations are much less obvious, and their boundaries may be drawn much less objectively than for the examples given above (Hughes 1981).

The idea of a local population is a straightforward idea, but it poses the question, How many local populations make up a natural population? In an extensive area, where do we draw the line? An arbitrary boundary with local populations interacting across it is not permissible, because it would create an abstraction for just the same reason that a single population is an abstraction. The only alternative is to seek a natural ecological barrier or some other barrier to the exchange of genes between populations.

In southeastern Australia the distribution of the blowfly *Lucilia cuprina* extends over more than a million square kilometers, and there is no obvious ecological barrier except around the margins of the distribution. There are other populations in other parts of the continent and in other continents. It is impracticable to work with such a large unit, but there is no reason why it should not be the paradigm for our concept of a "natural population".

This concept of the multipartite natural population is developed further in chapters 8 and 9. It is an integral part of our theory of distribution and abundance. This is not surprising considering the patchiness of most terrains (see, e.g., Elton and Miller 1954) and the highly specific demands that animals characteristically make on the sort of place where they will live (Andrewartha and Birch 1954, chap. 12).

The meaning for the multipartite "natural population" that emerges from this analysis may be summarized:

(1) It is the sum of a large number of interacting local populations.
(2) It is not abstract.
(3) It is not artificial.

There are two corollaries:

(i) As a consequence of (2) the natural population can be bounded only by natural ecological or genetic barriers. Consequently, in some species the only natural population that can be recognized is the whole species. In most if not all instances the smallest "natural" population will be too large to be sampled as a unit because it would be impracticable to take a sample that rigorously represented all the variability in the population. Of course ecologists are aware of this problem and know that their conclusions have to be qualified accordingly (sec. 8.4).

(ii) In the context of (3) we exclude experimental populations in the laboratory or field as natural populations but regard humans and all their works as part of nature. This means, for example, that the insect pests that owe their abundance to the great monocultures of commercial agriculture can be regarded as belonging to natural populations.

In seeking a meaning for "natural population" we have established a meaning for "local population" as well. This introduces a new idea, "the locality".

## 1.7  The Locality

The locality is conceived as an area that supports, or might support, a local population. The concept of locality is useful because it helps us to think about (a) the environmental risks experienced by the dispersive phase of the life cycle: a safe "landfall" is more likely when localities are abundant and densely distributed (sec. 8.411); (b) the stability in numbers of the natural population: a natural population that is distributed over many localities is likely to be more stable than one that is distributed over few localities (sec. 9.3); (c) the size of the natural population: a piece of country with many localities will support a larger population than the same area with few localities in it (sec. 9.2).

A locality might be obviously distinct from the countryside, as, for example, a carcass, a pond, or a tree. Or the boundaries might be distinct but the distinction may be so subtle that to perceive it calls for the esoteric art of the trained naturalist. We quote a passage from Ford (1945, 122) on this point:

We have three butterflies which are limited by geological considerations, being inhabitants only of chalk downs or limestone hills in south and central England, and they may reach the shore where such formations break in cliffs to the sea. These are the Silver-spotted Skipper, *Hesperia comma*, the Chalkhill Blue, *Lysandra coridon*, and the Adonis Blue, *L. bellargus*... The two latter insects are further restricted by the distribution of their food plant, the Horseshoe Vetch, *Hippocrepis comosa*, and possibly by the occurrence of a sufficiency of ants to guard them. Yet any of the three species may be absent from a hillside which seems to possess all the qualification which they need, even though they may occur elsewhere in the immediate neighbourhood. This more subtle type of preference is one which entomologists constantly encounter, and a detailed analysis of it is much needed. A collector who is a careful observer is often able to examine a terrain and to decide, intuitively as it were, whether a given butterfly will be found there, and that rare being, the really accomplished naturalist will nearly always be right. Of course he reaches his conclusions by a synthesis, subconscious as well as conscious, of the varied characteristics of the spot weighed up with great experience; but this is a work of art rather than of science, and we would gladly know the components which make such predictions possible.

The boundaries of an area in which the animals might live might be perceptible, yet the area might be too large to be conveniently thought of or sampled as a locality. A suitable statistical convention must then be devised to define the boundaries (secs. 2.39, 8.41) but each case must be treated on its merits, according to the demands of the hypothesis and the nature of the animal.

# 2
# The Web: Envirograms

## 2.0 Introduction

"Looking back", wrote Charles Darwin, "I think that it was more difficult to see what the problems were than to solve them". This chapter is about "seeing" problems rather than solving them. An investigator is likely to see problems by the light of the general theory that he has accepted (chap. 10). Our general theory of environment is outlined verbally in chapter 1. In this chapter we try to present the same idea graphically. When the theory in respect to any particular species (the primary animal) is presented graphically, in the form of an envirogram, the problems in the ecology of that secies seem to be illuminated and to stand out clearly.

## 2.1 What Is an Envirogram?

According to the theory of environment, activity in the directly acting components is the proximate cause of the condition of the animal, which reflects its chance to survive and reproduce. But the distal cause of the animal's condition is found in the web, among the indirectly acting components which modify the centrum. A modifier may be one or several steps removed from the centrum, and the pathway from a particular modifier to its target in the centrum may be joined by incoming pathways from other modifiers that may be behind or alongside the first one. To indicate the degree of indirectness in the action of a modifier, we call one that is n steps away from its target a "modifier of the nth order". The envirogram is a graphic representation of these pathways (figs. 2.01–2.06).

So the envirogram is concerned with causes: it is a dendrogram whose branches trace pathways from distal causes in the web to proximate causes in the centrum.

Early in the investigation the pathways have the status of hypotheses. The envirogram is not merely a description of unverified investigations. It is important to see the envirogram from the beginning as a guide to the most profitable experimentation. Each step contains a suggestion that is to be tested by observation or experiment in the laboratory or the field, but especially by manipulating populations in the field. As the investigation matures the envirogram assumes the status of an explanation (sec. 9.03). It may be presented as a plausible summary of the ecology of a particular species and be absorbed into the general theory of the distribution and abundance of animals.

Three advantages attach to the envirogram, especially at the beginning of an investigation. (a) Because the envirogram stems from the theory of environment, the whole environment is under scrutiny, not merely those aspects of it where biotic interactions might lead to competition. (b) Not only is the whole environment under scrutiny, but the theory also offers practical guidelines for its analysis (sec. 2.3). If the distinction between proximate and distal causes is strictly maintained (sec. 1.1), hypotheses become sharper and explanations more plausible. (c) In the early stages, as the envirogram takes shape on paper, the paper usually becomes too small. This experience is a salutary reminder to the enthusiastic naturalist (who should reside in every ecologist) to concentrate on the essential issues, to look first for key components in the centrum, then trace them back into the web. At the same time it is advisable to curb the natural enthusiasm for pursuing pathways to an unnecessary depth in the web. For example, when an entry is made in a pathway, this component does not necessarily have to invoke a centrum with a web of its own. Indeed, the policy is to retain only important components; or, when in doubt, an entry might be retained with a distinguishing mark to indicate that its ecology might be worthy of further investigation later. Unless such economies are practiced, pathways may become confusingly complex and defeat their purpose. A pathway may begin with a modifier that is n steps from its target. We know that the activity of this modifier must have an antecedent cause that might properly be placed n + 1 steps from its ultimate target, in the centrum. So why break into the series and begin at the nth step? The answer is not logical but practical or empirical. In practice it is rarely necessary to go beyond n = 2 or 3 to find an obvious starting point. All science is like this. The explanations stop short, but they can be pushed deeper when the need for further penetration has been conceived or conceded. In none of the envirograms in figures 2.01–2.05 are all four of the compartments of the centrum represented. The policy, to put in only what is known to be important, or suspected of being so, applies to the centrum equally with the web.

## 2.2    The Envirogram: Some Examples

We have tried to keep the following envirograms simple without doing injustice to any important component. The idea of distal causes in the web flowing in to activate proximate causes in the centrum is all-important. The detail of the structure scarcely matters; it is bound to vary as each operator pieces the story together in his own way. There are slightly fuller accounts of the ecologies of the rabbit, the spruce budworm, and the Queensland fruit fly in chapters 12 and 14, against which the pathways in figures 2.01, 2.02, and 2.06 might be checked; figures 2.03–2.05 are more fragmentary. They are included chiefly because they illustrate certain points. They must stand or fall by the brief summaries that accompany them. Some guide lines for making envirograms are offered in section 2.3.

### 2.21    The European Rabbit in Southeastern Australia.

An envirogram for the European rabbit, *Oryctolagus cuniculus* (plate 4), is shown in figure 2.01. The full study extended over five climatic regions (chap. 12). To do justice to the diversity of regional environments and rabbits' responses to them might require several envirograms. The story is condensed into one envirogram for

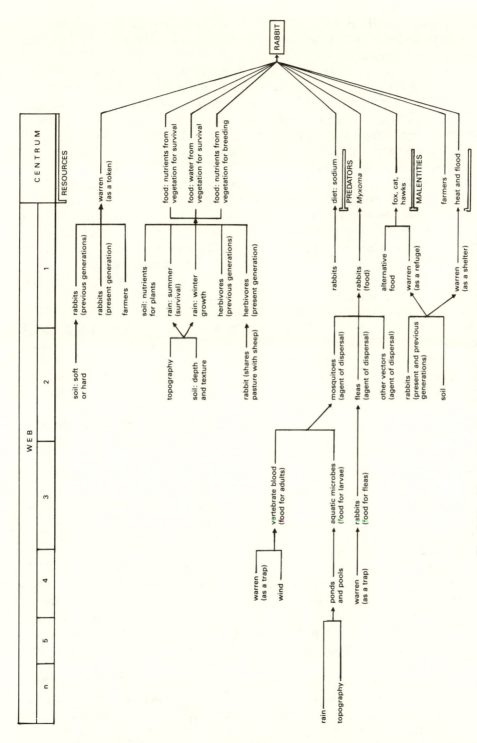

**Fig. 2.01**   The envirogram of the European rabbit in Australia.

convenience, but the condensation leaves the impression that the environment of the rabbit includes a greater number of important components than are likely to be found in any one region.

Looking at any envirogram, to move from left to right along the converging pathways is to trace the action in the environment as it complexly determines the animal's chance to survive and reproduce. To move from right to left, exploring the diverging pathways as they penetrate further into the web, is to retrace the progress of knowledge as the investigation matured. The reader may go either way, whichever is convenient.

Take, for example, the first pathway in figure 2.01. This pathway is discussed in sections 12.411, 12.414, and 12.441. One branch of this pathway arises in the web as a modifier of the second order. The warrens that were dug in hard soil by previous generations of rabbits may be inherited by subsequent generations. In the column of first-order modifiers this branch is joined by two other branches, and all three branches unite to influence the activity of the warren as a token. The meaning of token is defined in section 1.41, and tokens are discussed in section 3.3. When a farmer digs or rips out a warren he kills all the rabbits in it. So farmers also have a place in the centrum, as a malentity.

All the other resources are food in one form or another. The chief risks are likely to arise from deficiencies in the diet, chiefly water (survival season) or digestible protein (breeding season). During the survival season water in the vegetation, together with adequate nutrients, is important (secs. 12.32, 12.424). Drinking water is not important because this niche is not strongly developed in the rabbit (sec. 12.424). During the breeding season an adequate concentration of digestible nitrogen becomes important (secs. 12.31, 12.422). The activity of food in the environment of the rabbit can be explained by looking at the plants that the rabbit eats and at their environments. Rain (which provides water) and soil (which provides other nutrients) for the plants and various herbivores that are predators of the plants are all shown as modifiers of the first order ($n = 2$) in the envirogram of the rabbit. All these things occur in the centrum of one or another of the plants that provide food for the rabbit. Rain is chiefly important because plants can grow only when the soil is moist. The activity of the rain is modified by the topography and the water-holding capacity of the soil (sec. 12.423). Exploitation by rabbits and other herbivores including man in previous generations may have altered the type of vegetation available to the present generation of rabbits (sec. 12.427). In the present generation an excessive number of rabbits may cause an intrinsic shortage of food, or excessive grazing or exploitation by other herbivores may cause an extrinsic shortage of food for rabbits (see secs. 3.131, 3.132 for explanations of intrinsic and extrinsic). The pathway leading to sodium in the diet is brief because not much work has been done on it. It is important only in the subalpine region (sec. 12.425).

The virus *Myxoma* is the most important predator (sec. 12.431). The pathway that leads to *Myxoma* in the centrum has its origin in two modifiers of the fifth order. Rain and topography have an important influence on the distribution and abundance of pools and ponds of fresh water that support the larvae of mosquitoes, which are vectors of *Myxoma*. At the level of the second-order modifiers, this pathway is joined by a branch that relates to food for the adult mosquitoes, the females of which feed on the blood of rabbits and other vertebrates. This branch begins with wind, which blows the mosquito to where it might find food. At this point, among the modifiers of the second order, mosquitoes are joined by other vectors, especially the rabbit flea. The vectors carry the *Myxoma* to where it finds food—that is, to a rabbit. So wind modifies the activity (increases the probability) of food for the mosquito; and the mosquito modifies the activity (increases the probability) of food for *Myxoma*. The wind is a modifier of the fourth order, and the mosquito is a modifier of the second order in the environment of the rabbit. But neither the wind nor the mosquito (as a vector) enters the centrum of any organism in the environment of the rabbit. Vertebrate predators are discussed in section 12.432. In some of the more productive parts of the Mediterranean region there is an extremely subtle interaction between "alternate food" for predators and the persistent reduction in the abundance of rabbits that was initiated by the panzootic of *Myxoma* in 1950–52 and has, since then, been maintained over substantial areas by farmers. Rabbits are killed by farmers practicing various control measures (sec. 12.441), which makes the farmer a malentity. Warrens are destroyed by farmers, which makes the farmer a modifier of the first order in the pathway leading to the warren (as a token).

The final pathway leading to "heat and flood" is brief because this subject has not been much studied.

A component of environment becomes important only by virtue of its capacity to vary. In this context the capacity to vary from time to time has to be distinguished from the capacity to vary from place to place. For example, weather varies from time to time in the same place, while topography may remain essentially constant for many years. But topography may vary from place to place. So weather may be important when the spotlight is on temporal variability in abundance; but topography may become more important when the spotlight is on distribution. This point is illustrated several times in the envirogram of the rabbit. The modifiers of the fifth order in the pathway that leads to *Myxoma* provide a straightforward example: breeding places for mosquitoes are relatively abundant along rivers of the Murray-Darling system and their tributaries. The example that is provided by the modifiers of the first order in the pathway leading to food is more subtle: the succession in the vegetation has differed greatly in the regions (secs. 12.423, 12.427), and the regional abundance of the rabbit reflects these differences.

## 2.22 The Spruce Budworm

The ecology of *Choristoneura fumiferana* (plate 7) is summarized in section 14.1. Only the salient features necessary to appreciate the envirogram are given here. Virtually the whole of the boreal forest of Canada is the habitat where *Choristoneura* may be found. Extreme patchiness in the distribution of places that offer good food for *Choristoneura* is characteristic of the habitat. The habitat is immense, and patchiness is built into it on a grand scale. Not surprisingly, *Choristoneura* has responded by evolving extremely dispersive patterns of behavior (sec. 8.42).

Three stages in the life cycle are strongly adapted for dispersal (sec. 14.11). Those individuals that do not emigrate, as first- or second-instar larvae, from the place where they were born settle down, in the second instar, to begin eating and growing. They do not get another chance to migrate until, as adults, they develop wings. The top envirogram, figure 2.02, summarizes the ecology of such a local resident.

The pathways that converge on "food" are most important. The larvae of *Choristoneura* find good food in the flower buds, leaf buds, or young foliage (not more than two or three years old) of mature (forty years or older) trees of balsam fir or spruce. In the forest, no new tree can get a start unless an old one has died to make room for it; nor is it likely that a vacancy will remain unfilled for long. So to have a stand of uniformly aged trees covering a large area requires that, at some well-defined time in the past, some harmful "thing" must have swept through the area, killing all the trees. Trees of balsam fir or spruce mature in about forty years. A stand of mature trees requires that the destruction occurred forty years or more ago. Probably plagues of *Choristoneura* are the most important "agent of death" but wildfire, severe storms and man practicing clear-cut logging probably add their quota as well. In the envirogram of the resident *Choristoneura* the "agents of death" have been lumped together as a modifier of the first order, converging on "food". They have been itemized in the envirogram of the dispersing *Choristoneura*. It

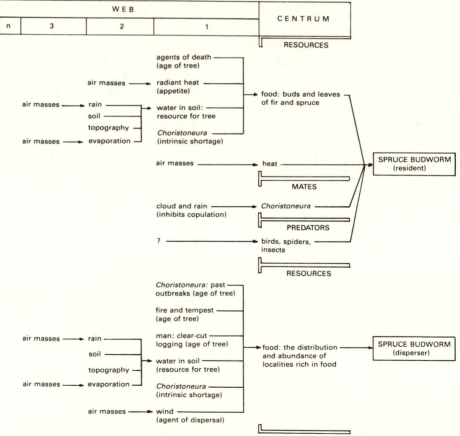

**Fig. 2.02**   The envirogram of the spruce budworm in Canada.

seems to have been established beyond reasonable doubt that a densely distributed abundance of mature trees is necessary but not sufficient to cause an outbreak.

There is strong suggestion that outbreaks have usually been preceded by a number of years (perhaps five or six) of anticyclonic weather. Anticyclonic weather is usually bright and dry. This relation has not been convincingly explained with critical experiments. A plausible hypothesis (sec. 14.211) is represented by the pathway that converges on "water in soil" in column 1 and passes on to "food", in the centrum. Water is a resource for the tree and therefore is a first-order modifier for the budworm. The pathway begins in column 3 with air masses (weather) modifying the amount of rain and the evaporation of water from the soil. Evaporation is one of four second-order modifiers that influence the activity of water in the soil. The depth and the texture of the soil influence the amount of water it can hold against gravity; the texture of the soil also determines what proportion of this water will be available to the trees. Topography influences the local distribution of the water before it soaks into the soil. The amount of water in the soil is important because, according to this hypothesis, a mature tree will produce an abundance of nutritious food for *Choristoneura* when the tree has been "stressed" for several years by a shortage of water. When the shortage of water is alleviated by a return to wet weather the tree will revert to its normal non-nutritious condition. This hypothesis represents a particular instance of a general theory of outbreaks developed by White (1974 and other papers; sec. 10.311). It seems reasonable to adopt this explanation for *Choristoneura* because (a) there seems to be a significant tendency for outbreaks to develop during spells of anticyclonic weather, (b) during anticyclonic weather rainfall is low and evaporation is high, and (c) the closely studied outbreak of 1949–57 came to an end after a run of wet summers, and it seemed that neither intrinsic shortage of food nor predators could be invoked to explain the final collapse of the outbreak.

There are two other branches to the pathway that converges on "food". (a) Radiant heat is a modifier of the first order because too little of it (dull, cloudy weather) will inhibit larvae from feeding; they may even be driven away from food to take refuge among the twigs and branches. (b) *Choristoneura* is a modifier of the first order because during an outbreak the trees in a locality may be severely defoliated and the consequent intrinsic shortage of food may be severe.

The only other resource worth mentioning is radiant heat, which penetrates the crown where the caterpillar is feeding, raising its temperature and so increasing its metabolism. The caterpillar eats more and grows more quickly when it is warm. Radiant heat in the centrum must not be confused with radiant heat in the web (see pathway converging on food). Anticyclonic weather, as determined by the movements of air masses, brings bright, warm days. And the open crown of a mature tree allows radiant heat and drying winds to penetrate. This pathway represents an alternative hypothesis (sec. 14.212), which is probably true, but the evidence does not suggest that it is either necessary or sufficient as an explanation of outbreaks. This pathway probably reinforces the influence of food, as outlined above, but it does not seem to be a key component. Two other pathways are included in the envirogram, one through predators and one through mates. Neither seems at present to be important.

The lower part of figure 2.02 contains an envirogram for a budworm that is dispersing, either a larva in the first or second instar or an adult (sec. 14.3). There is only one pathway in it: the only reward for dispersing is to find a new place to live where the food is good. The solitary pathway, converging on food, has virtually the same branches and the same components as the equivalent pathway in the envirogram for the resident. There are minor differences. The "agents of destruction" which are concerned with the age of the trees in diverse stands need to be spelled out, because the importance of the different sorts might vary geographically over the distribution of the natural population. Wind, as an agent of dispersal, was omitted for the resident (because local dispersal is taken for granted) but has been included for the disperser because all dispersal depends on wind. The branches that converge on water in the soil (a resource for the tree) assume a geographic perspective as the distribution of air masses, soil types, and topographic features come to be considered broadly. So the two pathways have much the same components, but for the disperser they are all being considered on the grand scale of the natural population. In the envirogram for the disperser, intrinsic shortage relates not to particular localities but to the total shortage that puts the whole natural population at risk (sec. 14.34). As an outbreak proceeds and the barren area of defoliated forest accumulates, the chance that a dispersing *Choristoneura* will make a good landfall becomes less and less, and the death rate among dispersers becomes higher and higher. Theory might safely predict that a continuation of this process would bring the outbreak to an end. Doubtless the logic is sound, but theory might have to assume the destruction of resources carried to an unrealistic limit in order to predict the extreme scarcity of *Choristoneura* that Morris (1963, 7) said existed between outbreaks. In the most closely documented outbreaks of 1949–57 the evidence suggested that something else overrode the intrinsic shortage of food to put an end to the outbreak while there were still substantial areas of good food left (sec. 14.211). That "something else" was probably an extrinsic shortage of food caused by changes in the activity of the pathway that reaches "food" through "water in the soil", a resource for the tree. The changed activity could probably be traced to changes in the distribution and pattern of air masses that caused a return of cyclonic weather (sec. 14.211, fig. 14.03).

## 2.23   The African Buffalo

Figure 2.03 presents an early stage in putting together an envirogram of the African buffalo *Syncerus caffer*. It is based on the ecology of the buffalo as we have summarized it from the writings of Sinclair (1977). The African buffalo needs green grass, and it cannot graze any that is too short (pathway converging on food, fig. 2.03). It also needs a regular supply of drinking water (pathway converging on water). The need for these two resources and the ability of herds of buffalo, acting in concert, to fend off the attacks of lions (first pathway under predators) are probably what restricts buffalo to riverine grassland or forest during the dry season. Seepage lines in forests provide enough water for small groups of buffalo but not enough to support a herd. As the dry season intensifies, the proportion of the population living on the riverine grassland increases, doubtless because of the assured supply of drinking water, and buffalo usually experience a shortage of food, not because all the long-bladed green grass has been consumed, but because

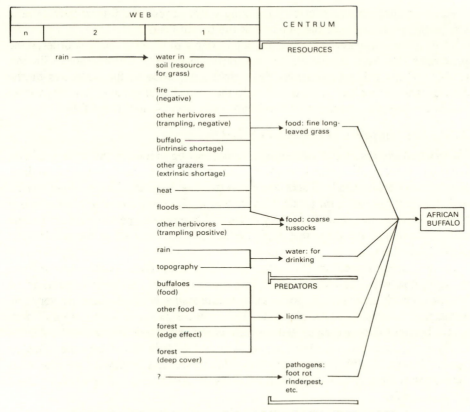

**Fig. 2.03** The envirogram of the African buffalo in East Africa.

it has turned into straw and does not provide an adequate diet. Even during the wet season few buffalo are found on the open plain because the grass, though it may be abundant, is usually too short, perhaps because it has been grazed by other herbivores—notably wildebeest but also topi, kongoni, waterbuck, and others. In the riverine plain the supply of food is sometimes reduced by flood or fire or trampling by other herbivores. In the long run the grasses grow better after flood or fire. Some coarse tussocks that grow close to the river may send out edible shoots after being trampled by elephants (eighth entry in first modifier of food). On a hot day the subdermal temperature of the buffalo may go above 40°C; then they will stop eating and cool off in a wallow (seventh entry in first modifier of food).

The herding behavior of buffalo is a good defense against lions. Nevertheless predation by lions accounts for 25% of deaths from all causes. The lions lie in ambush at the edge of the forest or among the tall tussocky grasses along the river. They complete the hunt and make the kill in the adjacent grassland. Deep in the forest the trees seem to provide a refuge from the lions. So in the envirogram deep forest represents "shelter" for the buffalo and the "forest's edge" represents "cover" for the predator (first-order modifier of lions). The former has a positive and the latter a negative influence on the primary animal's chance to survive and reproduce. Other predators include a number of pathogenic microbes and viruses; rinderpest and foot rot are perhaps the most important.

Rain is important as a source of drinking water (a resource for the buffalo) and soil water (a resource for the plants that the buffalo eats).

Most of the pathways in figure 2.03 begin with a modifier of the first order; none penetrates the web beyond the second-order modifiers. Unfortunately Sinclair (1977) did not ask the questions that might have allowed the pathways of the envirogram to penetrate more deeply into the web. It would have been interesting to see the extrinsic shortages of food and water investigated more fully.

## 2.24    The Limpets *Cellana* and *Patelloida*

Underwood and his students observed five interacting species of invertebrates in an intertidal habitat off the east coast of Australia, about the latitude of Sydney (Underwood et al. 1983). There were two grazing limpets, *Cellana tramoserica* and *Patelloida latistrigata* (plate 1); a barnacle, *Tesseropora rosea*; a predatory whelk, *Morula marginalba*; and algae (A.J. Underwood, pers. comm.). In this section we outline the shape that might be taken by envirograms of the two limpets (figs. 2.04, 2.05).

All five species occur on the surface of rocks. In sheltered places (where there is negligible wave action) the rock faces usually carry a dense population of *Cellana*, which graze on the spores and juvenile stages of algae; there are very few barnacles, *Patelloida*, or mature algae. The mature algae are absent from midshore levels because they are eaten before they can grow (Underwood 1980); whelks rarely eat *Cellana* (Moran 1980); barnacles do not settle in such calm water (Denley and Underwood 1979; Denley 1981); *Patelloida* settle, but they die because *Cellana* crushes them or eats their food, which is the spores and young stages of algae (Creese 1982).

A rock face in a place where there is moderate to strong wave action often carries a dense population of mature barnacles. Some of the spaces between the barnacles will support mature algae; some of the spaces and the barnacles themselves will support *Patelloida*, which graze on the spores and young stages of algae. There will

**Plate 1**   Intertidal animals depicted in the envirogram Figure 2.05. The three small hat-shaped animals on the diagonal that extends from the right lower corner are the limpet *Patelloida*. The two large limpets at the top middle are *Cellana*. The three nodular whelks at the top left are *Morula*. The others are the barnacle *Tesseropora*. Photo by R. G. Creese)

usually be few or no *Cellana*, and there may be some whelks. The *Patelloida* can persist there because the spaces between the barnacles are too small to support grazing *Cellana*, and the clusters of mature barnacles shelter the *Patelloida* from excessive wave action (A.J. Underwood, E.J. Denley and M.J. Moran, pers. comm.). Whelks eat large but not mature barnacles, but they prefer *Patelloida* (which are small) to barnacles (Moran 1980; Fairweather and Underwood 1983).

Some exposed shores where there is moderate to strong wave action are dominated by *Cellana* and other gastropods. Barnacles settle as usual on such sites. But if *Cellana* arrive in great numbers they "bulldoze" settling, and newly settled, barnacles, greatly reducing their number. If *Morula* are present they may eat many, if not all, of the remaining barnacles within a few months, thus returning the area to one dominated by *Cellana* and other gastropods. If settlement of barnacles is great and *Cellana* and *Morula* are not sufficiently abundant to dispose of most of them, the area may become dominated by barnacles.

The low shore areas, below those usually occupied by barnacles and *Cellana*, are largely dominated by macroalgae at most times (Underwood 1981; Underwood and Jernakoff 1981). These algae maintain a supply of spores to the midshore areas that support *Cellana* and *Patelloida*. There are also some patches of macroalgae in the midshore, particularly among adult barnacles where there are few limpets.

In figure 2.04 we try to select out those interactions that are relevant to the ecology of *Cellana*. If this is attempted at an early stage of a project, when some of the natural history has been studied, a tentative envirogram can be made. This will serve to focus attention on questions that are relevant to the ecology of the primary animal and may uncover questions that may not have been noticed.

In figure 2.04 a pathway that converges on "food" through "rock surface" occupies most of the envirogram. To get enough food the limpet needs access to a sufficiently large area of rock surface to graze over. The area of rock surface that is available to *Cellana* for grazing or for settling may be reduced if either algae or barnacles preempt it by growing to maturity on it. So, starting in the outer reaches of the web and working inward, among the modifiers of the third order *Cellana* appears twice, modifying "algae" as a predator and modifying barnacles as an aggressive malentity; *Morula* modifies "barnacles" as a predator, and "topography" (i.e., a large physical feature that might shelter a rock or group of rocks from waves) modifies "rock surface", making it more or less suitable for barnacles to

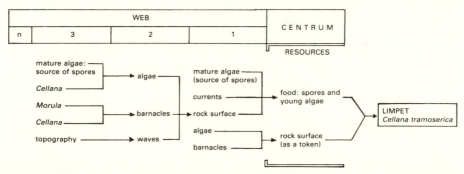

**Fig. 2.04** The envirogram of the limpet *Cellana tramoserica* on the coast of New South Wales.

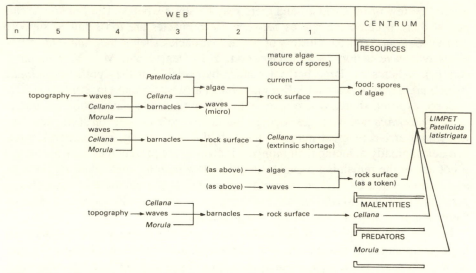

**Fig. 2.05**   The envirogram of the limpet *Patelloida latistrigata* on the coast of New South Wales.

settle on. There are three modifiers of the first order in the pathway to food. Mature algae are a source of spores for limpets provided they settle on a suitable rock face. Currents carry the spores to rock faces. So the area of suitable rock surface, beds of algae, and currents modify the activity of food for *Cellana*. The only other pathway of any importance leads to "rock surface" in the centrum as a token. The discovery of a suitable place to settle is a prerequisite for the morphological, physiological, and behavioral changes involved in settling and feeding; for example, the larvae do not settle on surfaces that are covered with algae. The distal branches of this pathway are the same as before. They would be redundant if they were filled in.

Figure 2.05 is an envirogram for *Patelloida*, the other grazing limpet in the group of five species studied by Underwood and his students. The two limpets have similar requirements for settling and feeding, with the exception that *Patelloida* can settle on adult barnacles and, being small, can continue to live there when mature. When these two limpets occur together, *Cellana* with its large radula crushes juvenile *Patelloida* and/or takes most of the joint stock of food for itself (Creese 1982). Thus *Cellana* is both an aggressive malentity and a harmful modifier of food in the environment of *Patelloida*. These interactions lead to some interesting variations in the envirograms. "Barnacles", which were twice a harmful modifer of the second order in the envirogram of *Cellana*, become a helpful modifier in the envirogram of *Patelloida*—of the second order modifying *Cellana*, which might rob *Patelloida* of its food, and of the third order modifying "waves" which might otherwise prevent *Patelloida* from staying in a place where it can rely on finding food.

These variations come about because (a) there is no way that barnacles could be said to preempt suitable rock surface at the expense of *Patelloida* (see above) and (b) rock surfaces that are dominated by barnacles offer *Patelloida* a microhabitat where it might find a refuge from *Cellana* and shelter from wave action (sec. 2.34). *Patelloida* is the preferred prey of *Morula*, which is shown as a component of the centrum in figure 2.05.

## 2.25 The Queensland Fruit Fly

The fruit fly *Dacus tryoni* (plate 2) is thought to have been indigenous to a belt of tropical and subtropical forest that stretched along the northern half of the east coast of Australia. If that were indeed the limits of the pristine distribution, the limits must have been set by food, not climate, because now that exotic fruits in orchards and gardens provide food where none was before, *D. tryoni* may be found in many different climates in habitats scattered over a large part of the eastern third of the continent. Abundance is greatest in areas with summer rainfall and in years when summer rainfall is high (sec. 14.2). Only in the central and western deserts and perhaps in southern Tasmania would the climate in itself be inhospitable enough to prevent the risk of an outbreak of *D. tryoni* (Meats 1981).

More than one hundred indigenous and exotic fruits have been recorded as suitable food for the larvae of *D. tryoni*. The larvae ingest the flesh of the fruit but probably obtain most of their nourishment from microorganisms in the fruit (sec. 14.2). The most important requirement is that there be a succession of fruits ripening throughout the period of the year when temperatures are suitable for breeding. In places where fruits are available it seems likely that food for adults, such as honeydew and secretions from leaves, will be abundant. If the temperature is warm enough for development, dormancy does not occur at any stage of the life cycle. Other species of animals are likely to eat the same fruits as the larvae, but probably only birds and mammals influence the amount available for *D. tryoni* (sec. 14.2).

In the original tropical and subtropical habitats fruits in the bush (now supplemented by those in orchards and gardens) provide a succession of fruit that lasts through all seasons. *D. tryoni* breeds continuously and has the potential to complete eight generations in a year (Meats 1981). However, there are usually fewer than eight generations because of lack of continuity of crops of fruit (R. A. I. Drew, pers. comm.). Farther south, in the areas where *Dacus* has become abundant only since European colonization, the succession of ripe fruit depends on exotics—for example, citrus, loquat, early stone fruit, late stone fruit, pome fruits, and figs. But there is no guarantee that consecutive stages in the succession will be geographically contiguous. It is more likely that the crop of ripe fruit that had nourished one generation may have matured and vanished by the time that the next generation requires food for larvae. It becomes necessary to find a new locality where there is suitable fruit. In the envirogram (fig. 2.06) the dispersing adult is distinguished from other stages in order to emphasize the importance of dispersal. Meats (1981, 152) commented on the dispersiveness of *D. tryoni*:

Each [local] "population" is essentially ephemeral because the host fruit for one generation is generally not available for the next. The progeny of one generation therefore disperses to find new host trees which may or may not be available in the vicinity. It is therefore possible to detect "populations" where in fact no breeding occurs. Fletcher (1974a,b) found that in areas of open bush where virtually no breeding takes place, the numbers of mature adults rose from zero in spring to a

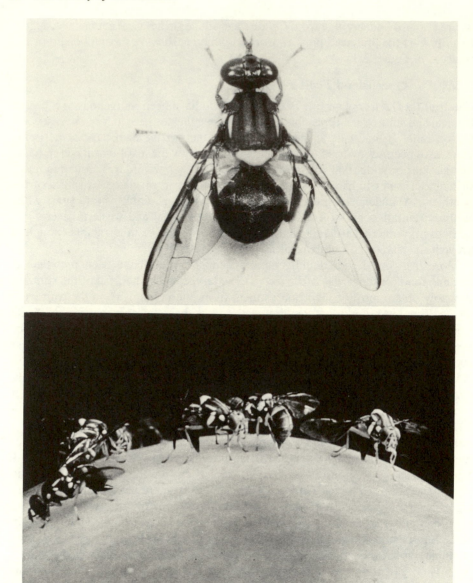

**Plate 2**   The Queensland fruit fly *Dacus tryoni: above*, adult male; *below*, adult females laying eggs into the surface of an apple. Usually no more than one female is seen on the surface of a fruit, but this photograph shows a caged population. (Photo courtesy of M. A. Bateman)

peak of 6256 ha$^{-1}$ in summer. He also found that mature flies could travel at least 24 km in 21 days. Since *Dacus* infests cultivated fruit it is quite likely to be transported [in the course of commerce] in the larval stage to virtually anywhere in Australia.

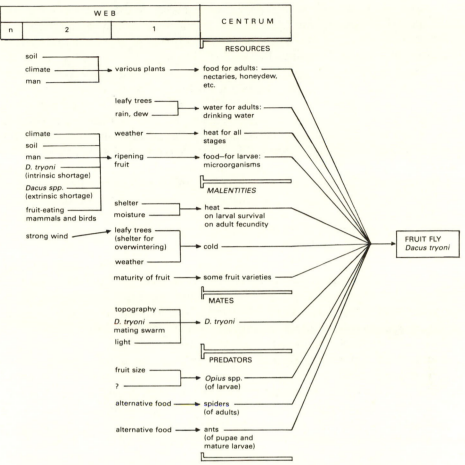

**Fig 2.06**  The envirogram of the fruit fly *Dacus tryoni* in eastern Australia.

Populations are therefore a series of short-lived propagules which are established after either natural or assisted migration. These propagules may establish and flourish consistently in some areas and only occasionally in others. In areas of consistent occurrence numbers may be consistently high, as on the tropical coast of Queensland (Gibbs, 1967) or they may rise and fall approximately a thousandfold, on an annual basis as in the coastal region near Sydney (Fletcher, 1974b). Otherwise they may be consistently present but never at high numbers as in east Gippsland, Victoria (O'Loughlin, 1964). In other areas they may be detected regularly each summer but apparently disappear from all but a few localities in winter (May, 1963; Monro, 1966). In yet other areas they may occur only in occasional summers (May 1963).

The tropical origin of *Dacus* is reflected in its response to shortage of heat. In the most southerly part of its range, in coastal eastern Gippsland, where summer is cool, *Dacus* usually completes only 2 generations in the full year. There is an unexplained need for water. Bateman (1968) observed that around Sydney, between November and April, the population declined unless there was at least 24 mm of rain per month. Adults, overwintering in leafy trees, benefit by drinking water from dew. So "heat" and "water" appear as resources in the centrum along with food. Heat is a malentity for larvae in fruit that happens not to be protected from the direct radiation from the sun. Heat is also a malentity for adults through its negative influence on fecundity (Bateman 1967).

As becomes its tropical origins, the fruit fly is highly vulnerable to moderately low temperature. The immature stages are not likely to survive the winter in any habitat where the prevailing temperature is too low to promote active growth or development. If the adults can find good shelter, such as in a leafy tree, some of them at least are likely to survive the winter if the mean daily minimum temperature does not fall below 2°C (Fletcher 1979). So we include cold as a malentity. The pathway in figure 2.06 that converges on cold indicates that strong winds influence the effectiveness of trees as shelter from cold. Winds can blow flies away from their overwintering quarters, especially when the temperature is below the torpor threshold. Fruit can be a malentity; most eggs laid into late season apples die as eggs or in the first larval instar. The inhibitory effect of these varieties declines a little as the fruit matures (Bower 1977).

We mention in section 4.211 that *D. tryoni* displays an elaborate mating behavior which is set off by decreasing light intensity at dusk. There is some suggestion that the flies are attracted to special parts of trees, usually on the windward side, where the males form a mating swarm as dusk approaches. Such behavior may be necessary to a species whose adults normally disperse after emergence without first mating (sec. 14.2). So we include "mates" in the envirogram.

Predators are important rarely and only locally. Ants can be important in preying upon pupae and mature larvae, and some species of native wasps of the genus *Opius* lay their eggs into larvae. The chance of a larva's being preyed upon is much greater in a small fruit than in a large one, where many larvae are out of reach of the wasp's ovipositor. In section 14.2 we discuss how these various components of environment interact to influence the distribution and abundance of *Dacus tryoni*.

## 2.3   Guidelines for Reading and Making Envirograms

With most species, differences in the life-styles of distinct stages in the life cycle raise the question, Will one envirogram suffice, or are several needed? With the rabbit (sec. 2.21, fig. 2.01), differences between the breeding stage and the survival stage are substantial. Also, the rabbit is distributed over a number of climatic regions in which different components may be most important. Nevertheless, for our present didactic purpose it is sufficient to condense all the variability into one envirogram. With the budworm (sec. 2.22, fig. 2.02), the requirements of the sedentary resident in a locality differ markedly from those of a dispersive migrant which is seeking to colonize a new locality. It seemed more realistic to make separate envirograms, but for convenience they have been put on the same page.

Most envirograms include interactions like the one that is represented in figure 2.01 by "warren" as a refuge, in the pathway through "fox". The warren might be said to modify "fox", which becomes a more effective predator when the warren is built in shallow sand. Alternatively, the warren might be said to modify "rabbit", which becomes more accessible as food when the warren is built in shallow sand. The choice is quite arbitrary. Another example is figure 2.01. "Wind" and "mosquitoes" in the pathway through *Myxoma* represent similar interactions. From one point of view the wind modifies the accessibility of food (vertebrate blood) to mosquitoes; from another point of view the wind modifies the mosquito's capacity to search for its food (vertebrate blood). Two steps farther along the same pathway, "mosquitoes" might be said to modify the ability of *Myxoma* to find its food. The ambivalence of such interactions comes from the inevitable reciprocity of the predator-prey interaction. But ambivalence of this sort sometimes occurs without reference to the predator-prey relationship. For example, in figure 2.02, in the pathway that goes through "water in soil" as a resource for tree, "topography" might be said to modify the activity of "water" (the amount of water in the soil around a particular tree) without influencing the *amount* of rain (as measured in a rain gauge). Or it might be said to have modified the amount of rain (in a particular place) because the topography was reponsible for diverting some of the rain that had fallen in the vicinity to or from that particular place. In the first usage "topography" is a second-order modifier, and in the second usage it is a modifier of the third order. But the meaning and the usefulness of the envirogram remain unchanged whichever usage is adopted. It is good practice to choose the one that seems most realistic and to avoid redundancy by taking care not to fill two steps in the pathway with the same component.

Another feature of an envirogram is that it contains living organisms, artifacts, excretions, and residues of living organisms, and also inorganic entities. These three classes of "things" are all used without prejudice. Nevertheless, when a component of environment is a living organism (whether it belongs in the centrum or the web of the primary animal), its activity will be determined by its environment; to investigate the pathway beyond this organism is merely to investigate its ecology. To explore any pathway beyond a step that is an inorganic entity may require any of a wide range of scientific specialities. But the distinction between proximate and distal causes is still useful (sec. 1.0).

## 2.31 The Distinction between Directly and Indirectly Acting Components

The distinction between proximate and distal causes (of the primary animal's condition) is the linchpin supporting the concept of the web. For euphony's sake we speak of components that act "directly" or "indirectly" giving these words a technical meaning narrower than what they have in literary or colloquial English. Directly acting is strictly reserved for those components of environment that are proximate causes of the primary animal's condition—that is no other cause can be seen to come between the directly acting component and the animal. When a limpet, *Lottia*, pushes a limpet of another species, *Acmaea*, from a rock surface that is covered by a fruitful algal "lawn" (on which limpets normally graze) so that *Acmaea* finds itself in the inhospitable sea where it finds no food and dies of starvation, the proximate cause of death must be registered as shortage of food. And *Lottia* must be registered as a distal cause of the death. Even though *Lottia* made physical contact with *Acmaea* and violently pushed it off the rock, *Lottia* does not rank as a directly acting component in the environment of *Acmaea*, because shortage of food came between the violence of *Lottia* and the death of *Acmaea*; *Lottia* contributed to the shortage of food but only indirectly caused the death of *Acmaea* (sec. 6.2). Similarly with the wind that blows the mosquito or the budworm to a place where it finds food. The mosquito, having been blown to a place where food is plentiful, may eat to repletion. The food is the proximate cause of the mosquito's repletion; the wind is a distal cause. Even though the wind physically impinges on the mosquito, imparting kinetic energy to it, the wind ranks as an indirectly acting component in the environment of the mosquito (secs. 1.0, 12.431). When reading or making an envirogram (or doing any thinking within the framework of the theory of environment), it is essential to make this critical technical distinction between directly and indirectly acting components of environment as the first step.

## 2.32   One Natural Object May Constitute More Than One Component of Environment

An envirogram for the tick *Ixodes ricinus* might feature several qualities of the sheep (sec. 3.131). In the centrum, among resources there is "sheep" because the sheep is food for *Ixodes*. In the web, one step removed from food, there is "sheep" again because the sheep is an agent of dispersal for *Ixodes*. In each instance, the component of environment strictly is a particular quality of the sheep that influences the tick's chance to survive and reproduce; the tick *is nourished* by eating the sheep (some of its blood), and the tick *is carried* by the sheep to a place where in due course it might find another meal. It is these passive verbs that identify the component of environment in the strict sense in which it is defined in section 1.1, in table 1.01, and in table 3.02. In other words, *the component of environment is known by what it does to the animal*. The sheep has more than one quality that impinges directly or indirectly on the tick.

Here we encounter a semantic problem. Our language does not have the abstract nouns to name all the qualities of natural objects and natural manifestations of energy that constitute an animal's environment. To push ahead without them is to fill the sentence with long phrases or whole clauses, whereas euphony and conciseness both demand a solitary noun. Fortunately the problem is not serious. In the present instance the two qualities of the sheep, "edibility" and "capacity for generating and transmitting kinetic energy" take the sheep into different positions in the envirogram. This is generally true of all components of environment, because any component is fully specified by its name (as a natural entity) and by its position in a particular pathway of the envirogram. Thus, by virtue of the contexts in which "water" appears in figure 2.01 the envirogram tells the reader that water might (a) nourish the rabbit itself, (b) nourish the plants that the rabbit eats, or (c) convert a barren hollow in the ground to a place where mosquitoes (vectors of *Myxoma*) might breed. In figure 2.05 barnacles, in one pathway, interfere with a malentity by inhibiting *Cellana* from settling on a rock surface in turbulent water. In another pathway mature barnacles reduce the turbulence close to the rock surface that might inhibit *Patelloida* from living and feeding there.

It sometimes happens that the density of the population of the primary animal influences the activity of components in its own environment. This kind of interaction occurs most often between the primary animal and components in its centrum, notably food and predators. But less direct interactions may also occur. For example in figure 2.01, in the column of second-order modifiers, sixth from the top, the component "herbivores (present generation)" includes sheep. We know (sec.

12.246) that sheep and rabbits share the same pastures. So the rabbit reappears in this pathway as a modifier of the third order, because it reduces the activity of food for the sheep.

The envirogram also includes interactions between modifiers that occur remotely in the web. In the ecology of the rabbit the same rain that fills ponds and thus promotes the breeding of mosquitoes might also promote the growth of plants that the rabbit eats.

Some readers, especially those who work with flowcharts, might argue that these interactions call for anastomoses between pathways. But such a concept is irrelevant to the theory of the envirogram. By observing the two rules that are set out in this section, that an entity may constitute more than one component of environment and that the function of any component is indicated by its position in the web; the envirogram comes logically to comprise every interaction that might be required to describe an ecology completely.

There is no environmental component that cannot be represented in an envirogram. There is no interaction that can be represented in a flowchart that cannot be represented better in an envirogram. As successive steps are added to its pathways the envirogram grows smoothly, consistently indicating the function of each component by the pathway that it is on and by the number of steps that it is removed from the centrum. In most flowcharts lines run hither and thither, adding to confusion as their numbers grow. Compared to the virtually insoluble complexity of the diagrams that occur, for example, in Barlow and Dixon (1980, 8, 9), the complete logical simplicity, the consistency, and the realism of the envirogram seem preferable. (see also sec. 9.01).

## 2.33 Dispersal

Wind as an agent of dispersal is prominent in the environment of *Choristoneura* (fig. 2.02). The evolution of diverse and extensive adaptations in physiology and behavior that allow it to exploit the wind show how important wind dispersal is to *Choristoneura*.

At first it might seem that wind is acting directly on *Choristoneura*, since the insect absorbs kinetic energy directly from the moving air. One might ask how this differs from the heat that is absorbed by a caterpillar basking in the sun or sitting on a twig that has been warmed by the sun. So why not classify wind as a resource? Because in this instance we know that movement itself is unimportant except in relation to the distribution of food and good places to live, whereas temperature, acting directly on the caterpillar, makes a big difference to the rate at which the caterpillar grows.

Agents of dispersal are important chiefly for small species or for larger ones that disperse while they are small. Some cling to larger, more mobile ones as a tick clings to a sheep or a snail clings to a duck. Others cling to logs or other flotsam in rivers or oceans or to mobile artifacts of other animals, especially man: various kinds of baggage, carts, and boats have long been dispersing some of the lesser animals; more recently automobiles and airplanes do so more rapidly. But no other agents of dispersal can rival currents of water and air. Hardy (1956, 13 et seq.) discussed the enormous number and variety of plankton that drift in the major ocean currents. Hardy and Milne (1937) and Johnson (1951) called attention to the surprisingly large numbers of small terrestrial animals that drift high in the air over fertile farmland.

Like the spruce budworm (sec. 14.11), all these animals show adaptations that allow them to exploit specific agents of dispersal. Dispersal is important to them because it increases their chance of finding food or a mate or of escaping from a predator or a malentity. The capacity to disperse has no selective advantage except in relation to one or other of these components of environment. Agents of dispersal always occur in the web. In ecological theory "dispersal" is a word that requires a complement. One must always ask, Dispersal toward what good thing? Or away from what bad thing?

## 2.34 Shelter

"Shelter" is also a word that needs to be qualified. One asks, shelter from what harmful thing? The answer can only be, From a predator or a malentity (table 1.01). So whatever provides shelter must be in the web, modifying a predator or a malentity.

A good den or burrow may protect its occupant from bad weather or wildfire if they are not too extreme. A log or a large stone, well embedded in the soil, may serve the same purpose for a host of small terrestrial animals. But the need for shelter of some sort is almost universal among animals. To embark on a list would be tedious.

## 2.35 Symbiosis: Mutualism

When an animal's food includes a residue (as defined in sec. 3.1), the animal, by its eating, does not enter the environment of the organism that made the residue. For example, the Australian mistletoe bird, *Dicaeum hirundinaceum*, does not, by virtue of *eating* the fruit of the mistletoe, enter the environment of the mistletoe as a predator.

The mistletoes of Australia, *Loranthus* spp., are parasites (predators) of a wide variety of indigenous shrubs and trees. The seed of *Loranthus* germinates on a branch of the host, and sucker-like roots penetrate to the host's vascular system. The fruit of *Loranthus* is a berry with a single hard-coated seed. The mistletoe bird, *Dicaeum*, is distributed throughout mainland Australia wherever *Loranthus* occurs. The bird feeds exclusively on the fruit of the mistletoe. The seed is not digested; it passes rapidly through the gut. It is likely to be excreted while the bird is visiting another tree. Because the seed is coated with an adhesive, it is likely to stick to any branch it touches. On a suitable branch the seed is likely to germinate and give rise to a new plant. The bird's activity in carrying seeds to new localities results in more mistletoe and hence more food for birds in future generations. How is this complex set of interactions to be represented in an envirogram?

The first step is to realize that the flesh of the mistletoe fruit and the seed inside it enter the environment of the bird in two quite different ways. Failure to recognize this leads to hopeless confusion in trying to work out the relation between the bird and the mistletoe plant. Only the fleshy part of the mistletoe fruit is in the centrum of the bird. All other relevant interactions occur in the web of the environment of the bird or of the mistletoe. The bird and the mistletoe benefit from the mutualism thus:

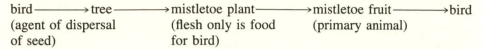

bird ⟶ tree ⟶ mistletoe plant ⟶ mistletoe fruit ⟶ bird
(agent of dispersal          (flesh only is food          (primary animal)
of seed)                         for bird)

Neither the seeds of the mistletoe nor any other vital parts of the plant are eaten, so the plant does not enter into the centrum of the bird. The mistletoe plant is the source of the fruit which is the food for the bird and so is in the web of the bird, as is the tree, which provides the food for the mistletoe. The bird must appear in the web of its own environment three steps removed from the centrum, because the bird is an agent of dispersal of the mistletoe. The bird is not a predator of the mistletoe even though it eats and is sustained by the flesh of the mistletoe berry. By eating of the edible flesh that surrounds the seed the bird in no way impairs the vitality of the plant. The pathway has the same form when the mistletoe is placed in the position of primary organism. A similar pathway, with the primary animal occurring in the web of its own environment several steps removed from the centrum, will be found in any envirogram that includes a mutualism. For example, a pat of cow dung contains nutrients that might nourish the grass on which the cow feeds. The nutrients might be lost or be recycled only slowly if the dung is left undisturbed, but they will be recycled rapidly when certain scarab beetles, by feeding on the dung, disperse and bury it.

In section 3.1 we distinguish between two kinds of food—living organisms and their residues. Excretions such as dung, honeydew, nectar, pollen, and the flesh of fruit (as distinct from seeds) are counted as residues along with carcasses and other organic refuse or litter. Mutualism parallel to that between *Loranthus* and *Dicaeum* occurs when honeyeaters, other sorts of birds, bees, flies, and other sorts of insects spread pollen from plant to plant while feeding on pollen, nectar, or the numerous small arthropods that commonly feed in flowers.

A variation on the same theme is seen in the relation between a cow or a wallaby and the microbes that live in the stomachs of such animals. The mammal provides nitrogenous food for the microbes by excreting urea from its blood into its saliva, and the microbes share the vegetable food, much of it rich in indigestible fiber, that the mammal swallows. The mammal adds to its diet of protein and digestible energy-foods by digesting the carcasses of the microbes that have died. Indeed, mutualism like this, that depends on the intervention of residues as food, is commonplace in nature.

Not all examples of mutualism depend on nonreactive food. In the classic example of the anemone that settles on the snail shell that shelters the hermit crab, it is said that the crab, acting as an agent of dispersal, improves the supply of food and perhaps oxygen for the anemone. And it is said that the anemone reduces the snail's risk of being taken by certain predators.

On analysis, all cases of mutualism follow the general pattern set by the mistletoe and the mistletoe bird in that each organism fits into the web of the other's environment but neither fits into the centrum of the other. The distinction between mutualism and the usual predator-prey relation between an animal and its living food is illustrated in the following segments of envirograms:

fox —————————→ rabbit
(predator)              (primary animal)

rabbit —————————→ fox
(food)                      (primary animal)

mistletoe bird ———————————————→ tree ———————————→ mistletoe
(agent of dispersal,                      (food)                         (primary plant)
modifier $n_1$)

tree ———————————→ mistletoe ———————————→ flesh of berry ———→ mistletoe bird
(modifier, $n=2$)       (modifier $n=1$)         (food)                          (primary animal)

## 2.36   Aggression between Different Species

In section 1.42 we mentioned that the environment of the ant *Myrmica rubra* contains a malentity and a modifier that can both be explained by the aggressive behavior of another ant, *M. scabrinodis*. The latter drives *M. rubra* from the most favored nesting sites and in the process may maim or even kill some. The violence done directly to the bodies of the victims is a malentity; but the enduring outcome of the aggression is that *M. rubra* must make do with inferior nesting sites where diverse components of environment are less satisfactory.

Such "double-barreled" aggression is commonplace among the social Hymenoptera (chaps. 6 and 7) and it is found in some other kinds of animals as well; but for the most part the chief or only outcome of aggression is that the victim is forced to tolerate an environment that is inferior in some respect, often but not always in terms of food.

## 2.37   Territorial Behavior

Territorial behavior is discussed in relation to food in section 3.121 and in relation to tokens in section 3.31. Territorial behavior in the rabbit is discussed in section 12.2, and that in the magpie in section 13.1. Here it is enough to say that the universal outcome of territorial behavior is to increase the probability that a certain component (or components) of environment will operate optimally for a favored section of the population no matter how deprived the remainder of the population may be. It is clear that this sort of aggression exerts its influence on the primary animal by modifying some other component of environment: the "territorians" have a place in the web of the primary animal. In this instance the mean represents the typical animal inadequately because territorial behavior separates the population into two very uneven subpopulations. Fortunately they can be easily recognized and taken into account separately as Carrick's (1972) observations on magpies showed (secs. 3.31, 13.1).

## 2.38   Simple Overcrowding

Overcrowding may be related to resources other than food and perhaps to components of the environment other than resources; in certain species, notably aphids, locusts, and armyworms, it seems that it can be related to the frequency of "jostling" or other sensual contacts with neighbors. But in this section we restrict the discussion almost entirely to overcrowding with respect to food. Overcrowding by "eaters" implies a local (or absolute) shortage of food (sec. 3.11). The eaters may belong to the same or different species. Overt aggression is ruled out; the most that is done in the interest of self or family is to strive to acquire or preempt as many mouthfuls as possible ahead of the neighbors.

In section 12.426 we mention that during the late 1940s when the rabbit was extremely numerous, before the outbreak of myxomatosis in 1950, in southeastern Australia, with rabbits feeding alongside sheep, the sheep got about 90% and the rabbits 10% of the shared pasture. These figures quantify one relationship that is represented by the sixth of the second-order modifiers in the branch of pathways that converge on food in figure 2.01.

When all the animals belong to the same species the reduction in food supply cannot be partitioned as above, but it can be measured in terms of how much food the overcrowded animals get at various levels of overcrowding. To measure the environmental component (usually food) is just as essential as to count the animals.

A good example of how to treat this sort of problem is given by Miller's analysis of overcrowding in *Choristoneura*. Miller (in Morris 1963) measured the total stock of food and the proportion of it that was used effectively (sec. 3.121) at different levels of crowding. He related the two quantities through a multiple regression (sec. 14.211).

## 2.39 The Locality

The concept of the "local population" was introduced in section 1.7 as an aid to studying the natural population, which is the reality. To think of a local population is to imply a locality that houses it. In section 1.7 we define locality as a place where a local population might live. Such a definition allows for localities that are empty but have a chance of being colonized as well as those that already support a local population.

We find the concept of the local population (living in a locality) useful in a number of specific contexts (e.g., secs. 3.121, 8.41, 8.5). The specific usefulness of the concept in making an envirogram turns on the ubiquity of dispersal among animals and the tendency for most species to specialize in dispersal during one stage of the life cycle. Even vertebrates which usually have a strong attachment to a "home range" while their attention is concentrated on growing and breeding usually have a stage in the life cycle, often as young mature adults, when this attachment is overcome by an urge to disperse and find a new place to live. With other kinds of animals specialization is carried further. One has only to compare the spiderling, dispersing on its gossamer, with the spider, lurking in place to snare its prey, or the planktonic mussel with the settled shellfish to appreciate the extremes of specialization that have evolved to fit animals for exploiting the chosen locality at one time and seeking a new one at another. The aphid even makes a massive amount of wing-muscle to sustain it during a long dispersal flight; then, after it has settled, the muscle is autolyzed and contributes to the nourishment of eggs.

Just as an envirogram should take account of the distinct requirements and risks associated with the age-classes in the local population, so it must take account of the differences between the sedentary and dispersive phases of the life cycle. For the dispersing animal the only risk that matters is the risk of not finding a locality. This risk will nearly always be determined largely by the distribution and abundance of localities and the activity of agents of dispersal, as, for example, in the lower part of the envirogram of the dispersing spruce budworm larva (fig. 2.02). den Boer (secs. 8.415, 8.42) would argue that natural selection will have balanced the dispersiveness of the animal against the scarcity of localities. Perhaps the extreme dispersiveness of the spruce budworm reflects a corresponding scarcity of localities in the forest between outbreaks, the scarcity arising through paucity of mature trees and prevalence of cyclonic weather.

In practice the boundaries of a locality may be drawn quite arbitrarily. Given that sampling methods are adequate, boundaries may be drawn to suit convenience. For such localities as a carcass, a pond, or a tree, the boundaries are obvious. The eye of a naturalist can often draw more subtle boundaries quite objectively. See the quotation from E. B. Ford in section 1.7.

But when the boundaries have been perceived the area may still be too large for convenience or for sampling. Statistical limits may have to be devised. den Boer (sec. 8.42), confronted by 1200 ha of heathland where many species of wingless carabid beetles lived, chose to regard as a locality an area surrounding a pitfall trap from which a beetle might come to fall into the trap. This definition sufficed for a study of "spreading the risk" that he had in hand. Each problem will need to be solved on its own merits.

# 3

# Resources

## 3.0 Introduction

Section 1.41 lists five resources. They include all the "things" that promote or organize a healthy metabolism—food, water, oxygen, heat, and tokens.

Food is interesting because it is variable and has many facets. Heat was discussed rather fully in Andrewartha and Birch (1954, chap. 6) and Andrewartha (1970, sec. 6.1), and we find ourselves with little new to say about it. The supply of oxygen is so reliable that we can find nothing interesting to say about that. On the other hand, the idea of tokens is new to us and we shall try to develop it.

## 3.1 Food

Animals occupy the second and higher trophic levels in the ecosystem, so their food must be organic, but it may be either the living organism itself or its residue. A residue might be a log or a carcass from a plant or animal, or any other sort of organism that has died. Or it might be an excretion such as feces, honeydew, or leaf litter from a living organism, or the flesh that surrounds the living seed of a fruit, or an artifact like honey from a bee.

The important ecological difference between "living" and "residual" food is that the former is reactive and the latter nonreactive (fig. 1.03). The primary animal that uses living organisms for food becomes a predator in the environment of the food-organism. It is this reciprocal relation between eater and eaten that is represented in table 3.02 (sec. 3.13). Residual food is not mentioned again in this chapter.

Unlike a treatise on animal husbandry, this book seems scarcely to mention the important subject of quality. We treat quality, in rather unusual contexts, as part of the discussion of extrinsic relative shortage of food in section 3.131 and in relation to the choice of food in sections 5.211, 5.221, and 11.312.

As a component of environment, food is important only when there is a chance that the animal will not get enough. As food becomes abundant and accessible, the probability of getting enough approaches one, and it cannot exceed one no matter how abundant the food becomes. So this section is all about shortages of food.

### 3.11 Absolute Shortage versus Relative Shortage

In Andrewartha and Birch (1954, 497) we coined the phrase "absolute shortage of food" for the condition where all the food in an area has been consumed and animals are dying of starvation because there is not enough food to go around. This condition is contrasted with the condition in which, though only a small proportion of the food has been eaten, nevertheless some animals are dying from starvation, and shortage of food sets a limit to the rate of increase in the population; we called this contrasting condition a "relative shortage of food". Relative shortage is discussed in section 3.13.

In 1954 we said that absolute shortage "is rather unusual in nature over broad areas, though it may happen sometimes as a temporary phenomenon in the smaller local areas of a larger population". We quoted Elton (1949, 16) in support of this opinion: "Indeed, so apparent is this that it has been for many years a convention among botanists to treat dynamic vegetation systems as though the animals were not having any influence upon the energetics of the plants at all; or only to bring in this idea when the inroads are of a very conspicuous kind, as with rabbit, or stock or deer grazing". We then went on to develop the idea of absolute shortage, using the results from experiments with blowflies that were forced to live crowded in a cage containing only a small carcass or a limited amount of meat for food.

Our theory about absolute shortage has not changed much since 1954. We still think it is a useful abstraction when it is applied to the idea of a local population. It is an abstraction for the same reason that the local population is itself abstract (sec. 1.7) and also because, in nature, the diversity of living places for food and the diversity of fitness among animals that search for the food make the extreme "egalitarianism" that is implicit in the definition of absolute shortage seem rather unrealistic even when the locality that might house the local population is small and relatively homogeneous. Furthermore, the idea of absolute shortage does not march with the realities of the natural population. Indeed, a true absolute shortage (all the food consumed) in a true natural population (the whole species) must have at least as low a probability as the extinction of the species.

The shortage of food that is characteristic of a natural population is a relative shortage, but this is not inconsistent with the presence of absolute shortages in some or many of the local populations at the same time.

In Andrewartha and Birch (1954, 624) we spoke of a "shortage of food in a relative sense" and on 491 we explained what we meant by this phrase. Under the heading "paradox of scarcity amid plenty" we wrote:

Most species are rare, relative not only to the places which they may occupy but also the stocks of food in the areas where they live. In other words, most natural populations (considering not small localized situations but the whole distribution or a substantial part of it) regularly consume only a small proportion of the food in their area. Territorial behavior may provide a partial explanation for this in those species which possess it; but it is lacking from many vertebrates, and it is absent or rare in the invertebrates. An obvious explanation, and one which may often fit the facts, is that some other component of environment—possibly, though not necessarily, predators—may hold the population in check. But this is not a universal explanation, for we may recognize certain sorts of situations in which a shortage of food may be chiefly responsible for the failure of the population to increase, even though, at first sight, food would seem to be plentiful, because only a small proportion of the total is consumed by the animals.

The essential feature of the relative shortage of food is identified in this passage: some, perhaps much, food remains uneaten, yet some animals, perhaps many, are dying from starvation. They are dying from starvation not because all the food has been eaten, as in the absolute shortage, but because they cannot find it in time. The food has proved inaccessible (or the animals' capacity to disperse and search has proved inadequate), perhaps because the food is widely scattered, or it is in a repugnant place or concealed in an unlikely place, or it is too dilute or has changed its condition with the passage of time, as when a food plant senesces and ceases to be nutritious, or for a variety of other reasons. Of course such changes in the condition of the food may also be responsible for an absolute shortage of food, not only locally but also over extensive areas.

## 3.12   Adaptations to Shortage of Food

Because animals mostly grow up in a local population, natural selection has equipped them to make the best of an absolute shortage of food. But because there are many reasons why it is often better to leave a local population than to stay with it, natural selection has also endowed most animals with an instinctive urge to disperse and with the means to conduct a search for food and a good place to live—the qualities that are needed to deal with a relative shortage. We discuss these two niches (professions) in sections 3.121 and 3.122 and also in section 8.42.

### 3.121   Fitness in the Face of an Absolute Shortage of Food

Any nonsocial animal that dies from starvation before it breeds might be said to have wasted the food that it ate. In Andrewartha and Birch (1954, 498) we called food that was eaten but not so wasted "effective food". Evolution seems to have followed two lines, one physiological and the other behavioral, that can be seen as mechanisms for maximizing the proportion of the total food that is used "effectively".

The physiological strategy is that of genotypes that can respond to a ration smaller than the optimum by curtailing growth and reaching the reproductive stage at a smaller size. It is an adaptation that is well developed in scavengers, but it is also found widely in diverse kinds of animals including man. For example, the blowfly *Lucilia sericata* lays its eggs on carrion at a specific (early) stage of decomposition. With plenty of food the maggots grow quickly, reaching a weight of about 60 mg before pupating. It is not essential for maggots to grow so big, and when food is scarce they may still pupate, provided they can attain a weight of about 25 mg (Ullyett 1950). When few eggs are laid on a piece of meat of limited size, a few large flies are produced; with a moderate number of eggs a larger number of smaller flies emerge. With increasing numbers of eggs laid on a fixed weight of meat, the number of flies emerging increases up to a maximum and then begins to decrease, eventually becoming zero. In Nicholson's (1950) experiments this maximum was reached when 25 females were allowed to lay their eggs on 50 g of meat. With increasing numbers, fewer and fewer maggots were able to get enough food to grow to the weight at which pupation became possible, until, with 150 females laying eggs on 50 g of meat, there were so many maggots that none got enough food and all died as larvae. All the food was eaten, but not one individual was contributed to the next generation. The ability to achieve the reproductive stage at 25 mg allowed the blowflies to reach a higher population density before beginning to lose effective food.

Behavioral adaptations for maximizing effective food all come under the general heading of territorial behavior. Several examples will illustrate the point.

The eggs of the bug *Nezara viridula* are laid in a "raft" of up to 90 barrel-shaped eggs standing on end, side by side. A single egg contains enough food for a larva of the predator *Asolcus basalis* to grow to maturity. The female *Asolcus* pierces the bug's egg with her ovipositor and places her own egg inside the egg of the bug. When more than one *Asolcus* egg is laid in the same *Nezara* egg it is usual for only one *Asolcus* adult to emerge, the oldest, and therefore the largest, *Asolcus* larva having eaten the others at an early stage. But according to Wilson (1961) it is unusual for a *Nezara* egg to have more than one *Asolcus* egg laid in it.

Wilson described how a female *Asolcus*, ovipositing on a raft of bug eggs, will examine each egg thoroughly with her antennae before either inserting or not inserting her ovipositor. If she pierces the egg she usually lays an egg in it (93% of occasions in one experiment). If she lays an egg she invariably marks the bug egg with a pheromone, by stroking the tip of her ovipositor back and forth over the top of the bug's egg. This mark is highly effective in preventing either herself or any other female from laying an egg in a bug egg that already carries an *Asolcus* egg. Table 3.01 gives the distribution of "stings" when seventy–five eggs of *Nezara* were exposed to either one or two female *Asolcus*. A sting was counted when a female was seen to pierce an egg with her ovipositor. On five occasions out of eighty-five a female pierced a bug egg without laying an egg in it. The values of $\chi^2$ at the foot of the table show no significant difference between the two empirical frequency distributions (one female compared with two females); but both of the empirical distributions were different (p < 0.001) from the theoretical Poisson distribution.

When more than one female was present on an egg-raft they took little notice of each other at first, but towards the end when most of the eggs had been marked one female usually became highly aggressive, driving all others away; she then continued searching on her own.

By virtue of the cannibalism of the larva, all the food was used effectively; none was lost with individuals that failed to reach maturity. By virtue of the female's

**Table 3.01**   Distribution of "Stings" in Seventy-five Eggs of *Nezara* Exposed to One or Two Females of *Asolcus*

| Number of Stings (x) | Number of Eggs (f) | | | |
| | Observed | | Expected on a Random Distribution | |
| | One Female | Two Females | One Female | Two Females |
|---|---|---|---|---|
| 0 | 1 | 2 | 23 | 21 |
| 1 | 66 | 57 | 27 | 27 |
| 2 | 5 | 11 | 16 | 16 |
| 3 | 1 | 4 | 6 | 7 |
| 4 | 2 | 1 | 2 | 3 |
| $\chi^2$ | Nonsignificant | | 86.5 | 55.4 |

*Source:* After Wilson 1961.

ability to recognize and obey a mark, the prospect of using all the food in the local situation was enhanced because eggs were not offered wantonly to the cannibals. And the female's readiness to fly away in the face of a little aggression as the absolute shortage in the local situation became imminent enhanced the chance that food in other localities would be found. The beauty of the system is that all these behaviors, while maximizing the numbers in the whole population, seemingly through cooperation, actually operated through maximizing the individual's chance of selfishly contributing her own progeny to the next generation. This is the level at which natural selection always works.

The female olive fruit fly *Dacus oleae* lays only a single egg in an olive fruit; after laying her egg she marks the fruit with "juice" regurgitated from her proboscis, which deters others from laying in the same olive (sec. 14.3). Similarly, females of the Queensland fruit fly *D. tryoni* are deterred from laying eggs in fruit that already has had eggs laid in it. The females alight on the fruit but do not stay; they depart, giving the impression that they have been repelled by something on the fruit, though it is not known what that substance might be (Gary Fitt, pers. comm.).

Territorial behavior in the Australian magpie *Gymnorhina tibicen* (plate 5; sec. 13.111) is superficially different from territorial behavior in *Asolcus* and *Dacus*, but fundamentally it is the same because it tends to ensure that any individual that gets any food will get enough to reach the reproductive stage in the life cycle. The magpie, which is about the size of a pigeon, is one of the commonest birds in Australia. Carrick (1963) studied the ecology of a population of magpies in several thousand hectares of farmland near Canberra (sec. 13.11) and found that the birds in this area comprised a hierarchy of groups. At one extreme were groups of two to ten birds which lived in and defended a "territory". The boundaries of the territories changed a little during the years as a result of vicissitudes experienced either by a group itself or by its neighbors. But, in general, once a group had established itself in a territory, it remained there for the duration of its existence as a group. The territory sufficed for all its requirements. The birds found all their food within its boundaries and reared their young there. The minimum requirement for a territory was a tree in which to build a nest, but most territories contained many trees. Carrick rarely found a bird from one of these groups that was undernourished, so apparently each territory contained abundant food for the occupants.

At the other extreme, a homeless flock comprised all those that were not members of a group with an established territory. The numbers of the flock were swelled each year, toward the end of winter, by the remnants of territorial groups that had abandoned their territories or been driven out of them. From time to time a group would originate in this flock and establish itself in a territory.

The birds in the flock were more likely to die from parasites, diseases, and other hazards than those in the territorial groups, and their supply of food was less certain, but the flock persisted because of the regular influx of young birds from the territories each year. On the other hand, the size of the flock had very little influence on the size of the breeding population, which seemed to be determined more by the innate "territorial" behavior of the birds than by anything that could be measured. Territorial behavior analogous to that found in *Asolcus* and *Dacus* is commonplace among insects in the group Parasitic Hymenoptera. Territorial behavior is also found in all the major groupings of the vertebrates, but in them it is more variable.

Insofar as the magpie forms a group to defend a territory throughout the year, for the duration of the life of the group, territorial behavior in the magpie is more typical of mammals than of birds. No vertebrate uses food so effectively as *Asolcus* and some other insects.

### 3.122   Fitness in the Face of a Relative Shortage of Food

To be good at dispersing and searching makes for fitness to deal with a relative shortage of food. In Andrewartha and Birch (1954) we devoted a full chapter to dispersal. Here we discuss dispersal briefly in section 3.132. See also section 8.42.

## 3.13   The Conditions of Relative Shortage

When the food is a living organism the animal that does the eating is a predator in the environment of the food. By its feeding the predator may reduce the supply of food for its posterity. This risk may be so small as to be negligible or so large that it dominates the ecologies of both species. Whether it is large or small will depend on certain aspects of both ecologies.

When the food is a residue, as it is with scavengers, this risk is zero, because the supply of food to subsequent generations is independent of the amount of food eaten in the current generation. The best way to summarize the interaction between the predator and its living food is in a 2 by 2 contingency table (table 3.02). We explore the consequences of this interaction in the next four sections.

### 3.131   The Extrinsic Relative Shortage of Food

The condition defined by cell 1 in table 3.02 has been caused not by the feeding of the predator but by some other component in the environment of the food. We call this an extrinsic shortage because it happens independently of the special predator-prey relation between the primary animal and its food organism. That is, the cause of the shortage is extrinsic to the particular relation, in contrast to the intrinsic shortage (sec. 3.132). In the condition of an extrinsic shortage of food many or most of the animals in a population go seriously short of food even though there would be plenty of food if only they could find it. The food may be plentiful or scarce, but in either case it is inaccessible.

The classic account of an extrinsic shortage of food in a large natural population occurs in a paper by Jackson (1936) on the ecology of the tsetse fly *Glossina morsitans*. The tsetse fly feeds by sucking blood, chiefly from ungulates. It needs to feed frequently, especially during hot weather, but it does not stay near its host after engorging, so each meal depends on the success of a new and independent search. If the food is sparsely distributed the fly may run a heavy risk of not scoring enough successes. Then it may die from starvation without contributing any off-spring to posterity—even though the blood in one antelope would be enough for many flies. Jackson pointed out that changes in the number of antelope occur quite independently of the number of flies feeding on them but nevertheless determine the rate of increase or decrease in the population of flies. In other words, the shortage of food for the flies might be alleviated or exacerbated, but the cause of the change might lie quite outside; that is, it would be extrinsic to the predator-prey relation between the fly and the antelope. Jackson gave his empirical observations and stated his inferences definitely and very clearly. Then, knowing that this was an

**Table 3.02** Analysis of Interactions That May Determine the Condition of Food for the Primary Animal When the Food Is a Population of Living Organisms

| Predation by Primary Animal on the Organism That Is Its Food or the Source of Its Food | Shortage of Food for Primary Animal | |
| --- | --- | --- |
| | Severe | Zero or Negligible |
| Zero or negligible | 1. Extrinsic shortage of food for primary animal (sec. 3.131) | 2. No shortage; permanent condition (sec. 3.133) |
| Severe | 3. Intrinsic shortage of food for primary animal (sec. 3.132) | 4. No shortage; temporary condition during one stage in succession (sec. 3.133) |

*Note:* Because the action takes place in a natural population, any shortage is essentially a *relative* shortage.

iconoclastic conclusion in the face of strongly entrenched contemporary theory, which placed absolute emphasis on "competition" and "density-dependent factors", he went on to drive his point home with a hypothetical numerical example which we paraphrase (see Jackson 1936, 886).

Suppose there are a hundred tsetse flies in an area which also supports just enough antelope to enable the flies to maintain their numbers indefinitely without increase or decrease. This does not mean that the total weight of blood is just adequate for the needs of a hundred flies, but rather that the antelope are just numerous enough to ensure that the average tsetse fly meets with food often enough to produce offspring at the rate required to match the death rate in the population. Suppose nine hundred new flies are introduced to the area. The fecundity of each newly arrived fly, depending as it does on the frequency with which the fly meets an antelope, will be just the same as that of the original inhabitants, and the population will continue to maintain itself at a steady level. This is because, even with the larger number of flies, there is still no shortage of food in the absolute sense. Jackson expressed the same point in different words:

It [food] may no more increase its action on a rising population of the tsetse fly than does climate, nor temper its severity towards a diminishing community. [In other words, food is not operating as a "density-dependent factor".] It seems that there is no pressure of numbers in the ordinary sense because there is probably no competition for food, and certainly none for shelter, as the writer's experiments indicate that the flies are so sparse that there can be no physical crowding even when the apparent density is comparatively very high.

Elsewhere in the same paper Jackson pointed out that there is no evidence that predators or disease influence numbers to any extent, and he concluded: "There remains the possibility that (at moderate densities at least) there are no dependent factors at all acting on tsetse".

Perhaps the best-documented account of an extrinsic shortage of food comes from the work of Milne (1949, 1951, and other papers) on the ecology of the sheep tick on some farms in northern England. The ticks *Ixodes ricinus* that live in the rough upland pastures of northern England feed mostly on the blood of sheep and only to an unimportant extent on that of wild animals. During most of the year the ticks remain inactive in the mat of vegetation near the ground. In spring they emerge from this shelter and climb up an exposed grass-stem. If, during the 9 or 10 days that they can survive in this exposed position they are picked up by a sheep as it brushes by, they cling to the sheep until they are engorged with blood. Then they drop off and crawl back into the mat of vegetation. They digest their meal and develop to the next stage of the life cycle. Each tick needs to feed three times during its life—as a larva, as a nymph, and as an adult. After the third meal the eggs are laid.

On one farm where Milne worked he estimated that a tick had about a 40% chance of being picked up by a sheep in any one year. The probability that a tick would be picked up three times during its life can be estimated as $0.4^3 = 0.06$. In other words a tick might have a 6% chance of getting enough food to allow it to complete its life cycle and lay eggs. This farm carried about one sheep to the acre. Had there been more sheep the ticks would have had a greater chance of getting enough food, since a tick's chance of being picked up depended only on the number of sheep in the area and their activity. It did not depend at all on the number of ticks in the area, because the feeding of the few that did get a meal made no difference to the number of sheep in the area. The shortage of food was an extrinsic shortage and a severe one; 94% of the ticks died from starvation.

Extrinsic shortage of food may take many forms. Many aquatic species feed by pumping through their mouth a continuous stream of water from which they filter out the small particles of food, discarding the water. The larvae of certain species of mosquito do this. If the concentration of food in the water falls below a certain threshold, the larva may not be able to filter water fast enough to get its minimum requirement of food and will die from starvation. This might happen even to a solitary larva, the sole occupant of a large pool that contained enough food to support many larvae if only it could be concentrated.

According to Main, Shield, and Waring (1959), *Setonix brachyurus*, a wallaby living on a small sandy island off the coast of Western Australia, becomes debilitated and many die from malnutrition during the long hot, dry summer. The main cause of death is shortage of food, chiefly protein. Yet they are living in a dense thicket of shrubs which they eat voraciously; indeed, the limit to their eating seems to be set by the extent to which the stomach can be distended. The shrubs do not appear to be obviously denuded. The contrast comes after the rainy season begins. During the mild, wet winter the shrubs put on new growth, which the wallabies prefer. For this brief season the wallabies get enough protein and water in their diet, and they prosper and breed. Like the mosquito larva, the wallaby could get enough protein out of the sclerophyllous vegetation that it must eat for most of the year if only the concentration of protein were greater. For the wallaby cannot afford a larger stomach, just as the mosquito larva cannot afford a more powerful pump. Natural selection always balances the ledger in this way.

A similar condition was observed by White (1966, 1969; see also sec. 10.311). White studied the ecology of the psyllid *Cardiaspina densitexta*, a small sap-sucking insect that lives on the leaves of the evergreen tree *Eucalyptus fasciculosa*, which is native to South Australia. It grows in places where the annual rainfall is between 450 and 650 mm, falling mostly in the winter, and the soil is mostly a shallow sand overlying an impermeable layer of clay-loam or limestone. Because of the soils they grow on and because it is their nature to produce a system of shallow roots, the trees are susceptible to the water regime. In the drier part of their distribution they are restricted to the moister localities; toward the wetter extreme they occur only in well-drained places.

Massive outbreaks of *Cardiaspina* occurred during 1914–22 and 1956–63. Between outbreaks the psyllids were scarce, being absent from many trees, in low numbers on many others, and abundant only temporarily and locally on solitary branches, solitary trees, or small groups of trees. White noticed that during 1914–22 and 1956–63 there had been the same sort of variation in the weather. The winters, normally wet in this region, had been wetter than usual, and the normally dry summers had been drier than usual. It seemed unlikely that such weather would influence the psyllids directly, but it was likely, given the sort of soil where *E. fasciculosa* usually grows, that this weather would be severe on the trees. During winter the shallow layer of sand where all the roots are would be waterlogged, killing the system of fine feeding roots. During summer there would be a shortage of water in the soil, and the trees would have few feeding roots with which to seek it. White (1969) suggested that in such a condition of "stress" the tree would mobilize nitrogenous building materials in the aerial parts and transport them in the sap to where they could be used to make new roots. The enriched sap would provide a source of nitrogen-rich food for the sap-sucking psyllids; on this rich food the female could lay more eggs and the newly hatched nymph could survive and grow. This performance is in marked contrast to the low fecundity and abysmally low survival rate of the young nymphs when their only source of food is the dilute sap that is characteristic of the healthy tree in its normal condition. This explanation has still to be verified empirically, but we accept it because it is plausible and because White (1978), in a wide search of the literature, found similar extrinsic shortages of food, which permitted an analogous explanation, in the ecologies of diverse animals living in diverse circumstances (sec. 10.311). Andrewartha and Browning (1961) described a number of other occurrences of extrinsic shortage. The phenomenon is indeed widespread and diverse. See also section 15.21 for extrinsic shortage of food for man.

### 3.132    The Intrinsic Relative Shortage of Food

The intrinsic shortage of food is defined in cell 3 of table 3.02. At first sight and from the anthropocentric viewpoint this sort of shortage looks like an extrinsic shortage. But the essential difference is that the primary animal has caused a relative shortage of food for itself by its own action as a predator in the environment of its food organism.

The classic example of an intrinsic shortage of food concerns the biological control of the scale insect *Icerya purchasi* on citrus and ornamental shrubs and trees in California (Quezada and De Bach 1973). The scale insect was an exotic pest on an exotic plant that was grown in one of man's artificial monocultures. The pest was reduced to low numbers and maintained in this condition by two exotic predators, the beetle *Rodolia cardinalis* and the fly *Cryptochaetum iceryae*, which were artificially transported across an ecological barrier and let go in California. The predators, by their feeding, created and sustained a severe relative shortage of food for themselves. As a consequence they became rare, at the same time keeping their food rare also. The shortage of food for the predators is an intrinsic shortage, because the cause of the shortage is intrinsic in the predator-prey relation between the predator and its food.

The biological control of *Icerya* in California is the classic example of an intrinsic shortage of food because it was the first documented case of the biological control of an insect pest and it is typical of all those that have been documented since. Notwithstanding popular opinion to the contrary, all the well-documented examples of intrinsic shortage of food for carnivorous insects have been artificially contrived by man as ventures in biological control. The animals experiencing the shortage of food are exotic, and the plants at the base of the food chain are exotics growing either as crops or as weeds associated with man's monocultures. It is conventional wisdom among ecologists that similar food chains generating intrinsic shortages exist among endemic species of insects living in places that have not been modified by man so extremely as have farms, but this opinion is not well documented.

There need be no doubt that the shortage of food experienced by *Rodolia* feeding on *Icerya* in California conforms to our definition of an intrinsic shortage because of the well-documented history of the original colonization by *Rodolia*, and also because the conclusion that was drawn from the early history has been confirmed more recently when orchards have been sprayed commercially with insecticides that kill *Rodolia* but not *Icerya*.

The evidence, as related by Quezada and De Bach (1973), may be summarized briefly: *Icerya purchasi* was accidentally introduced into southern California from Australia about 150 years ago. By 1890 it had spread and multiplied so much that it threatened to destroy the citrus industry. In 1888 a colony of 129 ladybird beetles, *Rodolia cardinalis*, from Australia were established inside a cage built over an orange tree; the tree was carrying a dense population of *Icerya*. By April 1889 the ladybird beetles had increased greatly and had destroyed nearly all the scale insects. The cage was removed, and within three months the *Rodolia* had spread through the orchard, destroying nearly all the *Icerya* there and thus making a severe intrinsic shortage of food for themselves. This sequence was repeated elsewhere in southern California; and ever since there have been few *Icerya* in this area. Smith (1939) explained this result in terms of the high powers of dispersal of the predator relative to its prey and also the high rate at which *Rodolia* multiplied. They multiplied rapidly not only because their food was plentiful and readily accessible, but also because they were favored by weather. Further evidence that the shortage of food for *Rodolia* is an intrinsic shortage comes from the use of insecticides. The predator *Rodolia* is more susceptible to certain insecticides than *Icerya*. Quezada and De Bach (1973) reported an outbreak of *Icerya* in a citrus orchard that was sprayed in the spring with methylparathion; *Rodolia* was not seen in that orchard until October, and the outbreak was not under control again until the end of the year. A number of other examples of intrinsic shortage of food generated by a successful agent of biological control were mentioned in Andrewartha and Birch (1954, sec. 10.321) and Andrewartha (1970, sec. 5.11). They might differ from *Icerya*, but only in detail.

The food, whether animal or plant, of a successful agent of biological control usually is patchily distributed. There are extensive areas within the distribution of the prey that are apparently favorable but are not inhabited; isolated colonies will wax and wane and perhaps die out after a while new colonies may be established where none was before and last for a while. But overall the food remains scarce and so, perforce, does the animal depending on it. We have observed this condition in species that are under strict biological control, notably the woolly aphis of the apple *Eriosoma lanigera*, which is the food of *Aphelinus mali* in Western Australia, and the sap-sucking bug *Nezara viridula*, which is kept under strict biological control by *Asolcus basalis* in South Australia. Similar patchy distributions have often been reported in the literature (Flanders 1947).

The same holds for the prickly pear *Opuntia inermis* in Queensland after its biological control by *Cactoblastis cactorum* (Dodd 1940). Today both cactus and moth occur over a huge area of Australia that is similar in extent to the original distribution of the cactus at the height of its abundance before 1930. But both cactus and moth now exist at densities that, relative to their former abundance, are quite low. In Andrewartha and Birch (1954, 87, 502, 658), quoting Nicholson (1947), we suggested that although *Cactoblastis* had great powers of increase and dispersal, it was unable to find and destroy all cactus plants because of the patchy dispersion of the cactus and the ready dispersal of its fruits. The cactus would be protected for some time from *Cactoblastis*, but sooner or later it would be found. While this is still true, it is one aspect of a more complex relation between cactus and *Cactoblastis*. Monro (1967) and Osmond and Monro (1981) have found that *Cactoblastis* moths do not disperse their eggs at random but concentrate on some plants and leave others free or relatively free, at least for a time. Some plants have so many eggs laid on them that the larvae experience an intrinsic shortage of food and very few survive. On other plants there is little or no intrinsic shortage of food, and the larvae survive to produce another generation of moths. Why some plants should be favored is not known, but Osmond and Monro (1981) have made observations indicating that the more susceptible plants are those that are near infested plants, that belong to a dense population, and that are fully green and succulent rather than yellow. The differences are subtle in nature, for a plant that is unattractive in one season may be attractive in the next. Whatever the causes may be, the outcome is that in every season some plants escape predation by *Cactoblastis*. Furthermore, there is always a reservoir of cactus in the form of underground tubers, broken-off cladodes, seeds, and seedlings in local populations of today that survive even heavy predation.

Whatever sort of animal the predator may be, if its food is likely to be scarce and patchily distributed we would expect natural selection to work toward making the predator dispersive and good at searching. Indeed, when we look into this matter we find a wealth of adaptations for dispersal in all sorts of animals and an urge to disperse that is sometimes so strong that to a casual observer it looks like a bad bet: individuals seem to take off looking for a new place to live long before the old one has become untenable through overcrowding or some other form of ecological succession. Often there is a dispersal stage in the life cycle whose behavior inexorably ensures that at this stage virtually every individual will abandon the place where it was born and seek elsewhere for the necessities of life.

In the sea the larger vertebrates traverse the oceans and often have well-defined routes linking the places where they spend different parts of the life cycle. Many invertebrates produce planktonic larvae that are especially adapted by their structure and behavior to exploit currents. In the air over the continents "aerial plankton" has been shown to contain not only adults of many kinds of small animals, mostly arthropods, but also first-stage larvae of larger ones. The larger ones may also disperse under their own power, taking advantage of winds usually closer to the ground (Johnson 1957; Johnson and Taylor 1957; Kennedy 1961). With species that practice territorial behavior, the often-surprising readiness to flee that is displayed by the loser is also part of this general syndrome. Not all dispersal can be attributed to the need to search for food or other resources, but the ubiquity of these adaptations for dispersal strongly suggests that the relative shortage of food (and other resources) is very much the rule in nature.

The distinction between extrinsic and intrinsic shortage of food first appeared in the classic paper by Howard and Fiske (1911). They were writing about the biological control of two caterpillars that were pests of forests. They called the extrinsic shortage "catastrophic mortality factors" that killed a constant proportion of the population independent of its density. They called the intrinsic shortage "facultative mortality factors" that killed an increasing proportion of the population as its density increased. They mentioned the extrinsic shortage only to discard it as unimportant, thus establishing a theory that has prevailed ever since. This theory places far too much emphasis on the condition that is defined in the bottom right corner of figure 3.01 and attaches far too little consequence to the condition defined in the bottom left corner of the figure.

Twenty years later Howard (1931) published *The Insect Menace* in which he documented the damage that insects did to commercial crops and herds. He suggested that many species that had become pests had been able to multiply to such numbers because of man's habit of growing crops of a single species densely crowded in contiguous farms distributed over wide zones or belts of country with similar soils and climate. He argued that these "monocultures", by providing extensive supplies of good, easily accessible food, allowed the insects to multiply to much greater numbers than they could otherwise have achieved. The implication was that the new practice had alleviated a critical extrinsic relative shortage of food that had prevailed under the old way of growing commercial crops. But the twenty years since the publication of Howard and Fiske's (1911) paper had been time enough for a theory that was based on "facultative" (density-dependent) factors and "competition" to become strongly established. So Howard's glimpse of a new and important idea (extrinsic shortage) was overlooked, as was indeed Jackson's (1936) elegant documentation and presentation of the same idea. How completely an established theory was able to obliterate the new idea is shown by the passage from Buxton (1955, 486) which we quote in section 9.34. Compare Buxton's explanation with Jackson's comment on the empirical evidence (sec. 3.131). That this theory is still influential is suggested by the passage from Krebs (1972, 351 quoted in sec. 9.34). A theory that is too strongly entrenched can hold up progress for many years. For further comment see Andrewartha (1961, 168; 1970, 156).

### 3.133   No Shortage at All

In the preceding two sections we have considered the conditions that are defined by cells 1 and 3 of table 3.02. To complete the analysis we must consider cells 2 and

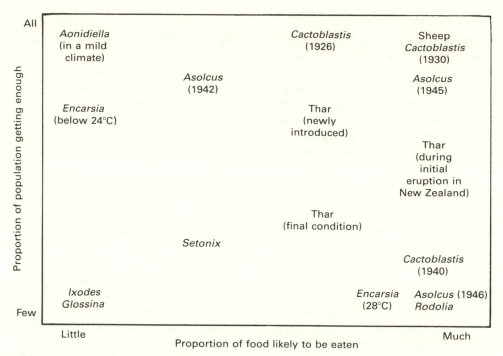

Proportion of population getting enough

All / Few

Little — Proportion of food likely to be eaten — Much

Aonidiella
(in a mild
climate)

Cactoblastis
(1926)

Sheep
Cactoblastis
(1930)

Asolcus
(1942)

Asolcus
(1945)

Encarsia
(below 24°C)

Thar
(newly
introduced)

Thar
(during
initial
eruption in
New Zealand)

Thar
(final condition)

Setonix

Cactoblastis
(1940)

Ixodes
Glossina

Encarsia
(28°C)

Asolcus (1946)
Rodolia

**Fig. 3.01**  Illustrating the conditions that may arise between a predator and its food in a natural population (i.e., between the animal and its food when the food is a living organism). The circumstances causing the particular condition for each species are outlined in the text. (Adapted from Andrewartha and Browning 1961)

4, which define "no shortage". Cell 4 must surely define an unstable and therefore temporary condition, because sustained heavy predation must cause a decline in the number of prey and therefore a shortage of food for the predator. This condition of temporary abundance of the feeding animal accompanied by heavy predation of its food was observed by Dodd (1936) while watching *Cactoblastis* establish itself as an effective agent of biological control of *Opuntia* in Queensland. We infer that a similar condition developed temporarily with initial upsurge in numbers of certain exotic species of ungulates as they established themselves in New Zealand (Caughley 1970; secs. 7.21, 10.32). We believe such a condition always occurs as a characteristic temporary phase in the succession that follows the establishment of a successful agent of biological control (fig. 3.01). We do not know, nor can we imagine, any other circumstances in which the conditions defined by cell 4 might be generated naturally except in the colonization of a locality where there is abundant food.

On the other hand, the condition defined in cell 2 is a well-recognized phenomenon in nature. It has been generally agreed among naturalists since Darwin that most species are rare relative to their food for most of the time (Andrewartha and Birch 1954, 22). If starvation is not a serious risk for at least one part of the life cycle, some other component must be keeping the animal rare relative to its food. Two examples are given in the top left corner of figure 3.01.

### 3.134 Conclusion

The conditions for relative shortage of food are defined in table 3.02 and illustrated in figure 3.01, which can be regarded as a sort of "map". The position of a species along the vertical coordinate depends on whether the "average" animal has a good or poor chance of getting enough to eat. The position of a species along the horizontal coordinate depends on whether the "average" organism in the population of food organisms has a good or poor chance to escape being eaten. If the environment changes, a species may appear in a new position. More specifically, those species that are likely to create an intrinsic shortage of food for themselves (e.g. "good" agents of biological control) when introduced into a place where food is abundant are likely at first to take up a position near the top of the "map"; later, as the supply of food is depleted, they move toward the bottom right corner. *Cactoblastis* is such a species (sec. 3.132); thar is another. Thar is an ungulate that was introduced into New Zealand. It went through an eruption like the one that is described in section 5.1 for moose on Isle Royale (Caughley 1970).

In terms of table 3.02, the bottom left corner of figure 3.01 houses species that experience an extrinsic shortage of food, the bottom right those that experience an intrinsic shortage; along the top are species that experience no shortage at all—in the top left corner the condition is natural and may be permanent; in the top right corner it is artificial (domestic sheep) or temporary (*Cactoblastis*; sec. 3.133). To hold a secure position in the top left corner, a species needs to have in its environment some component other than food that is keeping it rare relative to food. With *Aonidiella* on citrus plants in coastal southern California this component is a predator (De Bach 1969; sec. 5.227). With *Encarsia* it is heat (Burnett 1949; sec. 5.225). *Encarsia* is a small hymenopteron that preys on the whitefly *Trialeurodes*. At 24°C *Encarsia* breeds more slowly than *Trialeurodes* no matter how numerous the whitefly might become. But at 28°C *Encarsia* outbreeds and outdisperses its prey and so creates an intrinsic shortage of food for itself. This gives it a place in the bottom right corner of figure 3.01. The extrinsic shortages characteristic of *Glossina*, *Ixodes*, and *Setonix* are described in section 3.131.

## 3.2 Water

If free water is available most animals will drink; food always contains some water; some water is made from the respiration of dry matter, as oxygen from the air combines with hydrogen from the food; and some arthropods can gain water by taking in unsaturated air and drying it further before expiring or excreting it. On the other hand, animals that live in dry places lose water by respiration and excretion and by evaporation or osmosis through the cuticle.

To measure the activity of water in the environment, the ecologist must take into account the probability of loss as well as replenishment. The units of measurement must be chosen in relation to the mechanisms for conserving water in the body. The classic account of the adaptations for water conservation in aquatic animals was given by Krogh (1939), and that for terrestrial vertebrates by Schmidt-Nielsen (1964, 1975).

About replenishment there is not much to be said. Food may contain enough water without being supplemented by drink. Drinking water comes from rivers, ponds, springs and so forth; their distribution and abundance will be assessed in whatever way is appropriate to the particular study. The loss of water from the body and the risk of losing too much depend on the dryness of the places where the animals usually live and the adaptations that allow them to conserve their water against the forces of desiccation. It is instructive to look at these adaptations from the point of view of the engineering concepts of design, force, work and power.

## 3.21 The Dryness of Places Where Drought-Hardy Animals Live

The land snail *Helicella virgata* is common in southern South Australia, where the winters are mild and humid and the summers hot and dry. The snail is active during wet weather in the winter. During summer it aestivates, seeming to choose places that are especially hot and dry—the top of a fence post or the northern face of a stone wall. During five to six months of summer, while resting in this hot, dry place, the snail might lose 60–70 mg of the 400 mg of water it started with, but very few of them die. Meanwhile an open tank of water alongside the snails might have evaporated a meter.

The kangaroo tick *Ornithodorus gurneyi* lives in northern South Australia in arid rangeland where the annual rainfall might average 175 mm but is very erratic and the average annual evaporation from a free water surface is about 2 m. The tick lives in kangaroo "wallows", buried about 1–2 cm below the surface in the dust or in shallow cracks. When a kangaroo happens to use the wallow the tick may get a meal of blood, but this is infrequent. Meanwhile the ticks can stay alive without food or drink for months at a time. This performance should not be attributed entirely to the impermeability of the cuticle, because the tick, having lost a little water by evaporation, probably has the ability (like its relative the camel tick *O. moubata*) to absorb water from air when the relative humidity is 85% or more. Moreover, it can do this repeatedly. It will have the opportunity to do so because in the desert low temperature at night causes the relative humidity to rise even though the absolute humidity stays low (Andrewartha 1964).

Browning (1954) exposed immature nymphs of the camel tick *Ornithodorus moubata* alternately to 5% and 95% relative humidity at 25C. The ticks lost weight at 5% and regained it rapidly at 95% relative humidity. They continued to do this for 60 days without any change in the maximum weight that was reached each time. After 60 days the pattern was the same, but the maximum weight was lower; after 140 days they died, probably from starvation but certainly not from desiccation.

The kangaroo rat, *Dipodomys merriami*, lives in the desert in Arizona, where the temperature of the air above the ground may go as high as 45°C and that at the surface may be up to 70C; the relative humidity (R.H.) is about 5–15%. *Dipodomys* normally lives on a diet of air-dry seeds and drinks no water. According to Schmidt-Nielsen (1964), *Dipodomys* can maintain its water balance (66% of body weight is water) while living in air at 10% rh and 25C, eating only barley containing 4 g of water in 100 g of seed and drinking no water. Other small rodents living in the Australian and Arabian deserts are similarly drought-hardy (Schmidt-Nielsen 1964; MacMillen and Lee 1969). They all have long hind legs for hopping and long, pointed snouts. This shape has not been convincingly explained.

The flour moth *Ephestia kuhniella* can grow from an egg weighing less than 1 mg to a pupa weighing 16 mg (of which 10 mg is water) in flour that has been dried in an oven and kept in a closed container over concentrated sulfuric acid (i.e., in air of virtually zero R.H.). Even though it is eating such dry food and living in such dry air, the caterpillar can retain enough of the water of metabolism to maintain a water content of 64% in its body. This compares with 68% in a pupa that was reared in air of 70% R.H. on flour containing 14% water.

The brine shrimp *Artemia salina* can live in pools of seawater that have evaporated to saturation and have salts crystallizing out (Croghan 1958). In this very "dry" medium the brine shrimp maintains the normal water content of its body, and the potassium ions remain at their normal concentration in the hemolymph. The concentration of sodium ions and the osmotic pressure of the hemolymph increase about twofold; even so, the osmotic pressure of the hemolymph is only one-tenth that of the medium (measured as % NaCl).

The camel lives on the surface, exposed to the full severity of the desert sun. It must drink occasionally, but it uses water so economically, for a large animal, that it can go a long time between drinks. According to Schmidt-Nielsen (1964), there is an authentic record of a journey of 1,000 km made by camels in the Empty Quarter of the Sahara, where there is no drinking water (sec. 3.222).

All the above examples relate to animals in their normal active condition. As a general rule about two-thirds of their live weight would be water. Certain animals can assume a dormant condition in which they become extremely drought-hardy or cold-hardy. They seem to achieve this extreme hardiness, at least in part, by actively or passively discarding water from their bodies. In extreme cases they are left with only a small proportion of the water they would need in the normal active condition.

The larvae of *Polypedilum vanderplanki*, a small midge of the family Chironomidae, live in temporary freshwater pools that form in rock-hollows in Uganda. In their normal active condition water makes up about two-thirds of their weight. According to Hinton (1960), when the pool dries up the larvae dry up too, becoming virtually as dry as the dust in which they are lying. They survive the drought in this desiccated condition; when the next lot of rain fills the pool again, they replenish their tissues with water and resume development. When some dormant larvae were stored in air of 60% R.H. the water content of their bodies was about 8%. Some larvae that were stored in air-dry dust in the laboratory for thirty-nine months revived when they were placed in water and subsequently completed their development, apparently quite normally. This dormant condition is called cryptobiosis. It was first recognized in tardigrades and rotifers by Leeuwenhoek in 1701 (Hall 1922) and in nematodes by Baker in 1764 (Schmidt 1918). It is now well known that a wide range of eggs, cysts, spores, seeds, and other organisms in cryptobiosis are extremely resistant to drought and to high and low temperatures as well.

Many arthropods, and perhaps other sorts of invertebrates, may assume a dormant condition known as diapause (Andrewartha 1952). The diapausing egg of the grasshopper *Austroicetes cruciata* is about 2 mm by 1 mm. It weighs about 4.9 mg, of which 3.3 mg is water. While it remains in diapause it can survive the loss of about 2.5 mg, but it will die if it loses much more than this. In nature the eggs are laid in hard, bare soil about 1.5 cm below the surface. They remain dormant throughout the long hot, dry summer, develop during the winter (which is mild and rainy), and hatch in the spring. In a very dry summer some of the eggs die from

desiccation (Birch and Andrewartha 1942). It was estimated that a standard evaporimeter (a tank about 1 m in diameter sunk into the ground and kept full of water)
would evaporate 1 m of water while an egg was losing 2.5 mg. By the middle of
June (midwinter) diapause has disappeared and the embryo is developing actively;
the eggs absorb water from the damp soil and swell until they weigh about 6.2 mg.
They become much less drought-hardy. One lot of eggs exposed in air of 55% rh
at 20°C lost 2.2 mg in twenty days.

The adults of the beetles *Otiorrhynchus cribricollis* (Andrewartha 1933), *List-
roderes costirostris* (Dickson 1949), and *Leptinotarsa decemlineata* (Breitenbrecher, 1918) assume a condition of diapause with the approach of summer; the
water content of their tissues is reduced (in preparation for aestivation) to the point
where there seems to be virtually no water left. If they are examined by opening
the abdomen from above, the viscera can be seen completely dried out as a tough,
flat strip adhering to the floor of the abdomen.

As other species enter diapause, in preparation for either aestivation or hiber-
nation, the water content of their bodies is usually reduced, but not often so severely
as in these adult beetles. Animals in this condition can usually survive exposure to
extreme stress—not only heat and dryness but cold and other stresses as well.

## 3.22    Adaptations for Conserving Water

Physiology and behavior are interwoven in the stories that follow. Size has an
overriding influence on "strategy". Most terrestrial invertebrates are so small that
they must be more economical with water than even the kangaroo rat. They achieve
remarkable economies through adaptations in the cuticle, respiratory system, and
excretory system and by seeking shelter in burrows and elsewhere. The camel is too
large to make a burrow, so it must stay on the surface and counteract the heat and
dryness with a different set of adaptations that are appropriate to its size (see below,
sec. 3.222).

### 3.221    Shelter

A good burrow may provide adequate shelter for a small animal, even one that is
not drought-hardy. The trapdoor spider, *Blakistonea aurea*, which will lose water
rapidly and die from desiccation after a few days when forced to live in dry air at
30°C, survives 5–6 months of summer drought in South Australia by sealing the
mouth of the burrow with a silken lid and retiring to the foot of the burrow.
Above ground in the shade on a hot day the temperature may exceed 40°C, and the
relative humidity may be as low as 10%, but 20 cm below the surface at the foot
of the burrow the humidity remains above 90% and the temperature below 25°C.
The spider remains without food or drink for the whole summer. Apparently it is
safer for the spider to remain without food in a place where the air is moist than
to seek food, which would replenish its water, at the mouth of the burrow where
the air is likely to be dry (Andrewartha 1964).

The kangaroo rat *Dipodomys* weighs 36 g and is small enough to live in a deep
burrow. It spends the day in the burrow and emerges at night to forage for the
air-dry seeds that are its staple diet. According to Schmidt-Nielsen and Schmidt-
Nielsen (1950), during the day the temperature of the air near the ground varied
between 20 and 45°C and the relative humidity between 5% and 15%. During the

night the readings were 15–25°C and 15–40% R.H. The absolute humidity varied from 1 to 5 mg per liter outside the burrow and from 8 to 15 mg per liter inside the burrow. The temperature at the foot of the burrow was about 31°C and did not vary much from night to day.

Schmidt-Nielsen and Schmidt-Nielsen (1950) estimated the saving in water of respiration that *Dipodomys* achieved by staying in its burrow during the day. The body temperature of *Dipodomys* is 36–37°C. It was assumed that the expired air contained enough water to saturate air at 33°C; that is, 33.5 mg per liter. According to these figures *Dipodomys*, breathing the moist air of the burrow, saved 24% of the water it would have lost had it been breathing dry air at the surface.

An unmeasured but considerable amount of water was also saved by not sweating. By spending the day in the burrow and the night on the surface, *Dipodomys* spent all its time in places where the temperature was at least 5–6°C below its body temperature. Because it is small, it could dissipate sufficient heat by radiation and did not need to sweat to cool.

### 3.222   Design, Force, Work and Power

To remain healthy and active, an animal must maintain the water content of its body and the concentration of certain critical ions in the body fluid within a narrow range of what is normal for its kind. All terrestrial and freshwater animals and a few marine animals also maintain the total osmotic pressure of their body fluids constant within a narrow range. Such animals must do at least some work to maintain their water balance and those that live in truly dry places seem to achieve great feats. In considering how they do it design is important, as indeed it is in all machines that do work. Also, there is some insight to be gained by considering the force against which the work is done, the amount of work, and the time taken to complete it—that is, the rate of doing work, which is the power brought to bear on the project.

Design is important. None of the terrestrial invertebrates uses any water for cooling. Being poikilotherms, they do not need to; having a very small volume relative to surface, they cannot afford to. Neither do they use water to excrete nitrogen; uric acid is either excreted dry or, more often, stored in the body. These adaptations are not sufficient for those species that live in truly dry places. Most of them have also evolved a wax-covered cuticle that is highly resistant to the outward passage of water. Schmidt-Nielsen (1975, 412) listed a number of species for which the evaporation of water through the cuticle had been measured and expressed as $\mu$g of water evaporated per cm$^2$ per minute per mm of Hg of saturation deficit. They were, in $\mu$g, the tick *Dermacentor* 0.8, the mealworm *Tenebrio* 6, the locust *Schistocerca* 22, and the larva of the marshfly *Bibio* 900. These figures do not include water lost through respiration.

The tick *Ornithodorus*, weighing when fully hydrated about 900 $\mu$g, of which about 600 $\mu$g is water, lost about 5 $\mu$g per day while starving at 5% R.H. and 25C. This figure includes water lost with respiration. When $CO_2$ was added to the air at 5% R.H. the ticks lost water no more rapidly than before until the concentration of $CO_2$ reached 30%, which is about the concentration needed for anesthesia. Similarly, when nitrogen was added to the air ticks continued to lose weight at the normal rate until the concentration of nitrogen approached 98%, suggesting that the failure of the system when it came was due to lack of oxygen. These observations, together with the knowledge that the tick gains weight when starving in air moister than 85% R.H. suggest that water may be flowing into and out of the tick's body fluids at all humidities: in moist air the net flow is positive; in dry air the net flow is negative. If this is so, the observed loss of weight in dry air cannot be explained merely by reference to the permeability of the cuticle; the power that can be brought to bear on the process of water intake must be taken into account as well.

The partially desiccated larva of the mealworm *Tenebrio molitor* increases in weight while starving in air that is moister than 90%. Walters (1966) found that, in the starving larva of *Tenebrio*, the power (rate of doing work) that is devoted to water conservation was independent of humidity over a wide range of relative humidities but was closely dependent on temperature, which suggests that at any one temperature energy is being consumed for water conservation at the same rate independent of the relative humidity.

Unlike the insects and ticks, the snail *Helicella* seems unable to absorb water from humid air. As the air gets moister the snail continues to lose water, albeit slowly, until as saturation is approached it wakes up and crawls away. Once the snail has emerged from its shell it loses water rapidly unless it soon finds wet food or drink. Nevertheless, the retention of water in its body is an active process requiring energy derived from respiration (Pomeroy 1966). Forty snails were kept at 5% R.H. and 30°C and weighed daily until they died. The mean weight of a snail at the beginning was 592 mg, of which 407 mg was water. The mean daily loss of weight was 0.559 mg while the snail was alive and 19.830 mg per day after it had died. At the time of its death the snail had 271 mg of water in its body. It died 234 days after the experiment began. These figures allow us to calculate how much work the snail might have done on the assumption that it was using a pump to replace the water at the rate that it would have flowed out if the pump had not been working. We get this figure by subtracting 0.559, the daily outflow while the pump was working (snail alive), from 19.830, the daily outflow after the pump had stopped working (snail dead). The work done by a pump is given by the equation $E = Pv$ where E is energy, P is pressure, and v is volume of fluid pumped. We assume an osmotic pressure of 8 atmospheres for the body fluids of the snail, and we calculate an osmotic pressure of 4,000 atmospheres for the solution in equilibrium with a relative humidity of 5% from the equation $\log_e H = 4.6052 - 0.018P/0.0821T$, where H = relative humidity in percent, P = osmotic pressure in atmospheres, and T = absolute temperature in degrees centigrade (Andrewartha 1970, 122).

We find that $E = 0.5$ kilocalories, which requires the oxidation of about twenty times as much dry matter as the snail actually uses during aestivation. The analogy with a pump obviously does not hold. Yet all the evidence points to an active process:

1. Water conservation breaks down after death and in the absence of oxygen.
2. When experiments like the one just described were repeated at various combinations of temperature and humidity, the duration of life was independent of humidity but closely related to temperature.
3. The rate of loss of water depended on humidity, but even at the lowest humidity the snails seemed to die of starvation rather than desiccation.

The work done by the snail to conserve water during aestivation does not suggest "pumping"; perhaps it is more like "painting" as in restoring the waterproofing of a membrane. We do not know how the insects and ticks do it either, but the feces of *Tenebrio* and many other insects that can live in dry places are dried out very thoroughly in the rectum. The complicated structure of the tissues associated with the rectum suggest that they might be removing the water by a process that is analogous to pumping, at least with respect to the consumption of energy.

Birds and reptiles whose life-style depends on the sea and at least some terrestrial species that live in dry places excrete salt, chiefly sodium chloride, from a salt gland situated in the head or some other place. The concentration of brine coming from the salt gland may exceed the osmotic concentration of seawater. In one experiment with black-backed gulls that had been loaded with seawater through the mouth, ten times more sodium was excreted through the nose than through the cloaca. This design leaves the kidneys free to concentrate on excreting nitrogen. Nitrogen is mostly excreted as uric acid in birds and in reptiles that live in dry places, but in some aquatic reptiles the proportion excreted as uric acid may be less than 1%.

In mammals the kidney is designed to excrete both salts and nitrogen chiefly as urea. Compared with insects, birds, and most terrestrial reptiles, mammals buy energy with water, but they keep the price low by concentrating their urine. The kidney of *Dipodomys* can secrete urine that is osmotically more concentrated than seawater. The kidney of the camel can achieve only a much lower concentration (table 3.03), but the camel compensates for this disability by capitalizing on its large size. The adaptations that allow the camel to conserve water so well are, to say the least, unusual. The following brief summary of how the camel does it is taken from Schmidt-Nielsen (1964, 34 et seq.).

The camel is too large to seek shelter in an underground burrow. It often has to live in places that are much hotter than itself and where it is exposed to radiation from the sun and the hot ground. Because of its small surface relative to weight, the camel cannot dissipate enough heat by radiation; even when it is in a place that

**Table 3.03**   Pressure (in Atmospheres) against Which Certain Animals Work to Conserve Water in Their Bodies

| Species | Pressure | Empirical Information | Author |
|---|---|---|---|
| Rat flea | 910 | Gains water in air at 50% | Edney (1947) |
| Brine shrimp | 326 | Maintains body fluid in saturated brine | Crogham (1958) |
| Camel tick | 216 | Gains weight in 85% | Lees (1947) |
| Hopping mouse | 147 | Concentrates urine to 6.6 osmoles per liter | MacMillen and Lee (1969) |
| Kangaroo rat | 123 | Concentrates urine to 5.5 osmoles per liter | Schmidt-Nielsen (1964, 181) |
| Camel | 63 | Concentrates urine to 2.8 osmoles per liter | Schmidt-Nielsen (1964, 181) |
| Gull | 40 | Excretes concentrated solution of NaCl | Schmidt-Nielsen (1964, 181) |

is slightly cooler than itself, it must supplement radiation by the evaporation of sweat.

Compared with a man or a dog, a camel can lose more water from its body without dying. If a man loses more than about 10% of his original weight in water he becomes incapable of looking after himself. As he loses still more water his blood becomes viscous and no longer flows freely enough to transport heat from the deep tissues to the surface, where it can be dissipated. As the loss of water approaches 18% the temperature probably rises explosively, and death probably comes quickly. By contrast, a camel can lose water at least equal to 20% of its weight and still have its blood nonviscous and circulating freely, because the water that the camel loses comes largely from its tissues while the water that a man loses comes largely from his blood.

On a warm day a man will start to sweat as soon as his temperature exceeds 37°C; nor will his temperature fall much below 37°C even on a cold night. By contrast, the temperature of a camel, especially one that is dehydrated, will fall as low as 34°C during the night, and will rise to 40°C during the day before the animal begins to sweat. Schmidt-Nielsen (1964) estimated that an average-sized camel would require about 2,500 kilo-calories to raise its temperature through 6°C. To dissipate this much heat by evaporation would require 5 l of water. By a controlled departure from strict homoiothermy the camel can save 5 l of sweat a day. The camel's curly hair traps air that insulates the camel from heat radiating from the sun and the hot ground. Also, the fur ensures that when the camel is sweating the sweat will evaporate close to the skin where it will do most good.

We have seen that to consider the amount of work done on water conservation, or the amount of power devoted to it, may throw some light on the nature of the operation, but it is difficult to compare different species in these units: for example, how do we compare the energy required to make a resistant cuticle with that required to work a pump? On the other hand, whenever water is being moved from a place of low pressure to a place of high pressure, as, for example, when water is moved from concentrated urine into watery blood, or from concentrated seawater to dilute body fluid, or from dry feces or unsaturated air into body fluids, it seems plausible to accept the analogy of a pump and to compare the species with respect to the pressure against which they can work. Using the equation given above to convert relative humidity into osmotic pressure, we can compare the species that are brought together in table 3.03.

## 3.3 Tokens

The idea of "token" is defined in section 1.41. The resources in this category are less familiar than the resources we have discussed so far because tokens impinge on the animal through sensory receptors and neuroendocrine pathways that may be characteristic of each kind of animal. Perhaps the simplest example to introduce the topic is the hole in a tree that a great tit, *Parus major*, might recognize as a necessary prerequisite for a territory. The sight of the treehole impinges on the bird in such a way that it recognizes this as a place where it can reside and build a nest. And somehow the repetition of this message at the right time of the year stimulates the physiological and behavioral changes necessary for reproduction. It is this message that we recognize as the action of a token. On the other hand, the empty mollusc shell that the hermit crab recognizes as a place where it can make its home is not a token. It does not comply with the definition in section 1.41. The empty shell serves the crab merely as a refuge from predators and malentities and so is placed in the web.

As was explained in section 1.41, we use "token" in the context that the token signals to the animal that it is time to prepare for the next season or day or the next stage in the life cycle, or the reproductive cycle, or the cycle of diurnal behavior and so on. Light, especially the relative length of day and night, makes the most spectacular and most important tokens.

### 3.31    Preparing for the Breeding Season

In the vicinity of Adelaide, which has a Mediterranean climate, the grapevine grows during summer (October–November through April); during winter it is bare of leaves. The larva of the moth *Phalaenoides glycine* feeds on the leaves of the grapevine. It spends the winter as a diapausing pupa buried in or just below the debris on the ground. Diapause disappears during the cold weather of winter. The moths emerge about October and lay eggs on the foliage of the vine. The caterpillars measure the length of day and night and encode it, presumably in some part of the neuroendocrine system, as "long day". After the caterpillar has pupated, because the message that was encoded was "long day", a humeral message from an unknown source stimulates an endocrine organ in the brain (known as NSC, for neurosecretory cells) to organize spontaneous development of the pupa into an adult. The adult lays eggs, giving rise to a second generation of caterpillars. This generation of caterpillars, running into the autumn, encode "short day". As a consequence of this message, an endocrine organ situated in the subesophageal ganglion begins, as the caterpillar prepares to pupate, to secrete a hormone that acts on the NSC, putting this gland into a refractory condition. The same message as before "go ahead with adult development", is received by the NSC, but there is no response because the gland is in the refractory diapausing condition (Andrewartha, Miethke, and Wells, 1974). The consequences for the animal of perceiving the token "short day" go beyond the mere inhibition of the NSC. The caterpillar reaches maturity with more fat and less water than a nondiapausing one, and in this diapausing condition the pupa is drought-hardy and cold-hardy. Also, the behaviors are different: the diapausing caterpillar makes a much more thorough search for a good place to pupate.

Almost the same story could be told about the codling moth *Laspeyresia pomonella*, which is a pest of apples in the vicinity of Adelaide. With *Laspeyresia* the difference in behavior between diapausing and nondiapausing caterpillars is dramatic and easily observed. When the caterpillars are mature they leave the apple, and many of them pupate under the bark on the main trunk of the tree. The nondiapausing caterpillars seem in a desperate hurry to pupate; they will make do with the slightest bit of loose bark for shelter; they spin a flimsy cocoon and seem to pupate almost before they have finished spinning. When collecting them, most can be exposed and picked up with the fingers. By contrast, the diapausing caterpillars search thoroughly; nothing less than a firm crevice will do and they often improve it by boring into bark or even wood. They spin a dense, tough cocoon where they remain as diapausing larvae throughout the winter, pupating in the spring. When collecting them it may be necessary to dig them out with a penknife. By perceiving the appropriate token *Phalaenoides* and *Laspeyresia* ensure not only dormancy and hardiness for the duration of winter, but also a well sheltered place to spend the winter. Similar facultative diapause occurs widely among the insects and other terrestrial invertebrates. The ability to measure the length of day and to use it as a token also occurs in fish, reptiles, birds, and mammals. Perhaps it has been studied most widely in the birds (Lofts and Murton 1968).

After the breeding season or the molt, the gonads of birds usually regress and remain in a refractory condition until they are stimulated by a token, usually the length of day. The stimulus reaches the gonads through the pituitary. As with the insects, the full response to the token is broad. With the gradual maturing of the ovaries there are also other changes in physiology and behavior that are appropriate to the responsibilities of the breeding season. Birds that migrate before breeding lay down extra fat and assume a characteristic restless pattern of behavior that includes a tendency to repeatedly face in the direction that the migration will follow.

The Tasmanian mutton bird *Puffinus tenuirostris*, seems to use only length of day to time both its migration and its breeding: the precision that it achieves could hardly be won in any other way. Serventy (1967) recorded, for 20 years from 1947 to 1967, the egg laying of a colony of mutton birds that breed on a group of islands off the northeast coast of Tasmania. Every year the peak of the egg laying occurred on 25 or 26 November; and for the 20 years as a whole the range of egg laying extended over 13 days, from 20 November to 3 December. Moreover, one female, number 12378, laid her one egg on 24 November for 4 consecutive years, 1954–57. Apparently this timing has a powerful selective advantage which has not changed in recent times, because Serventy (1963) cited a historical record of eggs' being laid on these islands at the same dates 128 years ago. The precision of this timing is all the more remarkable because it follows a long migration from Asia, about the latitude of northern Japan. During the same 20 years the birds arrived on the island consistently during the last week of September. Such precise timing of both migration and breeding must surely rely on a very accurate measurement of the length of day; no other phenomenon would be precise enough.

Mammals, reptiles, and fish have also been shown to synchronize their life cycles with the seasons by measuring the length of day. Baker and Ranson (1932) did similar experiments with mammals. They found that the field mouse *Microtus* bred freely when exposed to 15 hours of light each day but virtually ceased breeding when this was reduced to 9 hours. By gradually increasing the length of day experienced by the ferret, Bissonnette (1935) was able to make it mature more rapidly, and he obtained animals that were sexually mature outside the normal breeding season. Similar results have been obtained by increasing the length of day experienced by hedgehogs, raccoons, a lizard, a turtle, certain species of fish (Marshall 1942; Bullough 1951), and domestic chickens (Tucker and Ringer 1982).

Animals that become reproductively mature in the autumn through the influence of light would have to respond to a decreasing photoperiod if they were dependent upon the photoperiod to time their cycles. Yeates (1949) showed that this was indeed so for sheep. He exposed sheep to an increasing length of day during autumn; the daily exposure to light was increased from 13 hours in mid-October (northern autumn) to 21 hours at the end of January. These animals became nonreproductive two months in advance of control animals. The day length was then reduced so that they experienced a decreasing photoperiod at a time of the year when they would normally be exposed to increasing hours of daylight. By the end of June these experimental animals were receiving only 5.5 hours of light per day. The first of them came into breeding condition in May, and the rest followed shortly afterward. Decreasing the photoperiod also hastened the arrival of reproductive maturity in the goat (Bissonnette 1941) and in trout (Hoover and Hubbard 1937). The trout were made to spawn in August instead of in December, the usual month for spawning in nature.

At least some of these experiments suggest that photoperiod (i.e., the relative length of day and night) might alone be sufficient to bring the animal into full breeding condition. On the other hand, the Australian magpie *Gymnorhina tibicen* seems to require something more than this from the environment. (We discuss the territorial behavior of *Gymnorhina* in secs. 13.1 and 3.121). The discussion is drawn from Carrick's (1963, 1972) 19-year study of a population of magpies that occupied 13 km$^2$ near Canberra. In 1958 there were 292 adult females; 189 (65%) of them were living in the "flock"—that is, they had failed to establish themselves in a territory. All the country that was any good for a territory had been claimed and was being defended by groups that between them included a total of 103 females. Of the 103 adult females in the territories, 19 failed to lay an egg or failed to rear an offspring. None of the females in the flock built a nest; almost certainly none laid an egg. The 84 females, out of a total of 292 (29%) that successfully laid eggs and reared young all came from good territories that were confidently and securely defended. A good territory contains plenty of trees, some of which are good for building a nest in, and plenty of food. Confident and secure defense of a territory usually depends on good leadership.

The gonads of the birds in the flock begin to enlarge in the spring at the same time as those of the birds in the territories, but in the flock the gonads stop when only partly developed and regress. In the good territories the gonads develop to maturity and the females lay eggs and successfully rear young. In the good, well-defended territory there are plenty of trees and good food, and there is peace and security. In the flock there is none of this; in a poor or poorly defended territory some of it is lacking. Something about a good territory, or life in a good territory, seems to constitute a token that brings to fruition the process that was started when the length of day heralded the approach of the breeding season.

Perhaps a similar explanation holds for the great tit, *Parus major*, and other territorial species. According to Kluijver (1951), the tits establish territories at the end of the breeding season. The original inhabitants are challenged by birds of the new generation that have reached maturity and by any immigrants that happen along. Prior ownership and experience count heavily, but some rearrangements and extensions occur. The territories are ignored during the winter but are claimed and defended again in the spring. No tit would defend a territory that lacked a place where a nest could be built—usually a tree hole. But Kluijver found he could increase the number of territories in a young woodland (where there were few tree holes) by putting up nesting boxes. The resumption of territorial behavior and mating in the spring reflects the length of day; but the territory with the tree hole in it is still necessary for completion of the reproductive cycle.

Conformists like the mutton bird or the tit that are well adapted to live in places where the weather and the seasonal flush of good food are reliable may do poorly where the rainfall is unreliable and largely independent of the season as measured by the changing length of day. The desert favors opportunists and nomads. Opportunists and nomads must make do with tokens that predict the near future. Even so, they may save invaluable weeks or months.

According to Frith (1957), the highly nomadic grey teal, *Anas gibberifrons*, has found such a token in the rising floodwaters of the great rivers of the Riverina in southeastern Australia. The rivers, usually in response to melting snow or rain in the eastern highlands hundreds of kilometers away, meander slowly across the virtually flat plains of the Riverina, filling billabongs, lakes, anabranches, and other backwaters as they go. The grey teal find abundant food for themselves and their young at the edges of the floodwaters, especially as they recede. We discuss the ecology of the grey teal in more detail in section 13.2. Briefly, the sight of the rising floodwater triggers the breeding cycle in the grey teal. The response is rapid; by the time the water begins to recede the first lot of teal have hatched a brood of ducklings.

According to Immelmann (1963), a group of black-faced wood swallows, *Artamus cinereus*, and the zebra finch, *Taeniopygia castanotis*, responded to falling rain. He watched the birds during a rainstorm that broke a 5-year drought near Alice Springs in May 1960. Some courtship displays began while the rain was still falling, and the birds copulated soon afterward. The first three nests were completed 6, 7, and 9 days after the rain. We are not told about the condition of the gonads before the rain. Rowley (1975, 189) suggested that in such obvious opportunists as the wood swallow the gonads may remain partially developed (after some earlier stimulus, perhaps day length), ready to respond quickly to the token that signals good breeding conditions in the immediate future. If so, the rain for the wood swallow would be equivalent to the supplementary token that we inferred for the magpie living in a good territory. Keast and Marshall (1954) found that during a severe drought the gonads of all species of birds examined in an area of desert in South Australia were completely inactive. Shortly after their examination 2 inches of rain fell in the area. When they examined the gonads two months later, thirty of the sixty species either had undergone spermatogenesis or had nested.

### 3.32 Preparing for the Daily Round

The "biological clock" that an animal carries in its body can measure and encode not only the length of day and night but also the time of day when light or darkness begins and ends. These clocks are normally set each day by the diurnal passage of the sun. Most clocks keep reasonably true time for several days, and a few exceptional ones operate, for as long as 2–3 weeks of continuous darkness. In most animals there is an innate rhythm of activity with a period of about 24 hours—circadian rhythm. One function of the "biological clock" is to keep this rhythm in step with the natural day and night.

In humans circadian rhythms are manifested in the regularity of our sleeping, eating, excreting and concomitant physiology. They are manifested as "jet lag" in those of us who fly across 180 degrees of longitude within a day or so. We adjust to local time after a few days' exposure to day and night in the new place. (For further examples see Moore-Ede, Sulzman, and Fuller 1982). Circadian rhythms are best known, at the level of natural history, among the vertebrates. Keepers of zoos, for the benefit of their diurnal visitors, train nocturnal animals to sleep during the natural night and remain awake during the natural day by blocking out the sun and switching the lights on at night and off during the day. The animals will then display all their usual nightly behavior during the artificial night. Ethologists, for the convenience of working during their natural day, sometimes play the same trick on nocturnal animals. Usually it is found that these animals, like the human traveler, adjust to the new regime after a few days' exposure to it. Evolution could hardly have done otherwise, because in nature the animals must keep in step with the daily changes in length of day that accompany the progression of the seasons.

Although circadian rhythms are more widely known from vertebrates, they may have been better studied in invertebrates. One nice study of the cockroach *Periplaneta americana* was done by Harker (1956). The cockroach is nocturnal; if it is living where daylight penetrates, it will remain quiet during the day and become active at night. But when a number of *Periplaneta* that had previously been exposed to the natural diurnal rhythm of light were kept in continuous darkness, they continued to be quiet during the day and active at night for about four days. After this their behavior lost its regular rhythm. When another batch was exposed to artificial light during the natural night and kept in darkness during the natural day they were quiet at night and active during the day, and this rhythm persisted for several days when they were kept in continuous darkness. By grafting a legless cockroach that had been conditioned for a certain rhythm onto the back of another that had lost its rhythm (by living in continuous light), Harker showed that the rhythm was caused by a hormone which, she said, originated in the subesophageal ganglion.

In the fruit fly *Dacus tryoni* there is a "clock" that can tell the time of day with remarkable accuracy after a lapse of twenty days from its last setting. Bateman (1955) showed that a rhythm imposed on one generation of the fruit fly *Dacus* was transmitted to the pupae of the next generation. He illuminated two batches of adult flies as follows: one batch was kept in the light from 9 A.M. to 5 P.M. and in the dark for the rest of the day; the other batch was illuminated from 9 P.M. to 5 A.M.. The larvae and pupae were kept in continuous darkness, and the adults of the next generation emerged in darkness. Most flies emerged during a period of 48 hours, but the emergences were not distributed evenly over this period. On the contrary, there were two peaks of emergence separated by 24 hours; the striking thing was that in each batch they occurred during that time of the day when the adults of the previous generation had been illuminated. We do not know how this message is conveyed across the generations and preserved through embryo larva and pupa before it is finally delivered at the right time. Similar sorts of clocks have been reported in certain species of *Drosophila* and we would expect them to occur widely among the insects.

### 3.33  Preparing for a New Life-style

The planktonic larvae of abalone are induced to settle and metamorphose almost exclusively on coralline algae. They respond to $\gamma$-aminobutyric acid, a neurotransmitter produced by the algae (Morse et al. 1979); see section 8.42 for further examples.

### 3.34  Things That Are Not Tokens

The honeybee *Apis mellifera*, when foraging or searching close to home, uses visual or olfactory landmarks. When it is far from the hive it is guided by celestial observations. It may use merely the position of the sun in the sky, or it may be guided by the direction and intensity of the polarization of the light reflected from the sky, which depends on the position of the sun. Because both methods depend on the sun, which moves, it is almost certain that the bee is also measuring the time of day, doubtless with a "biological clock" of the sort we inferred in the previous section (Lindauer 1961, 95).

Similarly, a bird close to home and in familiar country will use visual landmarks. But when it is far afield on a migratory or homing flight, mechanisms equivalent to a compass and a clock must surely be used as well. A number of explanations have been proposed, including the ability to read celestial angles and the stars (Mathews 1955; Griffin 1969). It is not for us to pursue the discussion of migration here. We raise the matter in order to point out that a landmark or a celestial "compass", even if it does involve measuring the length of day or knowing the true time of sunrise, has nothing in common with a token. These stimuli—the visual and olfactory landmarks of the bee and the bird—and other stimuli that at first might seem like tokens are in fact modifiers of some component of environment, usually a resource. They are placed in the web because they do not act directly on the primary animal. For example, the bird is already in a migratory condition. The token was responsible for that. The star (or other landmark) merely guides the bird toward its food, which is the directly acting component in this pathway.

The behaviors associated with courtship, including artifacts like nests or the bower and platform on which the male satin bowerbird does his courting, should like the pheromone released by the female gypsy moth, be regarded as an extension of the animal, not part of the environment.

## 3.4  Other Resources

Heat and oxygen are the only other resources that we can think of. For a discussion of heat see Andrewartha and Birch (1954, chap. 6). Oxygen rarely varies enough to be interesting except in polluted waters.

# 4

# Mates

## 4.0 Introduction

From the perspective of an animal struggling to contribute progeny to the next generation, it might seem that a mate is no less important a part of the environment than food. But from the perspective of an ecologist seeking to explain the numbers that he has counted, the important components of environment are those that vary most (sec. 1.31). Because ecologists inevitably study species that are abundant, they do not have much opportunity to observe sparse populations, where a shortage of mates is most likely to occur. This experience may well bias our opinion on the matter, because it seems that sparse populations are more typical than abundant ones. Certainly Darwin thought so when he wrote. "Rarity is the attribute of a vast number of species of all classes, in all countries. If we ask ourselves why this or that species is rare, we answer that something is unfavourable in its conditions of life; but what that something is we can hardly ever tell" (Darwin 1859, 319).

Smith (1935) also held this view with respect to insects. In Andrewartha and Birch (1954, 335), in a discussion of "too few animals of the same kind", we wrote:

In natural populations the reduction in the value of r associated with the increasing sparseness of the population may be carried so far that r becomes negative and the population proceeds to dwindle to extinction. There are two aspects to this phenomenon. One is the final extinction of a population which has been well established in an area but has been brought down to low numbers by natural vicissitudes or the deliberate destructiveness of man. The other is the failure of a small colony of immigrants to become established in a new area which is favorable in all respects except for the sparseness of the colonizing population. These two phenomena must be going on around us all the time, but they are mostly missed because they are so difficult to see.

Even though we rarely observe and even more rarely measure the disadvantages associated with a shortage of mates in nature, we can point to indirect evidence that this disadvantage has been important enough to provoke the widespread evolution of adaptations that serve to bring, or keep, the sexes together when numbers are few. Also, in a few instances, ecologists have learned how to create an artificial shortage of mates in a natural population. In the next section we refer briefly to one very successful empirical study of a shortage of mates. In subsequent sections we present a little of this indirect evidence.

## 4.1 A Shortage of Mates That Was Measured Empirically

Milne (1950) estimated the reduction in fertility that might be associated with a shortage of mates in a population of sheep ticks, *Ixodes ricinus*, on a farm in northern England. A mature female tick usually seeks a mate during April–May, and she will die without contributing progeny to the next generation unless she is fertilized during this season. At the beginning of the breeding season the ticks take shelter in the dense, moist mat of grass, moss, and decaying vegetation that usually covers the soil in this area. And the pasture is usually grazed by sheep—about two per hectare. The ticks are unable to crawl laterally through or over the mat; hence their chance of meeting a mate on the ground is negligible. They can and do climb to the tip of a protruding grass-stem whence they might, with luck, be picked up by a sheep that brushes by. A tick is likely to die from desiccation if it remains aloft for more than 4 or 5 consecutive days. Usually it returns to the mat to replenish its water (probably by absorbing water vapor from the nearly saturated atmosphere— sec. 3.222). This is a slow process, and a tick is not likely during the whole season to manage more than about 10 days on the grass-stem, lying in wait for a passing sheep. If it is not picked up during this time it dies without progeny. A female that has been picked up by a sheep has a good chance of mating only if a male is also picked up on the same quarter of the same sheep. The tick's capacity to crawl is so slight that two ticks of opposite sexes picked up at opposite ends or on opposite sides of the same sheep have little chance of coming together.

On one farm where he worked Milne estimated 150,000 ticks in the mat at the beginning of the season; during the breeding season 30,000 were picked up by sheep; 80% died without achieving even the first step toward mating.

The chance that any one tick will be picked up by a sheep depends on the number of sheep in the paddock but is largely independent of the number of ticks. Conversely, the number of ticks picked up by any one sheep depends on the number of ticks in the paddock but is largely independent of the number of sheep (at least within the bounds of the normal stocking rates of these upland farms).

Assuming that the sexes are picked up randomly and ignoring the risks associated with isolation by distance on the same sheep, in Andrewartha and Birch (1954, 338) we calculated the expectation, for different numbers of ticks per sheep, that a female would be matched by a male on the same sheep. The expectations ranged from 0 for 1 tick to 0.77 for 12 ticks per sheep (table 4.01).

Milne traced the historical "spread" of *Ixodes* through the upland farming country where he worked. It was very slow, sometimes taking decades to spread from one farm to the next or from one fenced paddock to another on the same farm. The few ticks that were carried through a sheep fence by hares, badgers, and other secondary hosts rarely, if ever, succeeded in founding a viable colony. A new colony was far more likely to be founded if a gap in a fence or an open gate allowed a flock of "ticky" sheep to invade a "clean" area. Milne did not measure the critical number of "colonists" below which the shortage of mates prevented the foundation of a viable colony. Presumably the threshold lies above the numbers that are likely to be carried by secondary hosts and below the numbers that might be carried by a flock of ticky sheep.

**Table 4.01**   Probability That a Female Tick Will Be Matched by a Male on the Same Sheep as the Number of Ticks Varies from One to Twelve

| Number of Ticks per Sheep | Probability That a Female Will Be Matched by a Male | Number of Ticks per Sheep | Probability That a Female Will Be Matched by a Male |
|---|---|---|---|
| 1 | .000 | 7 | .688 |
| 2 | .500 | 8 | .727 |
| 3 | .500 | 9 | .727 |
| 4 | .625 | 10 | .754 |
| 5 | .625 | 11 | .754 |
| 6 | .688 | 12 | .774 |

*Source:* After Andrewartha and Birch (1954, 339).

## 4.2   Adaptations That Bring, or Keep, the Sexes Together

Some species, notably among birds and mammals, that are likely to live through more than one breeding season may form a pair bond that lasts for life. No matter how they may wander or scatter from others of their species, they are assured of a mate while both live. The same end may be served by gregariousness or territoriality. These behaviors are also more appropriate to long-lived species, but they have been exploited by short-lived ones as well. Short-lived species, notably those that live through only one breeding season, like most insects, usually depend on more ephemeral behaviors. Attractants such as pheromones assume special importance (sec. 4.22). There is also a little evidence that some species of butterflies may be attracted to a "mating place" instead of, or in addition to, being attracted to an individual that is secreting a pheromone (Scott 1968, 1974; Shields 1967). Adults of the sheep-blowfly *Lucilia cuprina* in Australia are sparsely distributed until they congregate around a carcass for a protein meal and, in the case of females, for oviposition as well. The attraction of both males and females to sources of protein probably provides an important way for males and females to meet. Males feed on protein throughout their lives, though there seems to be no need for them to do so, whereas the female's eggs cannot mature unless she has a protein meal. Despite this means of getting the sexes together, 5% of fully gravid females found in the field have empty spermathecae, indicating they have not found mates. Having found a mate once in her life, the female does not need to mate again. One mating provides sufficient sperm for the two or three batches of eggs that a female is likely to lay in the course of her life, which is about two weeks (Foster et al. 1975; Vogt and Woodburn 1979).

### 4.21   Gregarious and Territorial Behavior

Territorial behavior is often related to resources (sec. 3.121), but this relation may not always suffice for a full explanation. For example, in a social group that defends a communal territory, as in *Oryctolagus* (sec. 12.2) or, more especially, *Gymnorhina* (sec. 3.31), this behavior assures the members of the group an adequate supply not only of food but also of mates. This dual function of territorial behavior is well exemplified in the American cottontail rabbit, *Sylvilagus floridanus*, which forms small groups much like those described for the European rabbit in section 12.2; but instead of defending a piece of ground the males defend the females, and the whole group is mobile but coherent (Marsden and Holler 1964). Nevertheless, to keep their tactics effective they need the confidence that comes from living in familiar surroundings. So their home range is strictly limited.

According to Ewer (1968, 87) this kind of social behavior is further developed in many of the large ungulates, with perhaps less attention paid to home range:

In many artiodactyls, seasonal movements of the herds in relation to changing food supplies make the holding of a stable breeding territory impossible. This, however, need not abolish "territorial" behaviour on the part of the males; instead of a definite area the group of females itself is defended, regardless of the fact that they are constantly on the move. We have already had an example of this in the impala, where the area over which the animals feed is too large, in relation to the degree of cover, for its defence as a spatial unit to be possible. The same principle occurs even more strikingly in the case of the blue wildebeest, *Connochaetes taurinus*. In the animal studied by Talbot and Talbot (1963) in Masailand, mating occurs when the animals are on the move from the open plains towards the bush. . . The size of the breeding groups is highly variable and may be anything from 2 or 3 to over 150 individuals. If the group of yearlings and females is large, then 2 or even 3 males may share it and co-operate in defending it against others, although showing no animosity to each other.

Ewer commented that this behavior ensures that "all females breed while there is a selection of males". A behavior that ensures that all females breed is obviously a useful buffer against the risk of not meeting a mate when the population is sparse.

One explanation for the evolution of gregariousness in species that are habitually rare or in abundant species that habitually migrate long distances may be that it ameliorates the risk from a shortage of mates. Waloff (1946) recorded the wanderings of many swarms of the locust *Schistocerca gregaria* in East Africa. The movements of the swarms were largely governed by temperature and wind. A characteristic response to temperature and wind made it likely that the locusts would be picked up, in a gregarious swarm, by ascending currents of air, and carried perhaps five to ten thousand meters above the ground, where a strong wind (its likely direction depending on the season of the year) might blow them perhaps hundreds of kilometers before a descending current would bring them to the ground again. There was no evidence that the wanderings led them, except by chance, to favorable places for breeding; but when, in the course of their wanderings with the wind, the locusts arrived at a place that was moist enough, they would develop to sexual maturity, copulate, and lay eggs. The advantage conferred on them by their gregarious behavior in keeping them together until they were ready to copulate is obvious. In other circumstances gregariousness may have other advantages, but it is a fact that many insects, fish, birds, and mammals whose usual way of life includes long migrations are gregarious, at least while the migration lasts.

## 4.211 Mating Places

Males of the antelope *Adenota kob* defend territories in a "territorial breeding ground" which attracts females in breeding condition. According to Ewer (1968, 86).

In the Uganda kob, *Adenota kob*, male territorialism is concerned purely with breeding and is unrelated to feeding (Buechner and Schloeth, 1965). Within the home range occupied by the females and juveniles a central "territorial breeding ground" contains the mating territories of the mature males. These territories are extremely small, only 15–30 metres in diameter, and there may be 30 or 40 of them in the territorial ground. Each male defends his territory and here the females come to him for mating. Territories are not all equivalent in value: the central ones are the most hotly contested and the ones in which most breeding occurs. In the Toro game reserve, where the observations were made, breeding occurs all the year round and any particular male may hold a territory for a very variable period— anything from a couple of days to what is described as "semi-permanent". In the latter case, the male periodically leaves the territory to feed or drink and then returns to it, which emphasizes the fact that the territory is concerned solely with mating. Leuthold (1966) has recently extended Buechner and Schloeth's work and has found that in addition to the males occupying the central territorial ground, there are always a number of others scattered about the periphery, each holding a larger territory some 100–200 m in diameter. There is a rough gradient with the territories in general smaller towards the middle of the range until finally they form the central clump which constitutes the territorial breeding ground. Although the peripheral males do defend their territories, there is little competition for them and the attention of the females is concentrated on the central area where almost all the mating occurs. Bourlière (quoted by Buechner, 1961), studying the same species in the Congo, failed to find this type of social organization. Leuthold's observations now make it clear that this was because the population density was too low to produce the characteristic clumping into a central territorial breeding ground.

Glover et al. (1955) described how a substantial and well-established population of tsetse flies dwindled to extinction after the vegetation had been cut down to destroy the special quality of certain little valleys called "kasengas" in that country. The study area covered 700 km$^2$; the area that was cleared in and around the kasengas amounted to about 2.2% of the total. During the decline of the population, and beginning well before the decline had progressed very far, Glover et al. (1955) found increasing numbers of nonteneral, uninseminated females in the routine samples. Such an event was unexpected: it had not occurred before during a number of years' routine sampling. It suggested, though the suggestion was not confirmed by critical evidence, that the flies congregated in the kasengas for mating. The Queensland fruit fly *Dacus tryoni* is known to congregate in special places for mating.

A series of studies on *D. tryoni* both in the laboratory and in the field have revealed a number of adaptations that increase the chance that a female will find a mate (Tychsen and Fletcher, 1971; Fletcher and Giannakakis 1973; Tychsen 1978; Smith 1979). Both male and female show sexual behavior only at the time of day when dusk occurs and at the low light intensity of dusk. The optimum light intensity for mating is about 9 1x. At the low light intensity of dusk male flies aggregate. Observations of fruit flies in cages placed over trees showed that males formed their aggregations completely before any females entered them. The aggregations formed on the windward side of trees. The aggregated males released a pheromone from a rectal gland; presumably the pheromone spread more rapidly through the tree when it was released on the windward side. In addition, each male stridulated by drawing the anal lobes of its wings across combs on its abdomen. The odor plume from the aggregation of males probably attracts females from some distance, and the large amount of pheromone released from an aggregation is probably more effective in drawing in females than pheromone released from males placed far apart. Each male defended a small area of leaf where he stridulated, and each female landed near a male and then walked up beside him. It seems that stridulation of the male is a short-distance attractant bringing the female to the vicinity of a male. The behavior of the male did not alter until the female entered his visual field. Mounting then occurred immediately. Aggregation of males does not seem to be obligatory for successful mating. Males have been observed stridulating alone on leaves of trees, and copulating pairs have been found with no other flies visible in the vicinity.

Tychsen (1978) found that the flies responded most strongly to stimulatory light (intensity of 9 1x) about or just before sunset, and he inferred that the fluctuation in the responsiveness of the flies was strictly controlled by a circadian clock (sec. 3.32). Smith (1979) crossed *D. tryoni* with a closely related species, *D. neohumeralis*, which is stimulated most strongly by light of intensity 10,000 1x. Because, in $F_1$, $F_2$, and backcross progeny, the intensity of the most influential light varied between the extremes represented by the two parents, whereas the narrowness of the "gate", that was opened by the circadian clock was the same for all the crossbreeds and both parents, Smith inferred that the same genes controlled the circadian mechanism in both species. The essential difference between the species was that *D. neohumeralis* was stimulated by light with an intensity of 10,000 1x and that their circadian clock made them maximally responsive during the middle of the day.

Despite the adaptations of *D. tryoni* that increase its chance of finding a mate, there are two periods in its seasonal cycle in temperate areas such as the environs of Sydney when the chance of a female's finding a mate must be quite small. Early in spring adults which have overwintered in sheltered areas fly out of their overwintering site in search of mates and fruit. The overwintering population is smaller than the population at any other time of the year, and the flies become widely separated when they move out of their overwintering sites. In addition, the first spring generation that arises from those females that have been fertilized is also smaller in number. Moreover, 80% of the newly emerged adults leave the local population in which they originated, and while they are becoming sexually mature they search for suitable new sites for laying their eggs. So a small population of newly emerged adults is left behind and a highly dispersed majority of adults is scattered far and wide. For both populations the chance of a female's finding a mate is probably small (Bateman 1977). The high dispersive ability of *D. tryoni*, which is so important for its survival in a patchy environment (sec. 14.2), becomes a liability when numbers are low and females need to find mates. Natural selection is always a compromise between different needs of the organism. Because populations in early spring are small and the flies at this time are immature and unmated, this would be the most strategic time to liberate sterile insects in any control program using this technique (sec. 4.32).

## 4.22   Pheromones

Chemicals that are released into the medium by one individual and influence the behavior of another individual of the same species are called pheromones by analogy with hormones, which carry chemical messages to diverse organs or tissues within the organism. Because pheromones must diffuse through, or be carried by, currents in air or water, they are usually volatile substances. Pheromones have been most studied among mammals and insects but they have been recognized in a wide range of other animals, and present indications are that birds and higher primates may be unusual in that they do not make much use of pheromones.

Pheromones are known to organize a wide range of behaviors, such as aggression, "friendship", courting behavior, and mating behavior. Our present concern is with those that promote aggregations that bring the two sexes together or keep them together when the time is ripe for mating. The actions of such pheromones fall broadly into three classes.

1. A pheromone that is secreted by one or both sexes causes those individuals that are ready to mate to congregate densely in one place. Species that mate more effectively in a crowd would profit from such behavior, and one might expect to find such behavior in a species that exploited a "mating place" like those mentioned in section 4.211.

2. A pheromone that is secreted by one or both sexes and serves to synchronize the mating behavior of all the members of a local or larger population. The most spectacular example in this category are the nuptial flights of the social insects, ants and termites. Not only all the sexually mature males and females from one nest but those from virtually every other nest for miles around will emerge for the nuptial flight on the same sultry summer evening. Weather is doubtless part of the trigger, but the precision and wide consistency of the response suggest a more specific stimulus as well. According to Hölldobler and Machwitz (1965), the females of the ant *Camponotus herculeanus* are stimulated to join the nuptial flight by a pheromone that is secreted by the males as they emerge from the nest.

3. One sex, nearly always the female, secretes a pheromone that will attract a male from a greater distance than he could perceive her with any other sense.

It is an old tradition among lepidopterists that male moths may be lured in good weather (warm, sultry, gentle breeze) by exposing a virgin female in a well-ventilated trap. For a long time naturalists, from Fabre onward, were skeptical whether this trick could be done with chemicals and a number of imaginative explanations not involving chemistry were put forward. Today chemistry forms the backbone of the explanation that we accept. From laboratory experiments we know that many different kinds of animals have extremely sensitive chemoreceptors (Shorey 1976, chap. 2). Estimates of the median response dose (RD 50) for males of the cabbage looper moth *Trichoplusia ni* varied from 8 to 800 molecules of pheromone per cubic millimeter of air, with a mean of about 80 (Sower Gaston, and Shorey 1971). It is common knowledge that dogs use a number of pheromones, including sex pheromones, and they have a very keen sense of smell.

We now know that many animals have mechanoreceptors that allow them to monitor the angle their bodies make relative to the direction of flow of the current, of air or water, that they happen to be in. Also, many animals can, with their eyes or some other receptor, monitor the movement of their bodies relative to certain fixed objects on the ground or in the heavens. An animal that possesses all three skills theoretically has the ability to follow a pheromone to its source, provided that the animal has the strength to run, fly, or swim against the current. The chemical stimulus organizes movement (travel); the mechanical stimulus directs the movement against the current; and the visual or other navigational stimuli allow the animal to correct for the vagaries of the current and to monitor the general direction of the current relative to the points of the compass. It is now generally accepted that this is the way that animals follow a pheromone trail in air or in water.

Theoretical models have been proposed from which it might be possible to predict the distance over which a pheromone trail might be followed to its source. But it is often not practicable to get good empirical estimates of the terms in the model. According to Shorey (1976, 31), applying such methods to data (and making some reasonable assumptions) for the cabbage looper moth gives estimates of not more than 100 m. With other species of insects the direct empirical method, marking males and recapturing them at a source of pheromone, has given estimates of not more than 10 km. But there seems to be a serious ambiguity intrinsic in this method: there is no way to tell how far the marked male may have traveled before it picked up the pheromone trail.

A natural shortage of mates may occur whenever any vicissitude of environment causes a population to become critically sparse. Even in a population that is not sparse, a shortage of mates may occur if some contemporary component of environment inhibits normal sexual activity. Man has exploited both these strategies in his attempts to create an artificial shortage of mates as a control measure against certain pests, mostly insects.

## 4.3   Artificial Shortages of Mates

### 4.31   Pheromones and Lures

It is well known that cue-lure, a mixture of 4-(p-acetoxyphenyl) butan-2-one and 4-(p-hydroxyphenyl) butan-2-one, will strongly attract males of the fruit fly *Dacus tryoni*. It is not known whether the female of *D. tryoni* secretes a sex pheromone. Nor do we have any other explanation for the attractiveness of cue-lure for the males of *Dacus*; but the empirical fact that it attracts them strongly is beyond doubt.

Bateman, Friend, and Hampshire (1966) postulated that an artificial shortage of mates for the females in a population of *Dacus* could be brought about by exposing a large number of traps baited with cue-lure and a poison. An extensive field experiment was done using small country towns as replicates. Most of the towns supported between three thousand and five thousand people, but the smallest had one thousand and the largest five thousand. There were twelve towns scattered over 8,000 km² of inland New South Wales. The countryside was used largely for growing wheat and grazing sheep. Localities (apart from the towns) that might support *Dacus* were few and far between, restricted for the most part to the few fruit trees that might be grown in the kitchen garden of a farmhouse. In the towns many gardens would support fruit trees; all the towns in the experiment had a good succession of fruit ripening throughout the summer. It was hoped that the towns would be sufficiently isolated from the adjacent farm gardens. In the event, this expectation was not fully realized, and so the results of the experiment lost some of their sharpness.

There were four treatments: (a) cue-lure and malathion; (b) protein hydrolysate (attractive to males and females) and malathion; (c) cue-lure, protein hydrolysate, and malathion; and (d) an untreated control. The lures were designated ML, PM, and ML + PM respectively. The traps were small blocks, about 5 cm$^2$, of Caneite (a porous artificial wood) that were painted with about 10 ml of lure plus poison and nailed up on trees, posts, and such, in a square grid with spacings about 50 m in both directions. The results were recorded by taking samples of ripe fruit at regular intervals throughout the summer and recording the proportion of the fruit that was infested with maggots. For analysis the towns were divided into three blocks. In each block each treatment and a control was represented once. The results for the second summer of the experiment are briefly summarized in table 4.02.

The results in the final column (the mean for the whole summer's sampling) are biased by the fertile females that flew in from adjacent farms toward the end of summer. The figures in parentheses (samples taken during January) represent the maximum response that could be recorded: treatment had been operating for 3 months of the summer, and no invasion had yet occurred. For our immediate purpose we are interested only in the towns that were treated with cue-lure—that is, treatments ML and ML + PM. We notice that ML alone caused about 50% reduction in infested fruit compared with the controls over the whole year and better than this on the figures for January. The protein hydrolysate lure was so effective that there was no room for improvement by the combined lure ML + PM, so this experiment did not test the hypothesis that cue-lure would be relatively more effective when there were fewer females attracting the males away from the trap. This consideration is of fundamental theoretical importance when a female sex pheromone is used as a male lure.

Shorey (1977) reviewed the use of pheromones in pest control. He suggested that the prior presence of large quantities of naturally produced pheromone, secreted by many females, might be one reason why the use of female sex pheromones as male lures had not been more successful against dense populations. When the artificial

**Table 4.02**   Proportion (%) of Fruit Infested with Maggots of *Dacus tryoni*

| Treatment | Block 1 | Block 2 | Block 3 | Mean |
|-----------|---------|---------|---------|------|
| ML        | 24 (5)  | 22 (18) | 13 (1)  | 20   |
| PM        | 6 (0)   | 14 (5)  | 3 (0)   | 7    |
| ML + PM   | 3 (2)   | 5 (1)   | 0 (0)   | 3    |
| Control   | 51 (62) | 34 (41) | 51 (63) | 43   |

*Source:* Bateman, Friend, and Hampshire (1966).

*Note:* The figures in the body of the table represent the means of samples taken throughout the summer from 5 December to 10 April. The figures in parentheses represent samples taken through January (midsummer). ML, Cue-lure and Malathion; PM, protein hydrolysate and Malathion; ML + PM, equal parts of ML and PM. The poison Malathion was present in the same concentration for all treatments.

pheromone is available in large quantities and is not too expensive, a better method might be to aim at disrupting communication between the sexes by flooding the area with pheromone and thus habituating either the receptors or the central nervous system, or merely confounding the message. As our knowledge of pheromones grows we may even discover chemicals (antipheromones) that will "block" or "mask" the natural pheromone.

## 4.32 Sterile Males

In Knipling's (1955) method of reducing or eradicating a population of insects that are pests, the object is artificially to create and maintain a severe shortage of males by flooding the population with males that have been sterilized by irradiation or by exposure to a chemosterilant. The shortage is not a real shortage because there is at first no reduction in the number of naturally fertile males. It is a statistical shortage: if the sterile males outnumber the naturally fertile ones by say twenty to one and are equal to them in libido, then the female has only one chance in twenty of mating with a fertile male, and 95% of the eggs will be sterile.

Knipling first demonstrated this method with a population of the screwworm fly *Cochliomyia hominivorax* on the island of Curaçao which has an area of 430 km$^2$. Large numbers of *Cochliomyia* were reared in the insectory and sterilized with X rays, using a dose that killed the sperm but did not otherwise reduce the sexual vigor of the flies. Fresh supplies of these sterile flies were released at weekly intervals to mix with the natural population. The sterile males would copulate with the wild females, thus stimulating them to lay eggs that were sterile. The number of flies in the natural population at the beginning was not known precisely, but certain assumptions were made and rough calculations done with the aim of having about ten times as many sterile as fertile flies to start with and of maintaining the same absolute number of releases for at least several generations. The theory was that if the early releases were sufficient to start a decline in the original population, then as time went on the effect of the sterile intruders should increase exponentially. In the event, weekly releases of 150 sterile males per square kilometer for 6 months were sufficient to exterminate *Cochliomyia* from the island. Subsequently the same method was used on a larger scale with equal success in Florida (Knipling 1960). The method has since been shown to work with a number of other species, and there have been certain minor refinements in technique and theory. But an attempt to eradicate *Cochliomyia hominivorax* from southwestern United States has not yet been successful because it has not been practicable to breed flies for sterilization that can match the genetic diversity and fitness of the target populations (secs. 11.2, 11.311).

## 4.33 Quarantine

Most developed countries maintain a sophisticated system of plant and animal quarantine. Plants or animals, or materials made from plants or animals, may be prohibited entry or permitted only after rigorous inspection. The aim is to prevent or discourage the entry, across an ecological barrier, of organisms that are known to be weeds, pests, or pathogens on the other side of the barrier or are suspected of having such a capacity on this side. If there is much trade or tourism across the barrier, quarantine may not achieve absolute prohibition, but by reducing the intake to a trickle quarantine may greatly reduce the risk that a viable colony will be established. There may be other risks associated with membership in a very small or sparse population or colony (Andrewartha and Birch 1954, 334), but shortage of mates is always an important one.

# 5

# Predators

## 5.0 Introduction

At every trophic level the food of animals must be a living organism or an organic residue (sec. 3.1). In the theory of environment, every organism that depends on living organisms for food finds its place in the bottom left cell of table 1.01 and is therefore a predator.

Table 3.02 depicts the rules governing the interactions between predator and prey which hold generally, whether predator or prey be animal, plant, fungus, microbe, or virus. In table 3.02 the interaction is viewed from the perspective of an organism seeking its food—that is, the predator is the primary animal. In this chapter we reverse the perspective, placing the prey in the position of primary animal. We seek to understand the influence of the predator on the distribution and abundance of the prey. This means, of course, that our study must embrace the ecology of the predator.

The circumstances in which a predator has a profound influence on the abundance of its prey are defined in cell 3 of table 3.02. There is not much more to be said about this interaction than has already been said in section 3.132, but we elaborate a little further in section 5.1. Cell 2 in table 3.02 houses all those interactions in which some component of the predator's environment other than food keeps the predator rare relative to its food supply. Cell 1 in table 3.02 defines the interactions in which the predator is kept rare relative to the food supply because the food, though present, is unavailable and so the predator has little influence on the abundance of its food. Cell 4 in table 3.02 defines the circumstances in which the predator has only a temporary influence on the abundance of its prey.

In section 5.2 we consider how the activity of predators is influenced by the prey (5.21) and by pathways in the web that lead through the predator (5.22). We conclude with an assessment of the influence of predators on the distribution and abundance of prey (sec. 5.3).

The emphasis in this chapter is on the influence of predators in natural populations. Despite the interest in predation by students of wildlife management and biological control, there are surprisingly few rigorous field studies of predation. This reflects the practical difficulty of making such studies and the scarcity of realistic models. Nevertheless, advances have been made in the past decade or so in understanding the role of predation in the distribution and abundance of animals in nature.

## 5.1   Feedback between Predator and Prey

When the predator causes an intrinsic shortage of food for itself (cell 3, table 3.02) there may be feedback between predator and prey. The predator influences the abundance of the prey and vice versa. Some aspects of the feedback between an herbivore and its prey (the vegetation on which it feeds) are considered in section 7.21. When the predator belongs to the third or a higher trophic level, a complex of feedbacks is a possible, though not a necessary, outcome of the relationship. Furthermore, the environment of a predator consists of far more than its prey, and the environment of the prey is more than its predator. When we consider the influence of the whole web of the environment of the prey and that of its predator, these other components may be overwhelmingly important to the virtual exclusion of the simple feedback such as are postulated in many mathematical models of predator and prey. With this in mind we consider the long term study that has been made of the moose and the wolf (plate 3) and the vegetation in Isle Royale National Park, an island of 540 km$^2$ in Lake Superior.

Before about 1900 there were neither moose nor wolves on Isle Royale. The moose, *Alces alces*, arrived in the early 1900s by swimming the 20–25 km from the Canadian mainland. The timber wolf, *Canis lupus*, arrived much later, about 1948. No other big-game animal or carnivore exists on the island. It has thus served as an isolated laboratory in which moose lived first without wolves and later with the wolf in its environment. It is fortunate that from fairly early on the possibilities of studying the ecology of the moose and then later the wolf were recognized. So we have a fairly continuous record of the history of these two populations and, to a lesser extent, the vegetation on the island.

By 1915 moose were well established on Isle Royale, with an estimated population of about 200 animals. Since then the number of moose has been estimated at irregular intervals by various people using different methods. Many of the early estimates are not much better than informed guesses. Since 1945 numbers have been estimated by aerial survey, and these estimates have improved with time. Figure 5.01 reflects both the number of moose and the increased accuracy of estimating this number.

The moose population increased from about 200 in 1915 to a peak population in the late 1920s and early 1930s. The estimates varied from 1,000 to 5,000, the correct figure probably being at the higher end of the range (R.O. Peterson pers. comm.). The population then declined to low numbers. This decline continued until about 1943, when there were possibly only a few hundred animals. Thereafter the population increased once more to about 900 in 1973, and it has since hovered between 900 and 700 animals.

Changes in the number of moose in the 1920s and 1930s can be interpreted with reference to the envirogram of the moose (fig. 5.02) derived from Peterson (1977 and pers. comm.) and Mech (1966), on whose accounts much of the following discussion depends. There are two key components in the environment of the moose; the pathway of the web that leads through food to the moose, and the pathway of the web that leads through the wolf (after 1948 when the wolf arrived on the island).

**Plate 3** *Top,* the timber wolf *Canis lupus; below,* the moose *Alces alces* being hunted by timber wolves. (Photos by R. O. Peterson; Isle Royale National Park, Michigan)

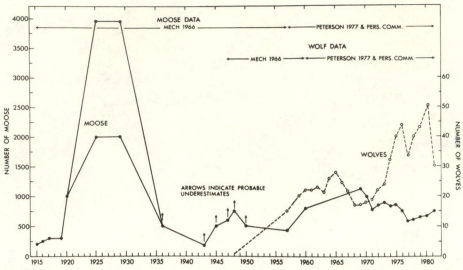

**Fig. 5.01** The numbers of moose and wolves on Isle Royale from 1915 to 1981. The graphs are based on sources indicated. Two sets of lines for the moose in the late 1920s and early 1930s indicate the divergence in views as to the real numbers during that eruption. The lower estimates are those of Mech (1966), the higher estimates are a mean of those of Hickie (5,000) and Murie (3,000), cited in Mech (1966) and considered to be the more likely estimates by R. O. Peterson (pers. comm.). According to Peterson (pers. comm.), the estimates for moose for 1936 to 1955 are probably underestimates; those for 1936 to 1943 were based on "impressions of fieldworkers conducting ground surveys of various sorts and are also likely to be low estimates"; those for 1945 to 1967 were based on strip counts of only parts of the island and have been corrected upward since they are probably underestimates. Mech's 1960 census was based on an aerial survey that was the only attempt to count all moose on the island. His 12% correction is probably too low in the light of subsequent experience and is therefore corrected upward.

The critical resource for the moose is suitable vegetation on which to browse. Moose characteristically browse on successional stages in the development of mature forests. For example, after a forest fire aspen and birch regenerate in the initial stages of recovery of the forest. This new growth is heavily browsed by moose. The mature forest itself provides little, if any, food for moose, since most of the leaves on the trees are out of reach. Some plants such as hazelwood and dogwood grow lateral shoots when browsed, so they can be browsed for longer periods than forest trees. Moreover, they are more tolerant of heavy browsing. In the two years before the peak population in the early 1930s, all species of plants that served as food during winter and several species that served as food during the summer were heavily browsed. This was so marked that one investigator predicted that many moose would die of starvation. This predicted intrinsic shortage of food actually began on an area of about 10% of the island about 1933, when 40 emaciated moose were found there. By 1934–35 "all browse was gone". By 1936 the population was down to about 400 animals, and it may have fallen even lower in the subsequent four years. The survivors in 1936 were said to be in such poor

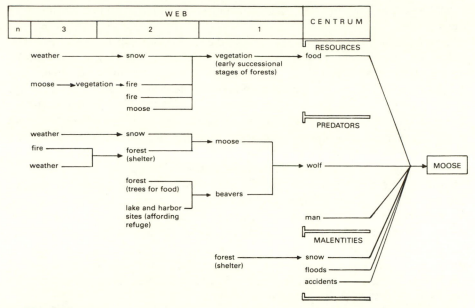

**Fig. 5.02** The envirogram of the moose *Alces alces* in Isle Royale National Park.

condition that the Michigan Department of Conservation live-trapped 71 moose and released them on the mainland in the hope that they would recover there. In 1936 fire destroyed browse on more than one quarter of the island. A few years after the fire, browse recovered and the herd began to increase, reaching about 800 animals in 1948. In the years immediately following "browse deteriorated", and an increasing number of dead animals were found. Available food again seemed inadequate to support the herd. By 1957 the population was reduced to 300 according to Mech (1966), though Peterson (pers. comm.) thinks that Mech's figures for 1945–57 are probably too low.

During the 1960s more food had become available, and the numbers of moose rose again. The population steadily increased to just over 1,000 in 1969; since then it has fallen to about 700 in 1981.

Wolves entered the picture in 1948 (fig. 5.01), but their numbers were not assessed until 1957, when at least 15 and possibly 25 wolves occupied the island (Mech 1966). Wolves subsisted almost entirely on moose in winter. At other times they included other animals in their diet, especially beaver when they became abundant.

The increase in numbers of wolves since they became established on the island (fig. 5.01) is real, but it also reflects the greater accuracy of more recent estimates. Initially, a single large pack of wolves hunted the entire island. About 1972 two large wolf packs were established, each occupying about half the island. The formation of two packs was associated with an an increase in the number of beavers. From about 1959 to 1973 the population of wolves fluctuated between 17 and 28. Between 1973 and 1976 their numbers doubled to 44. They consisted then of three large packs and several rather small ones. Wolves reached a peak of 50 in 1980. During the five years 1976–80 Isle Royale had the world's highest density of wild wolves. Then in 1981 their numbers fell to 30 wolves divided into five packs.

Each wolf pack travels about 11 km per day and 33 km per moose killed. The pack maintains a social hierarchy with one dominant (alpha) male and female. Mate-preference and restriction of courtship behavior among subordinates are thought to reduce the potential number of breeding pairs and often result in the birth of only a single litter to the pack. Not only do subordinate wolves have less chance of reproducing, they also have less chance of surviving, since they get the last share of any kill. When moose calves were abundant the pack produced more offspring. To some extent, then, the social behavior of a pack regulates the birthrate in relation to availability of food (Jordan, Sheldon, and Allen, 1967).

To what extent do the wolves on Isle Royale determine the number of moose? One possibility is that they have no influence on abundance of moose—for example, if they crop only moose that were going to die soon anyway from malnutrition, old age, or disease. According to Pimlott (1975), "In both simple and complex situations where human interference is minimal, it is evident that the effect of predation (by wolves) as a depressant on the population is minimised as a result of being concentrated on old animals and on juveniles which have not entered the breeding segment of the population."

On the other hand, evidence from Isle Royale suggests that wolves probably have reduced the number of moose that would otherwise be there. Between 1958 and 1974 Peterson (1977) found 837 dead moose. He considered that they had died from the following causes: wolf kill, 296; probable wolf kill, 230; malnutrition, 27; drowning in flooded streams, 27; accidents, 14, and unknown causes, 250. If "wolf kills" and "probable wolf kills" are combined, some 63% of moose that were found had been killed by wolves. Only a small proportion of those killed by wolves were recorded as undernourished.

Wolves do not kill randomly with respect to the age of the moose. Figure 5.03 shows the proportion of the total 837 deaths in the different age-classes alongside the assumed proportion of the different age-classes in the moose population. The highest proportions of deaths were among calves 1 year old (the commonest age group) and among adults 6–14 years old (much less common age groups). Moose aged 2–6 years were less vulnerable to wolves, though there were more of them than older moose. The importance of wolves as a cause of deaths in these two age-groups is further suggested by the life table of the moose (fig. 5.04), which is based on the age of the animals found dead. The survivorship curve takes a deep plunge in the first year of life and again after the seventh year.

Before 1970 about 13% of kills were in the age-classes 1–6 years; after 1970, 53% of kills were in these age-classes. Peterson (1977) associated the high predation rate after 1970 with malnutrition of moose early in life and greater vulnerability in deep snow. According to Peterson (pers. comm.), wolves were "surplus killing" moose that were undernourished in 1970. There were so many starving moose then that many died from malnutrition before wolves could attack them. Most healthy moose can protect themselves adequately from wolves; they are probably less vulnerable than smaller animals such as deer and caribou. Moose are in danger when undernourished and when away from protective refuges such as woods.

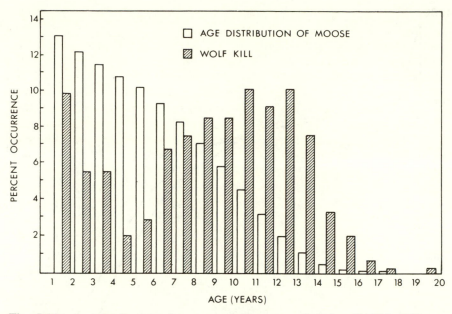

**Fig. 5.03**  Age distribution of moose and percentage of total wolf kill in different age-classes of the moose on Isle Royale. (After Peterson 1977)

However, since 1976, when wolves were abundant, few moose have been found dead from malnutrition.

The high population of wolves in 1976 was associated with high vulnerability of moose, and this in turn was due to the combined influence of successional changes in the forest, an increase in numbers of beavers, and severe winters, all of which are first- or second-order modifiers in the environment of the wolf. Heavy snow disables moose calves and they become more open to predation. The number of moose preyed upon by the wolf depends upon both the number of moose and the state of their nutrition. The fall in number of wolves from 50 in 1980 to 30 in 1981 is considered by Peterson (pers. comm.) to be due to a succession of winters since 1970 when food was short, calves were scarce, and the beaver population fell from 316 in 1974 to 122 in 1980. Already in 1980 3 dead wolves were found. From observations on the number of moose killed by wolves per day, Peterson (pers. comm.) estimated that in 1975/76, during the six months of winter when adult moose are most vulnerable to wolves, 44 wolves killed 187 moose (out of a total population of about 800). But during the 1979/80 winter, 50 wolves killed only 94 moose (out of a total of about 680). Hence when there was 1 wolf for every 18 moose, 187 moose were killed, but when there was 1 wolf for 13 moose only 94 moose were killed. Further, there is no correlation between the number of moose killed per day and the size of the wolf pack. A pack of 7 wolves and one of 20 wolves both have a maximum killing rate of 1 moose every 2.5 days. There seems to be no correlation between density of prey and predation rate or density of predators and predation rate. This surely points to the important influence that modifiers of various orders have on the "activity" of wolves (fig. 5.01).

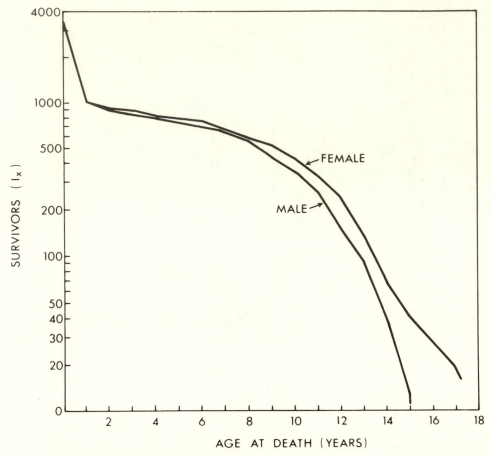

**Fig. 5.04**   The life table of the moose on Isle Royale. (After Peterson 1977)

The complexity of the environments of the moose and the wolf makes it difficult to determine what influence the wolf had on the abundance of the moose and what influence the moose had on the vegetation of the island. Caughley (1977, 131–32) contemplating the data up to 1975, considered that moose on Isle Royale are less abundant than they would be if no wolves were there. Peterson (pers. comm.), contemplating the data up to 1981, considers that the number of moose is determined primarily by the availability of food, which in turn depends very largely on weather. But moose losses to wolves become significant at times when moose, for reasons already given, become more vulnerable. These are not just moose that would soon have died from malnutrition or some cause other than predation.

What is the influence of moose on the vegetation? There is no doubt that the initial eruption of moose caused a shortage of food in the 1930s. Since the initial eruption, numbers of moose have been much lower, but browsing in these later years has had its influence on the vegetation, especially its composition. The density of trees in the upland forest has greatly declined, but average diameter has increased. This is due to browsing of the young firs by moose and to the elimination of mountain ash by browsing. White spruce, unpalatable to moose, has increased in density. The change in composition of the forest has probably reduced its susceptibility to crown fires, since there are fewer low conifers to bridge the gap between the ground, where most forest fires start, and the tree canopy (second pathway leading through vegetation to food, fig. 5.02). Browsing reduces or eliminates species favored for browsing. American yew, at one time abundant, now occurs only in places inaccessible to moose. Its place has been taken by thimbleberry, *Rubus parviflorus*, as the prevailing shrub in the understory. Proliferation of thimbleberry probably followed fires in the late 1800s, and the failure of American yew to replace it was the result of heavy browsing by moose on this palatable species (Janke, McKaig, and Raymond, 1978; Snyder and Janke 1976; Janke, pers. comm.). After fire the regeneration of aspen and paper birch is greatly retarded in some areas by browsing. Sometimes these trees are killed by intensive browsing following regeneration after fire. Some trees, like hazelnut, are kept in shrub form by browsing and probably provide more food for moose as a result of their low height (Peterson, pers. comm.). So there is no simple answer to the question, What influence do moose have on the vegetation in Isle Royale National Park?

How then can one say with any confidence as May (1973a, 322) has done, that "our [mathematical] model marches with the observed circumstances of an initially unstable vegetation—moose system and a subsequently stable vegetation-moose-wolves system?" These circumstances have not as yet been observed. Mathematical models are abstractions from nature. The theory of environment stresses that all important components should be considered and their activity analyzed. The feedback between predator and prey and prey and vegetation postulated by mathematical models, if it exists, is obscured by many other components of environment.

What would happen to the vegetation if there were no wolves on Isle Royale? We do not know. With certain combinations of components of environment the moose might increase, as they have at times in the presence of wolves (fig. 5.01), in which case the vegetation would no doubt change in composition as it has in those former episodes of increasing numbers. Further than that, little can be said with any confidence.

To find clear-cut examples of feedback between predator and prey, we must go to the literature on biological control of weeds and insect pests (sec. 5.3). In successful biological control the predator greatly reduces the abundance of the prey. In section 5.223 we discuss some suggestive examples from marine habitats as well. The existence of feedback between predator and prey does not imply the existence of an "equilibrium" or a "stable state" between predator and prey as postulated in deterministic mathematical models of this interaction. The real world is more complex than that. Frazer and Gilbert (1976) investigated the relation between the pea aphid *Acyrthosiphon pisum* feeding on alfalfa and its predator the ladybird beetle *Coccinella trifasciata*. Frazer and Gilbert counted the aphids and coccinellids in a field of alfalfa over only one season, but they made many studies on the ecology of predator and prey in contrived situations in the laboratory and in the field. From these studies they derived a "simulation model" of the numbers of predator and prey. The model predicted the number of aphids eaten, given the initial numbers and age distribution of aphids present on the alfalfa. A summary of how (in the model) the coccinellid hunts the aphids is provided by Gilbert et al. (1976, 135–36):

The coccinellid searches one plant after another. Each successive plant is chosen at random; there is no evidence that the coccinellid tends to move in any fixed direction up or down the row, and it often searches the same plant several times. Nor does it choose plants bearing many aphids in preference to those with few or none. If the predator searched more methodically, the model would have to be amended accordingly. The predator cannot detect the aphids at any distance; it finds them solely by running into them. An aphid may escape by falling off the plant when it feels the predator advancing towards it. If the aphid falls off a plant, it climbs on to a new plant chosen at random, i.e. without reference to the number of aphids already on that plant. Old aphids fall off more readily than young, presumably because they have less trouble finding a new plant. When the predator searches a plant, therefore, each aphid on the plant has a fixed probability of leaving the plant; but that probability varies according to the aphid's age. The predator has a certain probability of contacting each remaining aphid. That probability increases with the predator's hunger (defined as the biomass of aphid which the predator will currently eat to satiation), because hungry coccinellids search more anxiously than less hungry individuals. Once the predator has contacted an aphid, it has a fixed probability of capturing it; that probability again varies according to the age of the aphid, because older aphids can escape from the predator's grasp more easily than can young ones. When the predator eats an aphid, its hunger is reduced according to the biomass of the aphid. If the predator searches a plant without contacting any aphids, it spends a fixed average time on that plant, and its hunger increases accordingly. If it contacts an aphid but does not capture it, it searches the plant more thoroughly, and therefore spends a longer average time on the plant. The probability of contacting further aphids is consequently increased. If the coccinellid captures one or more aphids, the total time spent on the plant is increased by the time taken to eat the prey, which is proportional to the biomass of the prey.

The numbers of aphids and coccinellids counted on the plot are shown in figure 5.05. At the start of the season the number of aphids increased. After the coccinellids arrived the number of aphids fell. Low numbers then persisted through the time of maximum coccinellid numbers. Thereafter the number of coccinellids fell sharply, and the aphids increased again until the alfalfa was cut. These results have been repeated over several years (Gilbert, pers. comm.). Despite the evident correlation between the number of aphids and the number of coccinellid the evidence indicates that there was no sense in which the fluctuations of numbers of aphids and coccinellids conformed to any theoretical steady-state model of predator and prey. Other components of environment were more important in determining the trend in numbers of predators and prey. For example, temperature had a differential influence on the rate of predation and the rate of increase of the prey. It is true that for any density of aphids at any given temperature the model specified a density of beetles that would hold the numbers of aphids steady once the age distribution of the aphids had stabilized. But if temperature increased slightly, the model predicted that the coccinellids would begin to drive the aphids to extinction and would continue to do so, even if the temperature subsequently returned to its original value. Similarly, the model predicted that a temporary decrease in temperature would initiate an increase in numbers of aphids which the coccinellids could not subsequently halt. Including temperature in the model is a big advance on the classic models of the interaction of predator and prey. But we must remember that this model is still an abstraction, since it modeled predation abstracted from

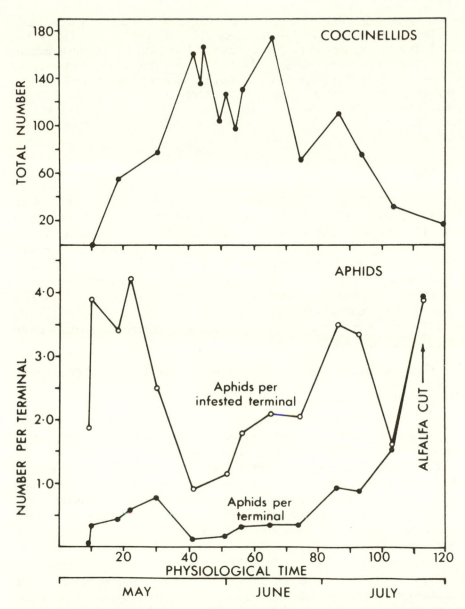

**Fig. 5.05** The number of pea aphids *Acyrthosiphon pisum* and their prey the ladybird beetle *Coccinella trifasciata* on an alfalfa plot during the growing season. Physiological time is calendar time transformed to "aphid time" because of the influence of temperature on the rate of development of aphids. (After Frazer and Gilbert 1976)

all of the environment except food and temperature for the ladybird beetle and predators and temperature for the aphid. The concept of a natural population comprising a number of local populations was also abstracted from the model. Nevertheless, based on the evidence provided on the interaction between the ladybird beetle and the aphid, Gilbert et al. (1976, 39) suggested the following generalization:

The conventional definition of stability, *viz.* a tendency to return towards a steady state, simply does not apply because no steady state is possible, even in theory. We must add that we know of no case where such an "equilibrium" has been demonstrated, except in the laboratory. . . . the predator-prey relationship sets no upper limit to the densities of either prey or predator, and no lower limit (except zero) for the predator; while the minimum prey density depends on both the temperature and the age-distribution of the prey. The truth is very simple: the predators leave the alfalfa field when the prey density has sunk so low that the predators cannot find enough to eat.

## 5.2    The Activity of Predators

In the theory of environment the influence of any component of environment on the primary animal is measured as the "activity" of that component. The probability that a rabbit will be eaten by a fox may depend upon the number of foxes that are hunting rabbits, or the number of rabbits, or the obstacles that may impede the hunt, and so on. As has already been discussed briefly in section 1.1, the activity of the fox might be represented by a factor for abundance of foxes multiplied by a factor for abundance of rabbits multiplied by a factor for success in hunting. In more general terms, the activity of a predator depends upon characteristics of the prey (sec. 5.21) and upon the web of pathways that lead through the predator to the primary animal (sec. 5.22).

### 5.21    Characteristics of the Prey That Influence the Activity of Predators

#### 5.211    Defense by the Prey

Plants and animals have evolved a variety of means of reducing their chance of falling prey to predators. Many plants produce substances that are neither helpful nor harmful to the metabolic pathways that are necessary for their normal growth and development. Some of these substances reduce the plant's palatability to phytophagous animals and may even make the plant immune to attack. Animals in their turn may coevolve the capacity to eat previously unfavored plants containing unusual biochemical substances. This capacity may require a degree of physiological specialization that limits their range of food. This and other advantages of specialization could lead to monophagy or at least to a reduced choice of plants used as hosts (Ehrlich and Raven 1965). The caterpillars of the moth *Depressaria pastinacella* feed exclusively on the umbels of the wild parsnip *Pastinaca sativa*. The umbels are rich in nitrogen; they also carry a higher concentration of furanocoumarins than any other part of the plant. The larvae detoxify the umbel tissue or are tolerant to its furanocoumarins and so they achieve a meal rich in nitrogen. They do not feed on the leaves and other parts of the plants that are low in nitrogen and in furanocoumarins. Many polyphagous insects feed on the wild parsnip, but not on the umbels, and so miss out on the most nutritious part of the plant. The price they pay for being generalized feeders is that the richest source of nitrogen in the parsnip is not available to them. The specialist is the sole user of the richest source of nitrogen and uses no other. Moreover, the specialist is attracted to the umbels by the furanocoumarins, which serve as the clue that identifies the parsnip. For the generalist feeders, furanocoumarins are both toxic and repellent (Berenbaum

1981). The arctiid moth *Seirarctica echo*, feeds on cycads that contain cycasin and a glycosidase. The larvae have special detoxifying mechanisms that protect them against the otherwise lethal components of their diet (Teas 1967). Buttercups (*Ranunculus* spp.) are distasteful to grazing animals and are rejected until other forage is gone. Several species contain protoanemonin, an irritant that can cause fatal convulsions in livestock. Nicotine and other alkaloids cause paralysis in aphids that feed on tobacco plants. Steroid cardiac glycosides in foxglove, *Digitalis purpurea*, can cause convulsive heart attacks in vertebrates.

Some toxic plant substances are less direct in their action. Tannins bind proteins into indigestible complexes which contribute to the growth-inhibiting properties of mature oak leaves for vertebrates and moth larvae (Whittaker and Feeny 1971). Larvae of the winter oak moth *Operophtera brumata* complete their development rapidly with a higher percentage surviving when fed on young oak leaves, which have a maximum content of protein and a minimum content of sugar and tannins. The feeding stage of larvae coincides with the time when young leaves are on the trees. When larvae are fed leaves that are two weeks older, so that the tannins have been laid down, they develop slowly, and many die (Feeny 1968, 1970). A concentration of only 1% tannin (dry weight) in oak leaves reduces the rate of growth of larvae of *Operophtera brumata* (Feeny 1968). The growth of larvae of *Heliothis armigera* is reduced when the concentration of tannin in the leaves it eats is as low as 0.1% (Chan, Waiss and Lukefahr, 1978). By contrast, four species of acridids were fed a diet containing up to 10% tannins without any deleterious influence on their growth and survival. Neither the amount of food that they ate nor the amount they digested was influenced by this high percentage of tannins, but concentrations of over 10% were deleterious. Evidently these acridids have overcome the plant's defenses (Bernays, Chamberlain, and Leather 1981).

Plants that accumulate alkaloids are distasteful to generalist browsing herbivores and can cause serious illness or death (Dolinger et al. 1973). Some insects that feed on alkaloid-rich species such as *Papaver* prefer the young leaves, which are relatively poor in alkaloids (Ehrlich and Raven 1965). Dolinger et al. (1973) have shown a relation between the alkaloid content of species of *Lupinus* and the availability of these plants as food for the butterfly *Glaucopsyche lygdamus*, whose larvae feed on developing and newly opened flowers of lupines. Local populations of lupines that flowered early relative to the flight of butterflies, and so had few if any eggs laid on them, had inflorescences containing a single alkaloid in low concentration. On the other hand, local populations of lupines that flowered during the flight period of the butterfly had inflorescences with a high alkaloid content. Nevertheless, accumulation of high quantities of alkaloid in the inflorescence was not in itself sufficient to prevent butterflies from laying many eggs on these plants, with subsequent predation by their larvae. A distinction exists between lupines that were heavily preyed upon and those that were least preyed upon. Plants heavily preyed upon had inflorescences containing nine alkaloids and showed little variability in alkaloid content. Plants least preyed upon had inflorescences that contained a mixture of three or four alkaloids, but the particular three or four were a selection from about ten possible alkaloids. Individual plants had a different selection, and the variability in total content was high. Larvae could be reared in the laboratory on plants of high but invariant alkaloid content with little mortality. The experiment of raising larvae on plants of variable alkaloid content was not done. Dolinger et al. (1973) interpreted individual variability in alkaloid content as a defense against the predator. It reduces the chance that strains of butterfly will evolve that can detoxify the poisonous compounds in their host plant.

The presence of highly toxic alkaloids in low concentration in some plants has been contrasted with the relatively low toxicity of tannins in high concentrations in other plants. Various investigators—for example, Feeny (1976), Rhoades and Cates (1976), and McKey (1979)—have suggested that these represent two different evolutionary strategies in defense by plants against herbivores. Which one is selected depends, they argue, upon herbivore pressure and the cost in energy to the plant of producing one or the other set of compounds. Since these are as yet unconfirmed speculations, we say no more about them.

Several trees, mostly gymnosperms, are known to contain analogues of the juvenile hormone of insects. These analogues prevent metamorphosis in some insects. Williams (1970) has proposed that these chemicals may be the plants'defense against predation by insects.

Two chemical compounds have been isolated from *Ageratum houstonianum* that depress the production of juvenile hormone and induce precocious metamorphosis and sterilization in several species of Hemiptera. In some holometabolous insects they cause sterilization and forced diapause. The biological action of these plant substances on insects is equivalent to removal of the corpora allata, the glands that produce the juvenile hormone (Bowers et al. 1976).

Certain nonprotein amino acids produced by plants are toxic to a wide range of animals. For example, canavanine, which is an analogue of arginine, is found in certain species of the Lotoideae, a subfamily of the Leguminosae. As much as 13% of the dry weight of the seed of the Neotropical legume *Dioclea megacarpa* is canavanine. It presents a formidable chemical barrier to all insects except apparently the bruchid beetle *Caryedes brasiliensis*. Not only can this beetle detoxify canavanine, but it converts it to canaline and urea which it uses as sources of nitrogen (Rosenthal and Bell 1979, 362).

Another probable defensive adaptation is discussed in section 10.331. It is commonplace for the tissues of perennials, especially woody perennials, to contain a low concentration of amino acids and proteins—too low to support a dense population of their normal predators, except perhaps during a brief annual season or during a period of unusual "stress" for the plants. Plants that are not highly nutritious may be protected from predation when the herbivore is highly dispersive and selects nutritious plants on which to feed or lay its eggs. Even the poor disperser that is unselective will tend to spare the less nutritious plants, since it will multiply slowly if it happens to find itself on such plants. Because it is a poor disperser, the larger numbers that multiply on nutritive plants will tend to stay there. However, in other circumstances we can imagine that herbivores that find themselves on less nutritious plants might have to eat more than they would on more nutritive plants, and so more of the less nutritive plants would be eaten (Moran and Hamilton 1980).

The so-called ant-acacias lack the bitter chemicals that make other acacias unpalatable to herbivorous insects. Instead the ant-acacias produce swollen thorns and specially modified leafbuds. Certain species of ants live in the swollen thorns and feed on the modified leafbuds. They also kill or drive off the herbivorous insects that might otherwise eat the acacia (Janzen 1966).

Various structural parts of plants have a defensive function. The large thorns of *Opuntia* are deterrents to large herbivores. Larvae of the moth *Lasiocampa quercus*, which feed along the edges of leaves of various trees, cannot eat sharply toothed leaves of holly (*Ilex*). But when holly leaves were cut so they had unbroken margins, the larvae ate them voraciously (quoted by Ehrlich and Raven 1965, 587). The tiny hairs on leaves and other parts of plants known as trichomes are defensive structures. They may be hooked or unhooked, glandular or nonglandular. Levin (1973) reviewed many examples in which herbivorous insects did less damage to the plants with the denser trichomes. Gilbert (1971) placed second- and third-instar larvae of butterflies of the genus *Heliconius* on the leaves of *Passiflora adenopoda*. The larvae could not move more than a few millimeters before their prolegs were securely hooked on the trichomes. The hemolymph drained from their wounds, and the larvae were dead within 24 hours.

The toxins and structures mentioned so far are characteristic of the plant whether it is being eaten or not. Some plants appear to mobilize chemicals and even structures in response to being eaten. The beetle *Epilachna tredecimnotata* feeds on cucurbits by cutting circles in the leaf, leaving only a few veins and bits of lower epidermis to hold the cut section in place. The beetle then feeds on the cut disk. The encircling takes about 10 minutes, whereas the complete feeding on the leaf disk takes from 1 to 2 hours. After feeding on a disk in the morning, the beetle crawls off the damaged leaf and does not feed again until the next morning. This behavior appears to be an adaptive response to the plant's ability to mobilize chemical deterrents and concentrate them at the site of the damage. Carroll and Hoffman (1980) found evidence of the accumulation of cucurbitacins in leaves 40 minutes after leaves were damaged by cuts. They could find no cucurbitacins in undamaged leaves. So the beetle makes its circular incision in the leaf with impunity, and by the time the plant has mobilized its toxins the circle of leaf is sufficiently cut away that these toxins do not reach it. On the other hand, an artificially cut leaf presumed to contain cucurbitacins inhibited feeding by the beetle. Despite the high toxicity of cucurbitacins to many insects and to sheep and cattle, some insects are resistant to them. Species of the genus *Diabrotica* not only are resistant to cucurbitacins but are also highly attracted by them to cucurbitaceae (Metcalf, Metcalf, and Rhodes 1980).

When leaves of potato or tomato plants are eaten by the Colorado potato beetle, *Leptinotarsa decemlineata*, there is a rapid accumulation of proteinase inhibitors in the leaves. Even leaves distant from those eaten show this accumulation. The proteinase inhibitors are released from wounded tissues and transported through the vascular system to other parts of the plant. Ryan (1979) has postulated that the proteinase inhibitors protect the plant by arresting digestive proteinase of the attacking insects, but as yet the conclusive experiments to test this hypothesis have not been done.

The autumnal moth *Oporinia autumnata* periodically defoliates large areas of the birch *Betula pubescens* in forests in northern Scandinavia and northern Finland. Haukioja (1980) showed that larvae reared on birch leaves that were damaged mechanically to simulate the feeding of larvae produced smaller pupae than those reared on intact leaves. Even more surprisingly, larvae fed on leaves from trees defoliated the previous year and even up to three years back produced smaller pupae than those that ate leaves from trees that had not been defoliated. The evidence suggests that leaves from damaged trees, even those damaged some years back, are less satisfactory for the nutritive needs of larvae than leaves from undamaged trees.

Haukioja interpreted this to mean that in response to predation the tree produces chemicals that defend it from the defoliating larvae. The larvae still eat the leaves of such trees, but if they produce smaller pupae they presumably become moths that produce fewer eggs and perhaps survive for a shorter time. These possibilities were not pursued.

According to Janzen (1979), African acacias make longer spines on shorter internodes after browsing of shoot tips by mammals. Janzen inferred from these observations that spininess is "expensive" for the plant and can be discarded with advantage if the plant is inaccessible to herbivores.

Animals also defend themselves against predators. They use a variety of defenses: sprays, bites, stings, mimicry, camouflage, and defensive social behavior. Among terrestrial animals the arthropods probably have the greatest diversity of chemical weaponry. There we find two types of defensive substances: those elaborated by special exocrine glands, and those contained in the blood, the gut, or elsewhere in the body. Bombardier beetles (*Brachinus* spp) spray predators with a secretion containing quinones which is the product of an exothermic reaction producing an audible detonation (Eisner (1970). The defensive secretions of some pentatomid bugs contain as many as eighteen compounds.

Many phytophagous insects incorporate in their bodies toxic chemicals from the plants that they eat. Rothschild (1972) listed forty-three species of insects in six orders that sequester toxins from their host plants. Sawfly larvae of *Perga affinis*, which feed exclusively on *Eucalyptus* leaves, have a diverticular pouch off the foregut in which they store an oily fluid that they regurgitate when attacked, a habit that has earned them the name "spitfires". The fluid consists of oils essentially identical to those in the *Eucalyptus* leaves on which the larvae feed. How the oils are extracted from the ingesta and collected in the diverticular pouch is not known. When tested against birds, mice, and three species of ants, the fluid was an effective deterrent. The defensive effect of the secretion is enhanced by the spitfires' habit of forming dense rosette clusters in the daytime (when they are not feeding) with their heads directed outward and secreting the offensive fluid synchronously (Morrow, Bellas, and Eisner 1976).

Most species of insects that incorporate toxins from plants in their own bodies do not concentrate them in special organs as does *Perga*. Toxic compounds from milkweed (*Asclepias*) are accumulated in their bodies by larvae of monarch butterflies. The toxins persist in the adults, making them distasteful to birds.

Some animals defend themselves or their offspring by appropriate defensive behavior. Morse (1980, 112–37) gives an account of a variety of methods. For example, prey may seek cover and hide. Birds that form social groups, on finding cover, may make low-intensity calls that muster the group but are inaudible to the predator. For example, the turkey *Meleagris gallopavo* makes soft singing notes. When the prey is not conspicuous, "freezing" may be a feasible alternative. This occurs frequently in flocks of birds, when one individual first spots a predator. Chickadees and titmice (*Parus* spp.) freeze immediately on hearing a call from another individual if they are in a leafless tree some distance from cover. Only by stalking or ambushing can predators approach cursorial prey, whose top speed is almost always well above that of the predator. Most prey are not vulnerable to a tiger unless it is initially within 20–30 m. The area of danger around a cheetah or a pack of wild dogs may be 100 m or more. Some prey mob their predators. Chaffinches and many other birds will mob a perched hawk or owl. Social animals give various kinds of displays to warn members of the group of danger. Plains-dwelling antelope and pronghorn have characteristic flash marks, that may function in this way (Morse, 1980, p.119).

It is well known that predators of large herding animals tend to stalk individual prey that have wandered from the group. The crowd is these species' defense. In the presence of wolves a herd of musk-oxen surround their calves and face outward in defense against intruders (Mech 1970, 242). Leopold (1933, 86) stated that herds of antelope of less than twelve to fifteen usually do not fight off wolves or coyotes as a herd. When such herds are attacked they usually stampede and scatter, so the loss is great. Herds of more than fifteen usually band into a defensive group and ward off attackers more effectively. In Andrewartha and Birch (1954, 345) we summed up, "In the presence of predators, an animal may have a greater chance of surviving if it is one of a large organized herd or flock than if it is in a small herd or solitary."

Invertebrates show elaborate responses to predators. Two species of limpets (*Acmaea* spp.) seek to evade species of starfish that normally prey on them, but they ignore those that do not (Phillips 1976). Many gastropod molluscs respond to predatory starfish with an avoidance response or an escape response (Feder, 1963). The female sawfly *Pseudoperga guernii* lays its eggs on the midrib of narrow leaves of *Eucalyptus*. It protects both the egg pod and the cluster of young larvae by standing over them in a defensive posture (Fox and Morrow 1981).

## 5.212 Size and Age of the Prey

Connell (1975) studied a population of the barnacle *Balanus cariosus* that initially consisted of classes aged 2, 4, and 10 years. Over a period of thirteen years predatory snails ate all young that arrived each year on most of the sites. At some sites not all new arrivals were eaten. If they survived to the age of 2 years, they became invulnerable to predation by all the common predators except the large starfish *Pisaster ochraceus*. Connell demonstrated the invulnerability of the older barnacles by protecting barnacles of different ages in cages on the rock surface. When he removed the cages or allowed predators to enter, all *Balanus* younger than 2 years were eaten. Older barnacles were seldom eaten.

The mussel *Mytilus edulis* is less vulnerable to attack by crabs when it reaches mature size. Only one species, a very large crab, *Cancer pagurus*, can break open large mussels (Kitching, Sloane, and Ebling 1959). Paine (1976) showed in particular study areas that *Mytilus* above a certain size were not preyed upon by the starfish *Pisaster*. He excluded *Pisaster* from an area for 5 years, after which it was allowed to reinvade. By this time the mussel shells had grown longer than 8 cm and could not be preyed upon by *Pisaster*. The African freshwater tiger fish *Hydrocyon vittatus* preys mostly upon fish less than 40% its own length. It cannot prey on larger fish because its habit is to swallow its prey whole. Fish longer than 15–20 cm are safe from predation. Jackson (1961) measured the length of 188 predatory tiger fish and the size of the prey in their stomachs. In no case were the prey greater than 50% of the length of the predator, and most were less than 40%. Small prey can of course be eaten by both small and large predators, but the habit of some prey of spawning in protective refuges keeps them, while they are juveniles, out of reach of predators.

We saw in section 5.1 that wolves on Isle Royale take mostly young moose (less than 1 year old) or old moose (more than 6 years old). Moose 1–6 years old are less vulnerable to predation because they can more readily fight off attacks and outrun wolves. Mech (1970, 239) wrote: "Because of the large size of the elk, the animal probably is almost as difficult for wolves to attack safely as is the moose, and no doubt most attacks on it are unsuccessful". Where prey that are easier to catch are available, wolves take fewer adult elk, though elk calves are killed because they lack the protection of size.

### 5.213   The Density of the Prey Population: The Functional Response of the Predator

When a predator has sufficient food and no other component of environment inhibits the rate of increase, the population is likely to increase. Solomon (1949) called this interaction the "numerical response". The predator-prey models of Lotka, Volterra and of Nicholson and Bailey rested on the assumption that the rate of increase would be entirely determined by the supply of food (i.e., the density of the prey–population; Andrewartha and Birch 1954, sec. 10.12). Solomon (1949) pointed out that the activity of the predators in seeking their food might also be influenced by the density of the population of prey. He called this interaction the "functional response".

Holling (1965, 1966) investigated experimentally the functional responses of a variety of predators. Some predators conformed to what he called type 1 functional response (fig. 5.06, bottom graph). The individual predator ate a constant proportion of the prey, as in the numerical response, but only up to a certain limit of density of prey, as in figure 5.06. Beyond that limit the proportion of prey that were eaten fell. The animals that responded in this way were mostly filter feeders such as brine shrimps and *Daphnia*. Presumably the filtering mechanisms take in a constant proportion of the water that flows past until the animal is satisfied, after which it stops filtering; hence the limit set to the numerical response. The functional response of the ladybird beetle *Adalia bipunctata* to different densities of the population of its prey, the lime aphid *Eucallipterus tiliae*, is a type 1 response (Wratten 1973). If predators with this type of response were to be effective in controlling increasing numbers of their prey, they would have to do so through increasing their numbers—that is, through a numerical response.

Most predators do not eat like filter feeders. Holling investigated the response of hunting predators such as corixid bugs, ditiscid beetles, and mantids. They all ate more of the prey as the density of the prey population increased, until it reached a plateau. Instead of the approach to the plateau being a straight line, as in the numerical response, it was curvilinear (fig. 5.06, middle graphs). Holling called this the type 2 functional response. The curve takes this shape because with increasing numbers of prey per unit area the predator spends more time "handling" and eating prey and consequently less time looking for it. Type 2 functional response is characteristic of many arthropod predators (Holling 1965; Hassell 1978, chap. 3).

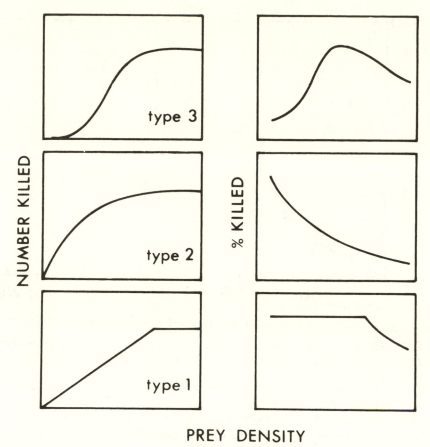

PREY DENSITY

**Fig. 5.06**   Three types of functional response of predator and prey. The number and the percentage of prey killed per unit time by a single predator are graphed against density of prey. (After Holling 1959)

There is yet a third type of functional response called type 3 response, which Holling identified when he did experiments on some vertebrate predators (fig. 5.06, top graphs). For example, Holling offered a deermouse, *Peromyscus leucopus*, various numbers of pupae of the sawfly *Neodiprion sertifer*, which are its preferred prey. The pupae were buried in sand so the deermouse had to search for them. Holling also provided an excess of an alternative and less-preferred food, dog biscuits. As the density of the sawfly population increased (fig. 5.07 and 5.06, top graphs), the predator captured more of them (and a greater proportion) up to a density of 129 sawflies per square meter, after which the response leveled off. Holling interpreted this response as follows. As sawfly density increased, the deermouse *learned* to search in the sand for pupae and became more proficient at it. At higher densities the falling off in the response of the predator was attributed to the combined effect of satiation and the shorter time left for searching after handling and eating so many prey. The learning experience provided by higher densities of prey may be an example of what Tinbergen (1960) called the "specific search image". He found that there was frequently a sudden increase in particular species of insects in the diet of the great tit *Parus major* soon after the particular

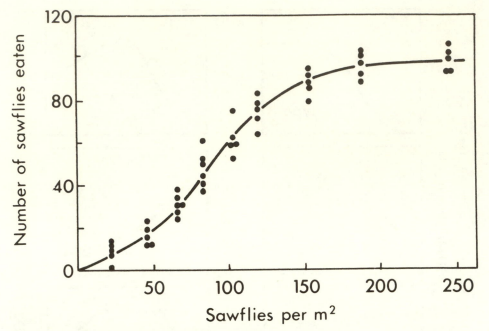

**Fig. 5.07**   The number of sawfly pupae eaten by a deermouse when an excess of an alternative food (dog biscuits) was provided. (After Holling 1965)

insect became abundant. There was a lag of a few days after the initial increase in density before the prey appeared in large numbers in the bird's diet. The increase appeared to be greater than one would expect simply from the observed increase in density of the prey. Tinbergen's observations suggested that a predator tends to overlook species of prey that are available only in low numbers. As the number of a particular prey species increases, so do the number of encounters of predator with prey. This leads to the predator's learning to concentrate on certain cues from the prey, such as size and color. The predator becomes skillful in picking out particular prey that are abundant, and so it takes a larger proportion than would be expected from random hunting. Type 3 functional response is apparently widespread among vertebrate and arthropod predators that learn to hunt better with experience (Hassell 1978, chap. 3). It is the only one of the three responses of predators that is "density dependent" in the sense that the intensity of predation per individual predator increases with increase in density of prey (up to a critical density).

A natural situation may be very different from a contrived one. In Andrewartha and Birch (1954, 344) we reported that when numbers of the grasshopper *Austroicetes cruciata* were high a grasshopper had a smaller (not a greater) chance of being eaten by a bird than when its numbers were low. This happened because toward the end of the outbreak food for the grasshoppers had disappeared (green grass and forbs had dried up) from everywhere except a few isolated localities that were, by virtue of their topography, moister than the surrounding country. The few remaining grasshoppers congregated in the few remaining places where they could still find food. And so did the birds. When the psyllid *Cardiaspina albitextura* was abundant on a high proportion of the foliage of *Eucalyptus blakelyi*, predatory birds ate a small proportion of psyllids. However, when psyllids were aggregated in high numbers on isolated patches of foliage so that most of the foliage had few if any psyllids, the proportion eaten was high. Again the birds were very effective in annihilating local concentrations of prey but ineffective in reducing the numbers when the population was dense and widespread (Clark 1964). Calvert, Hedrick, and Brower (1979) studied the mortality due to predation by birds of the monarch butterfly, *Danaus plexippus*, in its overwintering sites in Mexico, where the butterflies aggregate in enormous numbers and at very high densities in selected trees and groups of trees. In places the forest floor was carpeted with as many as 776 dead butterflies per square meter. Most of them had been killed by birds; two species of oriole and a grosbeak. These birds made repeated forays into colonies of butterflies, attacking hanging clusters. The number of butterflies killed by predation was inversely proportional to colony size. Large colonies had a far lower proportion of deaths from predation than small colonies. Calvert, Hedrick, and Brower (1979) suggested that predation is concentrated on the periphery of the colonies. Because of the relation between surface area and volume, the chance of being preyed upon will decrease with increasing size of colony. Intense packing of butterflies may indeed be an adaptation for gaining a safe position.

Pitelka, Tomich, and Treichel (1955) showed that predatory birds such as the skua and the snowy owl were able to congregate swiftly in localities where lemmings were abundant. However, the impact of these predators was greatest after the numbers of lemmings had fallen. By this time the predators had increased in abundance as a result of their earlier feasting. Likewise Keith (1974) and Keith and Windberg (1978) found that when snowshoe rabbits were at peak densities of about 7 per hectare, predators accounted for 5% of their deaths. When numbers had declined to 0.9 per hectare, predators accounted for 41% of deaths.

Besides taking an increased proportion of prey at low densities, at least some of the predators had also become more abundant by the time the prey were less abundant. According to Pearson (1966, 1971), predators of the vole *Microtus californicus* caught only 5% of the standing crop of voles when voles were at their peak densities, but as many as 88% when voles were rare. The predators in this case were feral cats, foxes, and skunks, which seemed very effective at hunting sparse populations of voles.

According to Newsome and Corbett (1975), plagues of the rodents *Mus musculus*, *Rattus villosissimus*, and *Notomys alexis* occurred in the central deserts of Australia after unusually moist weather. But long before the normally dry weather returned, and despite the continuing good seasons, their numbers fell. Newsome and Corbett noticed that the plagues attracted predators. Barn owls became abundant, as did dingoes, foxes and feral cats. In one habitat, as the members of *Rattus* fell the proportion of dingoes eating rats increased, suggesting that their specific search image for rats had been retained despite the rareness of the rats at that stage.

These observations may throw some light on what often appears to be a slow recovery of populations of small mammals after they "crash" from high numbers. Whereas the predator may have little influence on the prey when it is abundant, it may retard its rate of recovery after a crash.

The relation between the activity of the predator and the density of the prey may be further confounded by group behavior of the predator. An individual predator may have an increased chance of catching prey when there are many predators rather than few. In early spring when rivers along the Nile rise and water flows into channels leading to natural depressions, subadult crocodiles often form semicircles where the channel enters the depression. They face the onrushing water and snap at fish as they move toward them. Each crocodile stays in place, and there is no fighting over the prey. Any change in position would leave a gap in the ranks through which fish could escape. What might be a momentary advantage for one crocodile would be a net loss for the group (Pooley and Gans 1976). The activity of the predator in this case is a function of their characteristic behavior as a group, which increases the chance of finding prey.

Several authors have proposed that insect herbivores will increase most where their plant prey are most concentrated and that plants grown in polyculture are less subject to insect attack than plants grown in monoculture (Tahvanainen and Root 1972; Root 1973). In each instance the proposal is that herbivores concentrate on abundant prey, or else that the monoculture relieves an extrinsic shortage of food (sec. 3.131).

But this is not always the case. A preferred prey that is scarce relative to an alternative but less attractive prey may experience heavy predation, even to extinction (secs. 8.432, 10.22). Larvae of the fall cankerworm *Alsophila pometaria*, a polyphagous defoliator of canopy trees, hatch at the time of budbreak of scarlet oak and about ten days before budbreak of white oak. Moreover the larvae, after hatching, cannot survive more than two or three days without food. In a site dominated by scarlet oak but with some white oak, the larvae hatch as the buds of scarlet oak break and begin to defoliate the scarlet oaks. By the time the buds on the sparse white oaks break, the leaves of scarlet oak are becoming less palatable, and there is widespread dispersal of larvae to the palatable foliage of the white oaks, leaving the red oaks largely immune. In woods dominated by white oaks where there are a few scarlet oaks, only scarlet oak is in leaf at the time the eggs hatch. These few trees are defoliated by the time the dominant white oaks start breaking their buds. In each area the less abundant species of oak is exposed to greater predation in the spring than the more abundant species (Futuyma and Wasserman 1980).

## 5.22   The Web: Pathways That Lead through a Predator

The envirograms of chapter 2 show a number of pathways in the web leading through a predator to the primary animal. Some of these pathways become critical in determining whether a predator is a key component in the ecology of the primary animal. We now pursue some examples of such pathways.

### 5.221   Alternative Prey

On Isle Royale the wolves preferred beaver to moose, especially during summer. In the short term, predation on the moose was relaxed because they became an alternative but unpreferred prey. In the long term predation became more severe, especially in winter, because the wolves became more numerous as a result of concentrating on the more accessible food (sec. 5.1). A striking example of how predation-pressure on the scale insect *Aonidiella citrina* was increased by the presence of an alternative but unpreferred prey is discussed in section 10.22. The outcome can go either way, depending on whether the alternative prey is favored, but the interaction may be complex.

For example, the starfish *Pisaster ochraceus* preys preferentially on the mussel *Mytilus*, although it also eats a variety of herbivorous gastropods and chitons. These grazers influence the distribution and abundance of the filamentous red alga *Endocladia muricata*, itself an attractive site for the settling of mussels. Indeed, the density of settling mussels is highest on a substrate of *Endocladia*. After having settled on the filaments, the small mussels migrate down to the rock surface. As they grow larger they crowd out the *Endocladia* altogether. This happens when *Pisaster* is absent. However, in the presence of *Pisaster* there are fewer surviving mussels and the *Endocladia* bed may grow quite thick; *Pisaster* preys on both limpet and mussel. By preying on limpets the predator reduces the grazing pressure on *Endocladia*, which in turn favors the settling of more *Mytilus*. So the predation on the limpet by the starfish is doubly rewarding for the mussel. It reduces predation on the mussel and provides more *Endocladia* for the mussel to settle on (Dayton 1971; Paine 1969).

### 5.222   Vectors for a Predator That Is a Pathogen

When the predator happens to be a pathogen such as the *Myxoma* virus of rabbits or the bacterium *Pasteurella pestis* that causes "plague" in humans, the predator often depends on a third organism for dispersal—the vector. "Vector" means agent of dispersal. With *Myxoma* the important vectors are several species of mosquitoes and a flea (sec. 12.43). With *Pasteurella* the vector is a flea. In either instance the chance that the predator will find food depends on the abundance and activity of the vectors. So the vector is recognized as a modifier of the first order in the environment of the predator. The adult mosquitoes feed on the blood of diverse mammals and birds, and the larvae feed on microscopic aquatic organisms (fig. 2.01). The flea that is a vector for *Pasteurella* feeds on the blood of rats. So the ecology of a pathogen, especially in relation to the spread of the disease that it causes, becomes complex.

The gradual disappearance of plague from Europe had a number of causes including the possible evolution of lower virulence in the microorganism. Important among these causes was the cutting of the link between vector and pathogen. In this instance the prey was *Homo sapiens*, the alternative food for the predator was *Rattus*, and the vector was the flea *Xenopsylla cheopsis* and other species of fleas. At the time when the plague was common in Europe, buildings were mostly wooden and had thatched roofs. They provided suitable places for rats to burrow in, and the thatched roofs were ideal nesting places; they brought the rats and their fleas very close to the prospective prey of *Pasteurella*. Because of a shortage of wood, however, the wooden buildings were replaced by stone buildings and the thatched roofs by tiles. An important link between the vector and its prey was broken, and plague became much less common (McNeill, 1976, 172).

### 5.223   Predators of Predators

Many predators are themselves preyed upon which may effectively reduce the densities of their populations below the level at which they can exert much influence on the density of the population of the primary animal. So the predator

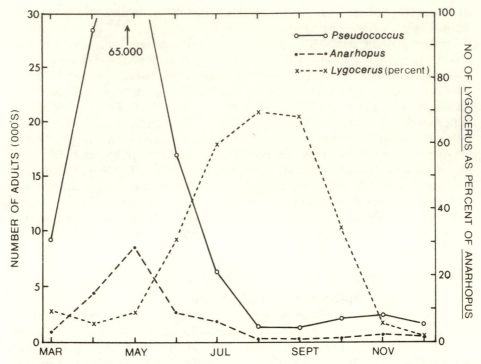

**Fig. 5.08**   Trends in the natural population of *Lygocerus* sp., *Anarhopus sydneyensis*, and *Pseudococcus longispinus* in a citrus grove in California. The first preys on the second, and the second preys on the third. (After De Bach 1949)

of a predator is recognized as a modifier of the first order in the environment of the first predator's prey. In the early days of biological control the influence of the predator of a predator was often overlooked, and the secondary predators were sometimes introduced along with the primary ones. For example, in Australia the scale insect *Saissetia oleae* is preyed upon by the encyrtid *Metaphycus lounsburyi*, which was introduced into Australia about 1925. It seems to have had little influence on the abundance of *Saissetia*, and this has been attributed to the presence of *Quaylea whittieri*, predator of *Metaphycus*, which had been introduced some time before.

The mealybug *Pseudococcus longispinus* in southern California has in its environment the encyrtid predator *Anarhopus sydneyensis*. *Anarhopus* is preyed upon by the calliceratid *Lygocerus* sp. The predator *Lygocerus* is handicapped by cold relatively more than its prey, with the result that *Anarhopus* is most numerous during spring and early summer (fig. 5.08). On the other hand, figure 5.08 also shows that *Lygocerus* was favored when a high proportion of the mealybugs contained larvae of *Anarhopus*, even though the absolute abundance of *Anarhopus* might be less. Thus *Lygocerus* presses most heavily on *Anarhopus* at the time when *Anarhopus* is pressing most heavily on *Pseudococcus*; this makes *Lygocerus* relatively more important in the environment of the mealybug than might appear at first sight.

It is partly because of complexities of this sort that authorities on biological control continue to disagree about the importance of predators of predators ("hyperparasites"). When the prey is living inside the body of another insect, as in the case of *Lygocerus*, the search for the prey becomes doubly difficult. This has been advanced as a reason for expecting the predator of a carnivorous predator to have less chance of being an important component of environment than a predator of an herbivore. (For further discussion see Andrewartha and Birch 1954, 471–473).

Interest in the predator of an herbivore has centered on the question, In what circumstances does release of the herbivore from its predator lead to overgrazing and hence an intrinsic shortage of food for the herbivore? For many years the deer and their predators on the Kaibab plateau in Arizona have been cited as an example. Reduction in numbers of predators was said to have resulted in more deer and the consequent overshooting of the carrying capacity, with great destruction of vegetation. However, Caughley (1970) has shown that other explanations account for the destruction of the vegetation.

Sea urchins are grazers of kelp on the west coast of North America. Sea otters eat sea urchins where their distributions overlap. Below a critical depth, sea otters do not prey upon sea urchins. Dayton (1975) found that on the shores of an island in Alaska the density of sea urchins increased below this depth (about 18–20 m). The cover of kelp also decreased below this depth. He postulated that predation of sea urchins by sea otters in shallow waters enabled the kelp to grow there. Although Dayton did not do the critical experiment, others (Kitching and Ebling 1961; Jones and Kain 1967; Paine and Vadas 1969) had demonstrated earlier that removing sea urchins from shallow water resulted in a thick growth of kelp, indicating that sea urchins were heavy grazers on the kelp. They also eat "drift algae". In the past sea otters were hunted commercially, but more recently this practice has been proscribed. As a consequence sea otters have become more numerous again, as have dense beds of kelp, and sea urchins have become fewer (Duggins 1980). It is said that in the past, when hunting of sea otters became intense, outbreaks of sea urchins usually followed (see Mann 1973). A similar relation between kelp, sea urchins, and the lobster *Homarus americanus* on the east coast of North America has been suggested by Breen and Mann (1976). There kelp is eaten by sea urchins, which are prey for lobsters which in turn are preyed upon by man.

The gastropod *Tegula funebralis* feeds on algae on the surface of rocks in the intertidal zone. Paine's (1969) study of the distribution and abundance of *Tegula* on a shore of the northwest Pacific coast of the United States showed something of the complexity of the relation of *Tegula* and its chief predator, the starfish *Pisaster ochraceus*. *Tegula* settles high up in the intertidal zone. In this area there is no predator. There *Tegula* develops for five or six years until it becomes sexually mature, after which it tends to migrate to a lower zone. There "food is apparently much more abundant", but there too is its main predator *Pisaster*. In this zone *Pisaster* consumes 25–28% of the population of *Tegula* each year. While *Pisaster* is the main predator of *Tegula*, *Tegula* is not its preferred prey. Usually there is alternative prey in the *Tegula* zone, which means that predation on *Tegula* is less than it otherwise would be. Paine believed that the presence of alternative prey reduced the chance that *Pisaster* might annihilate *Tegula* in this zone. But why should *Tegula* move away from a predator-free zone to one containing its predator? It is of course possible that as *Tegula* grows during its early life in the upper zone its need for food becomes greater than the algal beds in this zone can provide. To survive it may require thicker beds of algae, whatever the density of *Tegula* in the upper zone—that is, it experiences an extrinsic shortage of food in the upper zone (Paine 1971a). On the other hand, Paine (1969) postulated that the lower zone is favored by *Tegula* not only because there is more food there, but also because *Pisaster* prevents it from overexploiting its food supplies. At the same time *Pisaster* is not a major threat, since it prefers prey other than *Tegula*. This hypothesis could be tested by measuring survival and reproduction of *Tegula* with different densities of *Pisaster* in its environment. This does not seem to have been done, so the hypothesis remains an hypothesis for this example.

### 5.224   Mates of Predators

In chapter 4 we discuss "underpopulation". That is to say, there may be a critical low density below which a population may be unlikely to reproduce. In some cases this may be due to the risk that a female might not find a mate or to the inability of mates, when found, to become sexually active in a sparse population. Practical entomologists, using the method of biological control against insect pests and weeds, generally accept it as a sound maxim that if there are only limited numbers of a predator it is best to liberate them all in the same place and at the same time rather than to distribute them more thinly over a wider or more diverse area. Experience teaches that it is often difficult to build up, even in the one place, a density of population that will exceed the threshold required to give the colony a reasonable chance of becoming established. There are exceptions to this rule, which only serve to emphasize the complexity of the problem. We discuss it no further here. Other aspects are examined in Andrewartha and Birch (1954, 331).

### 5.225   Weather: Differential Influence of Weather on Predator and Prey

Burnett (1949) studied the influence of temperature on the greenhouse whitefly *Trialeurodes vaporariorum* and its predator *Encarsia formosa*. The relative influence of temperature on the birthrate and speed of development for the two species is shown in figure 5.09. The third component (length of life) of the innate capacity for increase ($r_m$) was also studied, and the trend with temperature was similar to that shown in figure 5.09 for the other two components. Between 27° and 30°C there was little difference in the life-span of adult females of *Trialeurodes* and *Encarsia*, but at temperatures near 20°C the whitefly lived longer and consequently laid many more eggs than its predator. Burnett did not present these data in the form of age-specific tables, so it is not possible to calculate $r_m$ precisely, but it is clear from inspection of the data that the relative changes in the value of $r_m$ with temperature are quite different for the two species: with *Trialeurodes*, $r_m$ would be at a maximum in the range 18–21°C; with *Encarsia*, in the range 25–28°C.

   When the two species were placed together in a greenhouse at 18°C, *Encarsia* increased so slowly relative to the rate of increase of its prey, despite the super-abundance of food (larval *Trialeurodes*), that it offered no check to the increase of its prey. When the temperature in the greenhouse was 24° or 27°C, the rate of

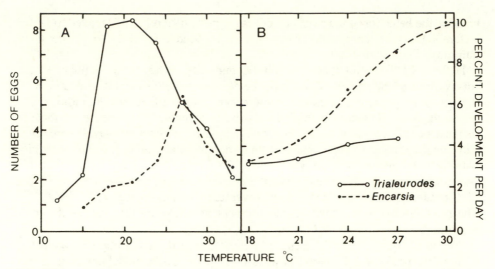

**Fig. 5.09** The influence of temperature on A, the daily rate of egg production, and B, the speed of development of a predator *Encarsia formosa* and its prey *Trialeurodes vaporariorum*. Note that the relative advantage possessed by the prey at low temperature disappears at high temperature. (After Burnett 1949)

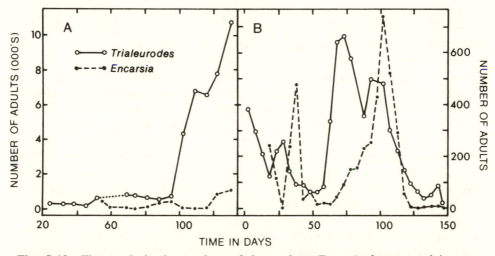

**Fig. 5.10** The trends in the numbers of the predator *Encarsia formosa* and its prey *Trialeurodes vaporariorum* when they were reared together in a greenhouse in which there were four tomato plants spaced several feet apart. In A the temperature was maintained at 18°C , in B at 24°C. Note that the predator did not multiply appreciably at 18°C, though there was plenty of food. (After Burnett 1949)

increase of *Encarsia* relative to that of *Trialeurodes* was so great that food for the predator soon became scarce (fig. 5.10). Local colonies of the prey were exterminated, but the population in the greenhouse as a whole persisted at a low density because there were always some prey that the predators failed to find. This was partly because *Trialeurodes* was constantly colonizing new situations and partly because the behavior of the two species was slightly different: they tended to prefer different parts of the tomato plants and to respond differently to gradients of humidity, light, and so on.

Hefley (1928) found that atmospheric humidity differentially influenced the survival rate of the hawkmoth *Protoparce quinquemaculatus* and its predator *Winthemia quadripustulata*. Relatively more prey survived at low humidities and relatively more predators at high humidities. Ahmad (1936) measured the net increase per generation in *Ephestia kühniella* and its predator *Nemeritus canescens* and found a differential response to both temperature and humidity. For example, with *Ephestia* the net increase was greatest, 78.3%, at 23°C; it fell to 42.1% at 15° and to 25.4% at 30°. With *Nemeritus* the net increase was also greatest, 38.7%, at 23°C but fell relatively more steeply at lower temperatures, being only 3.3% at 15°. The relative change in speed of development also favored the prey as the temperature fell from 23° to 15°C.

Sometimes quite subtle differences in climate are sufficient to tip the balance in favor of the predator or the prey. In section 5.1 we discuss a model that emphasizes the differential response to temperature of the pea aphid and its predator. The model predicts that at one temperature the coccinellid predator will drive the aphid to extinction; at another temperature the predator is quite unable to control its prey. We discuss a number of other examples in Andrewartha and Birch (1954, 469–473).

### 5.226    Refuges for the Prey

Gause (1934) described an experiment with *Paramecium* as the prey and *Didinium* as its predator in which *Paramecium* were protected from *Didinium* by a sediment in the culture tube. The prey retreated into the sediment, but the predator was unable to get in and eventually died of starvation. A more convincing laboratory model of a protective refuge was later demonstrated by Flanders (1948). His prey were larvae of the flour moth *Ephestia kühniella*. The predator was a braconid wasp. The two populations lived together in a cage on the floor of which was an open petri dish containing grain to a depth of 12 or 24 mm. The ovipositor of the braconid could probe to a depth of 6 mm. Those larvae that happened to be living between the surface and 6 mm were exposed to the predator. Those situated more deeply were secure. In every generation sufficient prey escaped to carry the population on. The adults were, of course, "immune".

The existence of protective refuges in natural populations is one reason amongst a number of others why prey and predators can continue to coexist. In Andrewartha and Birch (1954, 539) we described an example in which providing places where the scale insect *Saissetia oleae* might shelter from its predator *Metaphycus helvolus* resulted in fewer scale insects but more predators, thus making coexistence more secure. During five years of study of the cyclamen mite and its predator on strawberry plants, Huffaker and Kennett (1956) never observed the extermination of either predator or prey on any small group of plants or even perhaps on a single plant. The predator had difficulty in finding all the prey, especially those that were concealed in leaf crevices. And during periods of scarcity of prey the predator itself was to some extent protected by its ability to use honeydew as food.

Adults of the barnacle *Balanus glandula* in turbulent water on the shores of San Juan Island are distributed in a narrow zone near the highwater mark. In non-

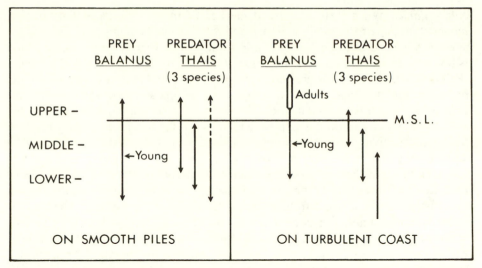

**Fig. 5.11** Distribution of prey *Balanus* and three predatory species of the snail *Thais* in the intertidal zone (upper, middle, and lower) of the coast of San Juan Island. (Connell 1970)

turbulent waters in quiet bays, and even in turbulent waters on pipes suspended from piers so that they do not reach the bottom, adult barnacles are found at all levels in the intertidal zone. The difference is due to the absence of predatory snails of the genus *Thais* in those places where adult barnacles occur (Connell 1970). These places provide a refuge from the predator. Elsewhere adult barnacles are absent because they have no refuge in their environment from *Thais* (fig. 5.11).

Connell (1970) placed stainless steel mesh cages over the rocky surface in turbulent waters to exclude predators to see if barnacles survived in the lower intertidal zones. They did survive for several years while the cages were present. Connell made observations on uncaged areas in these lower zones and found that over a nine-year study period every *Balanus* that settled there was eaten within two years after attachment. The settling barnacles in these zones were food for *Thais*. The failure of *Thais* to feed in the upper zone of turbulent waters is probably because it is unable to move up there, feed, and move down again with a period of high tide on the rough sloping surface. However, on smooth vertical piles of a pier *Thais* was able to crawl up to the highest intertidal zones, since the absolute distance was short. There were never any adult *Balanus* on these piles. Some 3 m away pipes that were suspended without touching bottom were covered with adult *Balanus*. In this case predators were unable to get to the pipes, since they do not have planktonic larvae and adults had no access from the bottom. The presence of *Balanus* at all intertidal levels in quiet bays was associated with rarity of *Thais* in those places.

Connell's studies leave no doubt that adult *Balanus* were virtually eliminated by predatory *Thais* in those places where they were accessible. Only the upper intertidal zone provided a refuge for the prey in the turbulent waters of the coast.

Much of the upper reaches of rocky shores of the exposed west coast of the United States is characterized by a distinct horizontal band of mussels, *Mytilus californianus*. When well established the bed of mussels virtually constitutes a monoculture, apart from their epifauna and its associated community. The boundaries of this band of mussels differ little from year to year; *Mytilus* can be found below the band to considerable depths, but there it is usually sparsely distributed (Paine 1974). The mussels are abundant in the "mussel band" and rare below the band because the band-area is a refuge from a predator, the starfish *Pisaster ochraceus*, which is abundant in deeper water. The starfish ventures up into the mussel-band on occasion, but exposure to air when the tide is out prevents it from being an effective predator there. Mussels below the band that have managed to survive predation by the starfish grow much larger and have a greater reproductive rate than those in the band, but the chance of survival there is small because of predation by the starfish. When a mussel manages to grow to a critical size without being preyed upon, it is safe from predation simply on account of size. The mussels in the band are small, almost certainly because they are exposed for long periods when the tide is out and they cannot feed. Below low tide they can feed at any time. Mussels appear to settle over a wide vertical expanse of the shoreline. They survive in high numbers only in the region of the high intertidal band or on the slopes of "submarine mountains" that are free from the starfish.

Many species of freshwater African fishes do not spawn in open waters where predators are common but use protected refuges where the juveniles can survive the early vulnerable stages of their lives. The cichlid fish *Tilapia macrochir* spawn in the swampy edges of lakes where dead mats of aquatic grasses protect the juveniles before they swim into open water. The mats of grass provide complete protection from predatory fishes as well as an abundance of insects and crustaceans and other food (Jackson 1961).

Errington (1967, 139) reported that the mongoose at first nearly exterminated the ground-nesting black-bellied tree duck in Jamaica. Fortunately for the ducks, a few of them adopted the habit of nesting in trees instead of at ground level. Thereafter the population of ducks began to gradually recover.

Craighead and Craighead (1969) estimated the number of meadow mice and hawks on nine sites in a study area of 30 km$^2$ in Michigan. Half the hawks observed were found in only six sites where meadow mice were abundant. Some other sites had high numbers of mice but did not have many hawks. The difference in the activity of the predator appeared to be due to accessibility of the prey. The activity of the predator increased (mice became more vulnerable) as a consequence of reduction in cover for mice owing to flooding and freezing and to overgrazing by mice, all of which are modifiers of the first and second order in the web of the hawk (mice being in the centrum of the hawk):

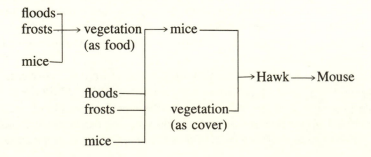

As cover disappeared mice moved to other places in search of food and cover. As the density of the mouse population fell as a result of predation, the activity of the predator did not seem to fall, because mice were still, even in their low numbers, highly accessible because of the destruction of their cover.

The muskrat, *Ondatra zibethicus*, and its chief predator the mink, *Mustela vison*, have been intensively studied (Errington 1943, 1963). The muskrat leads a semi-aquatic life. Wherever there is shallow water—along the banks of drains, around the shores of swamps, along the edges of ponds, and so on—it will construct its burrows and galleries, emerging from them to feed on grain and grasses; corn, *Zea mays*, is a favorite food. At all times, but especially during the breeding season, which corresponds approximately to summer, the muskrat shows a keen awareness of territory and will not tolerate crowding beyond a certain limit. So once the best and most secure places have been occupied to capacity by the more fortunate, the fiercer, or the more determined individuals, the others are either killed (especially the young), driven forth to seek homes in less secure places, or left with no homes at all. Some of these wanderers are taken by predators, but all or most of them are doomed to die, many predators or none.

The muskrat forms a staple food of the mink, *Mustela vison*. Mink sometimes bite their way into muskrat lodges and prey upon the resident muskrats, but this seldom happens. Muskrats living in territories in swamps are virtually invulnerable to mink. The muskrats that fall prey to the mink are those that other muskrats have driven out of the security of a home in a swamp to range far afield, or they are animals that have become handicapped by hunger or by bad weather which prevents their ready return to their refuges.

Errington also showed, during many years' study of the same populations living naturally in the same area in Iowa, that there was a close inverse relation between the density of the population at the end of winter and the rate at which it increased during summer. That is, the muskrats tended to increase more rapidly when they were few, so that the carrying capacity of the area was usually reached by the end of the season whether there had been few or many at the beginning. Since the mink determines neither the amount of cover nor the rate at which it becomes fully occupied, Errington concluded that predation had very little influence on the average abundance of muskrats in any particular area.

In his studies of the bobwhite quail *Colinus virginianus* in the central and northern part of its distribution, Errington (1945) concluded that predators were not important. The species is relatively tolerant of its own kind during winter, when coveys of fifteen to thirty birds, often comprising several family groups, come together and live in the shelter of low brush adjacent to fields and meadows which are the source of their food—insects and seeds. They are adept at seeking shelter at the slightest warning, so that if sufficient shelter is available their chance of being taken by predators during the winter is small. But when there are more quail than the area provides adequate shelter for, they are eaten by predators. The horned owl, *Bubo virginianus*, is the chief predator, but many lesser predators join in the hunt, depending on how accessible the food is. All the predators, including the owl, normally have many other sorts of food, so they do not press the hunt to the point where a quail is likely to be destroyed if it has adequate shelter.

During summer the quail breed, building nests in the shelter of grassy tussocks in the meadows. But by this time they have become highly intolerant of one another, and each male defends his chosen territory with the utmost vigor. This means that the rate of increase in a particular area during summer bears an inverse relation to the density of the population at the end of winter. These two phenomena—the relative invulnerability of well-sheltered quail during winter and the influence of the territorial instinct in determining fecundity during summer— make predators relatively unimportant. The analogy with the muskrat is close.

In the southeastern part of the distribution of the bobwhite quail, studies by Stoddart convinced Errington (1967, chap. 5) that predators had a greater depressive effect on quail there than seemed to be the case in his own study areas in the north-central part of the distribution. The most convincing explanation of this difference seemed to be that the north-central predators had such an abundance of staple prey to choose from, such as rabbits, several species of mice, and large insects, that they took quail mainly when quail were easily catchable. In the southeast the main staple food of potential quail predators was the cotton rat. When cotton rats were abundant predators depended upon them. When they were scarce, as happened every few years, predators might hunt quail more vigorously than they otherwise would. So the quality of a protective refuge for the quail is partly a function of the abundance of alternative prey for its predators. A refuge that provides immunity when there is abundant alternative prey may not provide immunity when alternative prey becomes scarce.

According to Errington's theory about predators of vertebrates, which was derived chiefly from his studies of muskrat and quail, predators of vertebrates are regarded as merely one of several "intercompensatory mechanisms" that together prevent the prey in any area from exceeding, except very temporarily, a characteristic level of abundance which is determined primarily by the amount of "cover", which Errington calls the "carrying capacity" of the area. The population may fall far below this level of abundance under the influence of other components of the environment; with muskrats, drought seems to be one of the most influential. But predators have little to do with this phase of the cycle. Using our terminology, we would say that Errington's theory placed little emphasis on the predators in the environment. He attributed most importance to the quality and abundance of shelter for the prey because these are more variable. The things that provide shelter appear as modifiers of the first order because they modify the effectiveness of the predator (sec. 2.34).

Errington's studies show that predators of some vertebrates do not necessarily reduce the abundance of prey, largely owing to territorial behavior in conjunction with the existence in territories of protective refuges. A direct approach to finding out if a predator influences the numbers of its prey would be to remove the predator and observe the consequences. Attempts have been made to do this. Predators of ruffed grouse, *Bonasa umbellas*, were removed almost completely in two places; nesting losses were reduced as a consequence, but the adult population did not increase (Edminster 1939; Crissey and Darrow 1949). This result supports Errington's conclusion. In section 5.3 we discuss circumstances in which vertebrate predators have influenced the numbers of their prey.

5.227  Intrinsic Shortage of Food for Predators

In the pathway leading to the wolf in the envirogram of the moose in figure 5.02, the activity of the wolf is shown as being modified by the number of moose in its environment. When there is an abundance of accessible moose the wolf may have an excess of food. Or there may be an extrinsic shortage of moose for the wolf owing to inaccessibility in dense forests. A third possibility is that few moose are available and accessible and the wolf eats them all. The wolf would experience an intrinsic shortage of food (cell 3, table 3.02). But this is all theory; there is no well-documented evidence that the wolf does in fact experience an intrinsic short- age of food, though there is little doubt that it experiences an extrinsic shortage from time to time. The existence of feedback between the wolf and the moose does not imply that the wolf has caused an intrinsic shortage of food for itself. There may be few moose because they have overgrazed the browse, which in turn could lead to fewer wolves. The shortage of food for the wolf in this case is not induced by the wolf. Only careful study can reveal whether an animal has an intrinsic shortage of food.

All the well-documented examples of intrinsic shortage of food for insect preda- tors have been artificially contrived by man as a venture in biological control. The animals experiencing the shortage of food are exotic, and the plants at the base of the food chain are exotics growing either as crops or as weeds associated with man's monocultures. Whether intrinsic shortages exist among endemic predators living in places that have not been modified by man remains to be demonstrated. The classic example of an intrinsic shortage of food concerns the biological control of the insect *Icerya purchasi* by two exotic predators, the beetle *Rodolia cardinalis* and the fly *Chryptochaetum iceryae*, which is described in section 3.132. Another example is the influence of *Cactoblastis cactorum* on its food the cactus *Opuntia* in Australia. *Cactoblastis* moved from a position of no shortage to one of intrinsic shortage of food, as illustrated by the change in position on figure 3.01 from the center top to right top and thence to bottom left. Similarly, though less completely, the thar, an ungulate introduced into New Zealand, began its history in New Zealand with no intrinsic shortage of food, but today much of the population experiences an intrinsic shortage (fig. 3.01; sec. 10.332).

## 5.228 Some Consequences of Sharing Food

The pea aphid *Acyrthosiphon pisum* in California used to have surges in numbers on alfalfa in both the spring and the autumn. After the accidental introduction of the blue alfalfa aphid *A. kondoi* into California, there were no longer surges in numbers of pea aphids in the autumn. This was in part because the blue alfalfa aphid increased the activity of the ladybird beetle predator *Hippodamia convergens* against the pea aphid. The blue alfalfa aphid has a higher rate of increase at low temperatures and so gets a head start in the autumn compared with the pea aphid. Having done so, it occupies the growing tips of the alfalfa so they are no longer available for the pea aphid, which settles for second best in the stems lower down. But that happens to be the preferred hunting zone of ladybird beetles, which cross from plant to plant at midheight. In the lower zone the pea aphid is ready prey for ladybird beetles which the blue alfalfa aphids avoid by being on the tips of the plants (Gutierrez, Summers, and Baumgaertner, 1980).

A predator may have in its environment another predator sharing the same prey. Conceivably the second predator may cause an extrinsic shortage of food for the first predator, or it may be a malentity. This eventuality concerns students of biological control from a somewhat different perspective. Their interest centers on whether the pest is best controlled by one or by many species of predator. According to Huffaker and Messenger (1976, 56), the empirical record on biological control supports the practice of introducing more than one predator. There has been no general indication that one has been harmed by the other, notwithstanding some reports to the contrary (sec. 7.22).

## 5.3   The Influence of Predators on the Distribution and Abundance of the Primary Organism

To assess the influence of a predator on the population of the prey, we need to compare the number of prey in the presence of the predator with the numbers that would exist in an otherwise identical environment in the absence of the predator. This is difficult to do, though we describe some examples of it. More usually, other means must be used to assess predators' influence on the numbers of their prey.

Examples of vertebrate predators that have little influence on the abundance of territorial prey are given in section 5.226. In Andrewartha and Birch (1954, 467) we estimated that Thompson's (1943) catalog of known insect predators of those insects and spiders that are of economic importance contained about 10,000 species. We chose at random from the catalog 20 insect pests known to be kept in check only by repeated application of insecticides. Without exception the catalog showed that each had several predators associated with it. The catalog made it clear that (1) an enormous number of species of insects prey on other insects, and (2) many insects remain numerous enough to be serious pests despite the presence in their environment of a number of predators. On the other hand, the mere existence of Thompson's catalog is tangible evidence of the success of "biological control" in a sufficient proportion of cases to justify expenditure of substantial sums of money. Krebs (1972, 372) estimated that 674 species of insect predators have been released in the United States and Canada in efforts for biological control of 154 pests of agricultural crops and forest trees. About 30% of the predators became established, though many of them did not control the pest; 21% of the pests were effectively controlled by the predators. The successes of biological control are outnumbered by the failures. Studies of natural predators of insects have often shown that predators have little influence on the abundance of their prey (e.g., Gutierrez, Morgan, and Havenstein 1971; Gilbert and Gutierrez 1973; Gilbert and Hughes 1971; Wratten, 1973).

However, some of the successes of biological control have been spectacular, leaving no room for doubt that in these cases the newly introduced predator has exerted a dominant influence on the abundance of the prey.

In more natural communities—that is, in those composed of indigenous species or at least where men have not deliberately introduced foreign predators—there has—until recently been no ready way to demonstrate the influence of the predator. With the invention and widespread use of chlorinated hydrocarbons as insecticides, it has become possible to gather experimental evidence on this point. Two papers, by Pickett et al. (1946) and Pickett (1949), drew attention to the selectiveness of certain insecticides and showed how their continued use resulted in the increased abundance of certain pests. De Bach (1946), taking advantage of this property of DDT, devised what he called "an insecticidal check method of measuring the

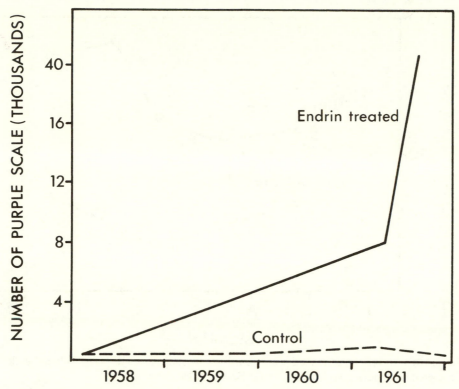

**Fig. 5.12** Increase in population of purple scale *Lepidosaphes beckii* on untreated trees and trees treated with Endrin, which is toxic to the predator *Aphytis lepidosaphes*. (After De Bach and Huffaker 1971)

efficacy of entomophagous insects" and subsequently used this method successfully with the mealybug *Pseudococcus longispinus* and the scale insect *Aonidiella aurantii* and other insects and their predators. Figure 5.12, for example, shows the increase in population of the purple scale *Lepidosaphes beckii* on trees treated with Endrin and those not treated. The predator responsible for keeping the scale insect at low densities in the control trees was the wasp *Aphytis lepidosaphes* (De Bach and Huffaker 1971). Another procedure is simply to cover foliage with insect-proof cages to keep the predator out. Figure 5.13 shows the effect of doing this with the scale insect *Aonidiella aurantii* to keep the population free from its predator *Aphytis melinus*. There was much greater mortality in the uncaged branches than in the caged branches.

Examples of successful biological control of insect pests and weeds, together with these striking demonstrations, leave little room to doubt that in these instances a predator is keeping its prey at very much lower numbers than the prey would maintain in the absence of the predator (secs. 3.132, 5.1, 10.23)

Among marine invertebrates there is evidence that some predators influence the distribution and abundance of their prey. Paine (1966, 1974) removed the predatory starfish *Pisaster ochraceus* from an area of shore extending through the intertidal zone to deep water. He removed the starfish by hand once a month for five years.

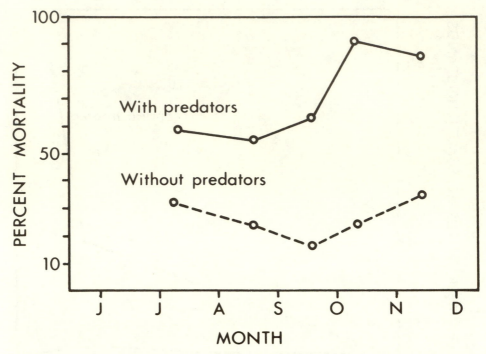

**Fig. 5.13**   Mortality of the red scale *Aonidiella aurantii* on English ivy with and without the predator *Aphytis melinus*. (After De Bach and Huffaker 1971)

Other areas were left undisturbed as controls. The abundance of algae and macro-invertebrates was estimated by transect counts at regular intervals. The control areas changed little compared with the large changes in the places from which the starfish had been removed. The barnacle *Balanus glandula* settled throughout much of the area where there was any settling space. Within three months it had occupied 60–80% of the available space. This was followed by *Mytilus* and the barnacle *Pollicipes* (*Mitella*), which crowded out *Balanus*. Eventually the area that had been cleared of starfish by hand at least once a month for five years came to be dominated by *Mytilus* and its epifauna, with scattered clumps of *Pollicipes*. Algae mostly disappeared. Chitons and limpets disappeared. The dense mussel band extended its lower boundary by about a meter in the first four years.

The extension of the *Mytilus* band and the colonization of lower areas by *Mytilus* were largely by migration of *Mytilus* from the band. *Mytilus* larvae settled in the area but were mostly eaten by other predators, especially two species of *Thais* and a small starfish (Paine 1976). The migrating larger *Mytilus* were invulnerable to this predation.

The removal of *Pisaster* was followed by a decrease in diversity of species below the *Mytilus* band (where *Pisaster* is normally an effective predator of *Mytilus*). The large number of species of algae and invertebrates became fewer with time until eventually a monoculture of *Mytilus* occupied the experimental site. Apparently in the presence of *Pisaster* the *Mytilus* population is so reduced below the *Mytilus* band that space becomes available for occupancy by other species. This experiment in Mukkaw Bay was repeated on Tatoosh Island with similar consequences and an even greater reduction in diversity of the fauna. Paine (1971b) did a similar experiment on an intertidal shore in New Zealand, with similar results. He removed the starfish *Stichaster australis* from a section of the shore. Its main prey, the mussel *Perna canaliculus* extended its vertical distribution by 40% of the available range. The number of invertebrate species fell from an original twenty to fourteen species. Furthermore, when the large brown alga *Durvillea antarctica*, which fully occupied a zone below the mussel band, was removed together with the starfish, the mussel extended its distribution to the exclusion of practically all else to occupy the zone formerly occupied by the large alga. The alga, if it gets there first, prevents the mussel from colonizing a place where it could otherwise settle and get the food (and other things) that it needs to grow and reproduce. So the alga is a modifier of the first order in the pathway that reaches the mussel through its food (and perhaps is a modifier in other pathways as well). The alga also reduces the supply of food for the starfish, so the alga appears as a modifier of the second order in the pathway that reaches the starfish through *its* food.

These experiments on mussels have also demonstrated that a certain group of species can live together in the same habitat when starfish are present but not in their absence. Starfish prefer to eat mussels. The starfish, pressing most heavily on the preferred prey, prevent the mussels from increasing to such large numbers or size that they crowd out the lesser species. When any one of these lesser species is put in the position of the "primary animal" the starfish appears in its environment as a modifier of the third order in a pathway that leads through food to the primary animal, as shown in the following segment of an envirogram:

$$Pisaster \longrightarrow Mytilus \longrightarrow \underset{\text{(for settling)}}{\text{Rock surface}} \longrightarrow \underset{\text{(as food)}}{\text{Plankton}} \longrightarrow \text{Primary animal}$$

It should be pointed out that Dayton (1971) carried out a series of experiments in which he also removed *Pisaster* from the same stretch of shoreline used by Paine (1966, 1974). In Dayton's experiments removal of *Pisaster* was not followed by a decrease in the number of species of invertebrates. Paine (1977) remarked that few mussels settled during Dayton's experiments and different organisms came to dominate the shore. The sequence of arrival of larvae of different species is apparently important in determining the outcome of the experimental removal of *Pisaster*. See section 7.11 for some more examples of how predators in the intertidal zone might influence the abundance of their prey.

Some experiments suggest that the numbers of fish may be influenced by their predators. In lakes and rivers young salmon are preyed upon by predatory fish and birds. Elson (1962) removed an average of 54 mergansers and 164 kingfishers a year for six years from a 15 km stretch of the Pollett River in New Brunswick. Some 8% of the fry of the Atlantic salmon *Salmo salar* that had been released into the river survived to 2 years old compared with 6% survival in three earlier years in which predators were not controlled. Most species of coarse fish doubled their numbers during the years in which predators were controlled. From 1935 to 1938 Foerster and Ricker (1941) removed by gill netting fish that were predators of young sockeye salmon, *Oncorhynchus nerka*, from Cultus Lake. The survival rate to 1 year of age of fry spawned naturally increased from 2% to 8% and that of fry released into the stream from 4% to 13%. Moreover, fish that had grown up without predators were larger than the average when they migrated.

We have already referred to the work of Errington, which suggests that predation has little influence on the abundance of birds and mammals that are strongly territorial (sec. 5.226). It is the territorial behavior of the prey, not the presence of predators that influences their abundance. He suggested that some of the higher ungulates that are preyed upon by canids might be exceptions (Errington 1946, 158).

According to Nelson and Mech (1981), most of the studies that have been done on wolves suggest that wolves do not usually deplete the populations of their prey. However, they cite two cases in which there is strong evidence that prey populations were severely depleted by wolves. In both cases the prey populations had been substantially reduced by deterioration of the habitat brought about by maturation of the forests combined with unusually severe winters. The additional heavy mortality of the prey caused by the predator appeared to aggravate the decline in prey numbers. Nelson and Mech (1981) and Mech and Karns (1976) provide data on one of these cases, namely, the predation by the wolf *Canis lupus* on the white-tailed deer *Odocoileus virginianus* in Superior National Forest in northwest Minnesota, a forest of some 5,300 square kilometers. The movements of deer and wolves and the number of deer killed by wolves were studied by various means such as radio-tagging of deer and aerial surveys during 1974–77. During this period both the number of deer and the number of wolves declined greatly (about 40% for the wolves) from what they had been during 1968/69.

In summer does and their fawns are more or less solitary. During summer the does are remarkably free from predation by wolves. Their solitary habit and the ease with which they can escape when there is no snow make this the safest time of the year for them. Most wolf kills then are of fawns. As winter approaches the does and surviving fawns move toward winter quarters, which may be up to 38 km from their summer quarters. In the process of getting there they form social groups. Nelson and Mech (1981) give reasons for supposing that social groups are better defended against wolves than solitary individuals in winter. Nevertheless, most of the deaths from wolf predation occur during the winter months. This, they think, is because the deer are under physiological stress owing to the cold and lack of food and because in deep snow their chance of escaping a wolf pack is much reduced. The deer that seemed to survive best were those that by chance found their winter quarters in the so-called buffer zone between wolf territories. Wolf packs in Superior National Park inhabit a mosaic of adjoining territories, each of about 125 to 310 km$^2$. Surrounding each territory is a strip about 2 km wide called the buffer zone, on either side of which wolves occur, but they rarely wander into the buffer zone. When they do so they appear insecure, and they try to kill members of neighboring packs if they meet them. There is evidence that deer that happen to be in the buffer zone are largely immune from wolves (Mech 1977).

Nelson and Mech (1981) estimated that 90% of the deaths that occurred during their study (in areas where none was killed by man) were due to wolves. Most of these deaths were among fawns, which are easiest for the wolves to catch. There seems little reason to doubt that at the time of this study, when deer were already declining for various reasons, wolves had a marked influence on the size of the deer population. Nelson and Mech's (1981) concern for survival of the deer was such that, as a result of their study, they recommended that deer be artificially fed in their winter quarters and that hunting be stopped.

Schaller's extensive study of the lion, *Panthera leo*, in Serengeti National Park in East Africa led him to conclude that the lion has little influence on the numbers of its prey (Schaller 1972, 400). Most of the lions in the park have wildebeest available for food for about one-third of the year. When the wildebeests migrate once a year en masse across the central plains of the park, only a small proportion of the lions are nomadic and follow the herds. When wildebeests are available they provide at least half the lions' food. On the other hand, predation makes little impact on the population of wildebeests. Of a population of 410,000 wildebeests lions took only 2–3% (Schaller 1972, 399). The proportion of wildebeests killed did not increase when the wildebeest population increased. Moreover, many of the animals killed were in poor condition and would probably have died soon had lions not captured them first. Hyenas were the only other predators of any significance for the wildebeests. They took about 2% of the wildebeest population and many, though not all of their prey, were debilitated.

While there is some evidence, which we have cited, that predators may influence the distribution and abundance of their prey, it is rare for a predator to exterminate its prey over the whole area of its distribution. When that does happen it is more likely to be on an island than on a continent. For example, the phasmid *Dryocococoelus australis*, endemic to Lord Howe Island, was eliminated within a few years by rats (Key 1978). Man has been blamed for the extinction of a large number of species, usually through destruction of their habitat. Human predation was important in the extinction of the dodo in Mauritius and the passenger pigeon in North America (Ehrlich and Ehrlich 1981, 215).

In laboratory experiments with predators and prey it is usual for the predator to exterminate the prey unless the prey is rendered to some degree invulnerable to the predator. Huffaker (1958) achieved this by giving both predator and prey varying opportunities to disperse. He used as prey the phytophagous mite *Eotetranychus sexmaculatus* and as predator the mite *Typhlodromus occidentalis*. The essential differences between his various models lay in the arrangement and number of oranges which served as food for the prey and of rubber balls the size of oranges. When few oranges were used, whether they were close together or sparsely distributed in space, the prey was annihilated by the predator in a matter of a month or so. But Huffaker did succeed in elaborating one experimental arrangement of his oranges in which both prey and predator persisted in a succession of three oscillations covering six months. The essential features of this design seemed to be the large number of oranges used (120 instead of 4), their regular arrangement on a flat surface, and, perhaps more important, the existence of barriers of Vaseline and other impediments to the mites' movements. This design must have reduced the dispersal of the mites, and it evidently gave an advantage to the prey by reducing the chance of a predator's finding it. The predators and prey in this experiment "played a game" of hide-and-seek that lasted some six months before all the prey were found. It is a nice laboratory model of the sort of situation that is important in natural populations of predator and prey.

The coexistence of predator and prey in natural populations is dependent upon many interactions between them that in turn depend on the environment of both. Some of the more important interactions are discussed in sections 5.2 and 5.3.

# 6

# Malentities

## 6.0 Introduction

A malentity is defined by three qualities: (a) The malentity acts directly on the primary animal, that is, it is the proximate cause of the animal's condition (secs. 2.1, 2.31). (b) The malentity is harmful to the primary animal; that is, the animal's chance to survive and reproduce decreases as the activity of the malentity increases (as in fig. 1.02B). (c) The activity of the malentity in the next generation of the primary animal remains unchanged (or it may be decreased) by an abundance of the primary animal in the current generation (as in fig. 1.03B or C). In section 15.23 we seem to have discovered malentities whose activity increases as the density of the population of the primary animal increases (fig. 1.03A). This reaction emerges because in the environment of man it seems that soldiers, terrorists, and man-made pollution should be counted as malentities. But our analysis of the environment of man is still elementary and tentative, so we let the definition of "malentity" in table 1.01 stand.

The concept "malentity" as it was first recognized by Browning (1962), under the name "hazard", has an air of "unfortunate accident" about it (the cow treading on the caterpillar or the bird flying into the television mast), and for this reason we call such "things" "stochastic malentities" to distinguish them from another class of malentities that we call "aggressive malentities". Whereas stochastic malentities may be living or nonliving, all aggressive malentities are living organisms that, directly by their overt aggression toward the primary animal, kill it or prevent it from living as well as it might live in their absence. Aggressive malentities are unlike predators in that their abundance does not increase as the primary animal becomes more numerous: (c) above. This is usually because the aggressive malentity does not use the primary animal as food, except in a trivial way. It profits from its aggression by displacing animals that are obstacles to its survival and reproduction, but it would have been better off had the victims of its aggression not been there in the first place, because it must use energy to dispose of them. By contrast, the predator gains energy from its activities against its prey. Both kinds of malentities fulfill unequivocally the conditions of figures 1.02B and 1.03B or C.

## 6.1 Stochastic Malentities

A sheep carrying a number of ticks might use a fence post, a tree trunk, or an upstanding rock as a rubbing post against which to squash or rub off ticks. The presence of such rubbing posts reduces the chance that the tick will survive and reproduce. A bullock walking on wet, heavy clay soil may leave deep hoofprints which may persist long after the bullock has gone. In wet areas the hoofprints may fill with water. Springtails (Collembola) often become trapped in enormous numbers in the surface film of such little pools. The hoofprints are malentities in the environment of the springtails.

Along much of the rocky shoreline of the west coast of the United States there is a distinct horizontal band of mussels, *Mytilus californianus*. When fully established this bed presents a continuous strip which is virtually a monoculture of mussels and their epifauna. Despite the apparent stability of this dense population, the mussels are in constant flux, with areas becoming cleared and recolonized, sometimes by other species, but eventually by these mussels again.

Harger and Landenberger (1971) showed that waves during storms shear off parts of the mussel beds, leaving patches that might be either very small or very extensive. They found that the loss of mussels through storms was related to the size of the beds: large beds lost relatively more mussels than small beds. In a heavy storm there is more chance that large sections of a large mussel bed will be ripped off or rolled up like a carpet. The risk is less for small beds. But the severity and frequency of storms is independent of the density of the population. So storms fit into the bottom right cell of table 1.01; they are malentities in the environment of a mussel. That a storm of the same severity is likely to kill a larger proportion of a dense population is not relevant to the diagnosis, but it is an interesting density relation to find in a malentity.

Another malentity in the environment of mussels is drift logs. Dayton (1971) has shown that many areas along the shores of the San Juan Islands and the outer coasts of Washington and Vancouver islands have large accumulations of drift logs. Many tidal organisms are killed by the battering of these logs. This process is important in providing space for settling of juvenile intertidal organisms. In areas that Dayton studied there was a 5–30% probability that any given spot would be struck by a log within three years. This estimate was made by measuring the survival rates of a cohort of nails embedded in the rocky substratum. In some places denudation of the substratum by drift logs was so great that sessile organisms were found only in crevices. Beds of *Mytilus californianus* were especially vulnerable to pounding by logs. Once the logs had displaced the beds, waves carried on with the work of clearing them away. One malentity led to another. Storms also kill more mussels if the beds have already been attacked by starfish (Harger and Landenberger 1971). In this case a predator's activity increases the activity of a malentity (water in the form of waves modified by a storm).

Of course, unfavorable weather is the most familiar and most important cause of stochastic malentities. Apart from the excessive accumulation of energy in water or air that accompanies storms and tidal waves, nearly all the malentities associated with weather are constituted by excessive accumulations of heat or water or by extreme deficiency of heat. Heat is rarely measured, but the flow of heat to or from the animal's body is inferred from measurement of temperature, which is more convenient. The responses of animals to extremes of weather were discussed in Andrewartha and Birch (1954, chap. 6); there seems no merit in repeating that discussion here. Examples are given in the case histories discussed in chapter 2.

Aggressive malentities have not hitherto been discussed, so we give more space to them.

## 6.2   Aggressive Malentities

Stochastic malentities might be called nonreactive because they conform to figure 1.03B. Aggressive malentities might be called reactive when they conform to figure 1.03C. In the language of table 1.01, a stochastic malentity is classified minus, zero and an aggressive malentity is classified minus, minus, or minus, zero. Both classifications end up in cell 4; so both are malentities. For the general purpose of chapter 1 it was convenient to confound them; for the present chapter it is convenient to distinguish them so that we can talk about malentities more particularly. The first question is, What must an aggressor do to be counted as a malentity?

Connell (1961) described aggression between two species of barnacles that seems to make one a directly acting component in the environment of the other. In the intertidal zone of a rocky shore in Scotland adults of the two species occupy two separate horizontal zones with only a small area of overlap. On the other hand, the juvenile members of the species from the upper zone are found in much of the lower zone. The species from the upper zone, *Chthamalus stellatus*, settled but did not survive in the lower zone. Connell (1961) investigated the hypothesis that *Chthamalus* was prevented by *Balanus balanoides* from growing up in the lower zone.

Connell (1961) transplanted nine pieces of rock bearing a known number of at least three year-classes of *Chthamalus stellatus* from the upper zone, where they normally live, to the lower zone, which is normally occupied exclusively by adult *Balanus balanoides*. After settlement of *Balanus* had ceased early in June and *Balanus* had reached densities of 49 per $cm^2$ a census of surviving *Chthamalus* on the transplanted pieces of rock was made by removing the rocks at low tide and making a map of all the barnacles on each of them. A line was drawn on each map dividing the area of each rock into two portions, each with the same number of *Chthamalus* on it. From one portion all *Balanus* that were touching or immediately surrounding each *Chthamalus* were removed with a needle. The other portion was left untouched. Censuses of *Chthamalus* were made every four to six weeks for a year. At each census *Balanus* were removed as before from the same portion of each rock.

In addition to the experiment with transplanted rocks, Connell found some areas where *Chthamalus* had settled naturally below midtide level, where *Chthamalus* usually do not occur. These areas were mapped and *Balanus* were removed as they settled, just as was done with the transplanted rocks. There were eight such areas in all.

In the absence of *Balanus* the survival of *Chthamalus* was high in all experiments. In the presence of *Balanus* there was progressive mortality of *Chthamalus* in six of the eight undisturbed sites and on all of the transplanted rocks. Overall, at the end of a year, about 80% of the original *Chthamalus* were dead. Figure 6.01 shows an example of the decrease in a number of *Chthamalus* in the presence of *Balanus* and the survival of *Chthamalus* when *Balanus* were removed.

Mortality of *Chthamalus* in the lower intertidal zone was due to interference by *Balanus*. Some grew over them, others undercut *Chthamalus* and lifted them off the

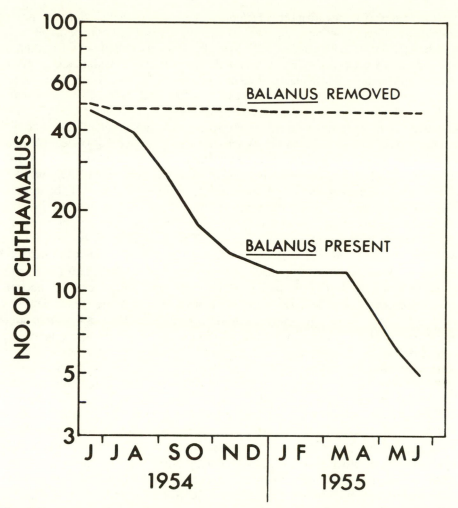

**Fig. 6.01** The effect of removing the barnacle *Balanus balanoides* on the survival of the barnacle *Chthamalus stellatus*. (After Connell 1961)

rock surface, others surrounded *Chthamalus* like a tall fence, preventing them from growing and eventually crushing them. These observations were made at each census. For example, in the case illustrated in figure 6.01 there were 47 *Chthamalus* at the start. Of these 42 died the following year: 24 were "smothered" by *Balanus*, 2 were undercut, and 1 was crowded out; 15 died from unknown causes. Many *Balanus* were needed to exclude *Chthamalus* completely from an area. The number of *Balanus* that settled varied, but *Balanus* grew more rapidly than *Chthamalus*, and a *Chthamalus* from last year's settling was still vulnerable to newly settled *Balanus*. It was unlikely that there would be a light settling of *Balanus* for two successive years. So the upshot was that *Balanus* destroyed virtually all the *Chthamalus* that settled and preempted much of the lower zone, thus reducing the chance that a *Chthamalus* might find a place. In effect, only those *Chthamalus* that reached the upper zone had a chance to survive.

When aggression toward a settled *Chthamalus* is directly lethal, *Balanus* must be counted as a directly acting component in the environment of *Chthamalus*. Because it is directly acting and because it fits into the bottom right cell of table 1.02, it must be counted as a malentity. We can be sure that smothering, crushing, and crowding out are direct causes of the death of *Chthamalus*. On the other hand, undercutting is an indirect cause of death. It exposes the naked base of the live *Chthamalus*, which probably in consequence dies of desiccation or is washed away by a wave and dies from some other cause. In its role as an "undercutter", *Balanus* is a modifier in the web of *Chthamalus*. When desiccation is the cause of death *Balanus* modifies the activity of moisture in the environment of *Chthamalus*. In other words, the influence of lack of moisture is aggravated by the activities of *Balanus*.

Does *Balanus* as a malentity get into the right column of table 1.01 by virtue of a zero or a negative reaction to the density of the victims of the population? As with all aggressive malentites there must be a chance that so much energy is used in aggression as to disadvantage the aggressor. This is unlikely to be the case with *Balanus*. However, insofar as *Balanus* settles on the top of mature *Chthamalus*, it makes an uncertain home for itself. After it has killed *Chthamalus* it remains attached to the shell of *Chthamalus*, which in its turn, sooner or later, becomes detached from the rock because of the death of its inhabitant. When the density of mature *Chthamalus* is high, more *Balanus* will settle on the tops of *Chthamalus*, and more will have shorter lives. The reaction of the malentity to the increase in abundance of the primary animal is negative. This does not happen with newly settled *Chthamalus*, whose surfaces are too smooth for *Balanus* to attach to. Connell (pers. comm.) knows of no case in which the local population of *Chthamalus* was so dense that it precluded *Balanus* from settling.

To destroy those *Chthamalus* that have settled is not the only harm *Balanus* does to *Chthamalus*. Many *Chthamalus* may be unable to settle because suitable places have been preempted by *Balanus*. Those that do not reach the upper zone may die from some cause other than violent aggression, perhaps from starvation. Even those that reach the upper zone may find life harder than it would have been had they been allowed to settle on the lower zone. These outcomes of aggression by *Balanus* are different from the violent death that was the fate of the settlers. In this context *Balanus* must be counted as a modifier in the web of the environment of *Chthamalus*. It is commonplace to find that the same thing represents more than one component in an environment (sec. 2.32). Connell's observations on *Balanus balanoides* and *Chthamalus stellatus* in Scotland were confirmed by Dayton (1971) and Connell (1972) for different species of the same genera on the Pacific coast of Washington, United States.

According to Brian (1956), in the west of Scotland two species of ants, *Myrmica rubra* and *M. scabrinodis*, sometimes occur together and sometimes occur separately. In woodland only *M. rubra* is found; in heath, heather, and felled woodland both are found, but in these places *M. scabrinodis* lives in short turf and *M. rubra* lives in longer grass-herb vegetation. This segregation within the heather or felled woodland is evidently not brought about by the ants' selecting of different places; it is caused, at least in part, by an active interference with one species by the other (Brian 1965, 91–92). Colonies are founded in summer. *M. rubra* are the first ants to arrive; their foundress queens seek warm, dry places; they form colonies in the warmest places, which may be wet but must not be waterlogged. *M. scabrinodis*, which arrives a little later, prefers the same sorts of warm, well-drained sites for its colonies. Some, or perhaps many, of the suitable sites have already been colonized by *M. rubra*. The second arrivals, *M. scabrinodis*, lay siege to the nests of *M. rubra*, remove the appendages of the workers, and devour their males. This, especially when associated with dry weather, is enough to cause *M. rubra* to evacuate its nests, and *M. scabrinodis* takes over the drier parts of areas previously occupied by *M. rubra*. What prevents *M. scabrinodis* from replacing *M. rubra* entirely in this region? The answer appears to lie in a difference of adaptations that gives *M. rubra* an advantage over *M. scabrinodis* in cooler, moister sites with longer vegetation.

While the fight is on some *M. rubra* die from the wounds inflicted by *M. scabrinodis*; by virtue of this violent aggression, *M. scabrinodis* becomes an aggressive malentity in the environment of *M. rubra*. But the enduring and most important outcome of the aggression is that *M. rubra* must live in a place that it does not prefer; presumably the place offers an inferior environment. The literature is not explicit on this point; perhaps good food is harder to come by or the weather is less favorable. Whichever component is important, the change in its quality has to be attributed to aggression by *M. scabrinodis*. In this context *M. scabrinodis* becomes a modifier in the web of the environment of *M. rubra*.

Whatever may be the disadvantage of the moist, shady places, it is felt more strongly by *M. scabrinodis*, because *M. rubra* usually wins the struggle for existence in this sort of place. As with the barnacles, each species comes to dominate the particular microhabitat that suits it best. The two cannot share the same microhabitat because of differences in their ecologies (sec. 10.2).

Aggression rather like what has been described for these two ants occurs fairly commonly in other ants and in bees. When the victim cannot find a microhabitat in which it can excel, the consequences for it may be more severe than for *M. rubra* (Wilson 1971, 447–52). The Argentine ant *Iridomyrmex humilis* has been successful in a number of places in exterminating native species of ants, as for example in the Mobile area of Alabama (Wilson 1951) and in Madeira (Brian 1956). Similarly, the introduced fire ant *Solenopsis saevissima richteri* is antagonistic to native species of ants in the areas that it has invaded in Alabama, Florida, and Mississippi. Since its spread, the native ants *Solenopsis xyloni* and *Solenopsis geminata* and the Florida harvester ant *Pogonomyrmex badius* have become scarce in the invaded areas (Wilson 1951). The ant *Oecophylla longinoda*, native to East Africa, has been destroyed in some coconut plantations by an introduced ant from India, *Anoplolepis longipes*, which replaces it in such plantations (Way 1953).

The native honeybee of Japan *Apis cerana* is contracting its distribution and is being replaced by the introduced European *Apis mellifera* both in apiaries and in mountainous areas where the bees live wild. *A. mellifera* monopolizes syrup dishes which both it and *A. cerana* have been trained to visit. *A. mellifera* is also said to be a more effective fighter and is able to prevent *A. cerana* from stinging it. *A. mellifera* attacks en masse colonies of *A. cerana*, which respond by evacuation. *A. cerana* destroys only weak colonies of *A. mellifera* which do not evacuate their nests when attacked. The spread of *A. mellifera* is aided by man's preference for this bee, which he protects in preference to the native one (Brian 1965, 91–92).

With all these species of ants and bees the aggressor seems to be both malentity and modifier. On the other hand, there are plenty of examples from throughout the animal kingdom of aggressors that are merely straightforward modifiers of one sort or another.

The limpet *Lottia gigantea* of the Californian and north Mexican coast, which lives on rock surfaces in the intertidal region, has been studied by Stimson (1970). Each individual limpet restricts its movements and grazing to a home range of about 900 cm². The home range is covered with a green algal film that develops only in the presence of the limpet and disappears if the limpet is removed. Evidently the limpet is necessary for the growth of the alga that serves as its food. The space occupied by the limpet and its food is unavailable to other grazers. If another grazer intrudes onto the grazing area of a limpet, the occupying limpet shoves off the intruder, including other members of its own species. For example, when a limpet, *Lottia*, found itself touching an *Acmaea*, which is also a grazing limpet, it stopped feeding, retracted its shell and lowered the forward edge, then suddenly thrust its shell and moved its foot forward 2 or 3 cm, striking and pushing *Acmaea*. Such shoves usually displaced *Acmaea*, which was then washed away by the next wave. If the first shove failed, the limpet would try again, and sometimes a limpet was seen to pursue the intruding *Acmaea* for up to half a meter within its home range before dislodging it. Not only are mobile intruders dislodged, but so are any sessile animals such as barnacles, mussels, and anemones that happen to settle within the home range of the limpet. The limpet scrapes them off.

Stimson (1970) did a number of experiments demonstrating that the presence of the limpet *Lottia* reduced the abundance of all other grazing animals on the rocky surfaces that it inhabited. He made counts of the numbers of the grazing limpet *Acmaea* inside the home range and outside the home range of *Lottia*. A typical result is shown in table 6.01.

**Table 6.01**  Number of Grazing *Acmaea* in 930 Square Centimeter Quadrats Inside and Outside the Home Range of the Limpet *Lottia*

| Location | Number of Quadrats | Mean Number of *Acmaea* |
|---|---|---|
| Inside home range | 6 | 4.2 |
| Outside home range | 12 | 21.0 |

*Source:* After Stimson (1970).

The populations of *Acmaea* were five times as dense outside the home ranges of *Lottia* as within them ($p = .01$). When *Lottia* were removed from areas of rocky shore the number of *Acmaea* increased compared with counts made in home ranges in which *Lottia* were present. Alternatively, when *Lottia* were transplanted and established home ranges in a new site, the number of *Acmaea* on the site decreased.

It is clear that by its aggression *Lottia* reduced the chance for *Acmaea* to survive and reproduce. There is no doubt that *Lottia* made direct physical contact with the bodies of *Acmaea* and pushed them. To an anthropocentrically oriented ecologist such behavior might seem to be direct action, but there is no evidence that the pushing or any other part of the aggression in itself harmed the *Acmaea*. We are not told, but it seems likely that the harm was due to starvation rather than wounding. If so, *Lottia* must be allotted a place in the environment of *Acmaea* as a modifier of food—not as a malentity.

In section 5.3 we mention that the mussel *Perna canaliculus* cannot settle on a rock face where the alga *Durvillea antarctica* is already established. In section 2.222 we mention similar examples among marine organisms, where the first arrival preempts a settling area for its own use. Latecomers seem, as a rule, not to be subjected to any overt violence; they simply cannot find a place to settle. They may find a place elsewhere, but they have at least lost the chance to exploit the resources that were to be found in the first place. Food is the most obvious reward for a successful settling. So the firstcomer is not a directly acting component in the environment of the latecomer; that is, it cannot be a malentity. It is in fact a modifier of all those things that happen as a result of settling.

Johnson and Hubbell (1974, 1975) found that many species of stingless bees in Costa Rica exhibit both intraspecific and interspecific aggressive behavior when they meet on flowers or at artificial baits. The loser in an encounter is forced to spend less time at the food source and gathers less nectar or pollen. At artificial baits the intensity of aggression rose sharply with increased sugar concentrations. Sometimes the aggression was so intense that the loser was killed. In another study Johnson and Hubbell (1974, 1975) watched the behavior of two species of stingless bees, *Trigona fuscipennis* and *T. fulviventris* which forage for pollen on *Cassia biflora*; *T. fuscipennis* forages in groups and tends to restrict its visits to large clumps of *Cassia*; *T. fulviventris* forages individually and visits more widely spaced or isolated plants of *Cassia*. If *T. fulviventris* happens to visit a clump of *Cassia*, it is displaced by the aggressive behavior of *T. fuscipennis*. This was observed frequently. Johnson and Hubbell (1975) had evidence that the one species had a preference for clumps of plants and the other for isolated plants but that plenty of mistakes were made, resulting in aggression. On this evidence it seems that *T. fuscipennis* is not a malentity but merely a modifier in the environment of *T. fulviventris*, like the limpet *Lottia gigantea* and like most species that practice interspecific aggression no matter what division of the animal kingdom they come from.

Interspecific aggression associated with defense of a home range has been reported in fish—for example, Gerking (1959), Ishigaki (1966), Low (1971), Thresher (1976), and Ebersole (1977). Low (1971) found that the pomacentrid reef fish *Pomacentrus flavicauda* defends its home range as a "territory"; it responded agonistically to thirty-eight species from twelve families but not to sixteen other species from six families. When it was removed from its home range there was an increase in the number of other species of algal-feeding fish in its former territories. Sale and Dybdahl (1975) studied eight species of pomacentrid fish that held home ranges throughout their juvenile and adult life and defended them from other species. Their borders were often contiguous and rarely overlapped. Sale (1978) showed that many of these species occupied similar spaces. When he experimentally removed individuals of the large species *Eupomacentrus apicalis* from their home ranges, this space was occupied by *Pomacentrus wardi*, a small species, and was held for at least a year against intruding *E. apicalis*. The presence of one species thus excludes another for considerable periods. There was much evidence of aggression but no suggestion of a malentity in these observations.

Similarly with birds. Despite the extent to which the distribution of birds is interpreted in terms of "competitive exclusion" (e.g., Cody 1974; Diamond 1979), there is little critical evidence of how the "exclusion" comes about. Most of these studies have been planned within the constraints of competition theory (sec. 10.21).

Birch (1979) found plenty of evidence of interspecific aggression in the defense of a home range by birds but virtually no suggestion of a malentity. For example, Pitelka (1951) observed that the male of the hummingbird *Calypte anna* was usually successful in preventing the hummingbird *Selasphorus sasin* from sharing its home range in Woolsey Canyon in California. From this and other observations Pitelka concluded that the presence of *Calypte* males in the canyon resulted in there being fewer *Selasphorus* than if they alone had occupied the canyon. Migrant rufous hummingbirds *Selasphorus rufus* arrive in eastern Arizona in late summer and establish feeding home ranges there on the way to their winter quarters in Mexico. They exclude other hummingbirds from these home ranges. They are intensively aggressive both to other members of their own species and to other species of hummingbirds such as *S. platycercus* (Kodric-Brown and Brown 1978). Further examples of such interspecies aggression in hummingbirds are given by Cody (1968), Stiles and Wolf (1970), Feinsinger (1976), and Lyon (1976).

Orians and Collier (1963) reported observations which showed that redwing blackbirds *Agelaius phoeniceus* resident in a marsh prevented the tricolor blackbird *A. tricolor* from occupying spaces that it would have used had the redwings not been there. Sixteen male redwings completely occupied the marsh by dividing it up into territories defended by each bird. Tricolor blackbirds made intermittent visits to the marsh but did not take up territories until nine of the redwings that were due to breed deserted their territories in the center of the marsh. Tricolors occupied the center of the marsh, leaving 7 territories on the periphery occupied by redwings. The male redwings on the periphery of the marsh harassed the tricolors of both sexes when they returned to the marsh to feed their young. The tricolors retained their own territories but did not succeed in establishing any additional territories later in the season.

Orians and Wilson (1964) and Murray (1971) reported further examples of interspecies aggression in birds, and M.L. Cody (pers. comm.) has said he knows of more than sixty species of birds that exhibit interspecies "territoriality".

So far as we are aware, the only field experiment testing the possible effect of one species of bird on the abundance of another was done by Davis (1973). He trapped juncos and golden-crowned sparrows for four years in a narrow field about 800 m long, bordered by trees and bushes. In this way he established that the juncos were twice as common as sparrows and occurred along the whole length of the field in areas with nearby cover close to water. On the other hand, the golden-crowned sparrows were concentrated in deep thickets at one end of the field. Juncos fed mostly on seed; the sparrows fed on newly sprouted annuals, buds and flowers. At times they flew together in mixed flocks.

In the fourth year, during two months' trapping over the whole area, Davis caught and removed ninety-seven of the sparrows. The proportion of juncos in the thicket area of the field rose from 60% to 94% over the next two months. Half the captured sparrows were later returned to the thicket end of the field, and the other half were released farther away. About half the latter reached the field in a week. Concurrently, the proportion of juncos in the thickets fell from 94% back to about the proportion that they had occupied in thickets before the removal of the sparrows. That the junco population rose when the sparrows were removed and fell when sparrows were returned suggests that sparrows excluded juncos from the thickets. Unfortunately there was no control area, and so other explanations are not excluded. Again, the aggressor is not a malentity but is in the web, on the pathway through resources.

Rather more experiments have been done with mammals, especially small rodents, but for the most part they have shown that one species reduced the distribution or abundance of another without throwing any critical light on how it was done. So there is not much in the literature that might help with this chapter. An incomplete and unreplicated experiment with *Microtus californicus* and *Mus musculus* suggested that *Microtus* was acting as an aggressive malentity in the environment of *Mus*. *Microtus* were observed to enter the burrows of *Mus* and disturb them, and it was thought they might have killed the nestlings (Delong 1966). On the other hand, the aggression of the chipmunk *Eutamias dorsalis* against *E. umbrinus* places it in the web of *E. umbrinus* because its aggression prevents *E. umbrinus* from getting food as readily as it would in the absence of *E. dorsalis*. There is no evidence that the aggressor directly harms its victim (Brown 1971). The same applies to four species of chipmunk that were studied by Chappell (1978). He found that aggressive encounters between contiguous species on a mountain slope contributed to the sharp zonation of the four species from the base of the mountain to its summit. In these encounters the chipmunks did not directly hurt one another, but they denied each other access to food.

All that can be said with any certainty at present is that in the mammals, as in the birds, interspecific aggression occurs quite often and that aggressors are often important in the environment of a victim as a modifier that reduces the activity of food. But there is little if any evidence of aggressive malentities. Perhaps they have not been looked for; the inspiration to look for them may not come from conventional competition theory. But perhaps they do not exist; perhaps most mammals, like most birds, learn to avoid an aggressor that is too strong for them.

## 6.3   Conclusion

Stochastic malentities occur more frequently and are more important than aggressive malentities. Most stochastic malentities are associated with some aspect of the weather; excessive heat or cold and excessive wetness are commonplace, but tempests and other meterological phenomena may contribute. In general, accidents of all sorts, including large ones such as volcanic eruptions and tidal waves, come into this class of malentity.

Living organisms, as when a sheep tramples a snail, or the artifacts left by living organisms, as when a bullock leaves deep footprints in waterlogged soil, may also contribute to stochastic malentities. But such things as the silken snare of a spider or the pit of an antlion are not malentities, because such artifacts are better judged as extensions of the animal which fit it for the life of a predator, like the claws and canine teeth of a cat or the talons of an eagle.

Aggressive malentities are less important; they are always living organisms. An organism that is an aggressive malentity is often a modifier as well. The aggression of a malentity is in itself sufficient to explain the concomitant response of its victim (the primary animal). But of course the aggression of a modifier is felt through the activity of an intermediate component, often food, that the modifier modifies. For this reason care and critical knowledge are needed to diagnose an aggressive malentity, to make sure an aggressor that is merely a modifier is not mistaken for one that is a malentity as well. The great majority of aggressors are merely modifiers, especially, it seems, among those groups of animals in which learned behavior is relatively important.

The behavior of an aggressor, including an aggressive malentity, is likely to have evolved in relation to a particular victim, because an aggressor that reduces the supply or the quality of a victim's food (or other resource) is likely to benefit to a corresponding degree from an enriched supply for itself. In this respect an aggressive malentity is like a predator.

But the benefit that may be gained by aggression along the same trophic level of the ecosystem is different from the benefit that is gained by predation across adjacent levels of the ecosystem. The benefit to the predator increases with the number of the prey up to satiation level; and it is unlikely that an excess of prey would harm the predator. On the other hand, the aggressor achieves a maximum return when it has destroyed or gotten rid of all the victims within its range. It may achieve this reward for the expenditure of less effort when the victims are few. When the victims are many the aggressor may exhaust its capacity for aggression without much result.

# 7

# Sharing Resources

## 7.0  Introduction

Controversy during the early decades of the twentieth century divided ecologists into two schools—one choosing to emphasize the importance of "biotic" components of environment, the other leaning more heavily toward the idea of "physical" components. There is no longer such a division in the ranks, but it seems to us that the continuing popularity of the idea of "competition" in current ecological theory, about both populations and communities, is a legacy from the neobiotic school that was prominent during the middle decades of the century and is still influential.

The idea of competition, as it has been conceived by ecologists, is unrealistic when tested against what really happens in natural populations and communities (secs. 9.4, 10.1). The unreality is most prominent in the area covered by the present chapter—that is, in relation to the interactions that occur between animals, of the same or distinct species, over resources.

The unreality takes two forms:

1. Ecological interactions have been seen where none exists. This error has been spurred by the tendency to overlook the fact that distinct species might live geographically close, perhaps displaying the same or similar niches, without being ecologically influential the one on the other. According to the theory of environment, such species need not be counted as part of the environment of the primary animal. Section 7.1 discusses the circumstances that might give rise to this condition.

2. There is a tendency to oversimplify. If a theory is beautiful for its logic or its intuition, it may come to be held so securely that it becomes invulnerable to challenge by "ugly facts"; the facts must be made to fit the theory and in doing so may lose some sharpness (secs. 9.0, 9.02, 9.03, 10.1). In sections 7.2 and 7.3 we catch a glimpse of the diversity of ways in which the activity of food or other resource in the environment of the primary animal may be modified by diverse other animals in the web. Section 7.2 deals with interactions that have traditionally been called intraspecific or interspecific competition. Such interactions are commonplace in nature; they are important in ecology no matter what theoretical background may be chosen for their study. But competition theory raises more problems than it solves (sec. 9.4). The theory of environment is more fruitful. In the envirogram the "other animals" of the same species is a modifier of a resource; the "other animals" of a distinct species is either an aggressive malentity or a modifier

of a resource. Usually it is food that is the proximate cause of the condition of the primary animal. Since aggressive malentities are discussed in section 6.2, section 7.2 is entirely about modifiers that either directly or indirectly reduce the activity of the food or other resource of the primary animal. Section 7.3 is about modifiers that either directly or indirectly increase the activity of food, or other resource, or mates, in the environment of the primary animal. Such interactions appear in the literature under various headings such as "symbiosis", "mutualism", and underpopulation (sec. 2.35).

## 7.1   Other Animals of the Same or Different Species May Share a Resource without Apparently Influencing the Distribution or the Abundance of the Primary Animal

An animal may share a resource with individuals of its own species without having its chance to survive and reproduce influenced by these other individuals. The principle is discussed in section 3.131. Table 3.02 shows, in relation to food, three classes of circumstance in which the chance to survive and reproduce is independent of the numbers of animals sharing the food (cells, 1, 2 and 4). The same principles apply to animals of different species sharing food with the primary animal. In this section we pursue the ways in which an animal may be kept rare in relation to a resource such as food.

### 7.11   Other Animals of the Same or Different Species May Be Kept Rare, Relative to a Common Resource, by a Predator

In the lower intertidal zone of the northwest coast of the United States, the barnacle *Chthamalus dalli* is unable to survive in the presence of the barnacle *Balanus glandula* unless the predatory snail *Thais* is also present. In the presence of the snail *Thais*, which preys on *Balanus*, the barnacle *Chthamalus* is able to grow and survive after settling (Connell 1970, 1972). *Balanus* is kept rare by the predator. By means of mesh cages Connell excluded *Thais* and was able to show that in its absence *Balanus* became sufficiently abundant to exclude *Chthamalus*.

In the absence of the predatory starfish *Pisaster ochraceus* the mussel *Mytilus californianus*, becomes the most abundant species in parts of the Pacific coast of North America. In the presence of *Pisaster* the population of mussels is so reduced that space becomes available for other species to settle. An original eight species increased to fifteen (Paine 1966, 1974). Even when there is no obviously dominant inhabitant such as *Mytilus* in the subtidal zone, the presence of predators may result in an increase in the number of species. For example, Virnstein (1977) studied the numbers of species and the abundance of macroinvertebrates in caged areas on the sandy bottom of Chesapeake Bay, with and without predators. The main predators were crabs and bottom-feeding fish. Without predators he found an average of eight species. With predators the number of species increased to fifteen after two months. In the presence of predators the abundance of all species seems to be below the densities at which one species influenced another's chance to survive and reproduce, and so more species can coexist. Further examples of this sort are given in section 5.3.

The predatory snail *Thais lapillus* is found in the intertidal zone of the New England coast where it feeds almost exclusively on the barnacle *Balanus balanoides* and the mussel *Mytilus edulis*. *Thais* is more common in some areas than in others, and there its prey is less abundant. These places typically are well protected from harsh wave action, and there is a dense algal canopy on the rocks. There *Thais* forages over a wide area of rock surface. By contrast, in areas exposed to strong wave action and without a dense algal canopy *Thais* is confined to crevices deeper than 10 cm. These crevices are also the only places in the exposed areas that are free of *Balanus* and *Mytilus*. Hence it appears that although *Thais* is present even in the most exposed areas they have little influence on the abundance of the prey in these places (Menge 1978a,b). In exposed places the prey are more abundant. This abundance seems to be due to the rarity of *Thais*, because when *Thais* was excluded completely by cages *Mytilus* became so common that it smothered and eliminated *Balanus*. When *Mytilus* and *Thais* were excluded by cages, *Balanus* persisted (Menge 1976). At least in some areas *Mytilus* is kept so rare by *Thais* that *Balanus* is able to survive.

In most examples of successful biological control of insect pests there is little reason to doubt that the predator keeps the prey scarce compared with its numbers in the absence of the predator (sec. 5.3). It is more difficult to assess the influence of predators in noncontrived natural populations of insects. For example, Varley, Gradwell, and Hassell (1973, 49) suggested a similar explanation for certain species of caterpillars that feed on oak leaves in England; they wrote:

There are many species of caterpillars which feed on young oak leaves in Spring. They must compete for food at least in those years when the trees are defoliated. We think that they coexist because each is regulated by specific parasites at such a low density that collectively they seldom run short of food.'

There is more reason to doubt the influence of predators on the abundance of prey when both are vertebrates (secs. 5.226, 5.3).

### 7.12 Other Animals of the Same or Different Species May Be Kept Rare, Relative to a Common Resource, by a Malentity

Four species of caterpillars feed on the leaves of *Pinus sylvestris* in central Europe. Taking all four species together, outbreaks that result in defoliation of pines and shortage of food have occurred fewer than ten times in sixty years. The four species, though closely related taxonomically, live on the same trees, and so far as can be determined they eat the same food, which is hardly ever in short supply. It is likely that weather as a stochastic malentity is one of a number of components of environment that keep the populations of caterpillars below outbreak levels most of the time (Andrewartha and Birch 1954, 449–65). Without long and detailed studies in the field it is difficult to determine to what extent this phenomenon is common among herbivorous animals.

Unfavorable weather is the most important cause of stochastic malentities. For example, it is bad weather in the intertidal zone along the California coast that keeps many animals rare relative to their resources (Connell 1975). More generally, weather is a major cause of rarity relative to resources (sec. 3.131).

### 7.13   Other Animals of Different Species May Share the Same Habitat and Certain Resources, but the Critical Resource Is Not Shared

Broadhead and Wapshere (1966) studied the distribution and abundance of two closely related psocid species, *Mesopsocus immunis* and *M. unipunctatus*. The two species can be found in plantations of larch and in natural mixed woodlands over a wide area of Yorkshire. They can be found living together in about the same numbers on the same tree at the same time. They live side by side in the same microhabitats on the tree. They lay their eggs at the same time, and the eggs hatch at the same time. They are both hosts to the same two predators. They both increase and decrease in number together from year to year. They were never observed to run short of food. The only shortage seemed to be suitable twigs on which to oviposit. Only a small proportion of twigs are suitable for oviposition, the rest being used by other organisms or being disturbed by branches or wind.

Despite the shortage of suitable oviposition sites, one species did not influence the abundance of the other because they preferred to oviposit in different sites. Eggs of the two species were distributed differently along a branch. On thin branches *M. immunis* deposited a greater proportion of its eggs in the axils of the dwarf side-shoots than did *M. unipunctatus*. Broadhead and Wapshere (1966) considered that an extrinsic shortage of available oviposition sites for each species kept the numbers of both species below the point at which the supply of food became short. There was no indication that one species encroached on the oviposition sites of the other. Had this happened, then doubtless one species would have influenced the abundance of the other, depending upon the extent of overlap. Broadhead and Wapshere (1966) obtained data that convinced them that the two species of psocids were not kept rare by their two predators. If this is the case, the two psocids are a nice example of two species' being kept rare by shortage of critical resources that they do not share to any significant extent.

We think of the twig as a resource because it seems that the presence of a special sort of twig is needed to evoke the "syndrome" of egg-laying in the psocid. So we call the twig a token.

### 7.14   Distinct Species Can Share the Same Habitat and Certain Resources Because "Refuges" Spread the Risk

The distributions of two species of blowfly, *Lucilia sericata* and *L. cuprina*, virtually coincide in the cool temperate parts of Australia. For a long time they could not be distinguished taxonomically. Now that they can be distinguished, their ecologies are known to be quite different in one particular respect which has been documented by Waterhouse (1947) and others. Both species will lay eggs on carrion, but only a fresh carcass will attract *L. cuprina*. Both species will also lay eggs on a living sheep, but only the maggots of *L. cuprina* can establish a colony on an uninfested sheep. Artificially, *L. cuprina* can be reared on a carcass if all other species of blowflies are kept out. In nature scarcely any *L. cuprina* survive the larval stage in a carcass because they are usually crowded out by the maggots of *L. sericata* and other species. In the living sheep the advantage is reversed: *L. cuprina* thrives but *L. sericata* does not; Waterhouse found that 90% of sheep that were infested with maggots yielded only *L. cuprina*, and the remaining 10%

yielded mostly *L. cuprina* plus a few *L. sericata* and other secondary species (Andrewartha and Birch 1954, 449–55). No component of environment other than food for larvae presses heavily or differentially on either species. So the two species "coexist", jointly exploiting other resources and each avoiding the same sets of predators and malentities. They are able to coexist because the habitat offers scope for specialization; it is heterogeneous with respect to one critical niche—food for larvae. The two species have specialized in this niche. Because of this difference between the two species it can be safely predicted that only one of them would persist in a habitat that was homogeneous with respect to food for larvae: in the absence of carrion only *L. cuprina* would persist; in the absence of living sheep only *L. sericata* would persist.

Such diversity of habitat (and the corresponding heterogeneity of niches) that allows two species to coexist, perhaps sharing a number of resources, has sometimes been explained by invoking the idea of a "refuge" (sec. 10.2). In this usage carrion is a refuge for *L. sericata* from *L. cuprina*; and living hosts are said to be a refuge for *L. cuprina* from *L. sericata*.

Underwood (1978) studied the herbivorous gastropods *Nerita atramentosa* and *Cellana tramoserica*, which live together on the same rocky surfaces of the intertidal shores of the east coast of Australia. In experimental cages the mortality of *Cellana* was greatly increased with increasing density of *Nerita*. Both feed on microalgae on the surface of the rock. Although they may share some food in common, the different structure of the radula in the two species indicates a difference in the sort of food that they scrape off the rocks. The densities of the populations in the experiments were similar to those that are found naturally in the intertidal zone. When *Nerita* was common *Cellana* was disadvantaged, probably by a shortage of food. Yet both species continued to coexist. Underwood thought that *Cellana* continued to share the same habitats in the intertidal zone with *Nerita* because *Cellana* can breed in the subtidal habitats where *Nerita* does not penetrate; *Cellana* also has a regular spatial dispersion that may have the effect of reducing its impact on its food supply. Also, both species are recruited from planktonic larvae, and this recruitment varies greatly from place to place and time to time; it is thus unlikely that high densities of *Nerita* would occur in all shores in every year. When the population of *Cellana* is reduced on any particular shore the empty spaces can be recolonized from elsewhere (sec. 8.42). So there are two sorts of refuge for *Cellana*. One is subtidal where *Nerita* never occurs, the other is a shore that, by chance, is not heavily populated by *Nerita*. Branch (1976) found that barnacles crowd out limpets on rocky surfaces, yet the two species continue to coexist. The limpets are replenished by migration and by larval recruitment from refuges lower on the shore where there are fewer barnacles.

"Foraging displacement" in birds may be interpreted on similar lines. A bird that is displaced from a source of food that is also available to another species will often find alternative, though less accessible, food. Morse (1967) described the relation between brown-headed nuthatches *Sitta pusilla* and pine warblers *Dendroica pinus* in longleaf pine forests in Louisiana where they are winter residents. They both participate in flocks of mixed species that form around chickadees and tufted titmice. When alone the nuthatches forage more heavily on trunks and proximal parts of limbs than they do when they are with pine warblers. When they are with the warblers they forage more on the distal parts of limbs and on twigs while the warblers forage on the proximal parts. Both species maintain this separation by aggressive displays.

The plain brown woodcreeper *Dendrocincla fulginosa* forages close to the ground when following swarms of army ants on Barro Colorado Island in the Panama Canal region. In the presence of the ocellated ant thrush *Phaenostictus mcleannani* they forage on higher or peripheral and less productive areas. They await opportunities to forage in their favored place if the ant thrush moves away. When the woodcreeper first occupies a ground-foraging site the ant thrush will occasionally dislodge it by snapping at it and fluttering its wings (Willis 1966). It seems reasonable to infer in such cases where one species displaces another in its foraging that the one is having some influence on the other's chance to survive and reproduce, though this might be difficult to measure.

Near the east coast of Australia, over a distance of some thousands of kilometers two species of tephretid fruit fly, *Dacus tryoni* and *D. neohumeralis*, can be found living in the same trees and breeding in the same fruits (Lewontin and Birch 1966; Gibbs 1967). The adults often live in the same tree and lay their eggs into the same fruit, and the two sorts of larvae often grow up together. We do not know whether the presence of one species of maggot in the fruit reduces the amount of food for the other species, but healthy mature maggots of both species often emerge from the same fruit. In short, the two species seem very much alike both taxonomically and ecologically, and they seem to coexist intimately over an extensive area. The relative abundance of these two species may vary greatly from time to time in the same place or from place to place at the same time.

Because most of the host plants produce ripe fruit for only a brief season, the fruit flies must rely on a variety of species to provide food throughout the year. As each locality becomes sterile the fruit fly must rely on its own dispersiveness to find a new and productive place for the next generation. In such circumstances the seasonal succession of fruiting and the geographic dispersion of the host plants (components that are extrinsic to density of the fruit flies' populations) are likely to be more important than intrinsic food shortages that might be caused by the feeding of an excessively dense population. It is known that *D. tryoni* has a strongly dispersive phase shortly after the adult emerges from the puparium. Fletcher (1973) found that 75% of adults that emerged in an orchard left the orchard during the first week of adult life. The dispersiveness of *D. neohumeralis* has not been measured, but it is likely to be about equal to that of *D. tryoni*. In the context of this section a vacant productive locality may be regarded as a refuge for either species—whichever one finds it first. It is a refuge only by virtue of its remoteness from members of the other species. This explanation of coexistence is consistent with den Boer's theory of "spreading the risk" (sec. 9.3).

### 7.15   True Biospecies That Are Taxonomically Distinct but Ecologically Similar May Share a Critical Resource Even Though No Refuge Is Apparent and There Is No Clear Evidence That the Species Are Kept Rare by Predators or Malentities

Starting with the hypothesis that species that are ecologically similar are more likely to be found living together in the same habitat than species that are ecologically dissimilar, den Boer proceeded to establish experimentally, for a fauna of carabid beetles comprising 149 species belonging to 41 genera, that (1) species of the same genus are usually more closely related ecologically than species from different genera; and that (2) relatively more congeneric species than disgeneric species would be found living together in the same habitat. (that is, relative to the total numbers of congeneric and disgeneric species in the region).

To verify the first point den Boer (1980) studied a sample of 16 genera with 3 or more species. All together 116 species were classified according to the following ecological characteristics:

1. Whether reproduction occurred in spring or autumn.
2. Whether the habitat was stable or unstable. "Unstable" was subdivided into "wet sites" and "sites disturbed by man". "Stable" was subdivided into "forests" and "others".
3. Whether the beetles were strongly winged or not; those that were not were mostly apterous or brachypterous.
4. The size of the beetle, which largely determined the kind of prey; four size-classes were recognized.

The results are summarized in table 7.01.

**Table 7.01** Proportion of Congeneric Species of Carabid Beetles That Shared Certain Ecological Characteristics Compared with the Random Expectation Based on the Whole Sample of 116 Species (Ignoring the Generic Classification)

| | Proportion of Species (%) | |
| --- | --- | --- |
| Ecological Characteristic | Congeneric Species | Expectation Ignoring Generic Classification |
| 1. Reproduce during same season | 84*** | 61 |
| 2. Share the same habitat | 53*** | 31 |
| 3. Similar capacity for dispersal | Nonsignificant | 75 |
| 4. Similar size | 77*** | 39 |

*Source:* After den Boer (1980).
*** $p < .001$.

For three of the four comparisons the null hypothesis that congeneric species are not different from the general average was disproved at $p < .001$. So we accept the alternative hypothesis that species within the same genus are usually more closely related ecologically than species from distinct genera. den Boer also found that, in the large taxonomically homogeneous genera *Amara* (21 species) and *Harpalus* (10 species) there was a highly significant clustering of ecological characteristics, but not so in another large genus, *Pterostichus*, in which the species were taxonomically heterogeneous.

The hypothesis that certain well-defined habitats would be favored by certain groups of species of the same genus was supported by the data in table 7.02, which show that 5 species of *Notiophilus* consistently inhabited a dry habitat where birch grew on windblown sand. During the same six years 5 species of *Pterostichus* consistently inhabited a moister habitat that was characterized by a mosaic of heath on better-structured soil. Furthermore, species of *Notiophilus* share resources in

**Table 7.02**   Numbers of Two Groups of Congeneric Species, *Notiophilus* and *Ptero-stichus*, Found During Five Years of Continuous Trapping in Two Distinct Habitats

| Species & Habitat | Number per Year | | | | |
|---|---|---|---|---|---|
| | 1961 | 1962 | 1963 | 1964 | 1965 |
| Small Birch Forest on Blown Sand | | | | | |
| *Notiophilus aquaticus* | 42 | 40 | 16 | 29 | 19 |
| *N. biguttatus* | 43 | 141 | 77 | 89 | 122 |
| *N. germinyi* | 26 | 30 | 16 | 15 | 20 |
| *N. palustris* | 3 | 15 | 9 | 10 | 9 |
| *N. rufipes* | 27 | 30 | 20 | 20 | 17 |
| Mosaic of Heath | | | | | |
| *Pterostichus versicolor* | 209 | 282 | 393 | 265 | 412 |
| *P. diligens* | 83 | 105 | 120 | 69 | 55 |
| *P. lepidus* | 31 | 22 | 102 | 139 | 70 |
| *P. niger* | 3 | 4 | 21 | 5 | 10 |
| *P. nigrita* | 8 | 6 | 7 | 10 | 13 |

*Source:* After den Boer (1980).

common which are quite different from the resources that are shared by species of *Pterostichus*.

The next step was to discover whether congeneric species were more likely to be found living together in the same habitat than species from distinct genera. The answer would be yes, if the following one-tailed null hypothesis were disproved: the number of congeneric species that are found living together in the same habitat is no greater than the number that would be expected if the choice of habitat were independent of the generic classification.

Den Boer had at his disposal 175 "year-samples" of beetles that were caught in pitfall traps at seventy-three sites. There were three traps at each site spaced 10 m apart. Trapping was continuous during periods that varied from one to nine years depending on the site. The grand total from 175 year-samples comprised 149 species divided between 41 genera. They came from diverse habitats that ranged from very wet (pools and banks) to very dry (windblown sand).

On the quite acceptable assumption that the repeated capture, during a full year, of two species in the same set of pitfall traps is sufficient evidence that the two species are living in the same habitat, all the data from the 175 year-samples could legitimately be used to test the null hypothesis.

The number of ways that a pair of objects may be drawn from a sample of N objects, provided there is no restraint on how the individuals may be paired, is N(N-1)/2. In the case of 149 species the number of ways is 11,026. If, however, N is divided into k groups of n; and the pairs may be constituted only by individuals drawn from the same group of n, then the number of ways that a pair may be constituted is given by:

$$\frac{\sum_1^k n_i(n_i - 1)}{2}$$

where $n_i$ is the number of individuals in the ith group of n. In den Boer's grand sample, the sum of 175 year-samples, there were 41 genera sharing 141 species unequally between them. The value:

$$\frac{\Sigma_1^k\, n_i(n_i - 1)}{2}$$

was found to be 582. So the number of different congeneric pairs that can be drawn from the grand total is 582. It follows that the chance that any one pair of species drawn at random will be congeneric is given by the ratio 582/11026 = 0.053. And the reciprocal 11026/582 = 18.9 measures the frequency with which congeneric pairs can be expected when drawing pairs at random from the grand total of species—that is, a congeneric pair is expected to turn up once in every 18.9 draws. den Boer called this statistic G.d. to stand for generic diversity. A smaller value for G.d. would indicate that species from the same genus are likely to cluster in the same habitat more often than would happen on the expectation that the choice of habitat was independent of the generic classification.

Because it was not possible to attribute precise probabilities to the results, and because the G.d. scale is unfamiliar, den Boer transformed the data to frequencies, which might more readily be judged by experience. He pooled the results from all the sites that were used each year, to reduce the comparisons to 10—one for each year and one for the grand total. Table 7.03 shows that the discrepancies between observed and expected frequencies are large and are highly consistent from year to year. There seems little reason to doubt their significance even though we cannot attribute a precise probability to them. We agree with den Boer that the null hypothesis has been confidently disproved. So we conclude that congeneric species are more likely to be found living together in the same habitat than are disgeneric species, and that the clustering of congeneric species in the same habitat will be

**Table 7.03**  Observed Number of Congeneric Pairs of Species of Carabid Beetles in a Number of Habitats around Wijster (Drenthe, The Netherlands) in Nine Successive Years Compared with the Number of Congeneric Pairs Expected If the Choice of Habitat Were Independent of Generic Classification

| Year | Number of Habitats | Number of Species | G.d. (on total) | Congeneric Pairs Expected | Congeneric Pairs Observed |
|---|---|---|---|---|---|
| 1959 | 17 | 83 | 17.4 | 258 | 366 |
| 1960 | 17 | 101 | 19.8 | 280 | 432 |
| 1961 | 21 | 106 | 16.3 | 412 | 560 |
| 1962 | 21 | 98 | 16.4 | 542 | 637 |
| 1963 | 17 | 98 | 14.4 | 553 | 604 |
| 1964 | 17 | 103 | 16.3 | 593 | 696 |
| 1965 | 24 | 103 | 16.9 | 642 | 827 |
| 1966 | 27 | 109 | 16.8 | 995 | 1,153 |
| 1967 | 14 | 105 | 17.6 | 843 | 1,015 |
| Total | 175 | 149 | 18.9 | 4,538 | 6,290 |

*Source:* After den Boer (1980).

*Note:* Results are summed over habitats for each year. G.d. is generic diversity. For explanation see text.

more pronounced than would be expected if the choice of habitat were independent of the generic classification. Because congeneric species tend to resemble each other ecologically (see above) den Boer (1980) concluded from the results of his own and Williams (1947, 1951) work that, "Species that are ecologically closely related will more often than not be found co-existing in the same habitat." We suggest, in section 10.2, that the converse is also true: when two species fail to coexist in the same habitat the explanation will usually depend on some critical difference in their ecologies.

Leaf-rolling hispine beetles of the family Chrysomellidae spend all their lives, except for the pupal stage, on the leaves of their host plants—larvae and adults live in leaves that they roll up and eat. Strong (1981) studied 156 populations from fifty-three localities (each of a few hectares) in Central America and Trinidad. There were twelve species of beetles in all—nine that belonged to the same genus, *Cephaloleia*, and three that each belonged to a separate genus. In no case was the population dense, and though many species might be feeding on the one plant and on single leaves, they ate no more than 0.5% of the rolled leaf that they lived in. Their host plants were all species of *Heliconia*. Each locality tended to have just one species of host plant. Seven localities had just one species of beetle, twelve localities had two species, thirteen localities had three, nineteen localities had four species, and two localities had five species of beetle. So it was usual to find a number of species of beetle feeding on the same plant and even on the same leaf. Far from there being any exclusion of one species by another, dense populations of different species of beetles occurred together more frequently than would be expected from random association. This tendency to aggregate was not the result of heterogeneity among leaves. Different species seem to be attracted to the same leaves. It seems they have very similar responses to stimuli from their environment which bring them together to profess the same niche. They conform well to den Boer's "co-existence principle" that "species that are ecologically closely related will more often than not be found co-existing in the same habitat."

Williams (1947, 1951) compared the faunal lists from a number of habitats with the faunal list of the whole region from which the habitats had been arbitrarily abstracted, on the criterion of homogeneity. The comparison was based on the number of congeneric species in the lists. On the twin assumptions that the countryside at large would offer scope for more niches than the single habitat and that the ecologies of congeneric species would be more alike than those from different genera, Elton (1946) had already suggested, in accordance with the theory of competitive displacement, that there would be fewer congeneric species in the lists for the habitats because competitive displacement would be more pronounced in the homogeneous habitat than in the heterogeneous countryside. Elton's analysis made no allowance for the size of the sample. Williams, adding a third assumption that the "theory of the index of diversity" (Williams 1944) holds for these data, postulated that in samples of the same size there would be no more congeneric species in the lists from the habitats than in those from the countryside at large, because species with similar ecologies would tend to congregate in places that offered scope for their niche. This is the sort of prediction that would emerge from the theory of environment. The data confirmed Williams' hypothesis. The conclusions that have been put forward by den Boer and by Williams have been supported by observations on other kinds of animals, but no other group of animals has been studied as deeply as the carabid beetles that live near Wijster. Case and Siddell (1983) summarize a more recent debate on the same subject, but in a different context.

Fraser (1976a,b) was unable to find any evidence that two species of salamanders, *Plethodon hoffmani* and *P. punctatus* that lived in the same places at the same time in eastern North America, influenced each other in any way. The young have very similar diets. The adults differ in size, and there are some differences in their food. *P. hoffmani* has a contiguous distribution with yet another salamander, *P. cinereus*; again there is no evidence that they ever ran short of their common food, though nesting sites might be limiting (Fraser 1976b). See section 6.2 (toward the end) for examples of different species of fish sharing the same habitat.

Similarly, it is usual to find many species of birds occupying the same habitat (Wiens, 1977). Pulliam and Enders (1971) found three to five species of finches living in abandoned agricultural fields in the southeastern United States in summer. With the autumn three new species arrived. Despite differences in the sizes of their beaks, there was virtually a complete overlap in the size of the seeds eaten by the various finches. Similarly, among desert rodents in the United States it is usual to find a number of species in the same traps in individual trapping stations. Heteromyid rodents eat seeds. Of six species investigated by Rosenzweig and Sterner (1970), there was no clear differentiation between species with respect to the size of seeds that were eaten, nor could the rodents be differentiated in terms of their efficiency in hunting and husking seeds.

The conclusions from the studies we have quoted in this section are diametrically opposed to the conventional wisdom of the "competitive exclusion principle" according to which ecologically similar species (and, it usually necessarily follows, taxonomically closely related species) necessarily exclude one another. It calls in question the often-quoted statement by Darwin (1859, 76): "As species of the same genus have usually, though by no means invariably, some similarity in habits and constitution, and always in structure, the struggle will generally be more severe between species of the same genus, when they come into competition with each other, than between species of distinct genera."

A critical analysis of this proposition is given in section 10.2, where coexistence and exclusion are discussed in more detail.

## 7.2 Other Animals of the Same or Distinct Kinds May Reduce the Distribution or the Abundance of the Primary Animal

It has been widely held that the distribution and abundance of a species depends on the extent of the "competition" that it meets from other animals of its own and other species in the habitat. The discussion in this section and the next (sec. 7.3) suggests that the distribution or the abundance of a species may be either increased or decreased by the activity of other animals of either the same or distinct species that modify the activity of food in the environment of the primary animal. But it would be a mistake to extrapolate from these results without giving due weight to the rest of the environment.

### 7.21   The Primary Animal May Experience an Intrinsic Shortage of Food That Is Caused by Too Many Other Animals of Its Own Kind

This interaction is discussed from a strictly environmental perspective in sections 2.38 and 3.132, and it is illustrated in figure 3.01 (bottom right corner). In this section we take a demographic perspective and consider how the rate of increase of the population is influenced by its density.

There has been no paucity of experiments concerning the influence of density on rate of increase in artificial populations in the laboratory. Much of this work was summarized in Andrewartha and Birch (1954, 351–397). The only examples from natural populations we could then refer to were from work on territorial birds, the vole *Microtus agrestis*, and the snowshoe hare *Lepus americanus*. Even now we have no information for natural populations of insects other than for predators that have been used in biological control (secs. 3.132, 5.3).

However, there have since been some studies on large herbivorous mammals (ungulates) that are relevant to this discussion. When an ungulate colonizes a region where there is abundant food and where other components of environment are favorable, its numbers increase. But the colony cannot increase forever. Any component of environment may become unfavorable to increase but the circumstance of special interest here is intrinsic shortage of food. The question we ask is to what extent the growth of a natural population of ungulates is likely to be inhibited by intrinsic shortage of food.

Information on the growth of populations of ungulates in the wild comes from two places. The reindeer *Rangifer tarandus* was liberated on three islands in the Bering Sea earlier in this century (Caughley 1976). In New Zealand eight cervids and three bovids were liberated between 50 and 120 years ago (sec. 10.32).

After reindeer were liberated on two of the Pribilof Islands in the Bering Sea, numbers initially increased and then declined. The peak was reached eleven years after liberation on one island and twenty-seven years after liberation on the other. According to Scheffer (1951), the decline was associated with depletion of lichen, which is an important source of food during winter.

Twenty-nine reindeer liberated on Saint Matthew Island in 1944 had increased to more than six thousand by 1963. Subsequently a crash reduced the population to forty-two animals in 1966 (fig. 7.01). In 1971 there were only twenty-eight females left. Klein (1968) associated the crash with shortage of food and with bad weather. By the time of the crash, lichens, a main component of winter food, were all but eliminated. The mean body weight of the reindeer fell dramatically as the population built to a peak. At the peak they were 40% lighter than when the population was climbing. The crash was a result of massive mortality and a decline in the number of pregnant and lactating females.

In New Zealand, populations of the newly introduced red deer characteristically erupted to high numbers and then declined abruptly. Caughley (1970) thought the decline was caused by a shortage of food as the deer overgrazed the pastures.

Likewise, the numbers of the Himalayan thar, a goatlike ungulate, quickly erupted at the place where it had recently been released. The eruption was followed by a decline. Before the thar were introduced tussocks of snowgrass formed a canopy, which was almost completely destroyed by grazing. It was replaced by turf-forming grasses that were more resistant to grazing. Caughley (1970) attributed

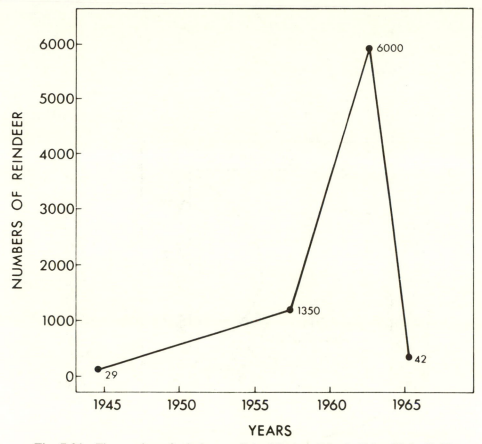

**Fig. 7.01**   The number of reindeer on Saint Matthew Island. (After Klein 1968)

the changes in the numbers of thar to changes in the supply of their food. At first food was abundant; as the population increased the thar first experienced an absolute shortage of food in a number of localities, then this changed to a relative shortage. The shortages were caused by the grazing of the thar—that is, they were intrinsic shortages. These events are discussed more fully in section 10.33.

When numbers are determined by an intrinsic shortage of food, the activity of food at any time is modified by the number of animals that have been eating it at some time in the past. Inversely, the number of animals eating at the present modifies the activity of food for the next and some subsequent generations or cohorts. This is true when food is living plants or living animals, because such food is not inert (sec. 3.01, table 3.02, cell 3). The relation between the amount of grazing (harvesting) and productivity (biomass of vegetation produced per unit area per unit time—e.g., a year) is complex. Figure 7.02A shows, in general terms, the growth in biomass of the standing crop as it matures without any harvesting. According to Caughley (1977) the general relation between biomass when harvested and the productivity that can be sustained (sustained yield) might be represented by a curve like figure 7.02B. According to this model, continued harvesting while the crop is young and immature gives a low sustained yield. The yield

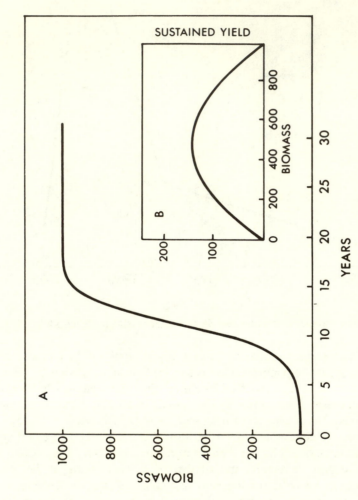

**Fig. 7.02** (A) Hypothetical growth in biomass of vegetation (or animals). (B) Productivity that can be sustained by harvesting at different biomass. (After Caughley 1977, 180)

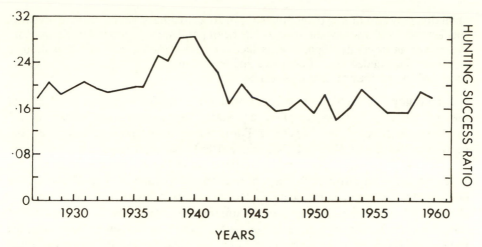

**Fig. 7.03** Relative number of deer in California as measured by "hunting success ratio" calculated from number of deer tags sold and number of deer reported killed. (After Caughley 1977, 19)

increases as cropping is delayed until the crop is about half-matured (inflection of the sigmoid growth curve); at this stage sustained yield is at a maximum. Of course this conclusion belongs to a model; it has been deduced mathematically from certain assumptions about the growth of plants. An empirical conclusion might be different.

The severity of the grazing influences the amount and quality of the food in another way. Most ungulates are selective grazers, so their grazing changes the quality of the vegetation as it is grazed, which was what happened after the outbreak of thar in New Zealand and moose on Isle Royale (sec. 5.1).

Further examples of outbreaks of ungulates followed by declines are provided by the growth of the sheep population in parts of Australia. In South Australia, Tasmania, and western New South Wales an initial eruption followed colonization. This was followed by a decline in numbers, after which the populations fluctuated around this lower level (Caughley 1976, 198–199).

The relative number of deer in California over a thirty-three-year span is shown in figure 7.03. This is one of the best-documented long-term studies of a large ungulate. The "hunting success ratio" gives an index of density in terms of catch per unit effort. This in turn is based on the number of deer tags sold and the number of deer reported killed by hunters. This is not, of course, an example of colonization in which one might look for an initial eruption and decline. However, there was a "surge of density" (Caughley 1970) between 1936 and 1939, presumably associated with a favorable environment for deer. The surge was followed by a decline that might have been due to overgrazing during the "surge", but it is also possible that components of environment other than food became unfavorable or that food became scarce for reasons other than the grazing of deer.

In reviewing the growth of local populations of ungulates following colonization, Caughley (1976, 198) concluded:

The eruption is the typical pattern of ungulate growth. It occurs in the presence of predators in the absence of predators, on continents as on islands, in areas of high plant diversity and of low plant diversity, in areas that have never seen the invading species as in areas where the species has been present throughout the Pleistocene. Whenever an ungulate population is faced with a standing crop of vegetation in excess of that needed for maintenance and replacement of the animals, an eruption and crash is the inevitable consequence.

A corollary of Caughley's conclusion is that it is unusual for ungulates to be kept rare in relation to their food supply by a predator, yet domestic cattle do not "erupt" in habitats where tsetse fly is endemic. Furthermore, the eruption of wildebeest and buffalo in Serengeti National Park since 1960 was associated with the disappearance of the viral disease rinderpest (Sinclair 1979). However, another herbivore may cause an extrinsic shortage of food for the primary animal. There is little doubt that the sheep in many local populations in Australia experienced an extrinsic shortage of food when rabbits were common. The reduction in the number of rabbits by the disease myxomatosis resulted in more food for sheep. Likewise, the removal of sheep from the Kaibab Plateau might have caused the famous eruption of deer there (Caughley 1970).

### 7.22    Other Animals of Different Kinds May Exclude the Primary Animal from the Habitat

The distribution of a species living alone may be broader than its distribution when it shares resources with other species. The principle is illustrated in figure 7.04 (from Andrewartha and Birch 1954, 427). A condition for this hypothetical model is that the opportunity for multiplication continues until space or some resource becomes limiting. It is a condition usually provided in laboratory experiments in which one species is crowded with another and food is replenished at regular intervals. There is no shortage of laboratory experiments with results conforming to this model, nor is there any shortage of attempts to interpret observations in

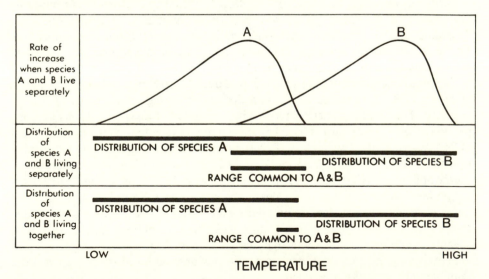

**Fig. 7.04**  Hypothetical diagram illustrating how two species that use the same resources may reduce each other's distribution with respect to some other component of environment—in this case temperature. (After Andrewartha and Birch 1954, 427)

nature in terms of this model. Yet direct evidence to support this as a model of nature is elusive. When we first published the model in 1954 we found our best examples from insects on carrion and invertebrates in sewage beds; we also cited just one example of interspecies territorial aggression between two species of hummingbirds. For the most part we found the evidence for many other supposed examples of the model unconvincing. As a model of nature it has been overworked, but many ecologists have continued to accept it as a general principle of nature. They have done so on the basis of theoretical and laboratory models, on indirect evidence such as "character displacement" or "niche partitioning", on the synchronous disappearance of one species following the appearance of another, or on indirect measurements of "competition coefficients" in species with overlapping distributions. There are numerous pitfalls in inferring from these sorts of indirect evidence that one species is in fact influencing the distribution and abundance of another. Connell (1975, 461) questioned the assumption underlying the widely held view of community ecologists that "competition is the sole or even the principal mechanism determining the area in which one finds a species as opposed to the potential area in which it can live." After reviewing evidence, mainly from plants and invertebrates, he concluded that "many species seldom reach population densities great enough to compete for resources, because either physical extremes or predation eliminates or suppresses them in the early stages." Wiens (1983a,b) comes to a similar conclusion about bird populations (see also sec. 9.4). Horn and May (1977) showed that one method of measuring "limiting similarities" may indeed be measuring something quite unrelated. Schroder and Rosenzweig (1975) calculated significant "competition coefficients" from data that measured habitat overlap of two rodents, only to find that perturbation experiments revealed no such "competition".

That one species of animal is observed to increase its numbers as another declines does not necessarily mean the two phenomena are causally related. The arctic hare *Lepus arcticus* originally occupied alpine, subalpine, and woodland habitats in Newfoundland. After the introduction of the snowshoe hare *Lepus americanus* about one hundred years ago, *L. arcticus* disappeared from the woodland, which was then solely occupied by the snowshoe hare. A popular explanation for the reduced distribution of the arctic hare was that in woodland it was "excluded" by "competition" from the snowshoe hare. The real story, said Grant (1972), is probably quite different. The arctic hare is now restricted to places that have ample cover for escape from predators. Furthermore, the principal predator, the lynx, is known to have increased in numbers after the introduction of the snowshoe hare. The disappearance of the arctic hare from the woodlands could be due to increased predation there by lynx whose numbers increased as a direct result of the availability of the snowshoe hare as prey. The arctic hare no longer occupies sections of subalpine and alpine habitats that have poor cover for escape from predators. Nor do snowshoe hares occupy these areas. A similar example from the insects is discussed in section 10.22.

According to Hassell (1976, 22), "Probably the closest we can get to observing Competitive Exclusion in operation comes from some examples of species replacement after very similar species have invaded the same area." He cites as an example what happened to three species of hymenopterous parasites of the oriental fruit fly *Dacus dorsalis* that were introduced into the Hawaiian islands (Bess, Van den

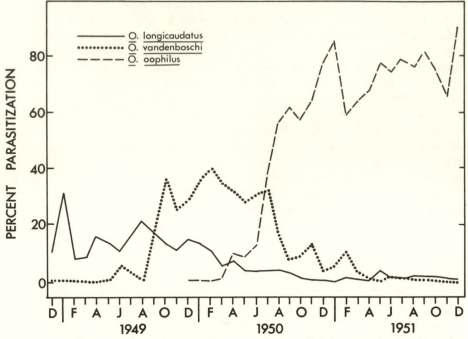

**Fig. 7.05**   Changes in the relative abundance of three parasites of the oriental fruit fly *Dacus dorsalis* on the Hawaiian island of Oahu. (After Bess, Van den Bosch, and Haramoto 1961)

Bosch, and Haramoto, 1961). However, the evidence in this case points more strongly to other interpretations. Figure 7.05 shows how *Opius longicaudatus* increased rapidly after its release on the island of Oahu in 1948. Suddenly it declined in numbers in the latter part of 1949, while *Opius vandenboschi*, which had been released about the same time, rose to a peak. However, by 1950 it was on the decline and a third species, *Opius oophilus*, which had been established in 1949, began a rapid rise in numbers. By late 1951 both *O. longicaudatus* and *O. vandenboschi* had nearly disappeared. According to Huffaker and Messenger (1976, 59), "currently *O. oophilus* is nearly the sole dominant." So the end point of figure 7.05 represents the situation today.

The successful survivor, *O. oophilus*, deposits its eggs in the eggs of *Dacus* in fruit. The other two parasites deposit their eggs in larvae. There is some experimental evidence that the presence of an egg of *O. oophilus* inhibits the development of larvae of the other two species (Clausen, Clancy, and Chock 1965). Apart from this there is no direct evidence that one species inhibits another. It is possible that the near disappearance of the two species of *Opius* is caused by the presence of another species. On the other hand, it is usual in the biological control of insects for a species to multiply for a few generations and then peter out. This is exactly what happened when *Opius concolor* was released in Sicily in an attempt to control *Dacus oleae*. It continued to multiply for several generations while releases were being made. It attained a substantial population, but when releases ceased its numbers fell, and after a couple of years it was rare (Monastero 1967). In the light of this example it is important to note that the decline of both *O. longicaudatus* and

**Table 7.04** Number of *Opius* Released in the Control of *Dacus dorsalis* on the Hawaiian Island of Oahu

| Year | Species | | |
| --- | --- | --- | --- |
| | *O. longicaudatus* | *O. vandenboschi* | *O. oophilus* |
| 1948 | 41,772 | 1,547[a] | |
| 1949 | 78,679 | 16,643[a] | |
| 1950 | 9,985 | 8,024 | 4,257 |
| 1951 | 0 | 200 | 937 |

*Source:* After Bess, Van den Bosch, and Haramoto (1961).
[a] The two species were not distinguishable in these years.

*O. vandenboschi* in Hawaii followed the cessation of releases (table 7.04). On the basis of this information it is reasonable to speculate that *O. longicaudatus* was never going to become established in any case, whether or not other species were introduced. The same might be said of *O. vandenboschi*, but with less conviction. Furthermore, Bess, Van den Bosch, and Haramoto (1961) stated that the three parasites prefer different fruit. Whereas *O. oophilus* was recovered from all fruits, the other two were recovered more frequently from mangoes and false kamani than from guavas, which are by far the commonest. So by their very habits the two unfavored species are attracted to less common fruits, while the favored species is attracted to the commonest. All together, thirty-two species of parasites have been introduced into Hawaii since 1947 for the control of *Dacus*, but only the three species of *Opius* listed above and three other genera have been recovered even in small numbers (Bess, Van den Bosch, and Haramoto 1961).

These species, like so many other introduced species, just never took —a fate that probably had nothing to do with the species that did. Further examples of species replacement that have been erroneously interpreted in terms of competitive exclusion are given in section 10.21. There is only one way to discover the extent to which species that share resources influence each other's distribution and abundance. The distribution and abundance of the species must be studied in the field, ideally before and after perturbation of one or other population. Mathematical models and laboratory models are abstractions from reality. To fail to appreciate that is to commit what Whitehead (1926, 58) called "the fallacy of misplaced concreteness". Models have their place in helping us simplify the complex, but in doing that we should heed Whitehead's further methodological warning to "seek simplicity and distrust it".

In natural populations as distinct from hypothetical ones, exclusion of one species by another is merely the extreme case in a gradient of interactions from cooperation to exclusion. Many species share resources and coexist (secs. 7.1 and 10.2).

In the next three sections we discuss three ways in which one species may exclude another from a habitat.

### 7.221 Modification of the Activity of a Critical Resource Such as Food

The Mediterranean fruit fly *Ceratitis capitata* has been reported as infesting fruit in the Hawaiian islands since early in this century. The Oriental fruit fly *Dacus*

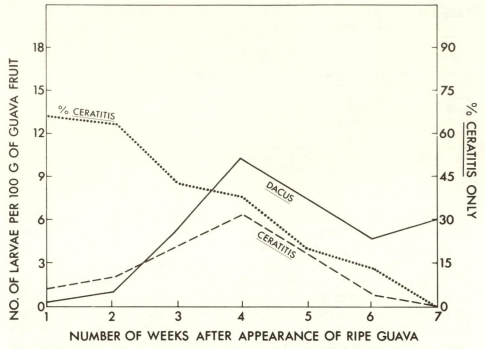

**Fig. 7.06**   Changes in the relative abundance of the Mediterranean fruit fly *Ceratitis capitata* and the oriental fruit fly *Dacus dorsalis* on the Hawaiian island of Lanai in 1950. (After Keiser et al. 1974)

*dorsalis* was introduced in the 1940s and spread rapidly. At the same time the distribution of *Ceratitis* contracted to peaches and some other fruits at high elevations and, at all elevations, to fruits such as coffee that *Dacus* rarely infests. At high elevations *Dacus* is rare or absent.

The infestation of both species in guavas in the island of Lanai in 1950 was studied by Keiser et al. (1974). Figure 7.06 shows the emergence of flies from guava fruits during the course of the six weeks spanning the first crop of guavas. In the first weekly collections *Ceratitis* constituted 66% and *Dacus* 33% of the combined infestation. Thereafter the percentage of *Ceratitis* decreased progressively until at the end of the season fruit contained only *Dacus*. In the first guava crop the following year *Ceratitis* was again in the fruit before *Dacus*, and again by the end of the season none emerged. After some years (which ones is not stated), only *Dacus* was found in guavas.

In section 10.21 we describe laboratory experiments which suggest that each species, as it oviposits, inoculates the fruit with a specific microorganism which the maggots require for food. The inoculum from *Dacus* seems to inhibit the inoculum from *Ceratitis*. In keeping with the theory of environment, we would say that one species modifies the activity of food for the other species.

The history of *Ceratitis capitata* and *Dacus tryoni* in Australia suggests that these two species would interact in the same way. *Ceratitis* was a major pest of fruit in the fruit-growing districts around Sydney from sometime during the 1800s until well into this century. Since about 1945 *Ceratitis* has not been found in Sydney or anywhere on the east coast of Australia. Its place seems to have been taken by *Dacus tryoni*, which is indigenous to tropical Australia farther north (Birch 1961). Because *Ceratitis* is no longer found in eastern Australia, it has not been possible to analyze the influence of one species on the other as was done with *Ceratitis* and *Dacus* in Hawaii.

In the intertidal region of Cape Banks near Sydney, the limpet *Cellana tramoserica* is abundant in a fairly dry part of the midtidal zone. Creese and Underwood (1982) found that *Cellana* excludes two other limpets, *Siphonaria denticulata* and *S. virgulata*, from this area. *Cellana* grazes by vigorously scraping the rock surface with a long, powerful radula. It is able to remove algal spores, diatoms, and very small microalgae from the surface. Because *Cellana* is abundant, it prevents algal spores from growing up to form an algal turf. On the other hand, both species of *Siphonaria* have radulae that are adapted to cutting rather than rasping. They are unable to feed effectively on microalgae. Because *Siphonaria* cannot graze as close to the surface of the rock as *Cellana*, they cannot obtain food from areas where *Cellana* grazes. On occasions when algae do grow to form a turf that *Cellana* cannot graze, *Siphonaria* comes in and grazes it. But as soon as the algae are closely grazed *Cellana* takes over and crops it to the surface rock. Thereafter, as long as *Cellana* is present there is no food for *Siphonaria*, except on the backs of *Cellana*, where *Cellana* cannot get it. The three species settle in this area from planktonic larvae, and all can live there as tiny limpets when their food requirements are very small. But if *Cellana* is present, before *Siphonaria* is more than a few millimeters in diameter it either dies from starvation or moves to another zone where algae are more abundant and where *Cellana* does not live. In terms of the theory of environment, *Cellana* modifies the activity of food in the environments of the two species of *Siphonaria*.

Dayton (1971) found that larvae of the mussel *Mytilus californianus* settle on the tips of the barnacles as well as on rock surfaces that are free from barnacles. When the mussels settle and grow on the tips of the barnacle, they come, after about two years, to cover the surface of the barnacle and thus prevent the barnacle's access to its planktonic food. Although no barnacle died of starvation from this cause during the two years that Dayton watched them, he expected that some would eventually do so.

The tapeworm *Hymenolepis diminuta* and the acanthocephalan *Moniliformis dubius* occur in the intestines of wild rats. Holmes (1961) infested rats that were free from parasites with measured numbers of the parasites, either one species at a time or both together. When both were together *Moniliformis* settled in a specific part of the gut and *Hymenolepis* occupied a section immediately posterior to the other; there was no overlap in the distributions. When either species was alone it occupied virtually the whole of the region of the gut that would be used by the two in combination. In the presence of the acanthocephalan, the tapeworm was shorter and lighter.

In another series of experiments in which the acanthocephalans were newly introduced into rats already infested with tapeworm, the tapeworm became displaced from the anterior region of the intestine. Tapeworms invading established populations of acanthocephalans caused no change in the distribution of the acanthocephalans (Holmes 1962); Holmes suggested that the acanthocephalans may reduce the availability of carbohydrates for the tapeworm.

Lock and Reynoldson (1976) provided evidence that the planarian *Polycelis felina* might inhibit the growth of populations of *Crenobia alpina* in spring-fed streams. Their food requirements are similar. When the food available was increased by artificially removing large numbers of both species, the lyoglycogen food reserves in the animals was increased, suggesting that they had previously been exposed to food shortage. Furthermore, a sudden drop in numbers of *P. felina* was followed by a rapid increase in numbers of *C. alpina*. These data suggest that *P. felina* reduced the chance of *C. alpina* to survive and reproduce by modifying the activity of its food. Further examples from Reynoldson's studies on planarians are given in section 10.24.

Some species of salamanders live close together but in distinctly different habitats. Such is the case with *Plethodon richmondi shenandoah*, which is restricted to areas of talus in Shenandoah National Park, Virginia, and *P. cinereus*, which inhabits soil outside the talus and does not penetrate into areas of talus, whether or not *P. r. shenandoah* is present (Jaeger 1970). From these observations one might infer that the distribution of *P. cinereus* is not influenced by *P. r. shenandoah*. On the other hand, it is possible that the restriction of *P. r. shenandoah* to talus is due to the presence of *P. cinereus* in the areas of soil. This is a hypothesis that could indeed be tested. Jaeger (1971) attempted to do that by enclosing areas on talus and soil. In one lot he placed one species alone, on another the other species alone, and in a third lot both species together. He counted survivors after twelve weeks. The results, although not completely conclusive, suggested that *P. cinereus* inhibited the survival of *P. r. shenandoah* in soil. Further experiments (Jaeger 1971), which were also inconclusive, suggested that *P. cinereus* modified the activity of food of *P. r. shenandoah* in soil, presumably by being more efficient in capturing prey.

In the 1870s and 1880s the blue tit *Parus caeruleus* spread rapidly 2,000 km west from Asia east of the Ural Mountains across Russia to the Baltic Sea. In so doing it invaded the range of the azure tit *Parus cyanus*. By the first decade of the twentieth century the azure tit had retreated a thousand or more kilometers eastward from the Baltic Sea, though it still overlapped the eastern part of the range of the blue tit. In the zone of overlap the two species segregated by habitat, the azure tit living in waterside thickets, the blue tit in upland forest. No one knows quite how one excluded the other, though it has been surmised that the retreat of the azure tit was due to its "poorer adaptation to forest than to waterside thickets" (Diamond 1978, 325). This could be interpreted to mean that the blue tit pre-empts resources in the forest without aggression. However, as Diamond (1978, 1979) pointed out, careful observations are necessary to confirm such a surmise. There are numerous examples of birds, excluding other species from their habitat by aggression (secs. 7.223, 6.2), but the necessary aggression may last only a week or two while territories are being established. If the observer is not there at the critical time it will be missed. Diamond (1978, 328) cited the example of wood thrushes, hermit thrushes, and veeries in Maine, which keep to mutually exclusive territories. They establish territories on returning from their southern wintering grounds. Wood thrushes fight the other two species for the first week after arrival, but not thereafter. The territories, which are marked out within the first week, are subsequently maintained without further fighting.

The fieldmouse *Peromyscus polionotus* and the housemouse *Mus musculus* occur together in old fields in the southeastern United States. Caldwell (1964) found that the numbers of each species fluctuated independently, suggesting that one did not influence the abundance of the other. However, in half-hectare enclosures he obtained different results. He released two pairs of each species in each of two enclosures (at different times). The numbers of both species rose initially, but the population of *Mus* soon fell, and it became extinct in one enclosure after five months (Caldwell 1964) and in the other after twenty months (Caldwell and Gentry 1965). From field observations of behavior he considered that *Mus* was much less adept at finding seeds in the sandy soil than was *Peromyscus*. That they could persist together outside enclosures he attributed to the ability of *Mus* to disperse when food became scarce or when other components of environment became unfavorable. Gentry (1966) counted the numbers of both species following invasion of a 4.5 ha cornfield that had been abandoned one year before. He found an inverse relation between the numbers of the two species, suggesting the possibility that one influenced the abundance of the other in this field, though he could find no direct evidence for this. In his observations the field populations were more like the enclosure populations that were studied by Caldwell.

It is generally agreed that the introduction of sheep, cattle, and rabbits into Australia probably caused the extinction of many native mammals. This was not because these exotic animals had needs that were similar to those of the native mammals. On the contrary, their requirements were in many cases very different. The reasons for the extinctions are not fully known. However, as Frith (1973, 109) remarked "There can be little doubt that alteration of the habitat by stock and the destruction of this cover was the major cause of the decline." In terms of the theory of environment, man, exotic livestock, and the rabbit modified (by reducing or destroying) certain resources or other components of environment that were necessary to the native mammals. Frith (1973, 108) cited some striking figures about the extinction of native mammals in the Riverina district of New South Wales. In the mid-nineteenth century this district had a rich marsupial fauna. One expedition recorded twenty-nine species. Today twenty-one of these species are extinct in the region. Except for two small marsupial mice, those that survived were either arboreal, aquatic, or extremely adaptable, like the echidna and the red kangaroo. Stock and rabbits changed the nature of the countryside. The vegetation changed. With overgrazing many native plants disappeared altogether, leaving denuded plains and sandhills in which the native mammals could neither find food nor make a burrow. Stock make resources unavailable to native mammals by simply destroying the resources. Not only did herbivorous marsupials become extinct (sec. 8.43), but so did carnivorous ones such as the western native cat *Dasyurus geoffroii*.

Not everywhere in Australia did the introduction of stock mean the decline of marsupials. The reverse is the case with the grey kangaroo *Macropus giganteus* and the red kangaroo *M. (Osthranter) rufa*, which are both widely distributed in areas occupied by sheep. Both kangaroos probably became more common after the introduction of stock. The causes are not well understood, but the presence of sheep means more watering places, and sheep probably changed the nature of the vegetation in favor of the two kangaroos (Frith 1973, 110).

### 7.222    Preemption without Aggression

Prior occupancy of a habitat by one species may result in exclusion of another species without defense by the one or aggression by the other.

When the anemone *Anthopleura elegantissima* completely covers a substratum in the intertidal zone, there is no recruitment of barnacles or other rock-clinging species (Dayton 1971). Anemones are not suitable surfaces for the settling of these organisms. When the anemones are numerous, all the available rock surfaces may be completely covered by them. Likewise if barnacles occupy the surface first they may prevent anemones from occupying the surface. Dayton (1971) described surfaces on the shores of San Juan Island on the Washington coast in which aggregations of the anemone *A. elegantissima* were distributed between clumps of the large barnacle *Balanus cariosus*. Dayton removed all barnacles from each of two areas one meter square. In one area barnacles occupied 45% of the surface and in the other 55%. He kept two control areas of 1 m$^2$ each from which *Balanus* were not removed. Figure 7.07 shows that the immediate outcome from removing *Balanus* was an increase in abundance of anemones compared with the control plots. This happened by both immigration and reproduction. During the four years when the numbers of anemones and barnacles were counted, the number of anemones in the control plots remained about the same. The initial increase in anemones in the experimental plots was not maintained. The numbers fell during hot summer months to a lower density than on control plots as a result of desiccation, but they recovered to reach high numbers each year during cool weather. Apparently during the hot summers the driest places are too dry for the anemones. There were more favorably moist places in the control plots because the rock face that was close to clumps of *Balanus* seemed to be moist enough for anemones.

Thus *Balanus* occupies different places in the web of the anemone in winter and in summer. During winter *Balanus* preempts places that might be suitable for the anemone to live; during summer *Balanus* adds water to a place that might otherwise be too dry for the anemone. Modifiers that, like *Balanus* during summer, are beneficial to the primary animal are discussed in section 7.3.

### 7.223    Preemption through Aggression

A species may exclude another from its habitat by overt aggression. A limpet may flip another off the rock surface on which it grazes. A bird may drive another away from its habitat. Some ants lay siege to nests of other species, killing some and driving others away. The animal that is aggressive to another and thereby directly reduces the other's chance to survive and reproduce is, in the theory of environment, an aggressive malentity in the environment of the primary animal. But when aggression is not the proximate cause of death or other disaster for the victim, the aggressor must be classed as a modifier. The distinction was made in chapter 6 and need not be pursued here.

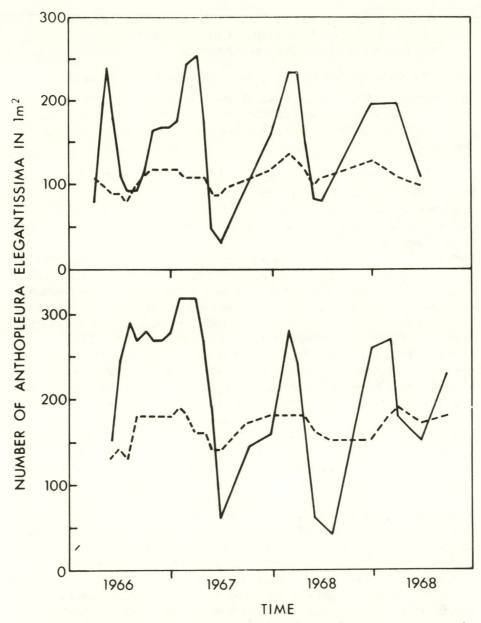

**Fig. 7.07** The density of the anemone *Anthopleura* in two plots one meter square after removal of 55 % (*top*) and 45% (*bottom*) of the cover of *Balanus*. Solid lines show density of *Anthopleura* when *Balanus* was removed. Broken lines show density of *Anthopleura* in the control plots where *Balanus* was not removed. (After Dayton 1971)

## 7.3   Animals of the Same or Distinct Species May Favorably Modify the Activity of Other Components in the Environment of the Primary Animal

### 7.31   The Same Species: Reducing the Risks from Underpopulation

Underpopulation in relation to mates is discussed in chapter 4. Underpopulation in relation to other components of environment was discussed in Andrewartha and Birch (1954) and Andrewartha (1970). We have nothing to add to those discussions.

### 7.32   Distinct Species: Symbiosis, Mutualism, Residues, and Artifacts

In symbiosis and mutualism one species increases the activity of a favorable component of environment (usually a resource and usually food) for the other. So each species has a place in the web of the other, but neither fits into the centrum of the other. This principle is discussed in some detail in section 2.35. We give more illustrations here. The barnacle *Balanus cariosus*, which provides a moist site where the anemone *Anthopleura elegantissima* can survive (sec. 7.222), has a place in the web of *Anthopleura*. The moisture it provides has a place in the centrum of *Anthopleura*. Likewise *Balanus cariosus* enables the snails *Thais emarginata* and *T. canaliculata* to survive on smooth vertical surfaces because the barnacles retain water on a surface much as water is retained in crevices (Connell 1972).

In the Serengeti plains of Tanzania zebras eat from the top of the herb layer, where the plant tissue has the highest cell-wall content. Topi and wildebeests eat the leafy layers below, which are higher in protein. Gazelles select high-protein fruits from the ground. These differences in eating behavior are imposed by differences in size and physiology (Bell 1971). The animals that feed at the lower levels have greater difficulty in finding their food when it is covered by the higher vegetation. In the now-famous migrations of game animals across the Serengeti plains in the wet season, the sequence of movement of animals corresponds to their sequence by the height of vegetation that they graze. First come the zebras, grazing on the tall stems, then the wildebeests and topi, and after them the gazelles. The grazing of zebras makes the preferred food of wildebeests accessible, and the grazing of wildebeests makes the preferred food of gazelles accessible. The same pattern of succession in grazing occurs in local areas where there are mixed herds. Each grazer modifies the food resource in a way that favors its successor.

Investigators before Gwynne and Bell (1968) tended to study the single species and its food alone, without following out the web of connections between its food and the other species of grazers, so they missed the important component of environment in the life of each grazer—other animals that modify its food, making the food more favorable.

PART 2

# General Theory of Population Ecology

# 8

# The Meaning of "Population"

## 8.0  Introduction

In this chapter we look for a realistic concept of "population" on which to build a realistic theory about the distribution and abundance of animals (chap. 9). To search for the best concept is no idle conceit, because the experiments that a scientist may devise and therefore the facts he may discover, as well as the explanations that he offers for them, depend on how he conceives nature. Perhaps the truth of this principle is more easily appreciated in relation to some discipline other than our own. In the time of Pythagoras the concept of number was bounded by the rational numbers; after his time the concept was broadened to include the irrational numbers. The essence of the change was that a symbol might be recognized as a number if it could be multiplied, even though it could not be added. According to Dantzig (1947, 99), the scope and usefulness of mathematics were greatly extended when the concept of "number" was revised in this way. In ecology the broadening of the concept of "population" to include the ideas of heterogeneity and luck seems overdue. Heterogeneity is noteworthy with respect to both the animal and its environment. Luck implies variability of environment that is independent of the density of the population.

In sections 8.1–8.3 we begin the search for a realistic concept of "population" by reminding ourselves of the sorts of facts (and their limitations) that emerge when a scientist attempts to count or estimate the number of animals as they are living naturally.

## 8.1  The Pragmatic Approach to Density

"Distribution" and "abundance" are virtually synonyms for "density". Beyond the limits of the distribution, density remains at zero (Andrewartha and Birch 1954, 5). Within the limits of the distribution pragmatism puts "density" into diverse contexts. A farmer might say that a population of herbivores was dense if there were enough of them to eat a substantial proportion of his crop. A huntsman or a fisherman might say a population of game or fish was dense if a reasonable bag could be filled in short time or with little effort. A conservationist might say a population was dense if it were destroying the habitat for some other species that he wished to conserve in a national park. The same ranger might say the population was sparse if he thought the risks of underpopulation were pressing heavily on it. A student of population ecology might say a population was dense if it were numerous relative to any component of its environment (sec. 9.1). A student of community ecology might say a population was dense if it were numerous relative to other species that are living in the same habitat. In every instance, the first step is to measure the density of the population or populations.

The fundamental unit of measurement for density is the number of organisms per unit area. With this information it is possible to satisfy any of the "interests" specified above. For example, a farmer might be chiefly interested in how much damage a plague of locusts might do to his wheat crop. The amount of damage will depend (other things being equal) on the number of locusts relative to the number of wheat plants. This statistic can be calculated from a knowledge of the number of locusts and the number of wheat plants per unit area. But "area" is variously specified for various ecological techniques.

## 8.2   Units of Measurement

The concept of "area" that enters into the concept of "density" has many faces. For example, in this book densities have been expressed as the number of rabbits seen at night by spotlight per kilometer of transect, the number of limpets per square meter of rock face, the number of scale insects per leaf of an orange tree, the number of aphids that were caught per hour in a suction trap suspended high in the air from a balloon, and so on.

When it is not practicable to see or to count every individual in the study area, numbers may be estimated by taking samples. Some sampling methods give estimates of the total population in the study area. Such estimates, as well as the full counts, are called "absolute" densities. However, it is often sufficient to know the factor by which one density exceeds another without knowing the actual number of animals in the area. Such estimates are called "relative" densities (Andrewartha 1970, 175). Estimation is often simpler for a relative than an absolute density; a relative density can often be estimated when it is impracticable to estimate an absolute density.

Any sampling device (trap, quadrat, etc.) that can be relied upon to discover a constant proportion of the population may be used to estimate a relative density. To convert the relative density to an absolute density it is necessary to know what proportion of the total population is likely to be discovered in the sample. Relative densities are more widely used in ecological studies than absolute densities.

For example, Parer (sec. 12.422), Carrick (sec. 13.1), and Myers and Parker (sec. 12.423) reported absolute densities because Parer and Carrick counted all the animals on their study areas and Myers and Parker estimated the number of burrows per square kilometer (knowing the area of the study area). But Cooke (fig. 12.01) and den Boer (sec. 8.42) reported relative densities, because Cooke estimated the number of rabbits per kilometer of transect and den Boer estimated the number of beetles per trap. Neither Cooke nor den Boer knew precisely what proportion of the population had been seen: Cooke had to assume that the proportion was constant from time to time; den Boer had to assume that the proportion was also constant from place to place and if he had wanted to compare species he would have had to assume that the proportion was constant from species to species as well.

## 8.3   Critical Sampling

There is only one rule about critical sampling. To take a critical sample it is necessary to give every individual in the population an equal chance of being taken in the sample. To achieve this goal it is necessary to eliminate bias from the dispersion of the sample sites (relative to the dispersion of the animals) and from the operation of the sampling devices (relative to different subclasses within the population). That is, all subclasses must be equivalent with respect to sampling (Fisher's "postulate of ignorance"; sec. 9.02).

Bias from the first source may be eliminated by dispersing the sampling devices (traps, quadrats, etc.) randomly over the whole area. This theoretically simple solution often turns out to be difficult to practice. Bias from the second source may be reduced by an improvement in technique or by devising special methods to match the behavior of the animal (Andrewartha 1970, 165–72). If serious bias still remains, the only course of action is to recognize and allow for it when drawing conclusions. One way of allowing for this source of bias is to draw conclusions about only those subclasses (in the population) that are adequately discovered by the techniques that are available.

If the data have been critically ascertained, the mean and the variability of the sample may be used to deduce the probability that the actual density of the population lies within certain limits. For any given level of probability the limits will be narrow (i.e., the estimate will be more reliable) when the variance of the sample is small. For any given size of sample, the variance will be small when the population is dispersed uniformly (conversely, the variance will be large when the population is dispersed patchily). For any level of patchiness the variance of the sample will be small when the sample is large relative to the whole population. In practice, the patchiness is usually an inherent feature of the problem and cannot be changed. Usually the only way to achieve a more reliable estimate is to increase the size of the sample. The degree of reliability that is acceptable depends on the purpose of the investigation.

Measurements of absolute density obtained by counting the whole population are good, but it is doubtful whether data that have been gathered by the method of mark, release, recapture can be regarded as critical (sec. 8.31).

### 8.31   Mark, Release, Recapture

In the method of mark, release, recapture a number of individuals are captured, marked, and released. The absolute density of the population is estimated from the proportions of the marked animals that are recaptured in subsequent samplings. A number of variants of this method have been suggested; they all depend on the assumption that, on the occasion of each sampling, the marked and the unmarked individuals have an equal chance of being taken in the sample.

There are many reasons why this assumption is rarely fulfilled when sampling in nature. We mention only two, which apply fairly generally to most kinds of animals. The latter parts of this chapter (secs. 8.4 et seq.) are also relevant.

1. Whatever the sort of animal or whatever the device used to capture it, some individuals in a population are more likely to be taken than others, not only on the first occasion but also on subsequent occasions. In a population of mice some

may seek out the trap and others may avoid it. In a population of butterflies some may be sluggish and some may be lively. In a population of snails some may be concealed where they are not likely to be seen, some may be fully exposed. And so on.

2. There is no plausible place to release a marked animal except the place where it was taken. If the marked animals must be released where they were taken, there is no plausible policy about the siting of recaptures. To cover the same ground would bias the result in favor of the marked ones unless the mark or the procedure of marking caused the marked ones to become unnaturally dispersive, in which case the bias would favor the unmarked ones. A similar argument holds if different ground is covered, or even if the catching sites are distributed at random.

The bias associated with (1) might be ameliorated by improvements in technique, but nothing much can be done about (2). The bias might disappear if animals "diffused" like the molecules of a gas, but it is well known that they do not. Most species are likely to disperse widely during the dispersal phase, but during that phase of the life cycle that is devoted to eating and growing or breeding they are likely to remain faithful to a "home-range" or, at any rate, to keep narrowly to the place where they happen to be. Neither behavior is likely to lead to the sort of mixing that might validate the estimates that are made by mark, release, recapture.

Practicing ecologists are accustomed to making do with sampling data that are not quite critical, because in ecology critical sampling is often difficult or even impossible (sec. 8.3). But it seems to us that with mark, release, recapture the gap between the critical and the practicable will nearly always be too wide to be acceptable. Students may read about the classical methods of mark, release, recapture in Jackson (1939) or in Dowdeswell, Fisher, and Ford (1940), and there is an elementary summary in Andrewartha (1970, 172, 198). For modern methods see summaries in Southwood (1966, 75–89) and Blower, Cook, and Bishop (1981, 44–73).

The foregoing discussion is intended as an introduction to the theoretical discussion in chapter 9. It is good to have a practical background when looking for plausibility in a theory. For the same reason we pass on now (sec. 8.4) to discuss the inveterate patchiness that seems to characterize the dispersion of animals in nature.

## 8.4   Patchiness

"Patchiness" may take many forms and have many causes. In this chapter we are especially interested in those aspects of patchiness that allow us to recognize "local populations" and the causes of variability between them.

### 8.41   The Concept of "Local Population"

den Boer's (1977, 1979) definition of an "interaction group" was based on the distance that a carabid beetle might travel during the nondispersive stage of its life. The definition implied that any individual in a group had a chance to breed with, or influence the breeding of, any other member of the group. Brown and Ehrlich (1980) and Ehrlich and Murphy (1981) studied local groups of butterflies of the genus *Euphydryas* and called these groups "demographic units". They defined "demographic units" as having "separate dynamic histories"; they distinguished such groups from demes, which are "separate evolutionary units". For example, they believe that the populations of *E. chalcedona* on Jasper Ridge consist of part of a single deme made up of a number of demographic units. The distinction seems to be that there is not enough movement of individuals from one demographic unit to another to significantly influence the numbers in "demographic units" but the movement could be sufficient to influence the genetic composition of demographic units. We use "local" and "natural" population partly to maintain our 1954 usage, but chiefly to emphasize that the "interaction groups" or the "demographic units" are really constituents of the natural population as we define it (secs. 1.6, 1.7). Two ideas are essential to the concept of "local population": (a) that the members of the group have a chance to influence other members of the group as mates and as modifiers of various environmental components, and (b) that a local population is likely to lose emigrants to other localities and gain immigrants from other local populations.

The concept of local populations with the individuals in each one responding to the particular environment of their own locality suggests that variability in local environments might lead to considerable variability in the demographic condition of local populations. There is scope for variability in all demographic statistics— life tables, age structure, birthrate, and deathrate. Given the possibility of a procession through colonization, growth, maturity, decay, and recolonization (fig. 9.03), there is the likelihood that the rate of increase, r, might be positive in some local populations and negative in others at the same time (fig. 9.05); whether one local population is declining or increasing may be largely independent of what the others are doing.

## 8.411  Variability Owing to Separation in Space

It is easy to see how spatial separation of localities may cause variability in local environments. Any component of environment associated with weather, soil, vegetation, other organisms and so forth, might vary, especially when the distribution of localities extends across distinct climatic, edaphic, or geographic regions, as is true, for example, for the rabbit (chap. 12), the grey teal (sec. 13.2), the spruce budworm (sec. 14.1), and the fruit fly (secs. 2.25, 14.3). The distribution of the fruit fly, *Dacus tryoni* extends along the east coast of Australia from northern Queensland to southern Victoria. In Queensland the climate is subtropical, and the habitat includes a wide range of fruits that ripen throughout the year. The fruit fly breeds continuously and usually achieves six generations in a year. It is abundant. In Victoria the climate is cool mediterranean; the habitat contains few fruits, which mostly ripen during summer. The fruit fly breeds only during summer and precariously survives the winter as an adult; it usually achieves one generation in a year, and its population is sparse. Similarly with all the other species mentioned above; the contribution of weather to the local environment might be inconsistent from one part of the distribution to another, not only on the average as predicted by climate, but also on particular occasions when extremes of weather might occur in particular parts of the distribution.

A good example of the disjoint influence of weather operating at two extremes of a large distribution comes from a comparison of the timing of outbreaks of *Thrips imaginis* in eastern and western Australia during the first third of the present century. An outbreak of *T. imaginis* is likely to occur during the spring of a year when there had been an early beginning to the wintry growing season followed by a warm winter (secs. 1.32, 10.12). A severe outbreak is likely to reduce the apple crop by as much as 60–80% of the expected yield (fig. 10.02). Table 8.01 shows that none of the six outbreaks that occurred from 1914 through 1930 coincided in the eastern and western parts of the distribution.

The outbreak that occurred in Western Australia in 1930 also demonstrated how vagaries in the local weather might operate on a much smaller scale. The apple-growing country in eastern Australia is distant from the apple-growing country of Western Australia by about 2,000 km. In Western Australia there are two apple-growing districts (centered on Mount Barker and on Bridgetown; fig. 8.01A), separated by about 150 km. In 1930 there were few thrips in Bridgetown, but a severe outbreak occurred in Mount Barker. The apple harvest in Bridgetown produced 20.9 million liters—about 95% of the predicted yield; at Mount Barker the harvest yielded 2.8 million liters—about 30% of the predicted yield.

For several days around 5 April the weather of Western Australia continued to produce a number of widely scattered thundershowers. Most of the rain fell in the arid interior region, which is usual when the weather is under the influence of a tropical air mass. On this occasion a trough of low pressure pushing down parallel to but inland from the coast probably helped produce one such rainy patch near the southwest corner of the continent. By a long chance the rain happened to fall on Mount Barker and to miss Bridgetown (fig. 8.01). The growing season for the food plants of *T. imaginis* began on 5 April in Mount Barker, but it was delayed for a further ten days in Bridgetown. The weather was still warm at this time of year. When "thermal units" were calculated by multiplying mean daily temperature by the number of days, it was found that at Mount Barker the twenty-six days in April after the "break" of the season contributed 406 day-degrees; but at Bridgetown the sixteen days between 15 April and 30 April contributed only 229 day-degrees. Subsequently there was no important difference in winter or spring weather in the two districts. A solitary, seemingly small accident in the autumn weather made the difference between a severe outbreak and no outbreak at all. See section 10.12 for an explanation of how such an event might influence the rate of increase in the thrips population.

A similar example from the ecology of the grasshopper, *Austroicetes cruciata* in South Australia was described in Andrewartha (1944). The northern and the southern boundaries to the distribution of the grasshopper correspond quite nicely to isohyets corrected for evaporation. But the distribution is not immutable: during a run of unusually wet winters the distribution extended toward the northern desert;

**Table 8.01**   Years in Which Severe Outbreaks of *Thrips imaginis* Occurred in Eastern and Western Australia

| Year | Eastern Australia | Western Australia |
|------|-------------------|-------------------|
| 1914 | + | |
| 1915 | | + |
| 1926 | + | |
| 1928 | | + |
| 1930 | | + |
| 1931 | + | |

*Source:* After Evans (1932).

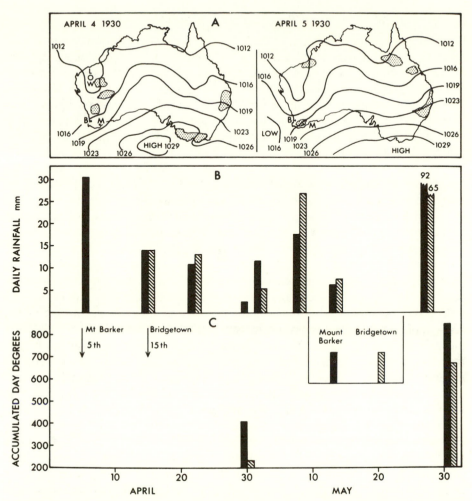

**Fig. 8.01** Vagaries in local weather that influenced an outbreak of *Thrips imaginis* in Mount Barker (M on map) but not in Bridgetown (B on map) in Western Australia in 1930. (A) Isobars indicate the distribution of isobaric pressure in Australia on 4 and 5 April. Shaded areas indicate patches of thundery rain. Patches of thundery rain were predictable from the isobaric patterns, but the odds against a patch that embraced Mount Barker but excluded Bridgetown must have been quite long. (B) The columns indicate the daily rainfall at Mount Barker (*black*) and Bridgetown (*shaded columns*). Once the season had "broken," the "follow-up" rains were adequate in both districts. (C) The end of the summer's drought, that is, the beginning of the winter growing season, is indicated by arrows. The day-degrees that had accumulated from the break of the season through to the end of April and of May are indicated by columns. The importance of this statistic in the ecology of *Thrips imaginis* is discussed in section 10.13 and illustrated in figure 10.01.

during a run of unusually dry winters it extended southward into a zone that is normally too wet for the grasshopper. In this way fluctuations in the numbers (and the concomitant risk of extinction; sec. 9.32) of an insect that is heavily dependent on the weather are strongly buffered against extremes in the weather (Andrewartha and Birch 1954, 8, figs. 1.01, 1.02).

## 8.412   Variability Owing to Topography

In addition to spatial separation of localities and local vagaries of weather, topography is a fruitful source of variability between localities. Topography appears prominently in the envirograms of the rabbit (fig. 2.01) and the spruce budworm (fig. 2.02); it is also important in the ecology of the grey teal (sec. 13.2). Variations in topography often result in the weather, as measured in the microhabitats of the locality, being quite different from that of the surrounding countryside, and rather different from one locality to another (Andrewartha, Davidson, and Swan 1938; Andrewartha 1940).

This point was nicely emphasized by Myers and Parker's (1965) study of the rabbit on an extensive study area in arid rangeland (fig. 12.02; sec. 12.423). The study area covered 670 km$^2$. Myers and Parker recognized five distinct "landforms" that were distinguished, among other qualities, by their characteristic topographies. In the "sand dunes" there was plenty of relief but not much opportunity for runoff because the soil largely absorbed the rain as it fell. But sand gives up a large proportion of its water to the plants before drying out to wilting point: plants growing on sand get more benefit from light falls of rain than plants growing on heavier soils. A consolidated sand dune is also a good place for a rabbit to make a deep burrow. The sand dunes were highly favorable to rabbits as long as the drought did not become too severe. In "stony hills" there was plenty of relief, and much of the rain ran off the hills and spread in shallow drainage-channels at their foot. During good times these drainage-channels, especially where they flooded out to form miniature "floodplains", provided plenty of good food. During the drought they provided virtually the only green food in the whole study area. Burrows that were dug into the sides of the hills provided good refuge from foxes (sec. 12.412) and were relatively immune to erosion when they were neglected (sec. 12.414). The other four landforms were markedly inferior. "Flat loams", "stony pediments", and "Mitchell grass plains" all lacked relief, and so there were few features that were well watered by runoff. In addition, the soil in the Mitchell grass plains was heavy clay and not much good for burrowing. During good times (1963–64), sand dunes, stony hills, flat loams, stony pediments, and Mitchell grass plains supported respectively 15, 7, 3, 2, 0 warrens per square kilometer. Judging by "active entrances" (sec. 12.414), there were more rabbits per square kilometer in the sand-dunes that in the stony hills, 58 active entrances per square kilometer compared with 42. At the height of the drought (1966) rabbits had died out completely everywhere except in the stony hills, where they were greatly reduced, 0.6 active entrances per square kilometer. After the drought (1969) the rabbits recolonized the stony hills rapidly and the sand dunes slowly, 52 compared with 8 active entrances per square kilometer. The explanation is explored in section 12.423. In the present context the important comparison is that the stony hills provided the best security during a drought. The sand dunes offered greater productivity during good times, but during drought rabbits became extinct there. All the other "landforms" were inferior at all times and fatally so during drought. It is clear from this study that the rabbits could not exist as a permanent population in an area that consisted only of sand dunes and landforms other than stony hills. The clue to understanding the distribution and abundance of rabbits near Tero Creek has come from the study of local populations in a diversity of habitats.

A somewhat analogous situation occurs with the mouse *Mus musculus* in wheatfields in Australia. According to Newsome (1969a,b), mice were abundant in 1961–62 in an area of about 10,000 km² in the wheat belt of South Australia; and during the preceding ten years there had been six other outbreaks elsewhere in Australia. During an outbreak the mice are extremely numerous in the wheatfield and may damage the crop severely. Nevertheless, if the wheatfield were the only habitat available to the mice there might not be any plague of them. Birch (1971, 117) summarized Newsome's explanation of the outbreaks:

Newsome (1969a) found that mice in wheat fields live a largely opportunistic existence. In most winters, the wheat fields are completely inhospitable for mice, especially when the soil becomes waterlogged and mice can no longer live in cracks and burrows in the soil. In summertime, the soil is too hard for burrowing and the mice rarely breed. If mice had only wheat fields to live in, they would not be able to maintain their populations. What usually happens is that wheat fields become colonized each year from more permanent local colonies in habitats, such as reed beds, that provide more suitable permanent homes for the mice (Newsome, 1969b). The numbers of mice in wheat fields were controlled by the supply of colonists, the suitability of soil for burrowing and the food supply, in that order; but it is not often that the soil is suitable for the establishment of colonies. The wheat fields, for the mice, are similar to the sand dunes for the rabbits. The mice survive and multiply best there in favourable weather, but in unfavourable weather they are unable to multiply and mortality is very high. The reed beds correspond with the stony hills for the rabbits, they were less favourable than wheat fields in good years and patchily distributed. In unfavourable years, however, they were the only sort of place where mice could survive and multiply.

Newsome (1965a,b) has reported a similar regime for the red kangaroo *Macropus (Osthranter) rufa*. The distribution of the red kangaroo covers most of Australia inland from the 38 cm isohyet (Frith 1973, 273). The climate is arid; there are frequent droughts. In the absence of drinking water (which was very scarce before men contrived artificial supplies for their livestock), the kangaroo survives the drought by choosing a diet that is rich in green herbage. It achieves such a diet, even through a prolonged drought, by choosing to live along the drainage channels that carry water down from the hills to the plain after rain. "Floodouts" are best, but the kangaroos exploit any depression that holds water long enough to wet the soil deeply. The perennial grass *Eragrostis setifolia* is an excellent indication that a particular topographic feature is likely to provide good food for kangaroos. The kangaroos feed at night and must shelter from the sun during the day. So in central Australia, where Newsome worked, the kangaroos tend to be dispersed along the ecotone of plain and mulga (*Acacia aneura*) woodland, but only in those places where a confluence of hills, to catch the rain, and plain, to receive the runoff, contrive to produce well-watered habitats that support *Eragrostis* and other useful species even during droughts. After heavy rain the kangaroos disperse into the mulga woodland, where good food is contiguous with good shelter. At this time the plains are also abundantly covered with good food, but the kangaroos ignore it, presumably because shelter is not conveniently close on the plain.

Key (1945) and Clark's (1947) study of the Australian locust *Chortoicetes terminifera* showed that the outbreak areas for this locust were characterized by their own specific topography. The solitary phase of *C. terminifera* can usually be found, often in very low numbers, throughout the southeastern two-thirds of Australia. A little breeding goes on between plagues, but the population would probably die out if it were not replenished by migrants from "outbreak areas" (fig. 8.02). During an outbreak swarms of gregarious locusts from the outbreak areas (perhaps reinforced by some breeding in "intermediate breeding areas") may be found throughout the distribution of the solitary phase. The gregarious invaders may, if the weather favors them, breed for one or several generations in the "invasion area" before most of them are killed, usually by unfavorable weather (Andrewartha 1940). The invasion area during a severe outbreak might cover more than a hundredfold the space that is occupied by all the outbreak areas put together.

The eastern outbreak areas are confined to a broad climatic region where annual rainfall, between 60 and 80 mm, is distributed fairly evenly between summer and winter and evaporation is such that, according to Davidson's (1936) method, the growing season might be expected to last from one to four months (see also fig. 12.04). According to Key (1945), the outbreak areas occupy about 16% of a large region (about $75 \times 10^4$ km$^2$) where the climate is tolerable but the soil and the vegetation favor the locust only locally. Localities where the topography, soil, and climate favor the locusts are dispersed irregularly through the outbreak area. According to Clark (1947, 10):

The Bogan-Macquarie outbreak area [2,500 km$^2$] has a flat topography. Practically the whole of the present surface is alluvial in origin. The slight local differences in level occurring throughout the area rarely exceed a metre. However, to them are related major changes in soil type and vegetation. The soils of the higher ground are compact. In general those at the lower level are self-mulching. The latter are regarded as a more recent alluvium than the former. The outbreak area as a whole consists of a mosaic of compact and self-mulching soils. Rarely can more than a few kilometres be travelled in a straight line in any direction without a major change in soil type being encountered.

The heavy self-mulching soils vary in texture from clay loams to clay, as well as in other features. The compact soils of the high levels vary in texture from sandy loam to loam.

All but the most compact of these soils were favored by *Chortoicetes* for oviposition. The nymphs, hatching from the eggs, might in dry weather find themselves short of food and shelter; but at other times, hatching after rain, they usually found in the low vegetation on these areas sufficient food and shelter to carry them through the first two or three instars. But this vegetation was likely to dry off quickly if no more rain came; moreover, the older nymphs and adults required more adequate shelter than the low, sparse vegetation that grew on the compact soils. Their chance of survival in most generations, especially during dry weather, depended on their being able to find more reliable sources of food and better places to shelter. The nymphs cannot fly, so they had to find their requirements within a few hundred yards of where they were hatched.

The characteristic vegetation growing on the heavier soils at the lower levels included perennial grasses such as *Stipa*, *Eragrostis*, *Chloris*, and *Danthonia*, the tussocks of which provided good shelter for the locusts. These species and others associated with them on soils of high water-holding capacity remained green and succulent for a relatively long time after rain and thus provided a relatively secure source of food for the locusts.

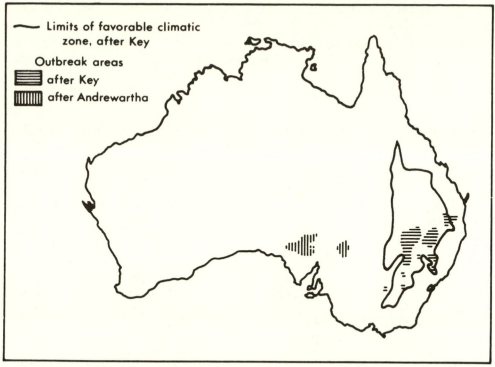

**Fig. 8.02** The distribution of outbreak-areas for *Chortoicetes terminifera* in eastern Australia and South Australia. (Modified from Key 1945 and Andrewartha 1940)

Wherever these two soils with their concomitant vegetations occur close together, especially if the boundary between them is abrupt, the locust is likely to find its best chance to survive during dry weather and multiply when the weather becomes more clement. This is especially true if the two soils that come into juxtaposition represent the extremes of their classes, because then the vegetation is likely to remain true to type even through extremes of drought. In the area that Clark studied these two soils were distributed as in a mosaic; consequently there were a great many places in this area where a locust would have a good chance to survive and reproduce—but these places were often separated by much ground that was barren from the locust's point of view. Such "outbreak centers" occupied only a small part of the whole "outbreak area".

Outbreak areas also occur in the arid rangeland of South Australia (2,000 km west from Bogan-Macquarie). The annual rainfall is only about 150 mm, but the typically arid topography of the drainage systems allows certain local features to receive more than their share of the water that falls as rain (Andrewartha, Davidson, and Swan 1938; Andrewartha 1940).

When the weather reverts to normal most of the locusts die. They become extremely scarce again except in the outbreak areas. In eastern Australia widespread plagues of *Chortoicetes* might be expected once in eight to ten years; in South Australia they occur about once in thirty to forty years. The difference in frequency of outbreaks reflects the difference in climate.

8.413   Isolation of Local Populations by a Finical Choice of a Place in Which to Live

In contrast to the highly dispersive locust, certain species of moths and butterflies seem to recognize, by some faculty that has not been explained, rigid boundaries to the locality where they live. In Andrewartha and Birch (1954, 519) we summarized Fisher and Ford's (1947) account of the moth *Panaxia dominula* as follows:

A colony of the moth, *P. dominula*. . . . occupied about 20 acres [8 ha] of fenlike marsh at Dry Sandford near Oxford. Part of the marsh was wooded, but most of it was covered by reeds and herbaceous plants, including comphrey, *Symphytum officinale*, which was the chief food of the larvae, and several sorts of nettles, which were also suitable for food. The marsh was bounded by woodland and agricultural land, into which the moths never seemed to penetrate, notwithstanding that they flew powerfully, were often observed circling around and above trees in their chosen area, and certainly were quite active in dispersing throughout the 20 acres [8 ha] in which they lived. There was another small area near Tubney, about $1\frac{1}{2}$ miles [$2\frac{1}{2}$ km] away, which was suitable for *Panaxia*, but Fisher and Ford found clear evidence that very few, if any, moths found their way from Dry Sandford to Tubney. . . Moreover, entomologists have collected in this vicinity for many years, and their testimony (reliable because *Panaxia* is a large brightly colored day-flying species) confirms that *Panaxia* is not to be found straying beyond the confines of the specialized [home range] where the colony lives.

According to Ehrlich and Murphy (1981), there are three distinct local populations of the butterfly *Euphydryas editha* on the Jasper Ridge Biological Reserve of Stanford University. Each is separated from the next by a few hundred meters but there is no physical barrier to dispersal. Despite the short distance between these local populations, there was very little movement of adults from one population to another. Fewer than 1% of the butterflies moved from one population to another in each generation. The populations seem to be kept separate primarily by the butterfly's reaction to its sources of nectar (D. D. Murphy, pers. comm.). Ehrlich et al. (1975) and Ehrlich (1979) have shown that the numbers of adults in these local populations at Jasper Ridge fluctuate independently of one another (fig. 8.03). One population (G) became extinct in 1964–65; the locality was recolonized in 1966–67; the population died out again in 1974–75, and by 1982 it had not been reestablished (Ehrlich, pers. comm.). The availability of food plants for larvae was similar in the three areas. In area G, however, nectar supplies were less; only one species of plant was a source of nectar for adults, in contrast to three species in the other area. In dry years in particular the presence of only one species as a nectar source in area G greatly reduced the chance of an adult's obtaining nectar. Adults are able to mature and lay eggs without nectar, but they lay many more eggs if they have fed on nectar. Ehrlich and Murphy (1981) considered that if the drought of the mid-1970s had continued one more year the other two populations of *E. editha* at Jasper Ridge would also have become extinct. During the drought of the mid-1970s three of four local populations of *E. editha* that were studied in the Sierra foothills became extinct. The fourth surviving population covered a larger area, and the plants there showed much greater variability in the time of onset of senescence. Peculiarities of the microhabitat where this population lived were presumably more favorable for an extended growing season for at least some of the plants. In other

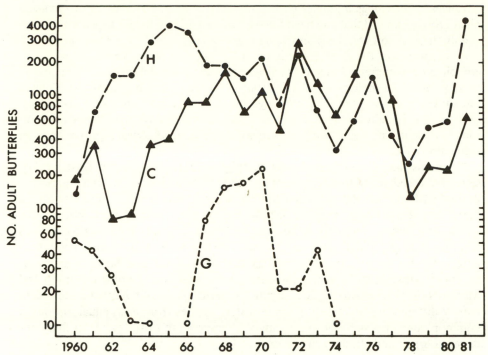

**Fig. 8.03** Numbers of the checkerspot butterfly *Euphydryas editha* in three local populations (H, C, and G) on Jasper Ridge, San Francisco Bay area. Note that the population at G became extinct in 1964–65, the locality was recolonized in 1966–67, and the population became extinct again in 1974–75. (After Ehrlich et al. 1975 and P. R. Ehrlich, pers. comm.)

localities populations became very rare during this drought. In general, localities where the larvae had access to perennial food plants tended to have more survivors than those with only annuals.

Studies by Ehrlich and his colleagues (summarized in Ehrlich and Murphy 1981) indicate how tenuous is the life of any single local population of the *Plantago*-feeding ecotype. Eggs are laid in batches on plants in early spring. After about two weeks they hatch. If the first-instar larva finds food it spins a web over part of the plant and remains in the web when not feeding. If there is no food within about 10 cm of where the eggs were laid, the newly hatched larvae starve to death. If the larvae find suitable food they continue to grow until the end of the third instar, when they molt and go into diapause. Diapause lasts throughout the summer and early winter; it is broken with the early spring rains when food again becomes available. Provided they find food, the postdiapausal larvae complete their development, pupate, and transform into adult butterflies. Many individual larvae fail to fit their lives into the growing season of their host plants. They die, either because the plant has undergone senescence between the time when the eggs are laid and when they hatch or because the larvae fail to reach the third instar during the period when green food is available. The chance that a larva will survive depends on how soon the plants senesce. At Jasper Ridge larvae feed on two host plants, *Plantago erecta* and *Orthocarpus densiflorus*. The vast majority of larvae that depend upon *Plantago* die before reaching diapause because the *Plantago* plants dry up as the season advances. A larva has a greater chance of survival if it finds itself on a *Plantago* plant growing on soil that had been tilled by gophers. Here the roots penetrate more deeply into the soil and the plants remain green longer. But by far the most likely way to survive to the third instar is to find a plant of *Orthocarpus* on which to complete development. Indeed, the size of the adult population in a given year is closely related to the density of *Orthocarpus* in the preceding year. These plants remain green longer than most *Plantago* plants. The dependence of *E. editha* on *Orthocarpus* is indicated by the fact that populations of *Euphydryas* around Stanford University are found only on serpentine soils; *Orthocarpus* grows only on serpentine soils.

"It seems likely" say Ehrlich and Murphy (1981), "that, for at least the last several hundred years, this ecotype of *E. editha* (subspecies *bayensis*) has persisted in the San Francisco Bay area as a shifting array of demographic units occupying the scattered patches of suitable habitat. Frequently, some of these populations would go extinct, eventually to be repopulated by migrants from nearby patches still maintaining populations' (see also sec. 8.431). These studies show clearly how unrealistic it is to seek to understand the distribution and abundance of a species except in terms of its local populations. To do otherwise, comment Ehrlich and Murphy (1981), is analogous to studying the performance of twenty thermostats in twenty different aquariums by pouring water supplies from each aquarium into a common container and then measuring the temperature of the water in that container. Ehrlich and Birch (1967) wrote:

A series of isolated populations with an array of different densities (including extinctions and re-establishment by migrants) may give the same superficial impression as a continuous population under rather tight "control". That is, to the casual observer, the species will be present each year. However, from the point of view of the way numbers change in nature, the two situations are entirely different.

The distribution of *E. editha* occupies a wide area in the western United States. In California alone there are more than twelve ecotypes which differ in the oviposition preferences of the adults and in the plants the larvae use for food. In most localities eggs are laid on just one species of plant, and the larvae tend to feed on a single species. However, some populations have one secondary host species or occasionally more or are polyphagous, regularly using several primary hosts. The life cycle of *E. editha* is synchronized better with some hosts than others. Which host serves best depends upon the climate and other features of the region where the populations live. A population of *E. editha* at Del Puerto canyon some 84 km from Jasper Ridge presented quite a different picture from those at Jasper Ridge. The main vegetation used for oviposition was not *Plantago* but *Pedicularis densiflora* and two other plants of the snapdragon family. *Pedicularis* is present on Jasper Ridge, but it has not been recorded as food for the larvae. At Del Puerto the larvae frequented defoliated patches of the food plant and then died of starvation. This never happened on larval hosts used at Jasper Ridge. Furthermore, nectar sources for adults were not co-located with *Pedicularis* at Del Puerto, hence adults were more vagile and moved considerable distances between nectar sources and oviposition sites (Ehrlich et al. 1975). The food plant used by *E. editha* in Gunnison County, Colorado, is *Castilleja linarifolia*, while throughout montane Nevada it is *C. chromosoma*. Yet both plants occur in both places. In Gunnison County a third host plant, *Penstemon strictus*, on which the larvae successfully mature is common, but oviposition and larval feeding are restricted to *C. linarifolia*. In drought years this plant remained greener longer than the others and was otherwise ecologically more suitable (Holdren and Ehrlich 1982). The implication of these observations is that local populations of *E. editha* have evolved to use the most suitable plants in each region of the distribution (Singer 1971).

### 8.414 Variability Associated with Asynchrony

Because the interaction of an animal and its environment is a dynamic process, the differential timing of ecological events within distinct localities must contribute considerably to the variability between localities. Much of this temporal component of variability is linked to the chance concurrence of erratic events. For example, the response of a population to an unusual frost or drought might depend on the age structure and also might alter the age structure differentially. One has only to reflect that the frost or the drought might act in either the centrum or the web of the primary animal to appreciate that such events may have large scope.

There is, however, an important component of variability between localities that is linked more systematically with the passage of time. It arises from the pre-established succession of birth, growth, maturity, and senescence that marks living organisms. The succession might be cyclical, driven by the succession of the seasons or by the innate life cycle of particular organisms, or there might be a number of other causes; but the timing might nevertheless be influenced by a random event that is more or less remotely situated in the web, as, for example, the outbreak of *T. imaginis* described in section 8.411. Insofar as these temporal changes are synchronized in distinct localities, they may not add much to the variability between localities, but when they are not synchronized, as for example, in *Euphydryas*, the contrary will be true. It seems that this temporal component of the variability between localities may add considerably to the heterogeneity between local populations.

For the fruit fly *Dacus tryoni*, only a place that offers ripe fruit will serve for breeding. Many host fruits have only a brief season, and usually each generation of adult fruit flies has to search for a new breeding place (sec. 14.2). In northern Queensland there are three other species of *Dacus* that commonly share the same locality, and sometimes the same fruit, with *D. tryoni*. They are *D. neohumeralis*, *D. jarvisi* and *D. kraussi*. G. Fitt (pers. comm.) has collected individual guava fruits from many neighboring localities in northern Queensland to find out the dispersion of flies in fruits. In any one locality all four species may be found in the same individual fruit. It is quite usual to find three species in a single guava. In one sampling Fitt collected three hundred individual guava fruits from twenty-nine sites; 25% of the flies in the fruits were *D. tryoni*. However, at two sites *D. tryoni* was the only species present and it was completely absent from eleven sites. Another 26% of the flies in the fruits were *D. jarvisi*. In some localities *D. jarvisi* was completely absent, in others it was the only species present. Likewise with *D. neohumeralis*, which constituted 37% of the flies in all the fruits collected; it was present in eighteen of the twenty-nine sites, being very rare in some localities and the only species present in others. There were even three localities in which no species of *Dacus* occurred in the fruits collected, despite their proximity to sites where flies were present. The overall picture of all four species is a very patchy dispersion both within fruits in a locality and between localities. Which species occurs in any place at any time probably reflects chance events during dispersal and the location of host fruits.

For the mosquito *Anopheles culifacies* the water that inundates a rice field is a good place for breeding. But such water is usually temporary. Only eggs that are laid while the rice plants are immature are likely to have time enough to mature. Macan and Worthington (1951) found that the locality ceased to attract egg-laying females once the rice plants had grown taller than 30 cm. Suitable places for the spruce budworm *Choristoneura* to breed occur where stands of spruce and fir that are older than 40 years have experienced several years of anticyclonic weather. Such places cease to be favorable when, as a result of an outbreak of budworm, most of the trees have been killed by overgrazing or when cyclonic weather has prevailed again for a number of years. The distribution of stands of trees of particular ages depends largely on the distribution of plagues of budworms in past outbreaks stretching back hundreds of years. The distribution and abundance of cyclones and anticyclones depends on the vagaries of various air-masses which form, drift, and decay over vast areas that transcend the area of North America.

In southeastern Australia, during the spring, the blowfly *Lucilia cuprina* will lay eggs freely on a freshly dead carcass of, for example, a sheep. If other species of blowflies are artificially excluded or happen naturally not to come in large numbers, the maggots will grow to a healthy maturity. But this is unusual. Usually other species, *L. sericata*, *Calliphora* spp., *Chrysomyia rufifaces*, and others come in large numbers, and all or most of the larvae of *L. cuprina* die young. Usually successional changes that are induced by these other species make the carcass uninhabitable for *L. cuprina*, but occasionally, when the other species happen to be fewer, more of *L. cuprina* survive to become adults.

Spight (1974) was able to identify thirty-nine local populations of the snail *Thais lamellosa* along 1 km of coast. The snail is relatively sedentary; all stages of the life cycle are completed on the shore, there being no planktonic stage. The habitat is patchy, so adjacent populations may live in quite different environments. During the five years of study the numbers in each local population changed greatly, and the changes were asynchronous. Some populations were reduced to low numbers and were probably saved from extinction only by the timely immigration of snails from other localities. Despite the large changes in size of local populations, the average number of adult snails along the kilometer of shore changed very little during the five years of study. Because Spight (1974) studied the snail as a multipartite population in a multipartite habitat, he was able to demonstrate key components in the ecology of the species that would otherwise have been missed altogether.

### 8.415   Demographic and Genetic Differences between Local Populations

There remain to consider only the ways in which the primary animal itself may vary between localities. Such variability may be either demographic or genetic. Demographic variation is universal because, with all other components of environment constant, birthrates and death rates will still vary with age structure; and the probability that a local population in nature will achieve a stable age structure is negligible. Genetic variation is also widespread and important.

Individuals within local populations and between local populations show phenotypic variation in their tolerance, preferences, and in other aspects of their lives that influence their chances to survive and reproduce. Spreading of risk by phenotypic variation is likely to be of most advantage in highly unstable environments. Where phenotypic variation reflects genetic variation between individuals, natural selection will shift the gene frequencies of the population as the environment changes and so further reduce the risk of extinction. This aspect of spreading the risk is discussed in detail in chapter 11.

Individuals within local populations and between local populations may show differences in tolerances, preferences and so forth, because they belong to different age-classes or development stages. The risk is spread when more than one age-class is exposed to environmental hazards, such as cold in winter, for example, the carabid beetles *Calathus melanocephalus* and *C. erratus* hibernate in both the larval and the adult stages (den Boer 1968, 169).

That concludes the discussion about variability between localities and local populations. In section 8.42 we consider the other essential premise on which the concept of the multipartite natural population rests, namely, that the universal dispersiveness of animals (secs. 2.22, 2.25, 13.2) has been bred into them by the universal need to found new colonies before the old ones die out (Andrewartha and Birch 1954, chap. 5). The discussion in this section shows that this premise rests on the well-established mutability of localities and the consonant precarious existence of local populations. To show that the innate dispersiveness of a species is quantitatively related to the dispersion or mutability of localities would strengthen the concept of the multipartite natural population (sec. 8.41). den Boer's experiments with carabid beetles, which we discuss in section 8.42, support the concept in this way.

## 8.42 The Dispersiveness of Species Relative to the Dispersion of Localities

The concept of spreading risk recognizes the importance of dispersal within the natural population that enables individuals to colonize new localities. We discussed dispersal at some length in Andrewartha and Birch (1954, chap. 5). Although there is a random element in some forms of dispersal, such as the drifting phase of planktonic larvae, there are also directional elements. Directional dispersal (searching) has been well studied, particularly in insects, birds, and mammals. However, it would be a mistake to suppose that it is absent even in animals that seem to be at the mercy of currents and winds. For example, planktonic life of the larvae of many sessile marine invertebrates comes to an end with an exploratory behavior when many potential substrates may be visited and tested. According to Doyle (1975), the duration of the exploratory phase depends upon the availability of suitable substrates on which to settle and metamorphose. In many species, the longer the search for the needed substrate the less demanding the larva is in its choice. *Spirorbis borealis* may eventually metamorphose on the surface film of water or without attaching to anything at all, but in its early exploratory phase it is very discriminating. In partially isolated tidal pools where the brown alga *Ascophyllum* is the only alga present, *Spirorbis* settles on the alga in profusion. A few meters away in exposed localities where *Ascophylllus* may be present but where *Fucus* is the dominant alga, it is *Fucus* that becomes heavily encrusted with *Spirorbis*. *Ascophyllum* has few or no *Spirorbis* settling on it in the turbulent waters. The reason is twofold. In turbulent water fewer *Spirorbis* choose to settle on *Ascophyllum*; and, those that do so have a low chance of survival because they do not seem able to cement themselves adequately to the surface of *Ascophyllum* in turbulent waters (Doyle 1975).

It is to be expected (in keeping with the principle that the niche is the animal's evolutionary response to its environment) that in a well-adapted animal dispersiveness will match dispersion of localities. den Boer (1977) took advantage of the recent widespread degradation of pristine habitats in the Province of Drenthe, The Netherlands, to test this idea, using as his material seventy-four species of beetles (Carabidae) that still inhabited the degenerate remnants of the habitats. den Boer noticed that when the pristine habitat had been such that it was likely to offer a dispersing beetle a new locality within walking distance of the place where it was born, the beetle was likely to be apterous or brachypterous. Species that lived in habitats where, even in the pristine condition, localities were likely to be widely spaced tended to have strong, functional flightwings.

den Boer argued that the less dispersive species were likely to have been more severely disadvantaged by the deterioration of habitats than the more dispersive ones, and he suggested that the extent of the disadvantage would be indicated by the proportion of the potential localities that were discovered and colonized. A species that is likely to find and colonize most of the localities soon after they become available might be counted as well adapted—not seriously disadvantaged at all. A species that has little chance of finding many of the vacant localities must be counted as ill adapted through lack of dispersiveness.

These arguments were based on three assumptions:

1. The search for a new place to live is going on all the time—an innate behavior that is independent of the density of the home population.
2. Any well-adapted species is endowed with dispersiveness that is commensurate with the likely accessibility of new localities.
3. This behavior has evolved because the quality of environment in the localities is likely to wax and wane more under the influence of successional, seasonal, and other such changes than under the influence of the density of the population itself (secs. 8.411–8.414).

To confirm den Boer's predictions lends support to the assumptions on which they rest and conversely, casts doubt on the opposite assumption—that dispersal from the local populations is under the control of some sort of "density-governing mechanism". To confirm den Boer's predictions also lends general support to the reality of the concept of "interaction group" (local population) as we use this idea.

We give a few brief details about the beetles, their dispersiveness, the extent of the degradation of habitat, the hypotheses that were tested, and the method of gathering and analyzing the empirical data before giving the actual results and the conclusions drawn from them.

The changes in the habitat had been profound. About 350 years ago a map of the vegetation of central Drenthe (covering about 600 km$^2$ around Wyster) showed that 3% of the total area waas occupied by farmland. The remaining 97% still carried either the pristine vegetation comprising deciduous forest and peat moss or a secondary mixture of heathland and blown sand which originated in an old predatory cultivation. There were ten small patches of farmland, clusters of farms, widely scattered through the pristine vegetation. By 1965 the position was quite reversed: 85% of the area was cultivated farmland and rather less than 8% was still carrying remnants of the original vegetation. There was only one large area (1,200 ha) of nature preserve consisting mostly of old heathland and pine plantations. The remainder consisted of small patches (none larger than 3 ha) of native vegetation scattered widely through continuous areas of cultivated farmland or pine plantation. The change may have been moving slowly in the same direction through the eighteenth and nineteenth centuries. Appreciable progress had been made by the beginning of the twentieth century; but most of the change had occurred since 1900, and most of it during the past few decades.

The dispersiveness of the species was inferred from the prominence of flightwings. At one extreme were species in which every individual had large, fully functional wings. At the other extreme were species in which all individuals were wingless. In between there was a wide range of brachyptery, with some species being dimorphic and others polymorphic with respect to brachyptery. If a carabid flies, it is likely to do so by day, during a period preceding the breeding season. den Boer used large sheets of glass with traps below them to catch flying carabids. In five years he caught 3,748 individuals from 26 species (class B below). He found that, in all the beetles that had demonstrated their ability to disperse by flying, the ratio of the area of the flightwing to the area of the elytron exceeded 2.0. He classified the 74 species according to this ratio putting them into the following classes:

Class A—25 species, wingless or strongly brachypterous, weakly dispersive.

Class B—26 species, ratio of flightwing to elytron greater than 2.0; strongly dispersive.

Class C—23 species, dispersiveness uncertain or intermediate because they were dimorphic or polymorphic or flightwings not so strongly developed as in class B.

The density of the local populations (interaction groups; sec. 8.41) was estimated by counting the beetles that were caught in pitfall traps. This was a good technique because Carabidae are predaceous beetles that seek their food, mostly other sorts of insects, by crawling on or through the litter at the surface of the ground, mostly at night. den Boer trapped only in the remnants of the original vegetation. His method was to sink three pitfall traps close together in a row and leave them at the same site for one or more years. He recorded his results as the number of beetles of each species counted at one site during one year. He trapped continuously for nine years, and by the end had accumulated 175 site-years. He caught more than 170,000 beetles and recognized 148 distinct species among them. But the 74 most abundant species between them accounted for 99.37% of the total catch. He analyzed the data from this group of 74 species. Each species appears at more than one trapping site; most sites yielded a number of species; all together den Boer had a total of 4,715 year-catches—a mean of 64 year-catches per species. Each year-catch for each species was counted as an interaction group (local population) of that species. den Boer (1977) described in some detail the precautions that he took to satisfy himself that the number caught in one site during one year was a good estimate of the relative density (sec. 8.3) of the population for that year at the locality that was defined by the trap at its center (Baars 1979). den Boer called the group of animals sampled by the trap an "interaction grouop". In this context the "interaction-group" comprises all those individuals that were born at a distance from the trap not greater than the distance through which an individual might move during the nondispersive stage of its lifetime. The "interaction group" might gain some immigrants from and lose some emigrants to other groups. These events were part of the interaction: they counted as "births" or "deaths".

den Boer explored two possible explanations for migration. If, as might be predicted by competition theory, emigrants left the locality only when driven out by overcrowding and/or immigrants arrived only in those localities that were not overcrowded, the density of the local populations should tend to cluster strongly around a modal density that is the maximum a potential disperser would tolerate. den Boer measured the relative densities (beetles per trap per year) of about 4,720 interaction groups. He inspected the result of this large sample and found no support for this hypothesis. There was no need to calculate precise probabilities. The failure of the hypothesis was obvious.

Alternatively, from the hypothesis of spontaneous (density-independent) dispersal outlined above, the expectation is for the logarithms of the population densities to be distributed more or less symmetrically about a mean which is likely to be less than the maximum density that the animals will tolerate. For a species whose dispersiveness adequately matches the dispersion of localities, the distribution of population densities (a statistic den Boer called DPS) is likely to be unimodal and more or less symmetrical as in figure 8.04A (solid line). A considerable proportion of the local populations are likely to be sparse because they have only recently been founded or refounded. Indeed, for the most adequately dispersive species the distribution DPS may be quite like the "normal distribution". For a species that has become inadequately dispersive because localities have become less accessible, there is likely to be a reduction in the proportion of recently colonized (or re-colonized) localities. Hence the distribution will be truncated as in

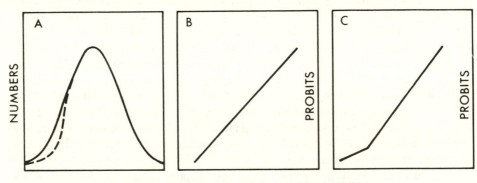

LOG DENSITY OF LOCAL POPULATION

**Fig. 8.04**    If the logarithm of the densities of the local populations is normally distributed, as in the solid curve in A, the same data, transformed to probits, will fall along a straight line as in B. If the frequencies of the lower values for log density are truncated as indicated by the broken line in A, the graph of the probits will show a point of inflection as in C. Provided the logarithms of the density are normally distributed, the contrast between B and C discriminates between adequately and inadequately dispersive species relative to the patchiness of the habitat. The transformation that den Boer (1977) invented (RPR) made the same comparison without depending on any assumption about the shape of the distribution of the statistical data.

figure 8.04A (broken line). When such distributions are transformed to probits they look like figure 8.04B or C.

Not all the distributions that den Boer had in his sample could safely be assumed to be like a normal, or truncated normal, distribution. But this is all right because the nonparametric statistic DPS measures the distribution of population size without requiring any assumption about the shape of the distribution. It was assumed that DPS would effectively represent the replacement of interaction groups relative to their survival time. den Boer referred to this concept as "Realization of population replacement", RPR. At the time (den Boer 1977) this assumption had not been verified in the field, but according to den Boer (pers. comm.), the survival times are of the right order of magnitude to support the assumption. If RPR (as measured by DPS) for species A is significantly less than for species B, we may conclude that species A has been more severely disadvantaged by recent changes in the dispersion of localities than species B.

$$DPS = 1 - \{\Sigma \ln(n_i + 1)\}/\{j \ln (N + j)\}$$

$n_i$ is the number of individuals in year-sample i.
j is the number of year-catches with $n_i > 0$.
N is the total number of individuals caught in all year samples.
To visualize the operation of DPS, think of the antilogarithm of

$$\{\Sigma \ln(n_i + 1)\}/j$$

as the geometric mean of $n_i$. Think of N as the arithmetic sum of $n_i$. Consider the changes in DPS as j changes from small to large. The lowest permissible value for j is 1—when there is only one population. So $(n_i+1) = (N+j)$ : DPS = 1 - 1 = 0; thus there is no replacement. At the other extreme, when j becomes large, the geometric mean of $n_i$ decreases relative to $(N+j)$. The ratio becomes small, and the value of DPS approaches 1 - 0 = 1. Replacement is fully realized.

The results of den Boer's analysis are given in table 8.02. den Boer was able to conclude, with considerable confidence, that the changes in habitat that had occurred in Drenthe had taken place too rapidly for the carabids to adapt to the change and that the less dispersive species had been more seriously disadvantaged by the changes than the more dispersive ones.

But for the purpose of our present discussion we can draw still another conclusion. That an experiment of this sort gave a strongly positive result, in keeping with den Boer's predictions, confirms the concept of the multipartite natural population. It suggests that the concept might be a good base from which to plan experiments about populations.

**Table 8.02**  Comparison between Dispersiveness and RPR (Realization of Population Replacement, Based on DPS—Distribution of Population Size) for 74 Species of Carabids in Drenthe, The Netherlands

| Dispersiveness | Number of Species | RPR | | Probability |
|---|---|---|---|---|
| | | Minimum | Maximum | |
| Group A, weakly dispersive | 26 | 0.4585 | 0.7834 | B > A < .001 |
| Group B, strongly dispersive | 25 | 0.6312 | 0.8109 | B > C < .001 |
| Group C, intermediate | 23 | 0.4522 | 0.7897 | C > A < .05 |

*Source:* den Boer (1977).

## 8.43   Lessons from Extinctions

There are semantic pitfalls to be recognized and avoided when discussing "extinction". den Boer's definition of "interaction group" (sec. 8.41) implies the largest grouping that still gives all individuals a reasonable chance to interact. Without some such safeguard the probability of extinction can be brought as close to certainty as desired by arbitrarily making the locality as small as may be required. The distinction between the "extinction of a species" and the "extinction of a local population" can also become a semantic myth, because the species is extinguished only with the extinction of the last local population. And the evidence from fossils is ambiguous because, even though the modern biospecies may seem quite distinct from the fossil palaeospecies, the palaeospecies might still be the ancestor of the biospecies. What can extinction mean in this context?

Nevertheless, the literature contains some useful accounts of extinctions of populations that might be considered to come within our concept of "local population" or den Boer's "interaction group" or Ehrlich and Murphy's "demographic unit" (sec. 8.41). A number of examples are mentioned in the next two sections.

### 8.431   Extinction of Local Populations While the Natural Population Carries on Securely

It is not known how many species of carabids were living in Drenthe one hundred years ago. den Boer (1968, 1979) thought it likely that some that were extant at the beginning of the century are extinct now and that some in his sample of seventy-four species might be in danger of extinction. The analysis in table 8.02 strongly rejects the explanation that extinction occurred because of the breakdown of some stabilizing process associated with the density of the population. On the contrary, the data support the view that the extinct species came to their end, and that the species that will soon be extinguished will come to their end, through bad luck—that is, through events that are not related to the presence or the numbers of the beetles themselves. In the present instance bad luck took the form of massive destruction of habitat by humans. An extrinsic shortage of something the animal needs is the commonest form of bad luck for nonhumans, and currently the great eruption of human populations is the commonest cause of such bad luck for most other species.

The following passage from Murphy and Ehrlich (1980, 319) describes the destruction of habitat for the butterfly, *Euphydryas editha* (subspecies *bayensis*) in the San Francisco Bay area. Figure 8.05 shows the total distribution of this species in this area.

It is now restricted to two localities, Jasper Ridge and Edgewood. It is extinct in eleven localities where it was known to exist previously. In two other localities its present status is uncertain. Large areas of habitat suitable for the subspecies have disappeared due to various causes. These include the building of a major freeway through Edgewood (EW), sub-division, construction and introduction of exotic species, and the combined effect of drought and livestock grazing at Silver Creek (SJ). As a result of these human activities the number of islands of habitat suitable for the subspecies is now greatly reduced and the distance between them has increased. The Edgewood population is threatened by a golf course and over the long term it seems unlikely that Jasper Ridge alone can maintain this ecotype.

According to Ehrlich et al. (1972), unusual spring weather climaxing in a late snowstorm in June 1969 in the subalpine area in Colorado damaged many herba-

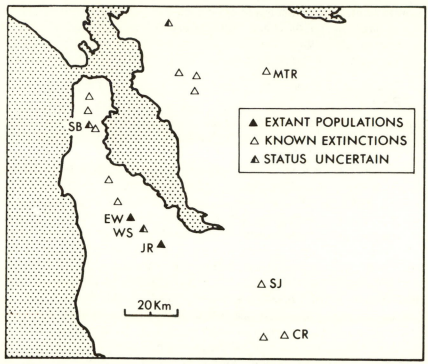

**Fig. 8.05** Local populations of the checkerspot butterfly *Euphydryas editha bayensis* within its entire distribution in the San Francisco Bay area: CR, Coyote Reserve; EW, Edgewood; JR, Jasper Ridge; MTR, Morgan Territory Road; SB, San Bruno Mountain; SJ, Silver Creek; WS, Woodside. The labeled localities have been actively studied by P. R. Ehrlich and his group. (After Murphy and Ehrlich 1980)

ceous plants and reduced populations of insects and small mammals. The storm also extinguished a population of the lycaenid butterfly *Glaucopsyche lygdamus* in this region. At the peak of the flight season in 1968 a census of two hundred inflorescences of *Lupinus caudatus* revealed a total of 233 eggs of the butterfly. In contrast, in 1969 an examination of three hundred inflorescences after the storm revealed only 6 eggs. Only one tattered female was observed during the time *Glaucopsyche* normally begin to emerge as adults in mid-June. The storm came at the peak of the flight time, and presumably many eggs had already been laid in inflorescences. These inflorescences were destroyed by the snow, and few of those eggs survived the storm. The local population of the butterfly was extinct in 1969. By 1970 the colony had not been reestablished. No adult was found and searching one hundred inflorescences yielded only 2 eggs.

Fatal "bad luck" may take many forms—for example, density-independent predation (sec. 10.22) or extrinsic shortage of food that is caused by density-independent predation of the food (*Oncopeltus* in section 8.432). We suggest in chapter 9 that the best defense against any sort of bad luck is to "spread the risk" (den Boer 1968) over a large number of local populations in as many and as diverse localities as possible. As we have seen above, to make efficient use of potential localities requires dispersiveness to match the dispersion of the localities. The urge to disperse which characterizes the behavior of most species is evidence that the multipartite natural population is a realistic concept. If the number and diversity of localities is large enough, the chance that fatal bad luck will visit all of them at once may become vanishingly small and thus account for the persistence of populations through periods that are properly measured on the evolutionary time scale.

### 8.432   Extinction of Local Populations Leading to the Extinction of a Natural Population

It seem that the last local population of the bug *Oncopeltus sandarachatus* on the island of Barbados, in the Caribbean archipelago, was on the verge of extinction when Blakley and Dingle (1978) studied it. During a month's thorough searching they found only five *Oncopeltus* on the island. The same search revealed only five plants of milkweed, *Asclepias curassavica*. Seeds of *Asclepias* are the sole food for *Oncopeltus*. Blakley and Dingle attributed the scarcity of *Oncopeltus* to the scarcity of *Asclepias*; the bugs were hampered by a severe extrinsic shortage of food.

The milkweed was scarce because it was heavily preyed on by caterpillars of a butterfly, *Danaus plexippus*; *Asclepias* is the preferred food of *Danaus* but when *Asclepias* is in short supply *Danaus* maintains its numbers by feeding on another milkweed, *Calotropis processa*, which is abundant on Barbados. *Oncopeltus* cannot feed on the seeds of *Calotropis* because the seed's coat is too thick and tough.

So when *Oncopeltus* ultimately becomes extinct on Barbados the proximal cause of the extinction will be an extrinsic shortage of food. Distal causes will be the predation of *Danaus* on *Asclepias* and the presence of *Calotropis* as an alternative but less preferred food for *Danaus*. Further back in the web of *Oncopeltus* there must be some component in the environment of *Calotropis* that allows it to flourish on Barbados but keeps it sparse or absent on most of the other islands of the archipelago. Both *Asclepias* and *Oncopeltus* are abundant on most of the other islands. The segment of envirogram that reflects this aspect of the ecology of *Oncopeltus* would look like this:

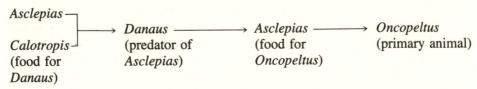

*Asclepias*
        → *Danaus*      → *Asclepias*      → *Oncopeltus*
*Calotropis*    (predator of     (food for      (primary animal)
(food for       *Asclepias*)      *Oncopeltus*)
*Danaus*)

The ecology of the prospective extinction of *Asclepias* on Barbados is like that of the extinction of *Aonidiella citrina* in southern California (sec. 10.22). But most extinctions would seem to be more like that of *Oncopeltus* on Barbados—due to an extrinsic shortage of some essential component of environment. Even so, *Oncopeltus* is not quite typical because, wherever one may look, the typical cause for extinctions seems to be the destruction of habitat by burgeoning populations of *Homo sapiens* (sec. 13.4).

Dilution and destruction of pristine habitats, such as den Boer studied in Drenthe (sec. 8.42), began in Australia in 1788. Even while the chief source of energy was horse, bullock, and man power, the impact of the European settlers (and the exotic plants and animals that accompanied them) on the primitive floras and faunas of this isolated continent was severe. But the destruction of habitat accelerated spectacularly as the twentieth century's technological revolution got under way. Rural, industrial, and urban development outpaced the naturalists: the documentation of the ecological changes that occurred is fragmentary. They have been summarized by Frith (1973), and the following brief resume of the extinction of Leseur's rat kangaroo (or boodie), *Bettongia leseuri*, from mainland Australia was taken largely from Frith (1973).

In 1788 and for some years after that date, *B. leseuri* was widespread and abundant throughout virtually the whole of the southern half of the continent. In the southern part of its distribution, where rainfall was sufficient to support arable farming, it was regarded as a pest; there is a record of a farmer in South Australia who shot 149 in one night. According to Frith, in certain parts of the rangelands of New South Wales where the soil is hard and stony the plains are still "sprinkled with the old warrens" of *B. leseuri*. In places the warrens are large and close together. They were probably just as numerous in the softer country, but erosion has destroyed the evidence of their presence. In 1896 an expedition through the semi-arid rangeland of the center reported that *B. leseuri* was probably the most abundant marsupial in that country. It was one of the most important accessory foods of the aborigines. In New South Wales the last specimen was collected in 1892, but some old residents can remember it as late as 1920. It was last seen in the arable belt of South Australia about 1910. In Western Australia the last sightings were made during the decade 1930–40, but as late as 1940 it was still present in fairly large numbers in the relatively undisturbed rangeland near the Everard Ranges in northern South Australia. In 1958 a few small, isolated populations were known to exist in this area and elsewhere in the arid central parts of the continent. Since then *B. leseuri* has not been seen in mainland Australia. A small population still exists on a small island off the northwest coast.

The whole of what was the distribution of *B. leseuri* is now exploited by Europeans. In the more humid parts there is arable farming dominated by a wheat/sheep rotation. In the more arid parts the rangeland is grazed extensively, with sheep dominating the parts where it is practicable to provide fairly frequent watering points, and with cattle ranging in most places where there are no sheep. The Europeans have been accompanied into almost all the places where they have settled by rabbits, foxes, cats and other exotic animals as well as by exotic plants, both crops and weeds.

According to Diamond and Veitch (1981), when New Zealand was first settled by Europeans, during the nineteenth century, there were seventy-seven species of indigenous nonmarine birds. Since then eight species have been extinguished and thirteen have become scarce. As in Australia, the explanation of these extinctions and near-extinctions seems to lie in the destruction of forest and the introduction of exotic predators by the settlers. The domestic cat was important, and so were a number of diseases.

According to Ehrlich and Ehrlich (1981, chaps. 6, 7), major extinctions of large mammals occurred in both Eastern and Western hemispheres soon after the arrival of man. Since then some extinctions have been due to predation by man (sec. 5.3), but, as Ehrlich and Ehrlich (1981) point out, most have probably been caused by change of habitat wrought by man and the species that man has introduced.

## 8.5 Conclusion: The Multipartite Population

In this chapter we have reviewed the methods for estimating the density of a population and recalled some of the philosophical limitations and technical difficulties associated with making such estimates. Populations in nature are inevitably scattered patchily over a heterogeneous terrain. The structure is not static. Ecological problems always contain a large element of dynamics. Patchiness makes it difficult to take critical samples or to calculate precise statistics, thus placing restraints on the testing of hypotheses. But patchiness itself places the greatest restraint of all on the formulation of theory. No general theory about the distribution and abundance of animals should have a chance of being accepted as realistic unless it takes full cognizance of the patchy dispersion of animals in natural populations.

The concept of the multipartite population that has emerged from our review seems realistic and workable. We use it for the empirical analyses that occupy part 3 and for the theoretical discussion that occupies the rest of part 2. The essentials of the concept are:

1. The natural population comprises many local populations.
2. Dispersal occurs between local populations but is nonexistent or negligible between natural populations because the natural population is surrounded by an ecological barrier.
3. Immigration to and emigration from a particular local population are regarded as equivalent to births and deaths.
4. The local population is the unit in which individuals interact with one another as mates or modifiers of various components in the environment of the primary animal. The risk of arbitrarily defining the local population too small (thereby magnifying the risk of its extinction) is avoided by thinking of the local population as the most extensive unit that can exist without denying its members, by its very size, a reasonable chance to interact with one another.
5. Even so, most local populations are doomed to extinction because most localities are ephemeral.
6. A locality is the distribution of a local population. A locality may constitute part of a habitat or comprise more than one habitat. This disjunction arises because the locality is objectively defined by the niches of the primary animal; the habitat is arbitrarily defined by the interests of an ecologist.
7. The risk of extinction for most natural populations is negligible, taken over a reasonable interval of time—say the life-span of an ecologist. This paradox (see 5 above) arises because the condition of the localities is so heterogeneous, with respect to time, that the chance that all local populations will face extinction at the same time is negligible. The risk of extinction becomes prominent only after the numerous local populations have become few.
8. Two exceptions to (7) might be:
   a. The physical event that so suddenly put an end to the climate of the Cretaceous era and ushered in the Tertiary, or other discontinuities that are further in the past.
   b. The current, apparently incurable outbreak of *Homo sapiens* which seems destined to cause the extinction of many nonhuman species before it reaches its zenith and crashes, like a plague of locusts.

The principle enunciated in (7) was the basis of den Boer's (1968) theory of "spreading the risk", which we discuss in chapter 9.

# 9

# Theory of the Distribution and Abundance of Animals

## 9.0  Introduction

The purpose of this chapter is to put forward a general theory about the distribution and abundance of animals. Theory is that step in the scientific method that is concerned with explanation. In this context "to explain" an empirical fact is to suggest a cause for it. In the present instance the empirical knowledge that requires a general explanation comprises the observed densities in populations of animals that are living naturally. Our theory is set out in sections 9.1 to 9.5. It is preceded by a brief description of the scientific method as we see it. Propounding a new theory or criticizing an old one is done better when there is agreement about the epistemological status of the familiar components of scientific method—models, hypotheses, experiments, explanations, and theories.

## 9.01  Models and Hypotheses

In population ecology "model" is the name for any symbolic manipulation of an idea or a fact that purports to copy or to predict the growth of a population. Mathematical models are not to be confounded with laboratory "models", which are populations that are reared artificially in a laboratory. Models may be manipulated by arithmetic, algebra, or simulation. The distinctions are not fundamental but expedient: simulation takes over when there is no practicable algebraic solution. Or a model may be presented graphically, as in figure 9.02.

Niven's (1967) mathematical model of a population of the flour beetle *Tribolium castaneum* is an example of a simple simulation model that opened the way for some new hypotheses by casting doubt on the currently accepted explanation of the empirical data. The real beetles were living in an artificially simple environment in a laboratory (Park 1948; Park, Leslie, and Mertz 1964). The beetles lived in a vial containing 8 g of flour at 29.5°C in air at 60–70% R.H. The beetles were counted and the flour was renewed every thirty days. One experiment lasted 2,100 days, about thirty-six generations. To develop from egg to adult, *Tribolium* requires about twenty days; a few adults lived nearly a year. The population (excluding eggs) fluctuated as in figure 9.01, solid line.

Niven conferred on the *"Pseudotribolium"* of her model an age-specific fecundity, an age-specific speed of development, an age-specific voracity (larvae and

*185*

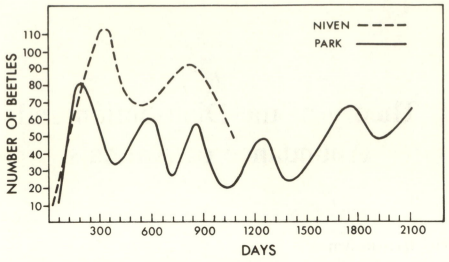

**Fig. 9.01**   Fluctuations in the number of *Tribolium castaneum* living in vials of flour at 29.5°C and 60–70% R.H. The flour was replenished and the insects counted every thirty days (after Park, Leslie, and Mertz 1964), compared with fluctuations in Niven's (1967) model, which was simulated on a computer without taking account of either conditioned flour or fluctuating gene frequencies.

adults eat eggs and pupae), and an expectation of life for the adult, all of which corresponded to estimates of these statistics that Park and his colleagues had made in experiments in which the results were independent of the density of the experimental population. The computer counted the population every day and recorded the result every thirty days. The results are shown in figure 9.01, broken line.

Park, Leslie, and Mertz (1964) had suggested that the major fluctuations, with a peak about every seven generations, might be associated with an accumulation of pollutants in the flour. They said the flour was "conditioned" by materials that were excreted by the beetles, especially when the beetles were numerous. McDonald (1963) suggested that genetic variations might be causing the fluctuations. Niven examined both these suggestions at once by leaving both of them out of the model. In the event, the model produced fluctuations that were at least as pronounced as the real ones, even though they did not agree with the real ones in phase and amplitude. The model had cast doubt on the existing explanations. The way was open for some new hypotheses.

The simple procedure that lies behind Niven's model of a laboratory population of *Tribolium* is extended to great lengths in the models that are recommended by Gilbert et al. (1976), Frazer and Gilbert (1976), and Barlow and Dixon (1980) for dealing with populations in nature. In these models the density of the population is traced through a computer (as Niven traced the population of *"Pseudotribolium"*) through successive units of time, occupying a generation or many generations, a season or many seasons, and so on. The simulated population begins with the density and age structure of the real population, which is also counted through the same interval of time. The pseudoanimal is exposed to the same "environment" as the real animal; that is, the environment of the real animal (those components thought to be important) has been measured and fed to the computer. This input can be made because the influence of the environmental components on the real animal has been determined empirically in independent experiments.

The real population is compared with the simulated one. If the agreement is close the environment and the niches (responses to the environment) of the real animal seem to have been judged correctly. If the agreement is not close the conclusion must be that the experimenter has made an error or errors. Perhaps an important component of environment has been overlooked; perhaps a niche has been mis-judged; perhaps a component of environment has not been measured accurately. So the next step is to revise the hypotheses or the techniques and try again.

The simulated population might be used to challenge a theory, as Niven (see above) did with *Tribolium* or as den Boer did with two species of carabid beetles (sec. 9.32). Or the simulated population may be used as the base for a population-management program (Hearn et al. 1981). Whether the purpose of the exercise is chiefly theoretical or chiefly practical, the prediction is more likely to pass the empirical test if the model is realistic.

For this reason we prefer the envirogram to the flowchart (sec. 2.32). The envirogram is more likely to focus attention on the critical components of environment which might be few (Andrewartha and Birch 1954, 558; Hearn et al. 1981). For this reason and because of its construction, the envirogram is more likely to give rise to hypotheses that can be tested within the framework of conventional statistics and thus have a probability attached to them.

Another sort of defect arises when the model is allowed to become so abstract that there is no way to test any of its predictions empirically (sec. 10.331). Then the temptation arises to treat the prediction of the model as if it had been verified empirically and thus fall into the error of the misplaced premise (sec. 10.12). Unfortunately, there is a powerful trend toward excessively abstract mathematical models in the modern literature on population ecology. One has only to read contributions from May (1973b, 1976), Holling (1973), Peterman, Clark, and Holling et al. (1979), Clark, Jones, and Holling (1978), Southwood and Comins (1976), and Roughgarden (1979) and follow up the references cited in them to realize how strongly this trend has been established. We cannot see many testable hypotheses emerging from such abstract models, despite the claim that "such models can be useful in suggesting interesting experiments or data collecting enterprises" (May 1973b, v). So we say no more about them. To a large degree many if not most of the more abstract modern models have been developed from the foundations that were laid by Lotka and Volterra, which we have already criticized on the grounds of unreality (Andrewartha and Birch 1954, 362, 406). A model has no valid function other than to give rise to testable hypotheses or plausible explanations (sec. 9.02).

## 9.02 The Distinction between Scientific Knowledge and the Prediction of a Model or an Hypothesis

Three nouns are used to name the logic that is used in science. "Deduction" works from the general to the particular, as in the syllogism or in pure mathematics; it is needed for modeling and postulating. "Inference", may be written instead of "deduction", but this is elegant variation because the *Concise Oxford Dictionary* defines "to deduce" as a synonym of "to infer": each word occurs in the definition of the other. "Induction" is the difficult word.

In the *Concise Oxford Dictionary* "induction" is defined as "to infer a general law from particular instances"—a definition that has changed scarcely at all during the three centuries since it was expounded by Francis Bacon in *The Advancement of Learning* and *Novum Organum*. Bacon grew up in a society in which Scholasticism dominated philosophy: it was generally held that deduction was the only procedure by which man could learn about nature (Russell 1946; Sherrington 1955; Bronowski and Mazlish 1960). Bacon strongly opposed this view and argued that knowledge of nature could be acquired only by experience of nature. He accepted observation, but he argued strongly for experiment. His experiments and writings on heat showed that he was aware of the need to contain his inquiries within the bounds of what we would now call an hypothesis (Russell 1946, 565 et seq.). So Bacon can justly be recognized as the founder of our modern concept that scientific knowledge advances by the empirical verification of hypotheses.

The Royal Society was founded in 1662, and, according to Bronowski and Mazlish (1960, 186), the founding fellows were inspired by Bacon: they were all enthusiastic experimentalists. In their enthusiasm for experiments they overlooked the need for hypotheses. They and their successors relied too heavily on what came to be known as "induction by simple enumeration". Later this procedure attracted the adjectives "Baconian" and "naive" (with derogatory semantic overtones). This was unjust to Bacon because, as Russell (1946, 564) wrote: "Bacon was the first of the long line of scientifically-minded philosophers who have emphasized the importance of induction as opposed to deduction. Like most of his successors, he tried to find a better kind of induction than what is called 'induction by simple enumeration'."

Russell went on to define "induction by simple enumeration" in what he called a parable:

There was a census officer who had to record the names of all householders in a Welsh village. The first that he questioned was called William Williams; so were the second, third, fourth. . . . At last he said to himself: "This is tedious. They are all called William Williams. I shall put them down so and take a holiday." But he was wrong; there was one whose name was John Jones. This shows that we may go astray if we trust too implicitly to induction by simple enumeration. (Russell 1946, 565).

The census officer was wrong to assume he had arrived at a generalization that was absolutely true. But a critic would also be wrong to assume that the census officer had learned nothing from his observations in the Welsh village. It is just this sort of uncertain knowledge (with its uncertainty rather precisely understood) that supports all the useful arts—agriculture, medicine, engineering, and so on. The theories that make these arts possible have been inspired by large numbers of particular experiences. The procedure is beyond doubt properly called induction. This lesson has eluded philosophers for more than three hundred years, perhaps because they have been seeking, with great singleness of purpose, for the way to make absolutely true empirical generalizations. This failure seems to be the chief of Popper's reasons for repudiating induction. He wrote: "Nothing resembling inductive logic appears in the procedure here outlined. I never assume that we can argue from the truth of singular statements to the truth of theories. I never assume that by the truth of 'verified' conclusions, theories can be established as 'true', or even as merely probable" (Popper 1959, 33). And he argued that "the procedure of [empirically] testing [hypotheses] turns out to be deductive." The argument seems to rest on a euphemism which implies that an hypothesis is a "singular statement" that can be verified or denied absolutely in a once-only experiment or observation. "Next we seek a decision as regards these (and other) derived statements [predictions from theory] by comparing them with the results of practical applications and experiments. If this decision is positive, that is, if the singular conclusions [predictions from theory] turn out to be acceptable, or *verified*, then the theory has, for the time being, passed its test; we have found no reason to reject it" (Popper 1959, 33).

This argument seems to rest on two misconceptions. (a) Even if it is proper to call the null hypothesis that is deduced from a theory a "singular conclusion" the procedure of holding the theory on trial while it is repeatedly and indefinitely challenged by successive "singular conclusions" seems to be indistinguishable from the procedure of induction by simple enumeration (see above) and equally vulnerable to the criticism of "infinite regress" which is said to mar induction (Popper 1959, 30). So long as the sample is less than the whole population there is a risk that new information may contradict what has already been discovered, as in Russell's account of a census in a Welsh village. (b) The results that come from experimentally testing a null hypothesis (or "singular conclusion") rarely if ever come out as an all or nothing yes or no. They come in the form of a statistical "population" of individual readings that have a statistical distribution from which it is usually possible to calculate a mean and a variance. If the sample is small or the distribution wide, the conclusion may be less reliable than when the sample is large or the distribution narrow. This procedure is also likely to degenerate into an infinite regress unless it is rescued by an appeal to the mathematics of probability.

While philosophers chased the will-o'-the-wisp of an absolutely true empirical generalization, scientists and mathematicians were exploring the possibilities of improving induction with the help of the mathematics of probability. According to Fisher (1958a, 1959, 26), the concept of precisely estimated probability found its way into scientific method from the art of gambling. During the seventeenth and eighteenth centuries the recreation of gambling was highly regarded by men of social prestige and power. They engaged the interest of some of the greatest mathematicians in the land to unravel the knotty problems that arose in gambling.

Much progress has been made with the application of the mathematics of probability to biological research during the twentieth century. Sir Ronald Fisher contributed prominently to this development. Fisher's (1959) pamphlet contains a critical but rather esoteric discussion of the conditions for scientific induction. We have attempted a brief paraphrase, in popular language, of Fisher's three conditions for scientific induction.

Scientific induction depends on the correct use of the mathematics of probability, which requires three conditions. (a) It is agreed in advance that the goal is a statement that is qualified by a precise probability—we are not seeking an absolutely true generalization (Fisher's "no subset can be recognized"—the "postulate of ignorance"). This condition marches with common sense and realism: scientists have long been in the habit of trusting their conclusions more confidently when their experience is deep. (b) The empirical facts that might be recorded belong to a class that is bounded by a prior hypothesis. (Fisher's "there is a reference set"). (c) The data must be homogeneously (sec. 10.11) and critically (sec. 8.3) ascertained

(Fisher's "the subject belongs to the set"). In other words, the generalizations of scientific induction may be stated with a precise probability attached to them, provided only that the hypothesis has been strictly stated before the experiment begins and that the data have been homogeneously and critically ascertained. With these improvements in the procedure of induction it is clear that the unattainability of certainty in an empirical generalization matters scarcely at all to science and scientific method. (No matter how deeply it may matter to philosophers). The current popularity in population ecology of models so abstract that it is difficult to see how a testable hypothesis might be deduced from them (secs. 9.01, 9.4, 10.331) is a more serious matter. How much weight and what epistemological status do the authors and supporters of such models attach to the conclusions that might be drawn from them? Does the popularity of such models mark the first step backward toward Scholasticism?

To return to the question that we started with, What is the difference between the "predictions" that stem from models and hypotheses and the "conclusions" that rest on experiments and other forms of practical experience? They are all statements about nature and therefore uncertain. The difference between them lies in the form of their logic and therefore in the source of their uncertainty. The "conclusion" is uncertain because it depends on induction. The "prediction" is uncertain because it has been deduced from a premise which depends either on overt induction or on inspiration. If the premise can be tested directly, by experiment or observation, it will assume the status of an hypothesis. It follows that the function of the deductive processes that are used to make a model or an hypothesis is to rearrange the premise into a form that is testable. If the test is objective, by experiment or observation, the result must be counted as a contribution to scientific knowledge. If the test is subjective, with the prediction being measured for its plausibility in relation to acceptable ideas about causal relations in its field, the result must be counted as an explanation, possibly a contribution to theory. Of course these two procedures are not mutually exclusive: one sort of test does not preclude the possibility or the desirability of the other (secs. 9.01, 9.03).

## 9.03   The Distinction between Scientific Knowledge and Scientific Theory

There is a useful convention that reminds us of one essential difference between scientific knowledge and scientific theory: scientific knowledge is always reported in the past tense; scientific theory is always described in the present or future tense. Scientific knowledge has to be reported in the past tense because scientific induction depends on particular events (or observations, etc.) that have happened in the past. Despite the work of Professor J. B. Rhine and his school, no scientist would claim to have empirical knowledge of the future. Another essential difference arises because scientific induction has nothing to do with the idea of cause. If event A is shown to have occurred after events B and C and B is shown to be associated with both A and C, then either B or C (or some other, as yet unknown, event D) might, with equal logic, have been the cause of A. The dilemma is that scientific knowledge of the past remains fruitless and barren unless it can be used to predict the future. Action that might be taken to influence the future manifestations of A is likely to be fruitful only if the cause of A is correctly understood. The decision (say) to prefer B over C or to neglect the possibility of D is subjective. The theories that make possible all the useful arts and crafts (agriculture, medicine, engineering, etc.) depend on wide and deep causal laws that are based, in the first instance, on scientific knowledge; but in the final analysis they consist of a subjective consensus that comes from pooling the wisdom and experience of many scientists and craftsmen. The development of one such theory—the theory of infection, which gradually superseded the theory of spontaneous generation—has been much discussed by historians of science; we refer to some of these discussions in section 10.1.

A simple imaginary example might help to clarify the point in an introductory way. In an experiment with an insect pest, one batch of caterpillars is reared in a warm incubator with a red door. An equivalent batch is reared in a cool incubator behind a blue door. The first lot grows more quickly than the second lot. There would be widespread agreement that the faster growth is caused by the higher temperature. This explanation might be stated as part of a quite general theory which, if true, might be widely useful. For example, a farmer living in the Southern Hemisphere might be well served by a prediction that certain pests (caterpillars, grubs, scale insects) will grow more quickly on the northern side of a tree than those on the southern side. Nevertheless, an inexperienced and ignorant observer might argue that it was the red door that caused the experimental caterpillars to grow more quickly. Because this alternative explanation can be tested by a practicable experiment, it must be accepted as rational scientific criticism. Yet no modern biologist would hesitate to accept the first explanation without bothering to test the alternative.

The general theory is too securely entrenched to be seriously challenged by such an *implausible* alternative. But the implausibility is purely a matter of subjective judgment: there is no way to attach a precise probability to an *explanation* as can be done with the *knowledge* that is gained by scientific induction.

Section 9.1 introduces a general theory that seems to offer a plausible explanation of the distribution and abundance of animals, as it is revealed by studying the densities of populations as they occur in nature.

Our theory rests on three ideas. (a) A rational analysis of "environment" which is defined colloquially in chapter 1 and formally in the Appendix; it is illustrated by the envirograms of chapter 2. (b) A realistic concept of "population"; the multipartite quality of a natural population is verified in chapter 8 and illustrated in sections 9.1 and 9.2, by figures 9.02 to 9.06. (c) The concept of luck as it relates to the multipartite population; den Boer named this concept "spreading of risk" (sec. 9.3).

## 9.1 The Local Population as It Appears Empirically

In section 8.41 we use the concept of "local population" to explain the ubiquitous patchiness that characterizes natural populations. For den Boer, working with carabid beetles (sec. 8.42), it was realistic to treat the catch from each set of pitfall traps as an "interaction group" because, for the beetles, walking was the preferred mode of travel while going about the daily round of living in a locality. Dispersal from the locality was a special behavior that the beetles usually assumed only once a year, just before the breeding season. Likewise, Ehrlich (sec. 8.413) was able to recognize distinct local populations in the butterfly *Euphydryas editha*, because the butterflies themselves were finical about their observance of a boundary which Ehrlich could not explain. But for the majority of species the boundaries are less well defined. It may not be practicable to sample local populations as such. Even so, the idea of the local population remains helpful in interpreting the results and theorizing about the natural population. We begin theorizing by considering a graphic model of a hypothetical (local) population from which time and space have been abstracted (sec. 9.11). We take one step toward reality when we restore time to the model (sec. 9.12). When time and space are both restored, the model approaches the natural population which is real.

The models in figures 9.02 to 9.04 arise from the theory of environment. The models in figures 9.05 and 9.06 arise from the theory of environment, to which has been added the idea of spreading the risk. They are expected to give rise to further experiment and observation that might deny or confirm or, more likely, modify the theory.

## 9.11   Model of a Local Population: With Time and Space Abstracted

In figure 9.02 the horizontal movement of the bouncing ball models the rate of increase in the density of a local population. In the diagram immigration is not distinguished from births or emigration from deaths (sec. 8.41). The risk of extinction is prominently illustrated, with two distinct routes to extinction (Milne 1957, 253). The influence of density is modeled by the slopes. In the middle, where the platform is a horizontal plane, density has a negligible influence on the horizontal passage of the ball. The slope on the right models the way that overcrowding may reduce the rate of increase; the slope on the left models the way that underpopulation may reduce the rate of increase. Milne (1957, 264) and Reynoldson (1957, 172) pointed out, and Nicholson (1957, 172) agreed, that the harmful causes of underpopulation were unlikely to be ameliorated by the absence of competition in a sparse population. This fact is modeled in figure 9.02 by having the two slopes separated by a horizontal plane.

To read the diagram, think of it not as a graph, but as a blueprint for a machine. A small sphere bounces up and down on a platform that is horizontal in its central part but curved at both ends. The sphere approaches the platform vertically. In the central part the sphere also leaves the platform vertically. But at either end the initial direction of the bounce is deflected from the vertical by the slope of the platform. The sphere is unlikely to land on the same place in successive bounces because, throughout the length of the platform, the sphere is influenced by horizontal forces that tend to push it toward the left or the right. The direction and magnitude of the forces that operate during a bounce are independent of the position of the sphere relative to the length of the platform. The forces for any bounce are drawn at random, relative to the position of the sphere along the platform, from a population of forces of known or knowable distribution. That is, any hypothesis that might be made about them is testable (Popper 1959, 40).

Each bounce of the sphere occupies a unit of time that is appropriate to the life cycle of the animal—say a generation. The place where the sphere strikes the platform indicates the density of the population, with high densities to the right. The slope of the platform indicates the influence of density on the rate of increase, $r$. Density is measured as number of animals per unit of area. In the sloping parts of the platform shortage of space is not as a rule the direct cause of the slope. It

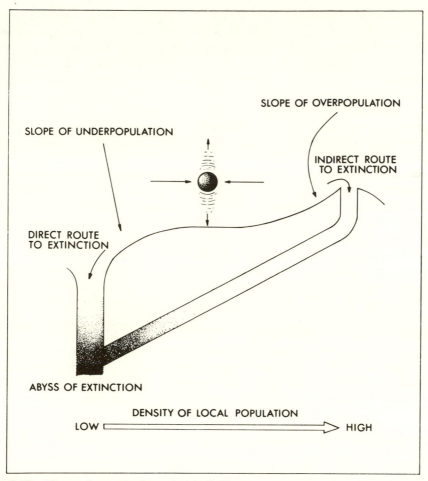

**Fig. 9.02**   The analogy of a bouncing ball illustrates two ways a local population may become extinct. The ball bounces vertically off the horizontal part of the floor, but at the ends the slopes impart a leftward bias to the bouncing. Each bounce marks one unit of time, or sometimes a generation. Between bounces, forces that are independent of density push the ball to the right or the left. The position of the ball along the horizontal axis indicates the density of the population. The population becomes extinct when the ball comes opposite the zero point and falls into the abyss of extinction. This is the direct route to extinction. The risk of extinction via the direct route is increased when the ball bounces on the slope of underpopulation (positive feedback to density). The risk of extinction via the indirect route is reduced when the ball bounces on the slope of overpopulation. Nevertheless, a strong run of right-pointing arrows from the density-independent "forces" might push the ball up and over the top. See comments on autocorrelation in section 9.32.

is more likely to be some component of environment such as food, mates, or predators that is related to area indirectly through the density of the population of the primary animal.

Extinctions may occur through either underpopulation or overcrowding (Milne 1957). The paradox of extinction through overcrowding occurs only in certain circumstances; it is likely only for animals with complex life cycles and generations that do not overlap or overlap only slightly. Certain insects come readily to mind—for example, locusts, armyworms, blowflies. In such life cycles, eggs or some other inactive stage in the life cycle may be accumulated quite innocuously in large numbers, beyond the capacity of the locality to produce food for them, once they have entered the growth stage of the life cycle. In the ensuing absolute shortage local populations of both predator and prey may be extinguished. During the initial eruptions of a newly introduced "successful agent of biological control" (fig. 3.01, *Asolcus*, *Cactoblastis*), many localities are depopulated in this way.

There are a few unusual animals for which the degree of proximity of other individuals of the same kind is critical, and this is measured as numbers per unit area or volume. Certain species of aphids produce apterous offspring when they are sparse and winged offspring when they are crowded (Lees 1967). Certain Acrididae grow longer wings and assume a gregarious, migratory behaviour when they are reared in a crowd (Faure 1932). The aphids respond to physical contact with other individuals of the same kind, perhaps through the cornicles; the locusts probably respond to visual and physical contacts. It has been suggested that certain small mammals also belong in this class (Chitty 1967), but that now seems unlikely (Andrewartha 1970, 75). In such species, if extinction of the local population comes through crowding, it seems to be due to an excessive emigration.

Extinction through underpopulation is less often recorded because it tends to be taken for granted. In terms of figure 9.02, extinction is likely to occur after the density-independent forces have by chance pushed the sphere onto the slope of underpopulation. The shortage of mates, vulnerability to predators, and other components that depend on density, but have a positive feedback to density, take over and drive the local population to extinction (Andrewartha and Birch 1954, 334).

### 9.12    Model of a Local Population: With Time Restored

Figures 9.03 and 9.04 represent one way of depicting local populations during part of their history. They depict only two aspects of the local population—its density and the richness of the locality in a particular resource such as food. No attempt is made to incorporate into these models the total environment of the animal in its local population (i.e., the envirogram), though of course both the density of the population and the resources available to it are influenced by the total environment.

The horizontal axes of figures 9.03 and 9.04 indicate time. For the curve, the density of the population is indicated along the vertical axis. For the symbols, the density of the population is indicated by the areas of the darkened part of the symbol. The total area of the symbol indicates the richness of the locality in a particular resource, say food or perhaps some other favorable component. Thus the proportion of the circle that is shaded indicates the density of the population relative to a resource such as food.

In figure 9.03 the first row of symbols indicates the condition of a local population that is likely to experience an extrinsic relative shortage of food (see the bottom left corner of fig. 3.01). The supply of food increases and then decreases, which is indicated by the area of the symbols increasing and then decreasing. But the supply of food does not respond, to any important degree, to the feeding of the primary animal. It increases and decreases under the influence of some other

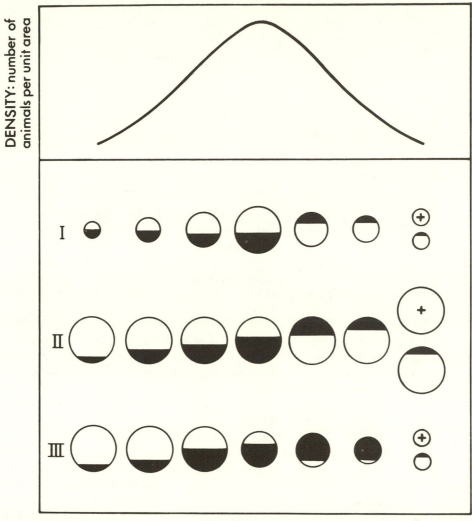

**TIME**

**Fig. 9.03** Model of typical cycles of growth of a local population in a locality. A cycle may proceed from colonization through expansion, maturity, decay, and perhaps extinction. The curve traces the rise and fall in density with time. Density is expressed as number of animals per unit area. The rows of symbols repeat the information from the curve for three distinct conditions in the local population. The area of symbol indicates the "carrying capacity" of the locality (say, the supply of food). The area of the shaded part of the symbol indicates the density of the local population. While the population is waxing the shaded area occupies the lower part of the circle; when the population is waning the shaded area occupies the upper part of the circle. The cycle might begin with: ⊕, a spent locality that is ready for recolonization; ○, a virgin locality that has not yet supported a population; ◐ A remnant from an old population that is now ready to enter another cycle of growth. The three rows of symbols represent for a local population: I, the condition of extrinsic shortage; II, The condition of no shortage at all; III, the condition of intrinsic shortage (table 3.02).

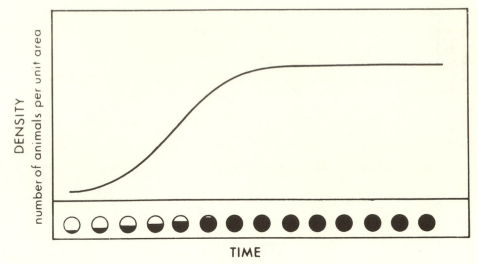

**Fig. 9.04**  Model of the colonization, growth, and persistence of a local population in a place where the numbers are chiefly influenced by an indestructible (i.e., reusable) component of environment. The curve and the symbols have the same meanings as in figure 9.03.

component in its own environment. A typical example is the influence of weather on the food of an herbivore. As weather ameliorates in the spring, the vegetation grows. With the approach of summer the vegetation dries out and is no longer highly nutritive as food for the herbivore (sec. 10.311).

In row I of figure 9.03 the resource, such as food, waxes and then wanes under the influence of weather—let us say as happens to the food of the wallaby *Setonix* (sec. 3.131). The density of the population of the primary animal is likely to increase as the quality of the food improves and to decrease as it deteriorates during the dry summer. Increasing the amount of low quality food in the summer does not alleviate the relative shortage of food in the locality for the primary animal. It needs a more concentrated food if that is to happen. (But of course the number of animals in the natural population is likely to increase if the amount of low quality food in the whole area happens to be increased by enlarging or increasing the number of localities; fig. 9.05).

The second row of symbols represents the condition of a local population that is not likely to experience any shortage of the resource at all (top left corner of fig. 3.01). The resource remains abundant and accessible, but the population waxes and wanes (area of darkened part of symbol) under the influence of some other component in the environment, perhaps a predator.

The third row of symbols represents the condition of a local population that is likely to create an intrinsic shortage of food for itself.

In all three rows the shaded area is at the bottom of the circle for a waxing population and at the top for a waning one. The age structure of the population is likely to differ in the two stages. The cycle might have one of three endings. The local population may be extinguished and the locality may remain in or return to a usable condition (circle with cross in it). The local population may be extinguished and the place where it had lived may remain unusable (not shown in the diagram). Or the local population may not be extinguished; after declining to a small remnant, it might start to grow again (circle with small remnant of black).

In figure 9.04 the curve is drawn to an asymptote; the circles remain the same size, and they remain fully shaded. This form of the model represents the sort of "stability" that may occur when the critical component of environment is "indestructible" in the sense that it is re-usable; for example, nesting sites such as holes in trees for certain sorts of birds.

## 9.2 Model of a Natural Population: With Time and Space Included

Figures 9.05 and 9.06 are best read as maps of the dispersion of a natural population. The dispersion of the localities is represented by the position of the symbols. The quantity of resource such as food is represented by the total area of the symbol, and the area of dark shading in it represents the density of animals, as in figures 9.03 and 9.04.

The purpose of these diagrams is not to portray how the total environment influences the density of animals. For that we need to consult the envirogram of the animal. Here we focus on the multipartite nature of the natural population. Local populations are more or less separated from each other. Likewise, the density of the population in each locality is more or less independent of the densities of the other local populations, as is the amount of resource available to the local population. We merely catch the condition of the natural population at one instant of time, as in a snapshot. In subsequent snapshots the picture would be different. These diagrams

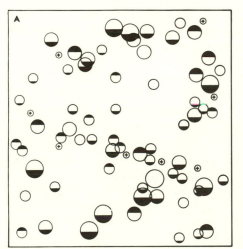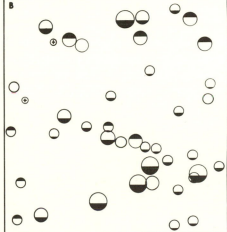

**Fig. 9.05** Model of extrinsic shortage of resource in both the local and the natural populations. Each diagram is a model (think of it as a map) of a natural population, which consists of local populations, the symbols for which have been taken from row I in figure 9.03. Localities that are suitable but unpopulated are shown as empty circles. The presence of symbols that have shrunk in size without any increase in the proportion of the symbol that is shaded indicates an extrinsic shortage of resource in both the local and the natural populations. The extrinsic relative shortage at the level of the natural population is greater in B than in A. For further explanation see the text.

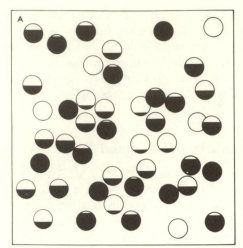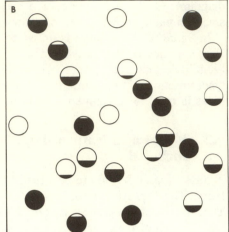

**Fig. 9.06**   Model of extrinsic shortage of an indestructible (reusable) component of environment in both the local and the natural populations. Each diagram is a model of a natural population that consists of local populations, the symbols for which are taken from figure 9.04, indicating an extrinsic shortage of a component of environment (say nesting places), not necessarily a resource. The population is denser in A than in B simply because there is a better supply of the indestructible component in A than in B. For further explanation see the text.

simply reinforce the verbal argument for the multipartite population (chap. 8) with a graphic illustration. Although they are abstractions from nature, they nevertheless illustrate one aspect of the natural population—its multipartite nature. This concept is fundamental to the theory of spreading the risk, which is discussed in section 9.3. A large number of different sorts of diagrams might be constructed along these lines. We have merely drawn two. In figure 9.05 the symbols are taken from row I in figure 9.03; in figure 9.06 the symbols are taken from figure 9.04.

Whereas in figures 9.02 to 9.04 the emphasis is on the interactions (between animal and environment) that are important to the animal during the nondispersive phase of the life cycle, in figures 9.05 and 9.06 the emphasis is on the risks that are associated with dispersing. In the nondispersive phase the emphasis is on surviving, growing, and breeding. In the dispersive phase the only important activity is to search for and find a new place to live. In figures 9.05 and 9.06 the contrast between A and B shows how the risk of not finding a new place to live may increase as the localities become more sparsely dispersed. The chance that a disperser will find a new place to live also depends on the dispersiveness of the species. This quality is important in the ecologies of agents (or prospective agents) of biological control (Andrewartha and Birch 1954, 474).

In figures 9.05 and 9.06 each symbol, including vacant localities, is about equally represented. The symbols are randomly placed with respect to each other and with respect to the coordinates of the "map". In nature the different sorts of local populations are likely to be dispersed more patchily than a random distribution (i.e., underdispersed) for various reasons, as might be imagined after reading section 8.4 and following sections. Some local populations are shown overlapping.

Although this is an outcome of their random distribution in the figure, it also corresponds to reality. Members of a local population may wander into the area of another local population to find a mate or may be influenced by some other component of environment there. In den Boer's terminology, local interaction groups overlap (sec. 8.42).

Figure 9.05 represents an extrinsic shortage of a destructible resource in both the local and the natural populations. It is built with the symbols from the top row of figure 9.03. The supply of the resource in each locality is likely to vary through causes that are not related to the density of the local population (see cell 1, table 3.02, and lower left corner of figure 3.01 for definition of extrinsic shortage). In addition, individuals that disperse away from a locality run the risk that they will not find a good place in which to live. This risk is also extrinsically determined, because the distribution and abundance and fruitfulness of localities does not depend on the density of the population. Whatever is causing the distribution, abundance and fruitfulness of localities, the extrinsic relative shortage at the level of the natural population is more severe in B than in A. A might represent a natural population of the wallaby *Setonix* (sec. 3.131) in a habitat that has many favorable localities. B might represent the natural population toward the edge of its distribution where favorable localities are few and far between.

Figure 9.06 represents extrinsic shortage of an indestructible component of environment in both the local and the natural populations. It is made with symbols from figure 9.04 indicating an extrinsic shortage of an indestructible component of environment such as nesting sites in the local population. Even with built-in security in the supply of resources in the locality, there remains, at the level of the natural population, the risk that a disperser looking for a new place in which to live may not find an adequate one. This risk is greater in B than in A because the resource is less abundant in B. The natural population in A might represent the great tit in broadleaf forests in Europe, while B might represent the same bird in a pine forest.

The use of "demographs" like those in figures 9.05 and 9.06 to visualize the concepts of "multipartite population" and "spreading the risk" was foreshadowed in Andrewartha and Birch (1954, chap. 14). den Boer (1968) gave the theory its name. In Andrewartha and Birch (1954, 660) we wrote:

The numbers in a natural population may be limited in three ways: (a) by shortage of material resources, such as food, places in which to make nests, etc.; (b) by inaccessibility of these material resources relative to the animals' capacities for dispersal and searching; and (c) by shortage of time when the rate of increase r is positive. Of these three ways, the first is probably the least, and the last is probably the most, important in nature. Concerning c, the fluctuations in the value of r may be caused by weather, predators, or any other component of environment which influences the rate of increase. For example, fluctuations in the value of r which are determined by weather may be rhythmical in response to the progression of the seasons (e.g., *Thrips imaginis*) or more erratic in response to "runs" of years with "good" or "bad" weather (e.g., *Austroicetes cruciata*). The fluctuations in r which are determined by the activities of predators must be considered in relation to the populations in local situations. How long each newly founded colony may be allowed to multiply free from predators may depend on the dispersive powers of the predators relative to those of their prey.

In the light of the meaning that has been given to "natural population" in this book, the distinction between (a) and (b) disappears. Andrewartha (1970, 156) condensed the statement above: "I conclude that the numbers of animals in natural populations may be explained by either (i) a shortage of time or (ii) a relative shortage of a resource." "Resource" is now defined more critically than in 1970 or in 1954. Insofar as a nesting site is a token that stimulates the animal to come into breeding condition, it is a resource (sec. 3.31); but insofar as the nesting site provides shelter from weather or refuge from predators, it is a modifier (sec. 2.33). There may be other material modifiers that are important. Shortages of such modifiers should be taken into account along with shortages of time and resources.

## 9.3   Spreading the Risk: An Outline of the Theory

The theory of spreading the risk recognizes:

1. That the natural population (den Boer's "population as a whole"; sec. 8.41) is the unit with which ecological theory must ultimately come to grips.
2. That the natural population is a multipartite population as we use this term in chapter 8.
3. That the smallest group into which the natural population can be realistically subdivided is the local population (den Boer's "interaction group") as it is defined in section 8.41.
4. That not only environments but also populations in localities characteristically vary in both space and time, for the reasons given in sections 8.411 to 8.415.
5. That particular local populations often do not thrive for very long; nevertheless, owing to the recuperative power of the locality and the dispersiveness of the animal (sec. 8.42), new colonies are being founded or old ones revived, on the average, fast enough to replace the losses and thus to keep the natural populations in being for very long periods of time. These periods are comparable with the periods that naturalists and paleontologists guess at for the duration of life for a natural population that is the whole species.

The theory of spreading the risk argues:

1. That the longevity of the natural population is merely an expression of the low probability that all the local populations will be synchronized to fluctuate contemporaneously to extinction either down the slope of underpopulation or up the slope of overcrowding in figure 9.02.
2. That "stability" in a natural population as measured by the range between the maximum and minimum densities that are likely to recur, or by the lack of slope in the linear regression of density on time, is another measure of the same probability as in (1).

The theory of spreading the risk explains this low probability in terms of the variability and asynchrony in changes in the rate of increase and hence the density of the local populations. Because rates of increase in the local populations vary asynchronously, they tend to cancel out in estimates of the variability of the natural population. The explanation depends upon the truth of the prediction that the variance of the natural population (taken as a unit) is less than the mean variance of the local populations. In symbols this proposition is represented by the inequality

$$\sigma^2 (\Sigma N) < \frac{\Sigma\sigma^2 N}{k},$$

where N is the density of a local population and there are k local populations.

## 9.31 Empirical Support for the Theory

The prediction above was tested by den Boer (pers. comm.[1]) for two carabid beetles, *Pterostichus versicolor* and *Calathus melanocephalus*, that were living in 1,200 ha of nature preserve in Drenthe, The Netherlands. There were eleven distinct trapping-sites; sampling was continued for nineteen years and nine of the sites were maintained continuously for twelve to seventeen years. At each site a set of three traps was left continuously in the same place. For *Pterostichus* the traps were considered to have sampled a circular area with a radius of 150 m; for *Calathus*, 50 m. According to Baars (1979), the traps gave a reliable estimate of the relative density (sec. 8.3) of the local population. The traps were serviced once a week. The total catch for the year extending from one breeding season to the next was taken as the relative density of the "interaction group" (local population) for the particular generation. The sum of the annual catches for all the trapping sites was taken as the relative density of the "population as a whole" (natural population). In this instance density was being expressed as number of beetles per unit area of ground (sec. 8.1).

Two criteria were used to compare the variability of the natural population with the mean variability of the local population.
1. The logarithmic range between the largest and the smallest density (LR).

$$LR = \ln N_{max} - \ln N_{min}$$

For the localities the mean LR was calculated by summing LR over all the localities and dividing by k, the number of localities.
2. The variance of the net rate of increase (LV). The net rate of increase (R) was measured by the quotient of the densities of the population for successive years (generations).

$$R = N_{t+1}/N_t$$

The variance of the net rate of increase (LV) was calculated in the usual way:

$$LV = s^2(R) = \frac{\Sigma(R - \bar{R})^2}{m - 1}$$

For the localities the mean variance of R was calculated by summing $(m_i - 1)s^2R$ over all the localities and dividing by $\Sigma_1^k(m_i-1)$; $m_i$ is the number of years in locality i. For the natural population; m = 19, k = 9–11. For the local populations: m = 12–17, k = 9.

---

1. We are deeply indebted to Dr. P. J. den Boer for allowing us to work from his unpublished manuscript and letters.

The hypothesis (sec. 9.3) predicted: (a) that the logarithmic range, LR, for the natural population would be less than the mean of the logarithmic ranges for the local populations and (b) that the variance of the net rate of increase of the natural population be less than the mean of the variances of the net rates of increase of the local populations. If we write N' for the density of the natural population, R' for the net rate of increase of the natural population, N" for the density of a local population, and R" for the net rate of increase for a local population, the hypothesis takes the form

$$LR(N') < LR(N'') \text{ and}$$

$$s^2(R') < s^2(R'').$$

The results are summarized in table 9.01.

The two species differ. The density of the population of *Calathus* in both natural and local populations is more variable than that of *Pterostichus* whether judged by LR or LV. Moreover, the reduction in LV from local to natural population is 65% for *Calathus*, whereas it is 85% for *Pterostichus*. By any of these criteria the risk is being spread more effectively for *Pterostichus* than for *Calathus*. One explanation for this difference might be that in the environment of *Calathus* the important components tend to fluctuate in unison, influencing all the localities in the same direction at the same time. Whereas in the environment of *Pterostichus* the important components tend to be out of step; there is less chance that either a bonanza or a disaster will strike all the localities at once. The natural population is relatively invariable because it is buffered by the variability of the local populations. If such buffering can be demonstrated for a relatively restricted population on 1,200 ha, it would be reasonable to expect more effective buffering over a wider, more diverse area—as indeed is characteristic of most natural populations. This is an important idea because the theory of spreading the risk accepts the buffered invariability of the natural population, which is a consequence of the variability between local populations, as sufficient explanation for the acknowledged longevity of natural populations. It is argued that to increase the variability between local populations

**Table 9.01**   Variability from Generation to Generation in the Numbers of the Carabids *Calathus melanocephalus* and *Pterostichus versicolor* Living on Kralo Heath

| Species | Logarithmic Range (LR) | | Variance of Net Rate of Increase (LV) | |
|---|---|---|---|---|
| | Natural Population | Local Population (mean of nine interaction groups) | Natural Population (variance of R') | Local Populations (mean of the variance of R" for nine interaction groups) |
| *C. melanocephalus* | 4.4379 | 4.8777 | 1.0507 | 3.2712 |
| *P. versicolor* | 1.0417 | 2.2559 | 0.0656 | 0.4252 |

*Source:* After den Boer (1981).

*Note:* Variability was estimated for the population as a whole (the natural population) and for nine of the interaction groups (local populations) that constituted the whole population. Variablity is measured by two criteria, logarithmic range and variance of net rate of increase per generation. See text for further explanation.

and thereby reduce the variability of the natural populations is to reduce the risk of extinction.

## 9.32   A Model That Elaborates the Theory

From the empirical base that he had established (sec. 9.31), den Boer (1981 and pers. comm.) used a simulation model to explore the idea that was developed in section 9.31. The model predicted the number of years (generations) that would elapse before the density of the population declined to a number less than one—that is, became extinct. The primary comparison was between the variance of the net rate of increase for the population as a whole and that for its constituent sub-populations. We call them "subpopulations" because, for the computer, den Boer did not retain the identity of the individual interaction groups. Instead, he consol-idated all 198 individual estimates of R″ for all the interaction groups for all the years into a single distribution from which he sampled at random, subject only to the restraints that are indicated under $\beta$ and $\rho$ in table 9.02. Similarly with the nineteen estimates of R′.

In rows 1 and 2 of table 9.02 $\sigma^2$ takes the value of $s^2$ as estimated for natural *Calathus* and *Pterostichus*. Because comparisons between columns 9 and 11 and 10 and 12 in rows 1 and 2 showed that populations of *Pterostichus* survived longer

**Table 9.02**   Computer Simulations of the Longevity of Populations of Two Carabid Bee-tles, *Calathus melanocephalus* and *Pterostichus versicolor*

| Input into the Computer | | | | | | | | Number of Years before N < 1 | | | |
|---|---|---|---|---|---|---|---|---|---|---|---|
| *C. melanocephalus* ($\sigma^2$) | | | | *P. versicolor* ($\sigma^2$) | | | | *C. melan-ocephalus* N, Deter-mined by Variance of | | *P. versicolor* N, Deter-mined by Variance of | |
| R′ | R″ | $\beta$ | $\rho$ | R′ | R″ | $\beta$ | $\rho$ | R′ | R″ | R′ | R″ |
| (1) | (2) | (3) | (4) | (5) | (6) | (7) | (8) | (9) | (10) | (11) | (12) |
| 1. 1.0507 | 3.0420 | E | 0 | 0.0656 | 0.4357 | E | 0 | 265 | 298 | 5,858 | 672 |
| 2. 1.0507 | 3.0420 | E | > 0 | 0.0656 | 0.4357 | E | > 0 | 170 | 162 | 10,127 | 295 |
| 3. 0.0656 | 0.4357 | E | 0 | — | — | — | — | 536 | — | — | — |
| 4. 0.0656 | 0.4357 | E | > 0 | — | — | — | — | 477 | — | — | — |
| 5. — | — | — | — | 1.0507 | 3.0420 | E | 0 | — | — | 894 | — |
| 6. — | — | — | — | 1.0507 | 3.0420 | E | > 0 | — | — | 415 | — |

*Source:* After den Boer (pers. comm.)

*Note:*

R′, net rate of reproduction for the population as a whole.

R″, mean net rate of reproduction for the nine interaction groups that constituted the population as a whole.

$\sigma^2$, variance of R′ or R″. (See table 9.01.)

$\rho$, The autocorrelation between successive values of R′ and R″. Two artificial values were used, 0 and >0.

$\beta$, the trend with time in the density of the population as a whole. E indicates that empirical values were used, *Calathus* −0.0807, *Pterostichus* +0.0142. For further information see text.

than those of *Calathus*, and because $\sigma^2$ was less and $\beta$ was greater for *Pterostichus* than for *Calathus*, den Boer investigated the relative importance of $\sigma^2$ and $\beta$ as follows. He put into rows 3 and 4 an imaginary *Calathus* which had all its own natural attributes except for $\sigma^2$, which was borrowed from *Pterostichus*, and into rows 5 and 6 he put an imaginary *Pterostichus* which had all its own characteristics except for $\sigma^2$ which was borrowed from *Calathus*.

Secondary comparisons included variation of the autocorrelation between successive generations (i.e., the tendency for successive values of R' or R" to be like each other) and variation in the regression of density on time. Autocorrelation, $\rho$, in the distribution of R' or R" was not calculated empirically but artificial values were obtained by taking what den Boer called "short runs" and "long runs". "Short runs" means simply sampling at random in these distributions; consequently the expectation for $\rho$ is zero. "Long runs" means choosing a number (in excess of random expectation) of consecutive values of R' or R" that were either positive or negative (ln1 $= 0$). den Boer thought that the empirical values for $\rho$ would probably lie between those he got with short and long runs. Thus in table 9.02 rows 1, 3, and 5 represent "short runs" and rows 2, 4, and 6 represent "long runs". The trend of N with time was measured as:

$$\beta = \frac{1}{m-1}(1nN_{(m)} - 1nN_{(o)})$$

The values of $\beta$ on nineteen years' results were $-0.0807$ for *Calathus* and $+0.0142$ for *Pterostichus*.

Throughout the exercise the mean net rate of increase was kept at the empirical values that had been estimated for it. So was the initial density, $N_{(o)}$. The density was estimated as a relative density. The absolute density was not measured precisely, but den Boer thought that *Calathus* might be at least ten times more numerous than *Pterostichus* and that the whole population of *Calathus* might be about $2 \times 10^5$.

The longevities predicted by the model are shown in the last four columns of table 9.02. Comparing the longevities of the population as a whole for both species, the predicted longevity for *Pterostichus* (variance of R' is low) is about nineteen times that for *Calathus* where the variance for R' is high. Comparing the longevity of the population as a whole with the "subpopulation", the difference is great, about nine-fold, for *Pterostichus* where the variance of R" is 6.6 times the variance of R' (table 9.01), whereas the difference is small for *Calathus* where the variance of R" is only three times that of R'. Clearly the variance of R seems to be important, but some of the differences in the longevities of the population as a whole may be associated with $\beta$.

The predicted longevity of the imaginary *Calathus* to which had been attributed the value of $\sigma^2$ that was appropriate to the natural *Pterostichus* was increased relative to that of the natural *Calathus* by a factor of 2.3. The predicted longevity of the imaginary *Pterostichus* to which had been attributed the value of $\sigma^2$ that was appropriate to the natural *Calathus* was decreased relative to that of the natural *Pterostichus* by a factor of 12.2. It seems that both $\sigma^2$ and $\beta$ were influential, but $\sigma^2$ more so than $\beta$.

In some runs that are not mentioned in table 9.02 the initial density, $N_o$, was varied widely while leaving the rest of the input at the empirical levels. To double the survival time of *Calathus* it was necessary to multiply $N_o$ by $10^3$. The survival time for *Pterostichus* was doubled when $N_o$ was multiplied by 16 for much autocorrelation and by 200 for no autocorrelation.

Extinction through overcrowding was abstracted by fixing an arbitrary ceiling and arbitrarily dealing with the next value of N. Several ways were tried without effectively altering the result.

The computer simulations especially emphasize the importance of low variability. It seems that low variability in R' might offset a substantial downward trend in N, for several hundred generations at least. One way for a natural population to achieve low variability in R' is to have local populations scattered through a variable terrain where environment fluctuates in such a way that R'' for particular local populations is often ranging in opposite directions at the same time. But the example of *Calathus* suggests that even a very great abundance of local populations may not add much to the stability of a natural population if the environment remains consistently the same in all the localities through all its fluctuations.

In general, the computer simulations predict that maximum survival time would be promoted by an environment that gave R' minimum variability, minimum downward trend with time, and minimum autocorrelation. The last condition carries the corollary that a randomly fluctuating environment is good for longevity. In other words, an environment in which a density-independent component (such as weather) is prominent is expected to promote a long-lived population better than an environment where a density-dependent component (such as an obligate predator which might generate much autocorrelation) is prominent.

den Boer (1981) reanalyzed these data using more realistic estimates for example, of autocorrelation. The new analysis predicted much shorter survival times, especially for *P. versicolor*, but in general left the preliminary predictions substantially intact. We quote den Boer's summary of this analysis in full (den Boer 1981, 39).

The survival time of small and isolated populations will often be relatively low, by which the survival of species living in such a way will depend on powers of dispersal sufficiently high to result in a rate of population foundings that about compensates the rate of population extinctions. The survival time of composite populations uninterruptedly inhabiting large and heterogeneous areas, highly depends on the extent to which the numbers fluctuate unequally in the different subpopulations. The importance of this spreading of the risk of extinction over differently fluctuating subpopulations is demonstrated by comparing over 19 years the fluctuation patterns of the composite populations of two carabid species, *Pterostichus versicolor* with unequally fluctuating subpopulations, and *Calathus melanocephalus* with subpopulations fluctuating in parallel, both uninterruptedly occupying the same large heath area. The conclusions from the field data are checked by simulating the fluctuation patterns of these populations, and thus directly estimating survival times. It thus appeared that the former species can be expected to survive more than ten times better than the latter (other things staying the same). These simulations could also be used to study the possible influence of various density restricting processes in populations already fluctuating according to some pattern. As could be expected, the survival time of a population, which shows a tendency towards an upward trend in numbers, will be favoured by some kind of density restriction, but the degree to which these restrictions are density-dependent appeared to be immaterial. Density reductions that are about adequate on the average need even not occur at high densities only, if only the chance of occurrence at very low densities is low. The density-level at which a population is generally fluctuating appeared to be less important for survival than the fluctuation pattern itself, except for very low density levels, of course. The different ways in which deterministic and stochastic processes may interact and thus determine the fluctuations of population numbers are discussed. It is concluded that some stochastic processes will operate everywhere and will thus *necessarily* result in density fluctuations; such an omnipresence is much less imperative, however, for density-dependent processes, by which population models should *primarily* be stochastic models. However, if density-dependent processes are added to model populations that are already fluctuating stochastically the effects are taken up into the general, stochastic fluctuation pattern, without altering it fundamentally.

### 9.33   The Theory Is Plausible and Meets Practical Needs

To be plausible, a theory must be consistent with the experience of practical men and with theory that is accepted in related branches of science (sec. 10.1). For population ecology practical experience might come from controlling a pest, harvesting a wild population or conserving a species in a natural community; related branches of science include evolution, ethology, and physiology. Part 3 might be regarded as a test (by such criteria) of our theory. There are summaries and explanations of the ecologies of eight well-known species. Four of them are pests; two occur in wild populations that are harvested for food or for sport; one is included because its ecology illuminates the literature about the conservation of nature. The eighth is *Homo*. In this company man must be classified as a pest that is nevertheless worthy of conservation. It seems that this species may have little chance of being adequately conserved unless he can first be recognized as a pest whose numbers badly need to be controlled—a problem that urgently needs to be tackled within the framework of sound ecological theory.

There is a particular application of the theory of spreading the risk that has to do with certain "facts" that are not so critically ascertained as those so far discussed. They relate to the inferences that might be made about the very long periods that most natural populations seem to persist. A natural population becomes extinct when the last of its local populations is extinguished. A local population may be extinguished in two ways (fig. 9.02). So the estimate that concerns us is the probability that every local population will proceed to extinction concurrently either down the slope of underpopulation or up the slope of overcrowding. Of course, it is not practicable to calculate a precise probability. But to contemplate the diverse sources of variability of environment between localities (chap. 8) and genetic variability between local populations (chap. 11), and the consequent extent of the buffering of the natural population, is to realize that there is little need to look beyond this buffering for an explanation of the longevity of natural populations. The longevity may be less than it seems at first sight because there is no way to tell how many generations ago the present species speciated from its direct line of ancestors. The period to be explained might well be shorter than it seems.

### 9.34   Criticism of the Theory of Environment

Most criticisms of the theory of environment (embracing the concept of the multipartite population and the theory of spreading the risk) take one of two forms. It may be said that these theories cannot or do not explain the "equilibrium" density, which implies an upper and a lower limit to the density of the population. This criticism may be explicit, or it may be implicit in a statement to the effect that only a density-governing mechanism can explain such limits. Or it may be said that the theory explains no more than fluctuations about a mean; it does nothing to explain the level of the mean about which the fluctuations occur, which is nearly but not quite the same thing. Several quotations follow that illustrate these points. Several critics emphasize the upper and lower limits; others emphasize equilibrium density; it is essentially the same criticism.

C. J. Krebs (1972, 351) wrote about the North Sea haddock:

This dilemma, the absence of any detectable relation between stock and re-cruitment, can easily be resolved on the theoretical level because it involves the question of population regulation. Some component of the vital statistics—births, deaths and dispersal—must be related to population density, in order to prevent unlimited population growth. As we saw in Chapter 11, population growth cannot be curtailed unless the net reproduction curve is depressed below 1.0 at high population densities.

Buxton (1955, 486) wrote about an insect:

It is difficult to see what factors may be responsible for regulating populations of *Glossina*. Unfavourable conditions of climate are presumably independent of den-sity and cannot regulate it. Competition for blood between individual tse-tse flies does not, we believe take place. We are left to suppose that the "enemies" (using that word broadly) of the fly or of the puparium must be responsible for the regulation. But it must be admitted that we have no evidence that any particular enemy becomes more numerous or effective at higher densities of the tse-tse population. We have indeed very little knowledge of the causes of mechanisms which prevent an indefinite increase of these insects when external conditions are favourable, but one must suppose that some such factor exists.

Southern (1979, 103) introduced a review of "small mammal populations" with the following statement:

If I had to point to one among the legion of ecologists who have argued that the stationary nature of most populations is the basic fact that demands investigation and explanation, I should choose David Lack. In his many papers and, pre-eminently in two of his books (1954, 1966) he has championed the theory that the numbers of animals are regulated by density-dependent mortality factors, i.e., those whose impact tends to cancel any departure from an equilibrium level up-wards or downwards as the case may be.

On page 105 Southern defined a "regulating factor" as "one that acts in a density-dependent way". Then, having observed that the numbers of two small mammals, bank vole and meadow mouse, which had been estimated twice a year for twenty-seven years fluctuated irregularly without a large or obvious trend with time, Southern wrote:

From the consistent way in which "what goes up comes down" we may conjecture that the system is a classical one with irregular fluctuations around different means for each species and this implies regulation.

J. R. Krebs (1970) wrote:

Analysis of the census data of the Great tit collected in Marley Wood since 1947

shows that clutch-size and hatching success are density-dependent. Mortality outside the breeding season is the key-factor responsible for annual fluctuations, and
may contain a weakly density-dependent effect. The density-dependent effects of
clutch size and hatching success are sufficient in themselves to regulate the population at the observed level. Thus it is not necessary to postulate any regulating
mortality other than that occurring in the breeding season.

In a general review of the biological control of insect pests, Huffaker, Messenger, and De Bach (1971, 19) wrote about the theory of environment:

We reject this hypothesis because: (1) it concerns itself mainly with *changes* in
density, largely leaving out the causes of the magnitude of the mean density; (2)
being concerned mainly with changes in density, the view ignores or denies the fact
of characteristic abundance, and provides no logical explanation of why some
species are always rare, others common and still others abundant, even though each
may respond similarly to changes in the weather, for example; (3) to accomplish
such long-term natural "control" it is presumed that the ceaseless change from
favourability to unfavourability of the environment in terms of species tolerances
can be so delicately balanced (a knife-edged balance) over long periods of time as
to keep a population in being without density-related stresses coming into play to
stop population increase at high densities, and without any lessening of these
density-related stressed coming into play at low densities to reduce the likelihood
of extinction.

To embark on a discussion of equilibrium density or limiting densities or of a
mean of a series of numbers that is not fully determined by the series itself is to risk
a tangle with metaphysics. Fortunately we can make all the points that are necessary
or useful without running that risk.

Sections 9.3 to 9.33 are about how the theory of spreading the risk might help
explain the risk of extinction for a natural population; section 9.11 and figure 9.02
emphasize the inutility of the idea of "competition" in relation to the idea of "lower
limits". There is no need to say more about "lower limits", except in section 9.4,
where we discover that the idea of a "density-governing" mechanism proves singu-
larly misleading about "lower limits".

Whatever meaning might be attributed to "upper limit", it must be expressed as
a density. So it must find a position along the horizontal axis of figure 9.02. There
seems to be no logical or empirical reason why an "upper limit" should not occur
somewhere along the horizontal part of the platform. Perhaps the critics have a
special interest in the slope of overcrowding. We can see no good reason why this
should be so. All that happens in this region is that the particular component of
environment that is under consideration becomes density dependent. There seems
to be no logical or empirical reason why this should exclude the component from
the theory of environment, which merely demands that the whole environment be
considered. It must be remembered that figure 9.02 applies only to a local population. We pick up the discussion again at this point in section 9.4.

The criticism that the theory of environment explains changes in density while
ignoring mean density seems irrational. The arithmetic mean is a statistic that is
implicit in any set of measurements. It becomes explicit when the particular values
are summed and divided by their number. To imply that any more information than
this is needed to define a mean is to invoke the metaphysical. We avoid entangle-
ment in a metaphysical argument by returning to a concrete example—Davidson

and Andrewartha's (1948a,b) study of a population of *Thrips imaginis*, which is summarized in section 10.12 and also mentioned in section 1.32. During fourteen years the population was adequately explained by a multiple regression of four terms which accounted for a satisfactorily high proportion of the variance, 78%. If any substantial component of environment (that was not significantly correlated with one of the terms that were already in the equation) had been left out, the agreement between hypothesis and empirical result would have been less complete. For example, the study area was a mixed urban-rural countryside. A large proportion of the area was occupied by buildings, roads, cultivated fields, and other artifacts that harbored no food plants. Had all this "preoccupied" land produced food plants, the thrips would doubtless have been more abundant; the mean of the fourteen years' counts would have been higher. Had such a change in the habitat taken place during the fourteen years' study, a fifth term would have been added to the equation to explain the trend with time in the numbers of thrips (and therefore the mean). As it happened, no such trend occurred. Had it been required to measure the influence of the density of the populations of food plants, the experiment might have been repeated in a suitably different habitat. This is just the sort of extension to an experiment that is likely to be suggested by the theory of environment. And the interpretation of the difference between the two study-areas can be plausibly made within the theory of spreading the risk.

## 9.4   Criticism of Competition Theory

When a feedback to density is put forward as the only plausible mechanism for "regulating" the density of a population (as it is in the quotations that begin sec. 9.34), it is nearly always clear from the context that the influence of crowding is assumed to be felt over the whole "population" and that the "population" responds as a homogeneous unit. Such an assumption might seem plausible when applied to one of den Boer's "interaction groups" (sec. 8.41), but not so when it is applied to the population as a whole of any widespread natural population. So competition theory is open to the broad general criticism that it is eminently unrealistic because it has not taken into account the multipartite structure of the natural population.

More particular criticisms arise when events in local populations are examined. The literature does not contain many accounts of the extinction of well-established natural populations, but, such as they are (sec. 8.43), there is no sign that the relaxation of a density-governing mechanism became important as the numbers declined. On the contrary, the last individual usually died (or the death of the last one seemed imminent) from an extrinsic shortage of habitat before the last of the habitat had been destroyed. It seems that there comes a stage, in the destruction of habitat, when the animal lacks the dispersiveness that is needed to cope with the ever-increasing sparseness of suitable places in which to live.

According to den Boer's experiments with carabid beetles in country where favorable habitat had been continuously depleted by the extension of arable farmland, there was no sign of a density-governing mechanism operating in the local populations, either to prevent the density from exceeding an upper limit or, by relaxing at lower densities, to slow the decline.

Others too have recently expressed their doubts about the conventional wisdom represented by competition theory. For example, Lawton and Strong (1981, 317) argue that:

Patterns in community structure of folivorous insects are similar to those observed in other groups of organisms; vertebrates, for example. However, we posit that interspecific competition is too rare or impuissant to regularly structure communities of insects on plants. We suggest that either the similar patterns occur for different reasons in different taxa or the role of competition is overemphasized as a cause of patterns in organisms other than insects.

Wiens (1977, 590) reported:

As we have learned more about the avifaunas of grassland and shrub-steppe habitats, the interpretation of our observations in the framework of competition theory has become increasingly awkward and forced. [He complained]: The basic premises of competition appear to be so fully accepted by ecologists that the theory is in danger of becoming a firmly entrenched dogma (Peters, 1976). [And he argued that]: The view of reality embodied in competition theory rests upon several assumptions of which four seem especially critical but are frequently left unstated or are set aside so that development of the theory may proceed unimpeded. [The four assumptions refer to (a) the intensity of selection that is implied by the theory. (b) equilibrium of the numbers of animals with respect to resources that are always fully utilized]. (c) it is not just the populations that are in equilibrium but their resource-utilization functions as well. (d) competition is the major selective force acting on resource-utilization traits.

Wiens doubted these assumptions largely on the grounds of variability in "environments" and the failure of empirical observations to confirm the results of predictions by the theory. Wiens concluded with a plea for long-term experimental studies that were not restricted by unrealistic assumptions. For further criticism of competition theory see chapters 7 and 10 and Dayton (1971), Caswell (1978), Price (1980), Connell (1980), den Boer (1980), Wiens and Rotenberry (1980), Wiens (1983a,b), Strong (1981), and Simberloff (1981).

Competition theory is unrealistic with respect to natural populations; it is untrue with respect to local populations of at least some species. Perhaps competition theory becomes a little more plausible, in respect to local populations, in the unusual circumstances of successful agents of biological control (sec. 3.131), but this is a very restricted field.

## 9.5   Conclusion

The theory of environment emphasizes that the whole environment should be taken into account. The concomitant theory of spreading the risk emphasizes that the multipartite natural population should be taken into account. As a consequence of following these precepts, we arrive at the theory that is summarized in the last paragraph of section 9.03 and propounded more fully in the subsequent sections of this chapter.

Before starting this difficult chapter, it seemed timely to review our concept of scientific method (sec. 9.0), because both the construction of our theory and the criticism of other theories depend on this concept. Sometimes ecological theory, especially that which leans heavily on modeling, seems to have laid itself open to the criticism that deduction has been allowed to weigh too heavily, without proper regard for empirical verification. To omit the empirical steps and to jump from hypothesis to explanation may be disastrous (sec. 10.12).

The temptation to discount the empirical evidence may be strong, especially when a general theory is strongly held, and known to be strongly held by many colleagues. For example, at least two of the authors whom we quote at the beginning of section 9.34 decided to stay with the popular theory despite their failure to find empirical support for it from the studies that they were reporting. It behooves the critical reader to evaluate such judgments for himself (sec. 9.02). And theories that rest too strongly on the logic of models without strong support from good empirical evidence should not be trusted too far.

# 10
# Some Controversies in Population Ecology

## 10.0 Introduction

The theme of this chapter is controversy, which serves to guide theory toward the truth. Controversy is virtually confined to explanation and theory (sec. 9.03). In any other part of the scientific spiral, disagreement is readily resolved empirically. But in explanation and theory much of the thought process is essentially subjective and is sustained by controversy, which is essential to healthy scientific progress. Population ecology has its controversies, some of which have been going on for a long time. Much of this argument has been fruitful when the rules of productive controversy have been observed. Some of it has led the contestants, or some of them, to retreat into dogma. It is important that the student become acquainted with current controversy. Regrettably, the art of controversy is too often neglected in the teaching of science.

In this chapter we discuss a number of controversies. In section 10.1 we discuss the controversy about the "regulation" of populations. In sections 10.2 and 10.3 we reexamine two concepts where the popular explanation has tended to congeal into dogma. One is the popular explanation for coexistence of species; the other is the popular explanation for outbreaks. We look at both in the light of the theory of environment and find new explanations that seem more plausible than the old ones. Before turning to these modern controversies, we look at an old one, about the spontaneous generation of life, that was led by Pasteur and Pouchet; we examine this controversy for the light that it throws on the nature of scientific explanation and theory.

## 10.1 Controversy

The argument between Pasteur and Pouchet was about what caused the putrefaction (including the population of "animalcules") that could be relied upon to develop in a flask of "broth" that had been boiled and left exposed to the air on a bench or table. Pouchet said the air carried a "vegetative force" that was absorbed by the broth and, in the presence of oxygen, gave rise to putrefaction, including the spontaneous generation of living animalcules. Pasteur said the animalcules had arisen from living ancestors that had been floating in the air and that the animalcules caused the putrefaction. Each protagonist devised his own experiments and repeated some of the other's, often with contradictory results, which was not surprising because they were working with different sorts of broth and therefore different sorts of organisms. We are not concerned with the details but only with the principles governing explanation. To keep the discussion simple, we mention only Pasteur's experiments and Pouchet's criticism of them.

There was a long series of experiments, each more ingenious than the one before. Always the particular hypothesis was that there would be putrefaction in a control flask but not in the treated one. When the hypothesis was confirmed, as it usually was, Pasteur's explanation was that the treatment had excluded the parent animalcules from the broth. Pouchet's alternative explanation was that the treatment had either destroyed the vegetative force or excluded it from the broth. Each time Pasteur devised a new treatment that seemed to meet Pouchet's immediate criticism of the last one. But Pouchet always had room to maneuvre; he had only to change his ground slightly to meet the new situation. Treatments included boiling the broth and sealing the neck of the flask in a flame, boiling the broth and allowing only "calcined" (i.e., strongly heated) air to enter it, sealing a number of flasks and then opening some in the parlor of a country inn and some on the top of a snow-covered mountain, and so on. The last of the series is famous for its elegance. The broth was placed in the flask. The neck of the flask was drawn out into a long, thin sinuous tube with a deep concavity in which, it was postulated, the parents of the animalcules would be trapped by gravity if the air moved through the tube slowly enough. The broth was boiled. The flask with the boiled broth in it was then allowed to cool extremely slowly. In the best of these experiments the broth remained free from putrefaction for a long time. Pasteur said that the infective animalcules had been trapped at the bottom of the "swan neck". To add to the evidence, Pasteur tilted the flask until the broth lapped the bottom of the swan-neck loop, then retreated into the flask. It worked: the broth, which had previously been free of putrefaction, quickly putrified. Pouchet withdrew from the public debate at this stage, and a committee of the Academy of Science agreed with Pasteur's explanation.

Some historians have placed much weight on the experiment with the swan-necked flask. For example, Hardin (1961, 225) wrote: "[Pasteur] designed a test of such simple elegance as to convince almost everyone". Elsewhere he wrote: "Pasteur answered all the objections by a classic experiment". Elegant this experiment certainly was, but it was no more critical than the ones that had preceded it. Pouchet had only to say that the "vegetative force" was trapped in the swan neck and Pasteur would have been back where he was before—in search of a new experiment. Elegance hardly seems a sufficient substitute for a distinctly new hypothesis. Hardin may have been partly right, but we think that certain other historians have gained a better perspective of Pasteur's explanation and how it has come to be accepted as it is today. For example, Nordenskiold (1928, 434) wrote:

The two antagonists were allowed to carry out their experiments before the French Academy of Science, and Pasteur at once succeeded in convincing some of its foremost members—Milne-Edwards, Claude Bernard and the chemist Chevseal. Pouchet likewise had his supporters and, especially among the scientifically educated and half educated public, he gained many adherents who regarded spontaneous generation as a philosophical necessity, indispensable for a natural-scientific explanation of the origin of life [in keeping with Darwin's theory]. Thus argument opposed argument and party faced party. In these circumstances the solution of the problem would never have become possible had not Pasteur been able to put his ideas into practice on a large scale. During the succeeding years he invented his well known methods of preserving milk by "Pasteurizing" it, of improving the manufacture of wine and beer by controlling the conditions of fermentation, of securing immunity from silkworm disease and chicken cholera by eliminating the micro-organisms that cause them.

And Conant (1951, 248) wrote: "We are here dealing with *converging evidence*, be it noted, for no single set of experiments by themselves would appear sufficient to answer the objection of his opponents, the believers in spontaneous generation". Conant went on to say that by the end of the nineteenth century the experiments of Pasteur had only historical interest so far as spontaneous generation is concerned because "bacteriology and biochemistry were narrowing the conditions of their concepts". "In other words", he wrote, "the advancing techniques and concepts of bacteriology placed any proponent of spontaneous generation in the position of having to specify what organism he claimed could spontaneously arise!"

Looking back, it is clear that the decision to accept Pasteur's explanation and reject Pouchet's was a subjective choice that had to be made by each critic according to his knowledge and understanding of the background to the whole controversy. The background consisted of experiments specifically aimed at the problem, experiments in related subjects, reputable theory in related subjects, and practical experience—especially success in agriculture, medicine, industry, and other relevant fields. It so happened that, as the background knowledge grew, Pasteur's explanation fitted in better and better; it was absorbed and enlarged in the process. Eventually Pouchet's explanation was robbed of all its plausibility. And so it must be with any scientific explanation (sec. 9.03).

There are two errors that commonly occur in controversy. We discuss the error of heterogeneously ascertained data in section 10.11 and the error of the misplaced premise in section 10.12.

## 10.11   The Error of Heterogeneously Ascertained Data

During the famous controversy about spontaneous generation, Pasteur did his experiments with broth that was made from infusions of yeast and sugar; Pouchet did his with broth made from infusions of straw and meat. It is now known that Pasteur's broths, being slightly acid, would favor non-spore-forming organisms, whereas Pouchet's broths, being slightly alkaline, would favor spore-forming organisms; and to kill the spores would require longer exposures to higher temperatures than Pasteur used in his experiments. The two sets of data were not comparable, but they were compared. Neither protagonist was inspired to suspect the difference or put it to an empirical test. So the controversy was more prolonged and became more bitter than it might have been. Claude Bernard (1957) described a number of delightful examples of heterogeneously ascertained data that led to unnecessary controversy among his physiologist colleagues at the Sorbonne. Although Bernard wrote in the middle of the nineteenth century, his book is still worth reading if only to remind the reader how easy it is even for able scientists to overlook the error of heterogeneous data.

In a more recent controversy, about how to explain the density of a natural population, Smith (1961) criticized Davidson and Andrewartha's (1948b) conclusion that the annual fluctuations in a population of *Thrips imaginis* were determined by components of environment that did not depend on the density of the population. He said the data did not support this explanation but did support an alternative explanation that was quite the opposite. Smith analyzed trends in the population during October–November (late spring), which were largely determined by the weather at that time, whereas Davidson and Andrewartha analyzed annual fluctuations about the mean of the maximum density, which were largely determined by the weather during the previous autumn and winter.

Because the two sets of data were distinct, having been ascertained differently, Smith's criticism of Davidson and Andrewartha's conclusion was irrelevant. But by analyzing a distinct subset of the data that were published but not analyzed by Davidson and Andrewartha, Smith made an interesting contribution to the ecology of *T. imaginis* (sec. 10.12). Such is the fruitfulness of controversy.

## 10.12 The Error of the Misplaced Premise

Unlike the hypothesis, which is a precise particular statement and therefore can be verified (or falsified) empirically, the explanation (or theory) is essentially a statement of cause and therefore cannot be confirmed (or falsified) by experiment but can only seem to be plausible (sec. 9.03). The error of the misplaced premise occurs when an explanation (or theory) seems so plausible that its truth is accepted with the absolute certainty that properly belongs only to the premise of a philosophical syllogism. Such certainty has no place in scientific method. For example, Smith (1961) reinterpreted Davidson and Andrewartha's data on *Thrips imaginis*. Davidson and Andrewartha (1948a) had published the mean daily number of thrips in their samples for September–December for fourteen years. Each year the population grew from low numbers in August to a maximum that was reached toward the end of November or early in December (southern summer). Smith postulated that the increase from October to November would be less in those years when the numbers were already high in October. To test the hypothesis, he subtracted the mean for October from the mean for November and correlated the difference with the mean for October. All calculations were done in logarithms. He found a large and highly significant negative correlation ($r = -.80$, $p < .001$). Such a correlation would be consistent with a population that was "regulated" by a density-governing mechanism. But Smith assumed that there is no other way such a correlation might be caused and so he fell into the error of the misplaced premise. In his own words: "They [the authors] have in fact demonstrated the degree to which the population conforms to a predicted level, hence the degree to which the population is regulated. They have no evidence for a density independent system. The regulatory system by which the thrips are held to levels well below that set by their food-supply remains unknown." (Smith, 1961, 406).

It seems as if Smith knew absolutely (as the major premise of a syllogism is held to be absolutely true) that a "density-regulating mechanism" is the only possible cause for a negative correlation between $N_0$ and $\Delta N$. $N_0$ is initial density, and $\Delta N$ is the increment in density in unit time. In this instance $N_0$ is the density of the population in October and $\Delta N$ is the increase in the population from October through November). Having established the correlation, Smith claimed to have established (it seems infallibly) that the population of *Thrips imaginis* is so regulated.

den Boer (pers. comm.) has called our attention to a purely statistical defect in Smith's conclusion. If the population sampled in November was independent from and denser than the population sampled in October, and if the rest is left to chance, the expectation is for a negative correlation. den Boer illustrated this conclusion: he took 14 pairs of numbers from a table of random numbers; he called the smaller number of each pair October and the larger one November; he calculated the nonparametric Spearman correlation between $\ln N_0$ and $(\ln N_1 - \ln N_0)$. He found $\rho = -.65$, $p = .01$. den Boer repeated this exercise with 17 pairs of logarithms taken at random from a normal distribution with a mean of 5 and standard deviation of 3 (which he said was more realistic). He found $\rho = -.52$, $p = .02$. den Boer commented that the populations in the study-area during October and November might achieve a measure of independence if the study area received many immigrants from and lost many emigrants to neighboring localities where the populations were out of step with the population in the study area and with each other.

However unlikely this explanation might seem in the present instance, it is one that Smith might properly have taken into account before claiming certainty for an alternative explanation. A good safeguard against such an oversight is to invoke the theory of environment and the concept of multipartite populations.

This theory also teaches that a scientific explanation needs to be supported by more than logic and can never be held absolutely. It must be judged by its consistency with other scientific theory and by its success in practical arts such as agriculture, medicine, and engineering. It can be held only tentatively, ready to be discarded if it fails these tests (sec. 9.03). Smith's explanation of the density of the population of *Thrips imaginis* seems not to be consistent with what has been published about the ecology of the species.

The population of *T. imaginis* that was studied by Davidson and Andrewartha (1948a,b) lived in a large garden set in a suburban area of houses surrounded by small gardens but abutted (on two sides) upon extensive fields, meadows, and wasteland. The sample-area was typical of the "refuges" that the population retreated into during summer and winter. The adjacent rural land and pockets in the gardens supported an abundance of spring-flowering plants, mostly annuals, that allowed the population of thrips to expand its distribution and increase its abundance during spring. During the fourteen years of the study the thrips consistently fell to low numbers during summer through winter and consistently increased to large numbers in the spring, but they were much more numerous in some years than others (fig. 10.01; sec. 1.32). During the twenty years immediately preceding the study there had been three major outbreaks whose severity could be measured by the reduction in the expected apple crop (fig. 10.02). There was no severe outbreak during the study; the next one occurred in 1950 (Andrewartha and Kilpatrick 1951).

The population was sampled by counting the number of adult thrips in twenty roses. Although the thrips did not breed in roses of this variety, they fed in them. Counts made from flowers of *Echium* and *Cryptostemma*, two annuals that are important breeding places for the thrips during spring, showed comparable trends. So it seemed that the roses might safely be regarded as "traps" that were attracting

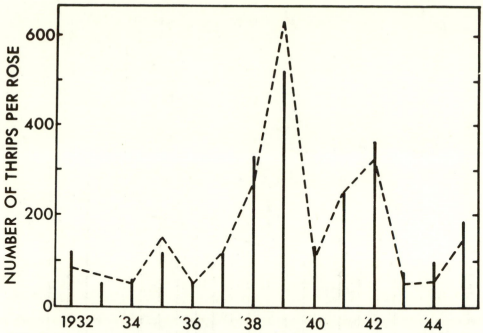

**Fig. 10.01**   The verticals represent the geometric means of the daily counts of thrips per rose for thirty days preceding the "peak" of the population. The curve represents the estimated values for the same qualities calculated from the appropriate regression equation given in the text. (After Davidson and Andrewartha 1948b)

an unknown but sufficiently constant proportion of the total population of adults, at least during the spring flush (Evans 1933, 1934, 1935). Because the chief purpose of the investigation was to find out what caused outbreaks in the spring (fig. 10.01; sec. 10.31), the size of the population as it was approaching its maximum was estimated for each of the fourteen years. The daily counts were smoothed by fitting a freehand curve to a 15-point running mean. The highest point on this curve was taken to indicate the date of the "peak" of the population. The geometric mean of the daily counts for the thirty days preceding the peak served to estimate the size of the population in the spring (fig. 10.01). These numbers were related, in a partial regression, to four quantities that it was thought might represent the weather that had caused them.

Because the catastrophic midsummer crash usually followed the peak of abundance fairly closely, it seemed likely that the dry weather that caused the crash might already have been at work before the peak was reached. Rainfall during spring (September–October) was included in the regression to represent this idea ($x_3$). It was thought that the distribution and abundance of food for the thrips while the population was increasing might be represented by a quantity that measured the date of the break of the summer drought and the atmospheric temperature during the growing period, thus reflecting the vegetative growth of food plants during autumn and winter ($x_2$). This idea is represented graphically by figure 10.03. It was also thought that temperature during spring might influence both the thrips and the food plants—both growing more rapidly in warm weather. Temperature during

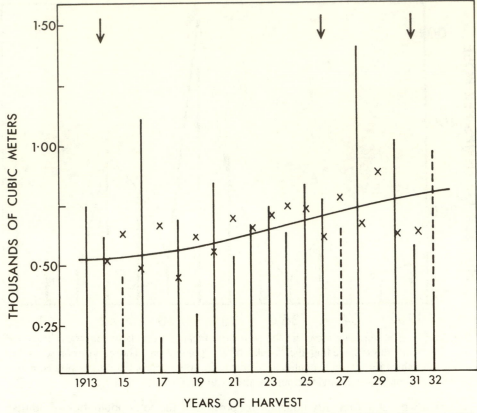

**Fig. 10.02** The relation between outbreaks of thrips in the spring of 1914, 1929, and 1931 (*arrows*) and the yield of apples from the state of Victoria. The actual yield is shown by the heights of the solid verticals. The trend with time has been smoothed by taking a three-year running mean (*crosses*). A freehand curve has been drawn through the crosses. Smaller crops than had been expected were harvested in the following autumn, 1915, 1927, and 1932. The deficiency is indicated by the broken verticals. Because the apple tends to bear a heavy crop in alternate years, the expected crop was estimated by measuring a deviation from the trend line equal to the deviation of the previous year but in the opposite direction.

September–October was included to represent this idea ($x_4$). The annual flowering-plants that supported thrips in large numbers during spring, inside and outside the refuges, usually spent the summer as dormant seeds (fig. 10.03). It seemed likely that such seeds would be more densely and widely distributed in the years following a season of abundant growth and profuse flowering, that is, after a year when $x_2$ had been large; $x_5$ was included to represent this idea. The authors considered whether $x_5$ might also represent the carryover of thrips from the year before, but they could find no evidence for this explanation. It seemed more likely that $x_5$ was influential because it represented the carryover of seeds from the year before. So four independent variates were calculated:

$x_2$ measured the accumulated warmth for the period that began when winter rain first broke the summer drought and ended on 31 August. It was based on the knowledge that the seeds of the food plants would germinate after the rain and that

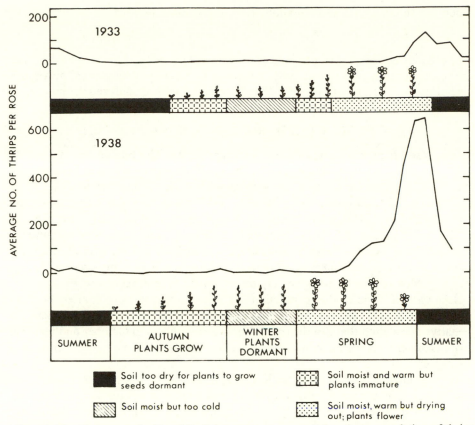

**Fig. 10.03** How an early "break" of the season may result in a large population of thrips in the spring. The curves represent the actual numbers of thrips counted in the roses in 1933 and 1938. The shading at the base of the curves illustrates the life cycle of the thrips' food plants. Food is scarce except when these plants are flowering. Also, the death rate among pupae is high while the soil is dry, before the "break" of the season and after the wet season ends sometime in spring. In 1933 the break of the season came on 11 April, and in 1938 it came on 19 February. In 1933 the total rainfall for September–October was 111 mm, in 1938 it was 51.4 mm. The summer drought began earlier in 1938. (After Andrewartha and Kilpatrick 1951)

the plants would grow faster in warm weather. So it was really attempting to measure the amount of growth the plants would have made by the beginning of spring. The value of $x_2$ was likely to be large when the break of the season came early and the maximum daily temperature after the break was high (fig. 10.03).

$x_3$ measured the total rainfall during September–October.

$x_4$ measured the temperature during September–October.

$x_5$ was merely $x_2$ repeated from the preceding year.

All the independent variates were calculated from the standard meteorological records kept at the Waite Institute. In mathematical form the hypothesis was

$$Y = a + b_2x_2 + b_3x_3 + b_4x_4 + b_5x_5$$

when the variates were written about zero, but it took the form:

$$Y = \bar{y} + b_2(x_2 - \bar{x}_2) + b_3(x_3 - \bar{x}_3) + b_4(x_4 - \bar{x}_4) + b_5(x_5 - \bar{x}_5)$$

when the variates were written about their means. The second form emphasizes that the regression analyzes fluctuations of y about its mean; y is the annual maximum size of the population. After the data had been collected and analyzed the numerical values of the coefficients were filled in:

$$Y = -2.390 + 0.125x_2 + 0.202x_3 + 0.187x_4 + 0.058x_5.$$

The result is expressed graphically in figure 10.01. The broken line represents the fourteen solutions to the equation; the columns represent the empirical estimates. The regression as a whole accounted for 78% of the variance of y. The hypothesis had been confirmed with $p < .001$. When the regression coefficients were divided by their standard deviations they took the values, from $\beta_2$ through $\beta_5$, 1.224, 0.848, 0.226, 0.511. In this scale they are comparable, and their magnitudes indicate the relative importance of each one's contribution to the overall result. So the terms in the regression may be ranked in the order of their importance, $x_2$, $x_3$, $x_5$, $x_4$, which is important for the argument that is developed below.

It is characteristic of this population that the decline that sets in after the peak quickly becomes a catastrophic crash to low numbers (Evans 1933). To explain the negative correlation that Smith reported between numbers in October and increase from October to November, it is necessary to consider this midsummer crash and its causes.

As the summer approached, the flowering of the thrips' food plants came to an end. A severe extrinsic shortage of food developed everywhere, especially in the temporary areas outside the refuges. The refuges were mostly home gardens that were irrigated during summer. The shortage soon became absolute everywhere except in the refuges and even in the refuges the shortage was severe and remained so until spring. At all times the shortage remained extrinsic. At the same time the soil dried out and the death rate among pupae from desiccation, increased (Andrewartha 1934; Evans 1934). Although these processes were most obvious after the "peak", they had also been operating, to greater or smaller degree, before the peak. We can make this inference because $\beta_3$ was the second most influential factor in the regression and also from Evans' (1935) observation that the peak usually occurred after the major food plants had ceased to flower (table 10.01). In other words, while the population was still rising the thrips were already dying from the same lethal components of environment that subsequently caused the catastrophic

**Table 10.01**   The End of the Flowering Period for *Echium* and *Cryptostemma* and the Date of Peak Abundance of *Thrips imaginis* in 1932, 1933, and 1934

| Year | Date of End of Flowering | | Date of Peak Abundance of *Thrips imaginis* |
| | *Echium* | *Cryptostemma* | |
| --- | --- | --- | --- |
| 1932 | 30 November | 12 November | 4 December |
| 1933 | 25 November | 3 November | 23 December |
| 1934 | 30 November | 8 November | 8 December |

*Source:* After Evans (1935).

midsummer crash. It remains to ask only whether these components were likely to be more active in those years when the population was large in October. If the answer is yes, we might have found a plausible explanation for Smith's negative correlations. Moreover, it is an explanation that carries no suggestion of any sort of feedback from density—that is, no suggestion of any "regulatory mechanism" in the sense in which Smith used that phrase.

The answer does seem to be yes.

The population was likely to be large in October when the season began early— that is, when $x_2$ was large. The simple (total) regression of the size of the population in October on $x_2$ accounted for 66% of the variance of y, p < .01 (Davidson and Andrewartha 1948b).

When $x_2$ was large $x_3$, which measured the rainfall in September–October, was likely to be small because $x_2$ and $x_3$ were negatively correlated (r = $-.63$; p < .02). This result was expected because Cornish (1936) had shown that during the ninety-five years 1839–1933 the duration of the wet season had been less variable than the date on which it began. This means that when the wet season started early it was likely to end early. This relationship would account for the negative correlation between $x_2$ and $x_3$. It would also account for the higher death rate from starvation and desiccation toward the end of the spring when the population had been large in October—and thus for Smith's negative correlations.

The influence of this pattern in the weather on the seasonal pattern of the death rate from starvation would be enhanced by the nature of the plants themselves. Annuals tend to set a period to their flowering despite the weather: when the flowering period began early it would tend to end early.

There is no suggestion that any of these quantities is likely to be influenced by the density of the population. Thanks to Smith's contribution to the controversy, the explanation that Davidson and Andrewartha offered for the density-independent control of the population while it was increasing toward its spring maximum can now be extended to cover the declining phase while the population is crashing toward its winter minimum.

There was another line of criticism. Kuenen (1958) pointed out that the very success of a regression (78% of the variance, p < .01) that dealt only with the growth of the population to its maximum each year implied that the nucleus, from which it began, was likely to be invariable; and, he said, this condition would be consistent with density-dependent control of the population during its minimum phase. Klomp (1962) supported this criticism, and Reddingius (1971) calculated a significant correlation between the logarithm of the maximum size and the ratio of the maximum to the ensuing minimum of the population.

Such an argument might be true of a regression that lacked the terms $x_3$ and $x_5$. In the present instance, when the influence of these two terms is taken into account the argument loses much weight. The term $x_5$ represents changes in the distribution and abundance of seeds of the spring-flowering food-plants through which the terms $x_2$ and $x_3$ make their influence felt. When the influence of $x_5$ is taken into account Kuenen's argument is turned back on itself because the significance of $x_5$ in the regression implies that the nucleus from which the population increases each spring should be at least as variable as $x_5$. Similarly with $x_3$. It was shown above that $x_3$ was the second and $x_5$ was the third most important term in the regression, that $x_3$ was negatively correlated with $x_2$, that in years when $x_2$ was large (because the winter rains were early) the summer drought was likely to begin early and that in years when the seeds germinated early in the autumn the plants were likely, independent of the weather, to senesce early in the summer. When the influence of $x_3$ is analyzed in this way we see that the regression predicts a relatively greater reduction in a large population than a small one. Such a prediction is consistent with Reddingius' correlation, but it is achieved by reference only to components of environment that are most unlikely to be density dependent. However, not too much weight can be attached to the statistics that were calculated by Kuenen or Reddingius because, although the reliability of the samples taken during the winter was not tested, they must have been less reliable than those taken during spring: the samples were small; the roses were scarce and so it was difficult to take a consistent sample; the thrips were less active and less reliable in seeking the traps; also, the age structure and sex ratio were different from those in the spring.

## 10.13   Conclusion

Controversy is an important arm of scientific method. Controversy is best conducted according to the generally accepted rules of scientific method—that is, with due care taken to preserve the distinct functions of hypothesis, experiment, and explanation and to maintain the overriding restraint that is imposed by the empirical arm. Experience is the final arbiter, whether experience comes by way of a formal, critically designed and controlled experiment or in some less formal way, such as the success of an industrial or medical process.

All the criticism that we have examined in this section was marred by the error of heterogeneously ascertained data. Pasteur and Pouchet used different kinds of broth; Smith confused two phases in the growth cycle of the population; Kuenen, Klomp, and Reddingius all overlooked important terms in the regression that they criticized.

The error of the misplaced premise marred the criticism of Pasteur, Pouchet, and Smith. With Pouchet and Smith the error seemed to arise from a reluctance to doubt the conventional wisdom of the day. An unfortunate consequence of this sort of error is that the protagonists may not become alert to the need to shift the ground of empirical inquiry. Witness how Pasteur's experiments, for all their increasing elegance, continued to be no more than variations on the same theme. And Smith seemed not to see the need for an experiment to test his explanation, which he seemed to hold infallible. Nevertheless, he made an empirical discovery that broadened the original explanation that he criticized.

Indeed, controversy is a robust arm of the scientific method: even though the major rules have been infringed, the stir that is made by argument is likely to focus attention on the need for more empirical knowledge, which brings the general acceptance of an explanation a little nearer. But the ultimate judgment, when at last the weight of empirical evidence leaves no room for a plausible alternative, is subjective. There is no epistemological magic by which a precise probability (far less a certainty) can be attributed to a causal explanation.

For sections 10.2 and 10.3 we have chosen two controversial topics that have interested population ecologists for a long time. We think that there is a positive contribution to be made to the explanation of "competitive exclusion" (which we prefer to call "coexistence") and of "outbreaks" by looking into these problems from the perspective of our theory of environment.

## 10.2   Coexistence and Displacement

Ecologists have been seeking to explain how species with similar ecologies cannot live closely together ever since Grinnell (1904) set the ball rolling by postulating that "it is only by adaptations to different sorts of food, or modes of food-getting, that more than one species can occupy the one locality". The ensuing discussion has given rise to one symposium (Anonymous 1944) and two informative controversies. The first controversy showed us how the same set of empirical data might lead to opposite explanations depending on whether emphasis was placed on the similarities or the differences in the ecologies of species (Elton 1946; Williams 1947; sec. 7.15). The second controversy showed us the philosophical difficulty of framing the right questions and the logical tangles that may arise from asking the wrong ones (Hardin 1960; Cole 1960a,b). All this discussion has taken place within the framework of competition theory, and Colinvaux (1973, 337) seems to have aptly summarized the majority view among modern ecologists: "Stable populations of two or more species cannot continuously occupy the same niche".

Nevertheless, there seems to be an element of paradox in this statement, because it is well known that in nature certain sorts of animals occur clumped together in distinctive habitats—marine animals in marine habitats, terrestrial animals in terrestrial habitats, tropical animals in tropical habitats, and so on. At least with respect to these broad categories, it is obvious that species are brought together by similarities and kept apart by differences in their niches. Might it be that this is a general law that holds throughout the whole spectrum of specificities from the broadest macrohabitat and the broadest macroniche down to the most specific microhabitat and most specific of microniches?

The paradox disappears when the question is examined from the perspective of environmental theory. From this perspective one naturally expects to find species with similar ecologies living together in the same sorts of places. If there is any question that presses strongly for an answer, it might be: How does it happen that species with different ecologies can sometimes live closely together? In this section we reverse Grinnell's hypothesis and find that in such cases the ecologies of the two species may be more alike than they seemed to be, or that the association of the two species in the same habitat may be less close than it seemed. In either case it makes good sense to explain success (in sharing the same habitat) in terms of critical similarities and to explain failure to do so in terms of critical differences in the ecologies of the two species. We have already introduced this topic in a different context in section 7.15. We now pursue it through sections 10.21 to 10.24.

But first the usage for "habitat" and "niche" must be made quite clear. For "habitat" we adhere strictly to the usage of Elton (1949) and Elton and Miller (1954). The boundaries of a habitat are as they have been arbitrarily proclaimed by an ecologist. The boundaries are usually chosen to contain an ecological feature that has a certain homogeneity with respect to the sort of environments it might provide for animals. For example a broad-leaved woodland is clearly distinct from a coastal heath. But such "macrohabitats" are heterogeneous. For example, within the woodland there will be "minor habitats", say trees of a particular species, and micro-habitats, say the space under loose bark on trees of a particular species. And so on. The habitat becomes more homogeneous as the boundaries are specified more minutely.

For "niche" we adhere strictly to the usage of Elton (1927, 63) and the extended definition spelled out by Odum (1959, 27).

The ecological niche, on the other hand, is the position or status of an organism in its community and ecosystem resulting from the organism's structural adaptations, physiological responses and specific behaviour (inherited and/or learned). The ecological niche of an organism depends not only on where it lives but also on what it does. By analogy, it may be said that the habitat is the organism's "address" and the "niche" is its profession, biologically speaking.

There is a "hand and glove" relation between the niche and the environment: the environment produces the external stimulus that gives the animal a chance to practice its niche (profession); the niche reflects the animal's intrinsic capacity to exploit or otherwise respond to its environment. As with habitat, so with niche it is helpful to think of the hierarchy of macro-, minor, and microniches. For example, the "profession" of predator might be exercised by an omnivore (many birds), a facultative carnivore (a ladybird that eats several species of scale insects), an obligate carnivore (a ladybird that eats only one species of scale insect), or a more specific obligate carnivore (a ladybird that eats only the second and third instars of one species of scale insect). And so on. Of two species that profess the same niche, one may do so more proficiently than the other.

In section 10.21 we describe how the larvae of two species of fruit flies that, in nature cannot share a fruit, such as a guava, can nevertheless live, on quite equal terms, in a dish of artificial medium because the medium neutralizes a difference between the species that, in nature gives one a consistent advantage over the other.

### 10.21   The Critical Difference between *Ceratitis capitata* and *Dacus dorsalis*

The total displacement of *Ceratitis capitata* by *Dacus tryoni* in eastern Australia and the precarious coexistence of *C. capitata* with *D. dorsalis* in parts of Hawaii are discussed in section 7.221.

According to Keiser et al. (1974), when eggs were laid naturally into a guava by normal healthy females of both species, it made no difference whether the two species laid eggs into the same end or opposite ends of the fruit, nor which species laid its eggs first (by half a day). The result was always the same: virtually not a maggot of *Ceratitis* survived to emerge, but *Dacus* survived and emerged normally. Seeking to identify the critical difference between the species, Keiser et al. did three more experiments that distinguished between the larval *Dacus* per se and something else, other than eggs, that the female puts into the fruit when she oviposits.

1. A female of *Dacus* that has been sterilized, in the pupal stage, by irradiation, will still go through the motions of ovipositing into a fruit. No egg is laid, but it seems as if something else (perhaps an inoculum of a particular microorganism) which interferes with the safe development of *Ceratitis* is injected into the fruit during the futile attempt at oviposition. Keiser et al. allowed sterilized *Dacus* to

"oviposit" into one end of a number of guavas. The healthy *Ceratitis* were allowed to oviposit into the same or the opposite end of each guava. When *Ceratitis* laid its eggs into the same end that had been "fouled" by *Dacus* the mean emergence of fully fed *Ceratitis* was 26 compared with 170 for those fruits where the eggs had been laid into the opposite end.

2. Gravid females of *Dacus* or *Ceratitis* will oviposit freely into a suitably perforated plastic cup. The eggs can be washed out of the cup with water. Eggs that had been collected in this way were used to set up a series of experiments in which fifty eggs of each species were artificially "oviposited" into each guava. In every other respect the design of the experiments was identical with those that had been done with healthy flies of both species ovipositing naturally into each fruit. It was expected that during the artificial collecting and "oviposition" of the eggs the harmful inoculum (whatever it might be) would become diluted. In keeping with this expectation, 26% of all emergences were *Ceratitis* compared with virtually zero in the other half of the experiment where all the ovipositions were natural.

3. Keiser et al. (1974) also did a large experiment in which the two species were reared on an artificial medium that was made essentially from commercial yeast and sugar in a base of wheat bran. At the beginning yeast made up about 3% of the gross weight and about 20% of the nutrient. At 28°C the commercial yeast would grow vigorously and would be expected to swamp any other microorganism that might have come in with the eggs. The eggs were collected in plastic cups as before and scattered through the medium. The ratio of *Ceratitis* to *Dacus* varied from 1:1 to 1:124; the total number of eggs also varied widely from wastefully few to densely overcrowded. At a few of the extremely high densities *Dacus* seemed to survive a little better than *Ceratitis*. But at all the realistic densities both species survived equally well at all ratios. In other words, the ratio of *Ceratitis* to *Dacus* for the survivors was not significantly different from the ratio for the eggs.

It is a pity that these elegant experiments stopped at this point, just when they were on the brink of identifying precisely the critical difference between these two species. However, there seems little doubt that the critical difference between *Dacus* and *Ceratitis* must be related to the food-eating niche. When the only food that is available in the microhabitat is a yeast that is foreign but acceptable to both species, any difference between them ceases to be critical: they become so alike, measured with reference to this narrowly defined niche, that both survive equally well. In nature, when a ready-made supply of yeast is not available the difference between them asserts itself; and *Ceratitis* cannot live in a place that is favorable in every respect except that *Dacus* is present. Knowledge of the precise nature of the difference is denied us because the experiments of Keiser et al. (1974) stopped short. Perhaps the story, when it is finally told, will run something like this. The fruits that *Dacus* and *Ceratitis* exploit do not contain a ready-made supply of microorganisms. Each species is adapted to inoculate the fruit with its own brand of microorganism. The inoculum that accompanies *Dacus* eggs inhibits the growth of the inoculum from *Ceratitis*; and the *Dacus* microorganism is not acceptable to the maggots of *Ceratitis*.

Because *C. capitata* is no longer to be found in eastern Australia, it has not been possible to conduct a similar series of experiments with it and *D. tryoni*. Gary Fitt (pers. comm.) has found a number of microorganisms associated with all stages of the life cycle of *D. tryoni*. Several of them were also found in infested fruit, but none was consistently found on the eggs in the close association that has been proposed for *D. oleae* by Hagen (1966). Fitt sterilized the surface of eggs of *D. tryoni* by washing them in sodium hypochlorite, which killed the microorganisms present. There was no microorganism within the egg. Larvae that hatched from these eggs could not grow in an artificial medium containing unhydrolyzed protein. They were, however, able to grow when the medium contained hyrolyzed protein. When *Enterobacter cloacae*, which are commonly but not always found in *Dacus* and infested fruit, were added to the medium the larvae developed normally. It seems that these bacteria and maybe other species, assist the larvae by hydrolyzing the protein in the medium. The larvae themselves produce no proteolytic enzymes. In view of the findings of Keiser et al. (1974) discussed above, it is possible that the disappearance of *Ceratitis* from eastern Australia (sec. 7.221) was the consequence of a similar interaction between *D. tryoni* and *Ceratitis* mediated by microorganisms.

In the next section we tell a story that is almost the converse of this one. Two species of scale insect, with prominently overlapping niches, that had closely shared a wide-ranging habitat for many years could no longer do so after the habitat had been enriched by the introduction of an exotic predator. A critical difference in the ecologies of the scale insects, which had lain latent and unknown, was activated by the predator. Both species of scale insect serve as food for the predator; but the predator prefers one of them. This one has been unable to survive in a habitat that contained both the predator and its alternative food.

## 10.22   The Critical Difference between Two Species of Scale Insect That Are Alternative Prey for a Facultative Predator

Before the publication of McKenzie's (1937) paper, it was not possible to distinguish between California red scale, *Aonidiella aurantii*, and yellow scale, *A. citrina*, from dead specimens in a museum. However, a good naturalist could tell them apart, with very few mistakes, from living specimens in the field. So De Bach, Hendrickson, and Rose (1978) were able to collect reliable information about the distribution and abundance of these sibling species in southern California going back to about 1890. Between 1890 and 1940 yellow scale was more widespread than red scale and numerous enough to be a serious pest in many localities. Its chief strongholds were in southern Ventura County and the southwestern corner of San Bernadino County. Red scale was also widely distributed and abundant, but, whereas yellow scale was recorded sparsely from most of the areas where red scale dominated, there was no red scale recorded from the major strongholds of yellow scale.

De Bach and his colleagues published a number of maps convincingly tracing the decline of yellow scale in the areas dominated by red scale and the spread of red scale into the areas that had been dominated by yellow scale. Some time between 1960 and 1970 the process was completed; and by 1970 the map contained not a single record of yellow scale but plenty of records of red scale where yellow scale used to be. Of course this does not necessarily mean that yellow scale was extinct everywhere, but it was scarce enough not to be mentioned in the official records.

Until recently insecticides had been widely used against both species of scales, sometimes aimed at exterminating the pest locally. Although economic control was practicable, extermination was not achieved. The maps show that, even in the

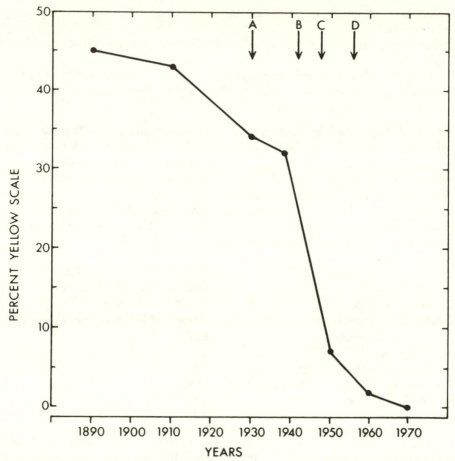

**Fig. 10.04**   The decline of yellow scale relative to red scale in southern California, showing the dramatic increase in rate of decline of yellow scale after 1940, and the dates when the following predators were introduced: arrow A, *Comperiella bifasciata* (Japanese race); arrow B, *Comperiella bifasciata* (Chinese race); arrow C, *Aphytis lingnanensis;* arrow D, *Aphytis melinus*. The data are derived from maps in De Bach, Hendrickson, and Rose (1978).

presence of the several predators that were present before 1940, yellow scale seemed to hold its own. A slight decrease in the abundance of yellow scale and a small increase in the distribution and abundance of red scale followed the introduction of the Japanese race of *Comperiella bifaxiato*, but considering the source of the data these changes may not have been significant. Soon after 1940 the relative abundance of yellow-scale began to decline rapidly, and its distribution contracted. At the same time red scale began to increase in numbers and enlarge its distribution, expanding into areas that had been predominantly or exclusively occupied by yellow scale. These changes were large and undoubtedly significant (fig. 10.04).

De Bach, Hendrickson, and Rose (1978) considered all the plausible interactions between the two species of scale that might have caused the extinction of yellow scale and its replacement by red scale. They dismissed:

1. extreme weather, including defoliating frost,
2. insecticides,
3. shortage of resources and physical crowding (the populations were not dense enough), and
4. simple predation (a number of obligate and facultative predators showed no sign, in the absence of red scale, of being able to depress the numbers of yellow scale toward any threat of extinction).

With all other plausible components of environment ruled out, De Bach and his colleagues concluded that the critical interaction between red and yellow scale must have taken place through a predator. It must have been one that was present after 1940 and one that would use red scale as a less preferred food in much the same way as *Ptychomyia* used *Eublemma* and thus pressed more heavily on *Levuana* (Andrewartha and Birch 1954, 486). According to De Bach, Hendrickson, and Rose, the following complex of exotic small Hymenoptera operated in the environments of red and yellow scale in southern California:

| Species | Introduced into California |
|---|---|
| *Aphytis citrinus* | Before 1900 (accidentally) |
| *Aspidiotiphagus citrinus* | Before 1900 (accidentally) |
| *Prospaltella aurantii* | Before 1900 (accidentally) |
| *Aphytis chrysomphali* | Before 1900 |
| *Habrolepis rouxi* | ? |
| *Comperiella bifasciata* (Japanese race) | 1931 |
| *Comperiella bifasciata* (Chinese race) | 1941 |
| *Aphytis lingnanensis* | 1948 |
| *Aphytis melinus* | 1957 |

The dates when the last four predators were introduced into southern California are marked with arrows on figure 10.04, which illustrates the change in relative abundance of yellow scale and red scale in southern California. It has been constructed from the maps published by De Bach, Henrickson, and Rose (1978). The symbols on the maps have been given arbitrary numerical values as follows: rare 1, moderate 3, abundant 9. These values have been summed, and the total for yellow scale has been expressed as a percentage of the total for both species.

The first four species have been present since before 1900 without producing the result that now has to be explained, and *H. rouxi* has never been found to eat yellow scale. It is known that the Japanese race of *C. bifasciata* will lay eggs freely into both red and yellow scales, but all the eggs that are laid into red scales are encapsulated and die. So red scale cannot serve as an alternative source of food for the Japanese race of *C. bifasciata*. The last two species between them, *A. lingnanensis* in coastal and *A. melinus* in inland regions, provide effective biological control of red scale over most of southern California. Both species will lay their eggs into yellow scale and can readily be reared in yellow scale in the laboratory. But *A. melinus* certainly and *A. lingnanensis* probably arrived too late on the scene to be charged with responsibility for anything more than participating in the final decline of yellow scale.

The Chinese race of *C. bifasciata* has been called the "red scale race" because, unlike the Japanese race, it can be reared in red scale in the laboratory. De Bach, Hendrickson, and Rose (1978) wrote:

About [1941] the introduced Chinese race of *C. bifasciata* slowly became established and distributed in the field on California red scale. . . . The Chinese race of *Comperiella bifasciata* is capable of using either scale as a host under laboratory conditions, but in the field, is found only in areas of pure California red scale or where the two host species occur on a single tree or in a single orchard.

The name "red scale race" for this species may be a misnomer and may have placed too much weight on its ability to make do on red scale when no yellow scale is about. Brewer (1971), working in South Australia, in an orchard where both species of scale occurred on the same tree and the only known predator was the Chinese race of *C. bifasciata*, collected 100 yellow scales and 200 red scales. He dissected them and found that 23% of the red scale had been parasitized and 66% of the yellow scale. Of the 55 eggs that had been laid into red scale, 91% were encapsulated and died; of the 66 eggs that had been laid into yellow scale only 6% had been encapsulated. In the laboratory he found that 85% of a total of 819 eggs that had been laid into mature red scale were encapsulated. When eggs were laid into nymphs of red scale 20 and 35 days old at 28°C, 26% of the eggs laid into the younger nymphs and 59% of those laid into the older nymphs were encapsulated. Despite its misleading name, the Chinese race of *C. bifasciata* seems to have a preference for and to achieve more success in the yellow scale than the red scale.

De Bach, Hendrickson, and Rose (1978) had much evidence that the decline in numbers and ultimate disappearance of yellow scale from places where it had long been established was usually preceded or accompanied by the penetration of red scale into these areas. They emphasized that there was no overcrowding or any other evidence that either species experienced any shortage of food. They explained the decline of yellow scale as "competitive displacement" of one "ecological homologue" by another. (The meaning of "ecological homologue" in this context is obscure. It must not be confused with the usage of Sigurjonsdottir and Reynoldson, in section 10.23 which is sharp and useful). De Bach, Hendrickson, and Rose thought that predators were important but they did not say how. For them, the idea of "competition" as implied by the name "competitive displacement" seems to have been sufficient explanation in itself. With our interest in the theory of environment, we would have liked to see the investigation carried to a more realistic conclusion in order to identify specific components of environment. Until the critical experiments are done the evidence must remain circumstantial, but, circumstantial though it is, the existing evidence points to an interesting conclusion.

From the observations of Brewer in South Australia it seems that the Chinese race of *Comperiella* may have a preference for yellow scale, and may achieve more success in it than in red scale. From the observations of De Bach, Hendrickson, and Rose in southern California it seems that *Comperiella* can maintain itself in reasonable numbers on red scale alone. The catastrophic decline of yellow scale set in shortly after the Chinese race of *Comperiella* had been released in California (fig. 10.04). If this evidence is accepted, we conclude that yellow scale died out because it could not survive in a habitat that also supported both red scale and the Chinese race of *Comperiella*. And we suggest that the predation-pressure on yellow scale became intolerably severe because the predators (a) preferred yellow scale and (b) could maintain their numbers (and therefore their pressure on the favored food) no matter how scarce the favored food became, because they had an alternative source of food in red scale. For other examples of alternative sources of food on the activity of a predator see sections 5.221 and 10.32.

We began this section by asking what condition must be fulfilled in order that two or more species may live closely together, in the same habitat, sharing the same food. Instead of a direct answer to this question, we seem to have found in this case history the reason why two closely related, ecologically similar species cannot coexist. It is because they differ from one another with respect to one critical niche—the capacity to evade or to tolerate predation by *Comperiella*. Red scale practices this "profession" more successfully than yellow scale. It is because of this critical difference between them that yellow scale cannot coexist with red scale in a habitat that includes *Comperiella*. Would yellow scale have a better chance to coexist with red scale if it were more like red scale? The evidence that emerges from section 10.23 suggests that the answer might be yes.

### 10.23   Two Monophagous Predators That Share the Same Prey

Most of the information in this section has been taken from Quezada and De Bach (1973).

The scale insect *Icerya purchasi* was accidentally introduced into California from Australia in 1866. Within a few years it had become widespread and abundant, and so bad was the plague that people said the whole citrus industry was threatened with extinction. The deliberate search for a predator in Australia and the introduction of *Rodolia cardinalis* (a ladybird beetle which is called the vedalia) into California in 1888 constituted the first attempt at the biological control of an insect pest. It has long been hailed as a classic example of biological control because of the speed and thoroughness of its success. The limelight fell on *Rodolia* because that was the one that was deliberately introduced; a dipterous predator *Cryptochaetum iceryae* came in unnoticed, an unobtrusive fellow traveler. In acclaiming *Rodolia*, the literature has been unfair to *Cryptochaetum*, which is now more numerous than *Rodolia* in the moist, mild coastal areas of southern California. None of those who witnessed the dramatic reformation of the citrus as they quickly changed back into healthy, virtually scale-free trees, doubted that the change was due to the predators that had been brought in from Australia. Today the local outbreaks of *Icerya* which consistently follow the local use of insecticides that kill the predators but not the scale confirm that early opinion beyond reasonable doubt.

The two predators are strictly monophagous, sharing the same prey (*Icerya*) over a large area, including everywhere that citrus is grown commercially in southern California. They have lived together, in this habitat, every since they were brought from Australia in 1888—for ninety years, which is time enough for more than three hundred generations. Both are efficient agents of biological control; either one could by itself, in a place where the weather favored it, probably keep *Icerya* scarce enough to satisfy the farmers. But, in southern California the two predators acting together are at least as effective as one would be acting alone, or perhaps more effective. It follows that each species creates an intrinsic shortage of food for itself and an extrinsic shortage for the other. Perhaps it would be more realistic to use the plural and say that between them the two species create an intrinsic shortage of food for themselves. The shortage must be regarded as persistent and severe because a high level of economic control has been consistently maintained for more than ninety years. The two species seem to be sharing the same niche. How do they achieve it?

There is an epistemological point to be made before we set out to find an empirical answer to this question. Many distinct species may share a heterogeneous macrohabitat, without critical interference, if each one lives in a distinct micro-habitat that allows it to exploit its own special niche, as is the case with the triclads of section 10.24 (see also sec. 7.1). As a corollary, two distinct species can be said to share the same niche only if it can be shown that this niche cannot be further partitioned into distinct microniches. Since a negative can never be "proved" empirically, we must as always be satisfied with the failure of an honest attempt to disprove it. With this point in mind we examine the way *Rodolia* and *Cryptochaetum* earn their livings by feeding on *Icerya*.

Quezada and De Bach (1973) recognized three climatic regions in southern California. The coastal region is humid and mild; citrus is grown largely on natural rainfall. The interior region is semiarid; citrus is grown largely under irrigation. The desert region is the most inland region. The climate is arid; the temperature ranges widely; little citrus is grown; *Icerya* and both the predators can be found, but there is evidence that *Cryptochaetum* at least is approaching the limits of its distribution in this region.

Throughout the whole of the distribution in southern California of *Icerya* its death rate from predation is undoubtedly high in every generation. Quezada and De Bach measured the age-specific death rate in a natural population of *Icerya* in the vicinity of Riverside (interior region) for four generations in the four seasons of the year (table 10.02).

Commenting on the results in table 10.02, Quezada and De Bach (1973, 661) wrote:

If we assume that an average of 200 crawlers is produced by each female, then 4000 crawlers were produced by the 20 females observed. The totals and percentages in [table 10.02] show that the per cent crawlers settled was less than 50 in the summer and less than 40 in the winter. During summer less than 50 per cent of the first stage larvae that settled attained the second stage. This mortality was largely due to vedalia attack. The second stage scales were subject to a heavier attack by the beetle, along with increasing *Cryptochaetum* parasitization. Only 3.2 per cent of the larvae which attained the second stage reached the third stage. This represents a survival of only 0.8 per cent of the 1,894 settled crawlers. At such low densities, the beetles dispersed, and a relatively large proportion of survivors from third instars (24.1 per cent) were able to produce egg-sacs in the summer. However this is only 0.37 per cent of the 1894 originally settled crawlers, which represents 99.63 per cent mortality after crawler settling. A similar pattern of mortality occurred in other seasons, with a very high proportion of scales destroyed by vedalia and *Cryptochaetum*, and the survival of a very low percentage of individuals that leave progeny.

The whole population could not sustain such a high death rate without declining rapidly. But 53–74% of the crawlers dispersed before settling. A few would have

**Table 10.02**    Age-Specific Survival Rate and Death Rate of *Icerya* during the Four Seasons of the Year at Riverside in the Interior Region

| Season | Stage in Life Cycle | Number of Scales | | Survival Rate | Death Rate (%) | |
| | | Begin | End | (%) | Age-Specific | Cumulative |
|---|---|---|---|---|---|---|
| Summer | Crawlers | 4,000 | 1,894 | 47.3 | 52.7 | 52.7 |
| | First instar | 1,894 | 903 | 47.6 | 52.4 | 77.6 |
| | Second instar | 903 | 29 | 3.2 | 96.8 | 99.2 |
| | Third instar | 29 | 7 | 24.1 | 75.9 | 99.8 |
| Autumn | Crawlers | 4,000 | 1,465 | 36.6 | 63.4 | 63.4 |
| | First instar | 1,564 | 415 | 28.3 | 71.7 | 89.6 |
| | Second instar | 415 | 37 | 8.9 | 91.9 | 99.0 |
| | Third instar | 37 | 8 | 8.1 | 91.9 | 99.9 |
| Winter | Crawlers | 4,000 | 1,063 | 26.5 | 73.5 | 73.5 |
| | First instar | 1,063 | 334 | 31.4 | 68.4 | 91.6 |
| | Second instar | 334 | 45 | 13.4 | 86.6 | 98.8 |
| | Third instar | 45 | 5 | 11.1 | 88.9 | 99.8 |
| Spring | Crawlers | 4,000 | 1,456 | 36.4 | 63.6 | 63.6 |
| | First instar | 1,456 | 374 | 25.6 | 74.4 | 90.6 |
| | Second instar | 374 | 35 | 9.3 | 90.7 | 99.1 |
| | Third instar | 35 | 4 | 11.4 | 88.6 | 99.9 |

*Source:* Quezada and De Bach (1973).

*Note:* Each generation began with the progeny from twenty egg sacs chosen at random. The figure of two hundred crawlers per egg sac was based on counts in the laboratory of crawlers emerging from a sample of egg sacs collected in the field. "First instar" means those that had settled.

been lucky enough to land on a food plant. Because they settle solitarily, they may have a better chance of not being found by a predator than those that stayed behind.

Quezada and De Bach estimated the intensity of predation on those that did stay behind by counting the number of *Icerya* that were eaten by each of the two sorts of predators during five consecutive generations (seasons) in each of the three climatic regions. All together 1,500 *Icerya* were tagged and watched; 100 were tagged in each region at the beginning of each season. The results are summarized in tables 10.03–10.05.

Because these scales had to be tagged individually, they may have been more isolated than the average scale, which is likely to be a member of a colony. For this reason the fifteen hundred scales may have been less at risk from predators than the average scale. Nevertheless, the weight of predation revealed in tables

**Table 10.03**    Causes of Death in Five Hundred Scales, *Icerya purchasi,* One Hundred in Each of Five Seasons, in the Coastal Zone

| Cause of Death | Autumn 1967 | Winter 1968 | Spring 1968 | Summer 1968 | Autumn 1968 |
|---|---|---|---|---|---|
| Old age, produced crawlers | 7 | 4 | 2 | 16 | 9 |
| Eaten by *Rodolia* | 9 | 3 | 4 | 10 | 12 |
| Eaten by *Cryptochaetum* | 78 | 85 | 91 | 72 | 76 |
| Unknown cause | 3 | 4 | 2 | 2 | 2 |
| Lost | 3 | 4 | 1 | 0 | 1 |

**Table 10.04**  Causes of Death in Five Hundred Scales, *Icerya purchasi*, One Hundred in Each of Five Seasons, in the Interior Zone

| Cause of Death | Autumn 1967 | Winter 1968 | Spring 1968 | Summer 1968 | Autumn 1968 |
|---|---|---|---|---|---|
| Old age, produced crawlers | 11 | 3 | 2 | 19 | 9 |
| Eaten by *Rodolia* | 50 | 48 | 16 | 42 | 58 |
| Eaten by *Cryptochaetum* | 22 | 43 | 78 | 32 | 26 |
| Unknown cause | 4 | 2 | 2 | 4 | 4 |
| Lost | 4 | 2 | 2 | 3 | 3 |

10.03–10.05 is so heavy that there need be no reasonable doubt that most of the heavy death rate of postcrawler stages that was recorded in table 10.02 was due to predation and that the prevailing low density of the population of *Icerya* in southern California is due to predation by *Rodolia* and *Cryptochaetum*. The figures in tables 10.03–10.05 also show that, despite the variation from season to season and from region to region in the number of scales eaten by each predator, the total number of scales destroyed by the two predators acting in concert varied scarcely at all from region to region, and not very greatly from season to season. Quezada and De Bach (1973, 657,661), commented on this point:

in any season, total mortality of the scales caused by its two natural enemies combined did not vary appreciably from one region to another. Statistical analysis showed no significant differences among total enemy-caused mortality in the three regions at any given season. Neither were the differences significant among the number of scales that escape attack in the three regions under the same circumstances. [And elsewhere they amplified this conclusion:] regardless of their relative dominance in the different regions, the two natural enemies combined cause similar mortalities of the scale. In the Riverside area, the average per cent of prey destroyed ranged from 74 per cent in the summer to 94 per cent in the spring. In the desert the respective figures were 74 to 90 per cent, and on the coast, from 82 to 95 per cent.

The consistency of the "predation-pressure" between regions is best explained by thinking that the number of scales that were left uneaten reflected the severity of the intrinsic relative shortage of food the predators could impose upon themselves. There is a corollary: because the "predation-pressure" remained constant independent of variations in the proportions of the contributions by the two predators, the two species must be equivalent to each other with respect to the severity of the

**Table 10.05**  Cause of Death in Five Hundred Scales, *Icerya purchasi*, One Hundred in Each of Five Seasons, in the Desert Zone

| Cause of Death | Autumn 1967 | Winter 1968 | Spring 1968 | Summer 1968 | Autumn 1968 |
|---|---|---|---|---|---|
| Old age, produced crawlers | 4 | 6 | 5 | 17 | 4 |
| Eaten by *Rodolia* | 88 | 76 | 78 | 74 | 91 |
| Eaten by *Cryptochaetum* | 3 | 13 | 12 | 0 | 0 |
| Unknown cause | 3 | 2 | 4 | 5 | 3 |
| Lost | 2 | 3 | 1 | 4 | 2 |

intrinsic relative shortage of food they can impose on themselves. In other words, it does not matter which species is doing most of the eating; it makes no difference to the threshold below which they cannot push the density of prey that is left untouched.

If the two conclusions we have just reached are true, namely (a) that predation is the component in the environment of *Icerya* that chiefly determines its abundance in all three climatic regions and (b) that the threshold below which the numbers of *Icerya* cannot be pushed by predation is the same independent of the relative abundance of *Rodolia* and *Cryptochaetum*, then we would expect to find that the density of the population of *Icerya* is the same in all three regions. This is just what Quezada and De Bach did find when they counted the *Icerya*.

Quezada and De Bach measured the density of the population of *Icerya* on citrus in sites chosen to represent the coastal region, where *Cryptochaetum* consistently and greatly outnumbers *Rodolia*, the desert region, where *Rodolia* consistently destroys more scale than *Cryptochaetum*, and the interior region where the same amount of scale is destroyed by each species. The density of the population was measured by counting the number of scales seen during a standard five-minute search of one tree. On the average 99.9 leaves were examined during five minutes, so the count was essentially the number of scales on 100 leaves. Twenty trees were examined on each occasion, and results were expressed as the mean for 20 trees. The results are summarized in figure 10.05. The range was from 0.70 scales per 100 leaves in the desert region in summer to 2.00 scales per 100 leaves in the desert region in winter. This is a strikingly slight population compared with the seething masses that were described before 1888, when the predators were first introduced.

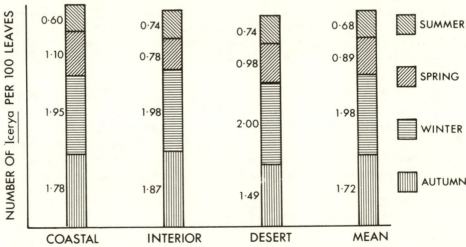

**Fig. 10.05**  The number of *Icerya* counted during five minutes' standardized counting (about one hundred leaves) in the three climatic regions of southern California during four seasons 1967/68. There is no significant difference in the densities of the populations of *Icerya* in the three regions even though *Cryptochaetum* is relatively more abundant than *Rodolia* in the coastal region and also in the interior region during winter, whereas *Rodolia* is relatively more abundant than *Cryptochaetum* in the desert and also in the interior region during summer. (After De Bach, Hendrickson, and Rose 1978)

**Table 10.06**  Relative Abundance of *Rodolia* and *Cryptochaetum* in Three Climatic Regions Expressed in Biomass and in Numbers

| Region | *Rodolia* : *Cryptochaetum* | |
|--------|---------|---------|
| | Biomass | Numbers |
| Coastal | 1:10.6 | 1:349 |
| Interior | 1:0.986 | 1:32.5 |
| Desert | 1:0.068 | 1:2.27 |

If the predators are responsible for the change, there can be no doubt about their effectiveness as agents of biological control, or about the severity of the intrinsic shortage of food that they sustain. In figure 10.05 the differences between the seasons may be significant, but the differences between the zones are small and not significant. This result confirms the conclusion, which we have already approached from other evidence, that *Rodolia* and *Cryptochaetum* are closely equivalent with respect to the qualities that make a predator a good agent of biological control. These are essentially qualities that relate to searching for food and exploiting it when it has been found.

There is still another inference to be drawn from the figures in tables 10.03 to 10.05. If we can assume that the biomass of the predators is proportional to the food that they eat and that one fully grown *Icerya* will be enough food for one *Rodolia* or eleven *Cryptochaetum*, we can estimate the relative biomass of the two predators from the relative numbers of *Icerya* that were eaten by them (table 10.06).

The second column of table 10.06 shows how the advantage that *Cryptochaetum* has over *Rodolia* in the coastal region is progressively reduced to a disadvantage in the inland regions. The gradient in relative biomass nicely matches the gradient in climate. To complement these figures there are the observations of Quezada and De Bach that *Cryptochaetum* seems to die out from the desert region each summer and recolonize it each winter—from at least 50 km away. And in the interior *Rodolia* tends to lose ground during winter and regain it during summer. It would be interesting and useful to have a physiological explanation for the difference between these two species with respect to their response to weather, but the experiments that might provide such an explanation seem not to have been attempted. On the other hand, we know rather more about the adaptations that make these two species so nearly equivalent in the qualities that make for a good agent of biological control—that is, a predator that can maintain a severe intrinsic relative shortage of food and still contrive to harvest a high proportion of the food in the area (fig. 3.01).

The fly *Cryptochaetum* lays its eggs into—and its larvae will eat—any stage of *Icerya* except the egg or the first larval instar. The flies are small, and as many as eleven may develop to full size inside one well-grown *Icerya*. The beetle *Rodolia* is larger. The larvae and adults eat *Icerya* from the outside; one *Rodolia* may need to eat several immature *Icerya* to grow to full size.

The scale *Icerya* tends to live in colonies, and it is usual to find either *Rodolia* or *Cryptochaetum* but not both in possession of a colony. In a recently found colony of *Icerya*, *Rodolia* will wander over the scales, browsing on them from the outside and making them unattractive to a *Cryptochaetum* that is seeking to lay eggs. Moreover, any eggs that are laid into whole scales are likely to be eaten by *Rodolia*, which readily eats scales containing eggs or young larvae of *Cryptochaetum*. On the other hand, *Rodolia* will not eat scales containing late-stage larvae or pupae of *Cryptochaetum*. Consequently, a colony of *Icerya* that was found first by *Cryptochaetum* and occupied long enough to produce some late-stage larvae may not be attractive to *Rodolia*. Quezada and De Bach observed on a number of occasions clear-cut examples of the operation of the rule that "possession is nine points of the law" and considered this a characteristic feature of the interaction between the two species. Thus, by two quite different behaviors, each species manages to "preempt" for its own use any colony of *Icerya* that it happens to have found first. Both species take about the same time to consume a colony, but neither is likely to destroy the colony down to the last scale.

Like many other ladybirds, the last *Rodolia* flies away from a colony of *Icerya* on which the beetles have been feeding before the last scale has been destroyed; the predator thus leaves behind a small remnant of survivors that can revive the colony, to the benefit of any *Rodolia* that might later be searching in its vicinity. *Cryptochaetum* achieves the same result by its feeding behavior. Larvae hatching from eggs that are laid into a mature adult scale that has begun to form its egg sac do not prevent the scale from laying some eggs. Because *Cryptchaetum* does not eat eggs or first-instar larvae, a small remnant of survivors in these early stages is likely to be left alive when the last *Cryptochaetum* flies away from the colony.

From the ability of *Rodolia* and *Cryptochaetum* to share an intrinsic relative shortage of food on equal terms we may infer, that they must be closely equivalent in their abilities to search for food, but we do not know how they do it. This aspect of their biology has not been closely studied. It may be that the difference in size holds a clue. For any given biomass a large number of small bodies may be best served by an involuntary random search; fewer large bodies may do better if there is a larger element of directed voluntary movement in the search.

The similarities in the ecologies of *Rodolia* and *Cryptochaetum* with respect to the food niche, predation on *Icerya*, are striking, especially in view of the fact that the two predators belong to distinct orders of the Insecta. The similarity (virtual equivalence in all the important characters that have been examined) is too striking to have occurred by chance: like all other adaptations, these qualities have been evolved because they carried with them selective advantages. In the first place each species has become tied, by strict monophagy, to one species of prey. Then, the struggle for existence revolved around adaptations that might allow one species to make better use of the prey than the other. So far the two species have emerged from this struggle virtually identical in all the qualities that pertain to a successful predator. As a consequence neither species has been able to oust the other, and the two have lived in intimate coexistence throughout southern California for rather more than three hundred generations. It may be that such an outcome is rather unusual for such a struggle. It may be more usual for the forces of the habitat to partition the niche into distinct microniches in which each species may specialize separately. But in this instance Quezada and De Bach (1973) found no partitioning of the food niche.

The most important difference between the two predators seems to reside in their response to weather. Judged by the biomass that each species can maintain (table 10.06, second column), *Cryptochaetum* has a large advantage over *Rodolia* in the coastal region; the two species are evenly matched in the interior region; and *Rodolia* has the advantage in the desert. Judged by the numbers each can maintain, *Cryptochaetum* has the advantage everywhere, but the same correlation with climate is still apparent (table 10.06). From an anthropocentric point of view these differences seem substantial but we must count them as noncritical because *Rodolia* and *Cryptochaetum* are to be found together in all three climatic regions and have been there for a long time. This seems to be a clear case of coexistence. The most plausible explanation is that they can live closely together like this because they are alike in sharing the same niche: the relative strength of those two species in this niche is such that each one can successfully share the niche with the other one wherever *Icerya* occurs in southern California (see tables 10.03 to 10.05). In other words, they profess the same niche with equal proficiency. This explanation has the advantage that it is consistent with the answer that must always be given when the question is, Why do different animals live together in the same habitat? They live together because they all share a niche that is critical for that habitat. Aquatic animals live in aquatic habitats because they profess the niche of taking oxygen from water; terrestrial animals live in terrestrial habitats because they profess the niche of taking oxygen from air. In the present instance *Rodolia* and *Cryptochaetum* profess the same very narrowly prescribed niche—predation on *Icerya*. They are so closely matched in this niche (and not so disparate in any other) that they can coexist in all three climatic regions of southern California.

Quezada and De Bach (1973, 678) put forward a different explanation. They concluded that, even in the interior region where the ratio of biomass was 1.0:0.99 (table 10.06), the observed coexistence was not real but only apparent; they implied that the same was true for the other two regions:

If the two natural enemies are true ecological homologues, their coexistence in the interior would be contrary to the principle of competitive displacement, which states: "different species having identical ecological niches (that is, ecological homologues) cannot co-exist for long in the same habitat. . . . We consider them as ecological homologues as defined by De Bach and Sundby (1963), for they have an identical food requirement, the cottony-cushion scale. We deduce the reason for their "co-existence" in the interior is that dispersal constantly occurs—the fly from the coast and the vedalia from the extreme interior or desert areas. Otherwise one or the other should be displaced.

This explanation rests on the error of the misplaced premise (sec. 10.12). It is also implausible in the light of the evidence.

1. The explanation is not general: it is not needed to explain the coexistence of *Rodolia* and *Cryptochaetum* in Bermuda. Bermuda is a small island in the Atlantic Ocean on about the same latitude as San Diego in coastal southern California. The climate is maritime; there is no scope for a gradient in climate such as occurs in

southern California. In Bermuda *Icerya* is chiefly a pest of *Pittosporum* but occasionally occurs on *Casuarina*; *Cryptochaetum* was introduced in 1948, some forty years after *Rodolia* had become established (Bennet and Hughes 1959). We are indebted to Dr. I. W. Hughes (pers. comm.) for the following comment on the present status of the biological control of *Icerya purchasi* in Bermuda:

The scale remains under excellent control primarily as a result of parasitism by *Cryptochaetum*. I have never looked critically at the relationship between the parasite and the predator, but from field observation it is apparent that *Rodolia* is the less common of the two control agents. Nevertheless, *Rodolia* remains well established and obviously an important influence in the control of the pest.

In Bermuda the two species coexist on much the same terms as in the coastal region of southern California without the possibility of any support by migrants from a distinct climatic region.

2. Of course both species are highly dispersive. Dispersiveness is a prominent feature of the food-finding niche in any good agent of biological control. But what is to be gained by abstracting dispersiveness from the makeup of these two species in order to predict that they would not be able to coexist if only they would not disperse—knowing that they do indeed disperse well? Such special pleading detracts from the plausibility of the explanation.

3. Quezada and De Bach's explanation implies a consistent and substantial net movement of *Rodolia* from east to west and a similar consistent and substantial movement of *Cryptochaetum* from west to east. Considering the distances involved, neither species seems large enough to be independent of the wind. The theory seems to demand at least a subsidiary explanation of how such complementary migrations in opposite directions might take place within the same system of winds.

There are some other examples of the powerful influence of competition theory in section 9.34.

In section 10.24 we tell of four triclads that seemed at first sight to be professing the same niche; but on closer examination each one was found (or inferred) to be specializing in a narrowly defined niche of its own.

## 10.24  Four Triclads That Specialize in Narrowly Defined Niches

The following account is taken from the work of T. B. Reynoldson and his colleagues. An entree to this work may be made through the following papers and their bibliographies: Reynoldson (1966, 1975) and Sigurjonsdottir and Reynoldson (1977). We restrict the discussion to the four species that were studied most thoroughly: *Polycelis nigra*, *P. tenuis*, *Dugesia polychroa*, and *Denrocoelum lacteum*.

Critical differences between these species allow each one to preempt its own particular microhabitat. Each one is superior to all the others in its ability to exploit a particular sort of food. So we refer to this relation between the animal and its food as a niche; Reynoldson used the term "food refuge", which we think is misleading (sec. 10.2).

These four species of triclads live in well-shaded shallow fresh water, in ponds or in the littoral zones of freshwater lakes, where the banks are not too steep or undercut by the water. Stony beds are preferred, and in such places the triclads live on the underside of the stones, where they also find their food. The richness of the food supply both in variety and in quantity depends largely on geology. Hard rocks give rise to chemically poor water, and the productivity of such lakes is low. Soft rocks give rise to water that is rich in chemicals and lakes that are highly productive. Shape, depth, aspect, and, as usual, the activities of humans add their influence. But the concentration of calcium in the water is quite a reliable guide to the heterogeneity of these habitats—that is, to the variety and richness of triclad food that they contain. Reynoldson (1966) measured the calcium in the water of 107 lakes and also measured the standing crop of small metazoans—chiefly worms, crustaceans, and snails—in the places where triclads live. He found that the biomass and the number of species of both triclads and their food were correlated with the amount of calcium in the water (table 10.07).

In rich lakes, with more than 10 mg of calcium per liter, where food is diverse and abundant, the four species were often found living together, and *P. tenuis* was usually the most abundant species. With one minor exception *D. polychroa* and *D. lacteum* were not found in lakes with less than 5 mg of calcium per liter, where the food was obviously less diverse (Mollusca and Crustacea had virtually disappeared). Reynoldson thought that in these lakes the *Polycelis* species, being better adapted for living on the remaining sorts of food (mainly worms but perhaps also some insects), caused an intolerable shortage of food for *D. polychroa* and *D. lacteum*, which they could not alleviate by turning to an alternative sort of food. Of forty-one lakes with less than 5 mg of calcium per liter, twenty-seven contained *P. nigra* alone, twelve contained *P. tenuis* alone, and in two both species occurred together. Reynoldson could see no consistent difference in the lakes to account for the distribution of the *Polycellis* spp. in them, but there is less than one chance in fifty that they were distributed in this ratio by chance. It seems likely that these two species also differ from each other by having specific food niches and that in only

**Table 10.07**  Distribution of Triclads and Their Food in Lakes of Varying Richness as Indicated by the Calcium Content of the Water

| Calcium (mg/1) | < 5 mg/1 | 5–10 mg/1 | > 10 mg/1 |
|---|---|---|---|
| Food: molluscs | Trace | + + | + + + |
| Food: crustaceans | Trace | + + | + + + |
| Food: total | + | + + | + + + |
| *Polycelis nigra* | 75 | 40 | 35 |
| *P. tenuis* | 20 | 100 | 100 |
| *Dugesia polychroa*[a] | 0 | 20 | 80 |
| *Dendrocoelum lacteum* | 0 | 20 | 60 |

*Source:* Compiled from Reynoldson (1966).

*Note:* Total food was measured as biomass of standing crop ranging from 0.4 to 6.7 g collected per hour. The distribution of molluscs and crustaceans was scored as the proportion of lakes in which *Lymnea* and *Hydrobia* or *Asellus* occurred: the range, for lakes in which there was more than 5 mg of calcium per liter, was from 20% to 83% for the molluscs and from 10% to 83% for *Asellus*. The data from different lakes were not strictly comparable. The pluses may be read as a quantitative estimate—low, moderate, or large numbers. The abundance of triclads was scored as the number caught in an hour: the range was from 0 to 160.

[a] Until 1968 two sibling species were included under the name *Dugesia lugubris*, but it now seems likely that most of the populations were dominated by *D. polychroa*.

two of the forty-one unproductive lakes was the habitat heterogeneous enough to support the most favored food of both species of *Polycelis*. On the other hand, in the productive lakes the habitat remained heterogeneous enough to support all four species of triclad, each one exerting its preference for its own sort of food, even though as the season advanced food became so scarce that the triclads usually lost a lot of weight.

Shortage of food is likely to recur after every breeding season, and triclads are well adapted to it. According to Reynoldson (1966), an adult *P. tenuis* that was starved for eight months shrank to 0.5 mm from an original length of 8 mm. Such shrunken individuals will grow into normal adults again when given access to food.

To gain direct evidence of the food preferences of triclads in the field, Reynoldson and Davies (1970) collected triclads from twelve lakes and tested the contents of the gut against sera that contained specific antibodies against Oligochaeta, Gastropoda, and the crustaceans *Asellus* and *Gammarus*. All together 1,837 triclads were examined. One set of results, based on 624 positive reactions, is summarized in table 10.08.

A positive result indicated the nature of the last meal, but only if it had been eaten recently enough to be still found in the stomach. Some individuals gave positive results to two or even three antisera, indicating that they had recently eaten more than one kind of food. Of the 624 positive results 155 came from *Polycelis nigra*, 156 from *P. tenuis*, 157 from *Dedrocoelum lacteum*, and 176 from *Dugesia polychroa*. For each species of triclad the sum of the positive responses to each kind of food is expressed as a percentage of the total of positive responses for that species of triclad. The kind of food with the highest percentage was thus identified as the preferred food. And the "advantage" listed in the last column was calculated by subtracting the score that was achieved for the same food by the most similar species. For example, 58% of the meals identified for *Dugesia* were Gastropoda, so this was its "food niche"; 17% of the meals that were identified for *P. nigra* were also Gastropoda, and neither of the other two species ate a higher proportion of Gastropoda than this; so *P. nigra* was the species most similar to *Dugesia*, and the "advantage" to *Dugesia* over *P. nigra* with respect to *Dugesia*'s food niche was

**Table 10.08**   Food Preferences of Four Species of Triclads

| Species of Triclad | *Asellus* | *Gammarus* | Oligochaeta | Gastropoda | Most Similar Species | Advantage |
|---|---|---|---|---|---|---|
| *Dugesia polychroa* | 16 | 4 | 22 | 58 | *P. nigra* | 41 |
| *Dendrocoelum lacteum* | 63 | 23 | 14 | 0 | *P. tenuis* | 38 |
| *Polycelis tenuis* | 25 | 8 | 57 | 10 | *P. nigra* | −11 |
| *Polycelis nigra* | 8 | 7 | 68 | 17 | *P. tenuis* | 11 |

*Source:* After Reynoldson and Davies (1970).

*Note:* Columns 2 to 5 show the percentage of meals made from each species of food. The most preferred food is recognized as the "food niche" of the species. See text for explanation of columns 6 and 7.

41%. Further sampling confirmed these results but failed to discover any reliable difference between *P. nigra* and *P. tenuis*. Yet these species are not ecologically identical, because it is possible to predict with moderate certainty which one will be found in an unproductive lake by measuring the concentration of calcium dissolved in the water. And it is almost certain that calcium is not acting directly but merely as an indicator of the heterogeneity of the food in the habitat.

Extensive laboratory experiments were done that largely confirmed and explained the results from the field. Triclads, either with one species at a time or with two or more species sharing the same small dish, were offered various species of food, either one species at a time or several at the same time. In some experiments the response to prey that were hale and whole was compared with the response to prey that had been slightly wounded or "stressed" by being entangled in a thin smear of vaseline. Responses were recorded by watching the triclads and recording their behavior, by measuring their growth (either positive or negative), and by measuring the consumption of food. It turned out that *Dugesia* was almost restricted to gastropods, *Dendrocoelum* preferred crustaceans, especially *Asellus*, and the two species of *Polycelis* preferred worms (Oligochaeta) but also would eat *Asellus* and some snails, especially if the prey were wounded. All species of triclads were more strongly attracted to prey that had been wounded, stressed, or restrained. Although the experiments were extensive, they failed to reveal any consistent difference between the two species of *Polycelis*. For any species, its preferences were more than mere preferences: to have access to the food that it preferred seemed to convey an advantage to the triclad that was competing with another species for food that was in short supply. This advantage was especially evident in the results of an experiment designed as a modification of the model proposed by de Wit (1971) for measuring interspecific competition in plants (Sigurjonsdottir and Reynoldson 1977). In one leg of the experiment, six lots of ten triclads comprising six different proportions of *Polycelis tenuis* and *Dugesia polychroa* were confined in a small glass dish and offered either snails, oligochaetes, or both together. In the other leg of the experiment, a mixture of *Polycelis tenuis* and *Dendrocoelum lacteum* were offered *Asellus* or Oligochaetes or both together. The treatments were continued for twelve weeks. The supply of food was not sufficient to promote positive growth: the triclads in all the treatments lost weight during the experiment, indicating that for most of the time the food was in short supply. Four of the six treatments confirmed the results of the earlier experiments. That is, when two species of triclad were confined on one species of food, which was the most favored food of one of them, this one increased its biomass relative to the other. When two species of triclads were confined on two species of food, which comprised the most favored food of both of them, the ratio of their biomass did not change during the experiment—both did equally well. In both experiments it made no difference to the result which species outnumbered the other at the beginning of the experiment.

An apparent exception occurred in one series of experiments in which *Lumbricillus* (Oligochaeta) was offered as the sole food for *Dendrocoelum lacteum* and *Polycelis tenuis*, mixed in various ratios. Sigurjonsdottir and Reynoldson (1977) summarized the results of these experiments:

In the regime with *Lumbricillus* as the sole food, the several lines of evidence all suggest that *Dendrocoelum* and *Polycelis* were able to co-exist at all the mixtures tested and maintain an equilibrium close to the original biomass ratios. It would appear that under the experimental conditions the two species were behaving as ecological homologues. The input-output ratios for biomass lie around the equilibrium line and it seems clear that the slope of the regression line will not differ from unity.

In other words, there was no critical difference between *Dendrocoelum* and *Polycelis* in the way they exploited *Lumbricillus* as food or with respect to any other niche within the confines of this experiment. This result was unexpected because, according to table 10.08, Oligochaeta was the most recent meal for 58% of *Polycelis tenuis* but for only 14% of *Dendrocoelum*: it was the most favored food for *P. tenuis* but only third preference for *Dendrocoelum*. On the other hand, according to the criteria of table 10.08 *P. tenuis* was the species that resembled *Dendrocoelum* most closely.

The experiments of Sigurjonsdottir and Reynoldson did not include a trial between *P. tenuis* and *P. nigra*, presumably because, despite extensive and profound investigation in the field and in the laboratory, no food preference had been found that might distinguish these two species. Nevertheless, the distribution of *P. tenuis* and *P. nigra* in lakes of varying calcium richness (table 10.07 and Reynoldson's comment on these data) suggests that such a difference does exist, even though it has not been identified.

The conclusion to be drawn from this study is that each of the four species of triclad is highly specialized with respect to its ability to exploit a narrow food niche. Each species can, and often does, eat a variety of food, but in times of prolonged or severe food shortage the species that is likely to survive (at the expense of the others) is the one whose most favored food is present. When two or more species survive a shortage of food while living together in the same place, it is because the most favored foods of both (or all) the species are present. More generally, there are critical differences between the ecologies of these four species that prevent any two of them from sharing the narrow food niches that are likely to be practiced during times of food shortage in the habitats where they live. The evidence (in tables 10.07, 10.08, and relevant parts of the text) suggests that the interactions between *P. tenuis* and *P. nigra* and between *P. tenuis* and *Dendrocoelum* conform to this generalization even though the critical difference between *P. tenuis* and *P. nigra* has not yet been identified and, in one laboratory experiment, *P. tenuis* and *Dendrocoelum* seemed to be able to live together on one species of prey.

## 10.25   Conclusion

We follow Elton's usage strictly: the niche defines the animal's profession. So we say in section 10.21 that the larvae of *Dacus* and *Ceratitis* both profess the same niche—feeding in guava. They differ with respect to the symbiont that provides food for them. The difference allows *Dacus* to dominate *Ceratitis*. In artificial medium the difference was eliminated: the contest was equal; both species professed the niche of eating artificial food with equal proficiency, and coexistence seemed to be perfect. In section 10.22 *Aonidiella aurantii* and *A. citrina* professed the same niche—to withstand predation by a particular predator but with different proficiency. They had coexisted for scores of generations while this difference between them remained latent—that is, in the absence of the predator. When the difference was activated i.e., in the presence of the predator—*A. citrina* was soon extinguished. In section 10.23 two predators with different life-styles, because one was a fly and the other a beetle, professed the same highly specific and seemingly homogeneous niche—predation on *Icerya purchasi*. According to all the ecological criteria by which we have been able to judge these two predators, they seem to profess this particular niche in remarkably similar ways and with remarkably even proficiency. They have been coexisting for hundreds of generations and seem likely to go on living together indefinitely. In section 10.24 four triclads in good times or rich places seem to share the same niche—feeding on various freshwater invertebrates. But when food was scarce, especially when the variety was small, each species of triclad manifested the ability to specialize in using its own favored food to the detriment of those others, whose favored food was absent or scarce. But the two species of *Polycelis*, with very similar ecologies, were most successful at living closely together in a homogeneous habitat.

Our review of these modern studies of coexistence and displacement, together with those that are reviewed in section 7.15, leads to two useful generalizations: when two distinct species that seem to depend on the same niche fail to live together in the same habitat, the critical difference between the species can usually be traced to one's being better adapted than the other to profess that particular niche. Conversely, when two distinct species seem to depend on the same niche but nevertheless continue to live together in the same habitat, there may be two explanations. It may be that (a) their ecologies and adaptations can properly be said to be the same with respect to this particular niche. That is, the closer the contest the better the chance that the two species can live together. This rule includes the special case where success in a niche is assured because some other component of environment keeps both species rare with respect to opportunities to exploit the niche; for example, the condition of the species that occupy the top left corner of figure 3.01. Or (b), the animals may have succeeded in partitioning the niche in a way that scientific inquiry has not yet revealed, as perhaps with the two species of *Polycelis* (sec. 10.24).

Finally, we think that our general conclusion, especially the first leg of it, has been implicit in much of the conventional discussion of "competitive displacement" but it has not hitherto been explicit. Too much of this discussion has been conducted under that misleading concept, competition. The theory of environment leads to an explanation of the empirical facts that is more realistic and more plausible. The facts themselves are not disputed.

## 10.3   Outbreaks

Great, often sudden increases in the numbers of such well known pests as locusts or mice are known colloquially as plagues or outbreaks. During the height of an outbreak of locusts or mice their numbers might exceed by several orders of magnitude those that persist, usually in refuges, between outbreaks. There are many other species whose outbreaks are just as spectacular as those of locusts and mice—spruce budworm (sec. 14.1), thrips (sec. 10.12), the pink gum psyllid (sec. 10.311), the grey teal (sec. 13.2), and the European rabbit (chap. 12), to mention a few that happen to have found their way into this book. These species are remarkable only in the magnitude of their oubreaks. If a more modest criterion were chosen to specify an "outbreak", say, an upward fluctuation that exceeded the inter-outbreak mode by perhaps a factor of ten or twenty, most species might have to be counted as likely to produce an outbreak occasionally. Probably the only ones that could be safely counted out are those that conform to the paradigm that we (Andrewartha and Birch 1954, 23) set up with a hypothetical population of bees that made nests in auger holes in fence posts. The auger holes were reusable, and in each generation (so favorable was the rest of the environment) every hole was found and used by a bee (see also fig. 9.06).

In nature, herons *Ardea cinerea*, storks *Ciconia ciconia*, and the great tit *Parus major* seem to conform to this paradigm. A ceiling to the number of breeding pairs seems to be set by the number of reusable nesting sites, which does not vary much. And bad weather, which might be the most likely cause of a "crash" in numbers, is not likely to reduce their numbers to less than half of the ceiling for herons and storks or one-fifth for *Parus* (Colinvaux 1973, 362). A variety of seabirds that nest on small offshore islands and certain strongly territorial animals like the Australian magpie (sec. 13.1) might be added to the list. But the full list might not be large. Most species must be counted as having a chance of an occasional outbreak, so outbreaks are commonplace. The question is, What causes them? It might be best, at least for a start, simply to inquire into the causes of fluctuations without setting any arbitrary limit on how big a fluctuation might have to be before it may be called an outbreak.

It was shown in section 10.12 that fluctuations in the size of the population of *Thrips imaginis* might be predicted by four sets of meteorological data. These components of weather were chiefly important as modifiers that influenced the supply of food for the primary animal, but certain aspects of weather also acted directly on the thrips, either as a resource other than food or as a malentity.

The theory of environment suggests that any component of environment might be influential, but experience teaches that the pathways that lead through food are likely to be the important ones. Another reason for concentrating attention on food in this section is that the chief purpose of this chapter is to compare competition theory with the theory of environment as a source of explanations for controversial issues in population ecology. Competition theory must inevitably rely on some form of "density-governing mechanism", and that leads almost exclusively to the predator-prey interaction. So the discussion of outbreaks in this section is restricted to those that are related to the predator-prey interaction. This restraint excludes relatively few species. Apart from the species that have specialized in nonreactive food—organic residues, artifacts, excretions and so on (sec. 3.1), every animal is a predator (sec. 1.42) and virtually every animal is prey as well. The herbivorous predators prey on plants, and the carnivorous predators prey on animals. It is pointless, for a general ecological analysis, to distinguish between herbivorous and carnivorous predators.

There are only two logical alternatives to explain an outbreak in terms of the predator-prey interaction—an unusual relaxation either of a food-shortage or a predation-pressure. A useful distinction can be made between intrinsic and extrinsic shortages of food. By analogy a similar distinction can be made for predation-pressure (sec. 10.33).

## 10.31   Relaxation of an Extrinsic Shortage of Food

In the Canadian Arctic, major outbreaks of the lynx *Lynx canadensis* recur about ten times in a century and major outbreaks of the fox *Alopex lagopus* recur about twenty-five times in a century. Records of the fluctuation in abundance of these two species can be traced back as far as the early days of trading in their pelts by the Hudson's Bay Company—at least back to the middle of the previous century. The fox eats chiefly lemmings and other small rodents. The lynx depends on the hare *Lepus americanus*. From early days it was known, from the reports of professional trappers, that outbreaks in the numbers of the prey of the lynx or the fox usually coincided with or preceded the peaks of abundance in the numbers of predators (figs. 10.06, 10.07). But these outbreaks have still not been fully explained.

In the "logistic" theories of Lotka and Volterra changes in the intensity of food-shortage and predation-pressure are by definition equally responsible for the predicted "coupled oscillations" that appear in the model (sec. 10.331, fig. 10.08). In the model, both the shortage of food and the "pressure" from predators are what we would now call "intrinsic" (sec. 3.131). Early discussions of the outbreaks of lemmings, foxes, hares, and lynx tended to follow the popular logistic theory. Elton (1942) was the first to recognize that the outbreaks and crashes in the numbers of the carnivorous fox and lynx were the result of an extrinsic abundance of food followed by an extrinsic shortage. In other words, neither the origin of the out-breaks of the herbivorous lemming or hare nor the subsequent collapse of the outbreaks could be attributed to "predation-pressure". For a modern summary of these discoveries, with historical leanings, see Colinvaux (1973, 483). We are left with the need to explain the outbreaks of these small herbivorous mammals without leaning heavily on the relaxation of predation-pressure.

There has been a spate of research and theorizing about the causes of the outbreaks, and more particularly the crashes that follow outbreaks, in the population of voles, mice, and lemmings. Despite the progress that has been made, we are still without a satisfactory explanation for the outbreaks. Food has usually been considered and then put aside. There seems to be a consensus that food cannot be important: because (a) Most of the adults that are present when the outbreak reaches its maximum and the crash is about to begin appear to be well fed and in good condition; the numbers crash chiefly because these apparently well-conditioned young adults fail, throughout the whole breeding season, to produce viable young. (b) The population crashes while, to the onlooker, there still seems to be plenty of food about (Krebs 1964).

Nevertheless, Andrewartha (1970, 82) and Colinvaux (1973, 488), after re-viewing the literature, speculated whether food had been sufficiently explored, not so much the quantity of food as its quality. The trained eye of a naturalist might estimate quantity with confidence, but quality may be more elusive (Newsome 1967). The close synchrony between outbreaks and crashes of lemming populations over the vast area comprising the 8 regions represented in figure 10.06 suggests that weather may be influencing the timing of the outbreaks. And weather is quite likely to operate through the food of an herbivore (secs. 10.12, 10.311, 12.31, 12.422). Mullen (1969) discovered one way in which weather might cause an outbreak of lemmings but not, it seems on present evidence, through its food. The breeding

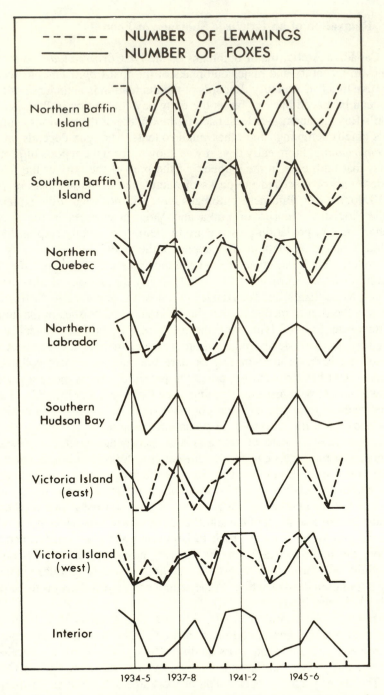

**Fig. 10.06** Changes in numbers of lemmings and arctic foxes and years of major migrations of snowy owls (shown by the vertical lines). Note how consistent the curves are for all eight districts of the large Hudson Bay territory. Also, the curves for predator and prey lie close together but with a slight but fairly consistent tendency for the curve for the prey to "lead" the curve for the predator. The coincidence in the migrations of snowy owls with the "peak" in the numbers of foxes is also convincing. (After Chitty 1950)

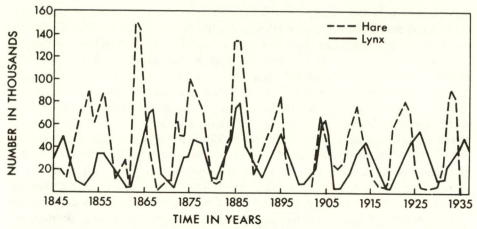

**Fig. 10.07** The numbers represent the numbers of skins of hare and lynx traded by the Hudson's Bay Company. Assuming constant trapping effort, the curves represent the size of the populations of hares and lynxes. As in figure 10.06, the curves lie close together, though there is a slight but fairly consistent tendency for the curve for the prey to "lead" the curve for the predator. (After Colinvaux 1973)

breeding season for the lemmings occurs in the period July through August, but it may be long or short. Mullen found that the breeding season was longer, because it began sooner, when there was a burst of unusually warm weather during the period May through early June. Mullen suggested that the burst of warm weather was what we would call a token because it triggered the animal's change to the breeding condition.

The European rabbit in Australia can breed only when its diet includes green growing herbage. So the breeding season is restricted to winter and early spring because the herbage dries off and changes to straw or becomes highly sclerophyllous with the approach of summer. Among the new generation, only those will

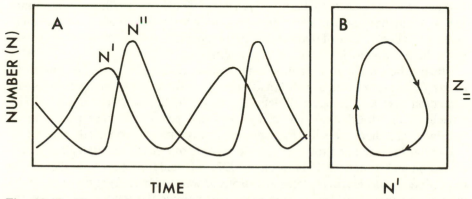

**Fig. 10.08** The coupled oscillations that are intrinsic to the predator-prey interactions according to the theories of Lotka and Volterra. $N'$ is the prey, $N''$ is the predator. A, numbers of predator and prey plotted against time; B, numbers of prey plotted against numbers of predators. In the Lotka-Volterra equations, when $dN'/dt = 0$ and $dN''/dt = 0$ the loop in figure A becomes a point (the "singular point"). At this point all oscillation ceases. (From Andrewartha and Birch 1954, 414)

survive that were born early enough to have matured (i.e., to weigh more than about 1 kg) before the vegetation loses its quality (secs. 12.31, 12.422). Outbreaks are likely to develop when there has been a run of good breeding seasons.

Perhaps thrips (sec. 10.12), rabbits, and lemmings all have this much in common—that to develop an outbreak they need a run of unusually long breeding seasons extending through a number of successive generations. With thrips and rabbits it is clear that weather is the ultimate cause of the outbreak. For both of them, weather determines both the beginning and the ending of the breeding season. The cause and course of the outbreak are entirely due to the changes in the severity of an extrinsic shortage of food. With thar (sec. 10.32) it seems likely that outbreaks might develop through favorable weather acting on their food. But it seems likely that the outbreak would come to an end through an intrinsic shortage of food. With the lemming, not knowing how outbreaks begin or how they end, we can say nothing about the relative importance of extrinsic or intrinsic shortages of food except perhaps this: if the end of the outbreak is caused by an intrinsic shortage of food it will be a shortage of some cryptic component of the diet, not a straight-forward shortage of edible material. In section 10.311 we discuss deficiencies in diet, but they are all associated with extrinsic shortages. We do not know of an authentic example of an outbreak that has developed through the alleviation of, or has come to an end through, an intrinsic deficiency of diet (but see secs. 12.424, 12.426).

### 10.311   The Special Importance of Diet

It was largely for the sake of logical completeness that Andrewartha and Browning (1961) included "deficient diet" in the class of relative shortages of food. But the example we happened to quote (of the wallaby that browses on a shrub which remains sclerophyllous and deficient in essential nutrients for most of the year except for a brief flush of nutritious growth during the winter) has turned out to be almost typical of the large array of examples of deficient diet that have now accumulated around this idea. The discussion in this section depends largely on the writings of White (1969, 1970a,b, 1971, 1973, 1974, 1976, 1978, and pers. comm. 1982). White has suggested that temporary alleviation of an inadequate diet might be the most likely cause of outbreaks of herbivores and, through them, of organisms in the higher trophic levels of the ecosystem.

The psyllid *Cardiaspina densitexta* is virtually restricted to the pink gum *Eucalyptus fasciculosa*, which is a small tree. It used to grow in open woodland formation over a large but well-defined area of South Australia. Farms now encroach on the distribution, but many trees and patches of trees persist in the farmland, and there are still large areas of nearly pristine woodland to be found.

Popular literature and folklore contain records of a massive outbreak of *Cardiaspina* on pink gum from 1914 through 1922. It is said that many trees died before the outbreak subsided; many trees that survived were cut back from the north so that the tree remained lopsided, with most of its canopy to the south of the trunk; this appearance is characteristic of trees that have been severely infested with *Cardiaspina*. The next oubreak lasted from 1956 through 1963.

During the long interlude between outbreaks, the population reverted to its usual sparse condition. Most trees supported no psyllid. Occasionally part of a tree, a whole tree, or a group of trees would be found infested with moderate numbers of psyllids. Very occasionally, and always in special circumstances, a tree or a group of trees would support numbers equivalent to those that are found during a widespread outbreak. For example White (1969, 907) reported one such instance, not for *C. densitexta* but for a related species, *C. retator*, which occasionally develops similar outbreaks on *E. camaldulensis*.

One such [occurrence] was a very localized outbreak of *Cardiaspina retator* Taylor on *E. camaldulensis* growing along the banks of the Torrens River in the hills behind Adelaide. This occurred during the long dry spring and summer of 1963–64, and was confined to trees growing on the edge of a lake formed by a weir across the river. Daily records of the Water Supply Department showed that this lake was kept full for an unusually long period during the winter of 1963, drained early in the spring, and left empty all through the ensuing summer. When the lake is full many of the affected trees have their roots flooded, and when it is empty the water level is some 20 feet lower, leaving the trees literally "high and dry." [White went on to anticipate the argument that we develop below.] This very localized and short-lived outbreak would appear to have arisen as a result of a sudden and extreme increase in stress of the trees attacked, caused by the artificially prolonged winter flooding followed by the prolonged summer drought. This is very similar to the natural rainfall pattern associated with general outbreaks of psyllids on *E. camaldulensis* in the Adelaide area.

There is no dormant stage in the life cycle of *Cardiaspina*. Development is slow during winter, but it does not cease. There are three generations in a year. The young adults of the spring generation are highly dispersive; their behavior makes it likely that they will be picked up by ascending currents of air and carried to join the aerial plankton. In contrast, the adults of the summer and autumn generations are highly sedentary. If they happen to be displaced from the tree where they were born and grew up, they will do their best to regain it.

Whether the adult on a tree is a resident of long standing or a newly arrived immigrant via the aerial plankton, it is highly selective about where it will settle down to feed. It feeds by inserting its mouthparts through a stoma into the phloem. It prefers the north face of the tree to the south face, a "sun leaf" to a "shade leaf"; a mature leaf is preferred almost exclusively to an immature one or an old one past its climax; and a place near the midrib in the basal third of the leaf is strongly preferred to any other part of the leaf.

Because for the female feeding is punctuated by egg laying, the eggs are laid in much the same place as the parent feeds. The distribution of the eggs is more clumped than that of the actual feeding sites because the female, without moving far from where she was feeding, nevertheless seeks a "ridge" or an obstacle against which to place the egg. For the first eggs to be laid on a leaf the midrib will do, but subsequent eggs tend to be placed against other eggs that are already in place. The nymphs tend to settle close to where they hatch, but they do wander, especially when crowded, so the distribution of the nymphs becomes less clumped than that of the eggs. But the nymphs either settle in the place where their parent chose to feed or they find a similar place for themselves. One must infer that such a strong and reliable pattern of behavior has evolved in response to strong and consistent selection pressure. It seems reasonable to postulate that the reward for this behavior is a good supply of food for the female while she is producing eggs and for the nymph while it is growing. It is difficult to imagine a shortage of sap. So it seems reasonable to interpret "good supply of food" as food of better quality than can be found in any other sort of place. White investigated this idea.

White had noticed repeatedly that nymphs that happened to settle in atypical places—on trees other than pink gum, on the south side of pink gum trees, on shade leaves, even on the apical instead of the basal third of a "sun leaf"—would die young. They showed signs of having died from malnutrition. Their mouthparts were normally placed through the stoma into the phloem, as if they had been sucking sap to the last. They had built at least part of the lerp which they normally build to cover the body. The construction of the lerp suggested "business as usual" up to the time of death. There was no sign of physical violence or disease. Because the raw materials for the lerp are carbohydrates which occur in the sap in excess of nutritional requirements, the malnutrition could not be attributed to these constituents. White thought the deficiency might be nitrogen in soluble form, probably as amino acids.

White confirmed the starvation hypothesis (but not the specific deficiency of nitrogen) by allowing newly hatched nymphs to settle on leaf-disks cut from mature leaves and floated on distilled water. Half the disks were exposed to dull light (200 lx) and the other half to bright light (7,000 lx). By the twelfth day 50% of the nymphs on the disks in the dull light were dead; by the twenty-eighth day 98% were dead. About 75% of those that died under the lerp with the mouthparts still in place showed the same signs of death by malnutrition as had been noticed for nymphs that had settled in the "wrong" place in nature. On the disks under bright light 90% of the nymphs were alive and apparently healthy after twenty-eight days. White concluded that as a result of the active photosynthesis that probably went on under the bright light the sap was nutritious in all the ingredients the psyllids required for healthy growth. Under dull light, in the absence of active photosynthesis, the sap was deficient in at least one essential constituent. But the dead psyllids had been building lerps, apparently quite normally. So the deficiency was not likely to be among the carbohydrates.

A closer look at the sort of place where an adult *Cardiaspina* that is living during an interval between outbreaks will elect to feed and lay its eggs suggests that the psyllid has chosen a place where the products of photosynthesis are likely to be concentrated. Photosynthesis is likely to be most active in mature "sun leaves"—sunlit leaves on the northern face. Because the products of photosynthesis are transported from the leaf through the phloem vessels in its stem, they are likely to be concentrated in the phloem vessels in the basal part of the leaf as they converge on the stem. Even so, a nymph that settles here, in the best place it can find, may be exposed to a high risk of death by malnutrition because even the best place may not be good enough when the tree is experiencing the environment that prevails between outbreaks. At such times the distribution of colonies of psyllids is extremely patchy, and in most colonies the death rate among juveniles is high. Most trees support no psyllid, probably because the chance of finding nutritious food in them is too low. This consideration led White to look at the environment of the pink gum.

Assume, in order to build a hypothesis, that the critical deficiency is in soluble nitrogen, probably specific amino acids. What variation in the environment of a pink gum is likely to lead to a concentration of amino acids in the place where a psyllid is most likely to look for it—the basal third of a mature sun leaf on the north face of the tree?

In the Mediterranean climate of South Australia, rainfall is the most likely candidate; and the fine feeding roots of the tree are the tissues that are likely to be extensively damaged by an unfavorable water regime. The smaller roots might be damaged if the soil remained waterlogged for too long; they would stand in urgent need of repair with the approach of the summer drought.

To appreciate the influence of rainfall we must take the soil into account. Through the heart of the distribution of pink gum the mean annual rainfall varies from 470 mm to 540 mm. The soil is shallow; a layer of sand overlies a layer of clay which overlies a layer of impermeable limestone. The layer of limestone increases the risk from waterlogging; the shallowness of the soil and the lightness of the top layer imply a small capacity for storing water and thus, an increased risk of desiccation during the summer drought. The northern limit to the distribution occurs at about the 450 mm isohyet, and the trees grow in light soils in localities that are topographically moister than the general countryside. The southern limit occurs at about the 700 mm isohyet. The trees grow in soils that are well drained but not too deep; water that is stored deeply would be out of reach of the shallow rooting system of a pink gum. White said that the pink gum is very sensitive to its water regime, and he commented (White 1970c, 15):

This sensitivity takes the form of a low tolerance to seasonal differences in soil moisture and is largely the result of its shallow-rooting habit accentuated by the prevailing soil type and the marked seasonal distribution of the rainfall. . . . Thus in the north the distribution is limited by summer drought at a point where rainfall in dry years is not sufficient on these desert soils to provide enough stored water to maintain the species through the hot summer (Specht and Rayson, 1957). In the south the distribution is again limited by summer drought because here winter rainfall makes all but the high, dry, sandy sites too wet for pink gum; but on these dry ridges in dry summers the soil dries out to the extent that only the deep-rooting *E. baxteri* can survive.

White sought a statistic that would measure the influence on this sensitive tree of variation in annual and seasonal rainfall. Because, as we have seen, within the distribution of pink gum, differences in soil and topography tend to compensate for regional differences in rainfall, it is best to convert deviations from the mean rainfall for each district to the scale of the standard normal deviate. This transformation allows comparisons to be made across regions as well as across years in the same region. A "stress index" (SI) was defined:

$$SI = d_w - d_s$$

where $d_w$ stands for deviation from the mean winter rainfall and $d_s$ stands for the deviation from the mean summer rain. The relevant winter precedes the relevant summer. The deviations are expressed in the scale of the standard normal deviate. Because $d_w$ will be positive for an abnormally wet winter and $d_s$ will be negative for an abnormally dry summer, the index will be large and positive when a dry summer follows a wet winter.

The biological rationale of the index is that during a wet winter the soil may become waterlogged and the feeding roots may be drowned; the consequent short-age of water (and nutrients) may injure the whole tree, including the aerial parts; the damage may be aggravated by a dry summer. The unusually large supplies of nutrients required to repair the injuries are likely to be made and mobilized in the leaves, especially the mature sun leaves on the northern face. As the nutrients are collected for general transportation, strong concentrations of them are likely to develop in the vessels of the phloem as they converge on the stem, that is, in the basal part of the leaf. On this rich diet the psyllids will lay more eggs, and the young nymphs will survive in large numbers where previously most of them had died from malnutrition. The stress index for Keith, a town in the heart of the distribution of pink gum, is shown in figure 10.09. Both of the known outbreaks were associated with high values for the stress index.

Accepting the theory that lies behind the stress index, it is possible to compare the condition of the psyllid during an interlude between outbreaks with its condition at the beginning of and during an outbreak. While the stress index remains negative or low the psyllids remain scarce; they are to be found mostly in small and patchily distributed colonies. Most of the trees support no psyllid. Doubtless some of these empty trees will not have been found by a wandering female with eggs to lay,

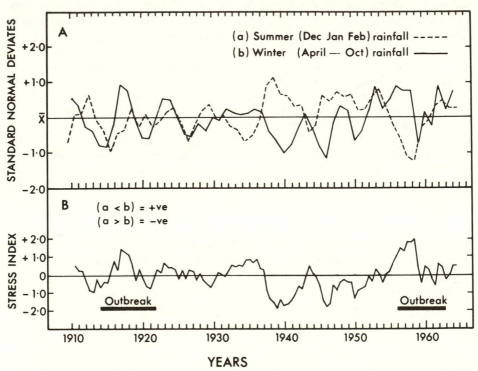

**Fig. 10.09**   Outbreaks of *Cardiaspina densitexta* on *Eucalyptus fasciculosa* at Keith, South Australia, in relation to "stress index." (A) Curve a (*dashed*) shows the summer rainfall plotted in units of the normal deviate about its mean, and curve b (*solid*) shows the winter rainfall similarly plotted in units of the normal deviate about its mean. Both curves are three-year running means. (B) The curve shows the stress index, which is the algebraic sum of b minus a. The stress index is positive when a is less than b, that is, when the summers were dry relative to the preceding winters. (After White 1969)

because they are scarce also. But it seems unlikely that many of these trees would have offered an immigrant colonizer a diet rich enough to give her a chance to establish a colony. In other words, a dispersing female has only a vanishingly small chance to find a place where she can found a new colony. Nevertheless, by chance a few trees, or parts of trees, scattered about the woodland, perhaps in atypical and therefore unfavorable places, have been "stressed" sufficiently for them to support a colony.

When the weather pattern changes and the stress index begins registering a sequence of high values, a rich concentration of nutrients develops in the sap, and most trees come to offer a rich diet to any psyllid that lands on them. The established colonies increase rapidly, and the luck of the dispersers changes dramatically. Their chance of landing on a pink gum may still not be very large, but for those that do the chance to found a colony has changed from close to zero to close to one. The outbreak is under way, and the psyllids remain numerous until the stress index reverts to its normal low value again.

Toward the end of the 1956–63 outbreak a number of species of predators became numerous. There was no evidence that they had caused or were likely to cause a decline in the outbreak before the stress index reverted to low values.

This explanation of outbreaks of the psyllid *Cardiaspina densitexta* on pink gum, *Eucalyptus fasciculosa*, attributed the outbreak to a relaxation of an extrinsic relative shortage of food. The shortage takes the form of an excessive dilution of an essential nutrient (possibly nitrogen) in the sap of the tree. The shortage is relaxed when the tree is "stressed". The immediate cause of changed condition of the tree is an unfavorable water regime which comes about through the persistence of an unusual pattern in the rainfall. So the ultimate cause of the outbreak is weather, measured as seasonal rainfall, which is a modifier of the first order in the envirogram of the tree but a modifier of the second order in the envirogram of the psyllid. The weather occupies two slots in the envirogram of the psyllid because, for the tree, excessive water during winter is a malentity and sufficient water during summer is a resource. The explanation seems plausible except for one serious gap; we still lack critical evidence that the food shortage is due to an intolerable dilution of soluble nitrogen in the sap. However, a good deal of circumstantial evidence from other species points in this direction.

Many plants, especially those that grow in reliable climates, have evolved a seasonal rhythm in their life cycles that allows them to spend most of the year with only dilute concentrations of soluble nitrogen in their tissues. In this condition they offer little prospect of a nutritious diet to any herbivore that might seek to eat them (McNeil and Southwood 1978). This defense against prospective predators breaks down when the plant is growing vigorously or producing fruit because at such times soluble nitrogenous compounds are made in large quantities in the mature leaves and transported to the growing points. For the same reason the progression of the leaves from immaturity through senility tends to weaken this defense. The vigorously growing immature leaves consume more than they produce; the mature leaves produce more than they consume; the senile leaves put back into circulation the soluble nutrients that can be retrieved from their decadent tissues. Wherever a stock of soluble nutrients accumulates, especially soluble nitrogenous compounds, it is likely to be exploited by herbivores. There are many adaptations. Adaptations by herbivores to exploit are countered by adaptations by plants to protect (sec. 5.211).

White (1974, 1978, and pers. comm. 1982) wrote about this subject too, but he gave more space to the sporadic sort of outbreak like the one that he had studied in *Cardiaspina*. He reviewed the literature and found, chiefly among insects, birds, and mammals, an impressively large array of examples in which: (a) An herbivore was kept rare by an extrinsic relative shortage of food. (b) The shortage seemed, or was thought by the author, to take the form of an extreme dilution of nitrogen. (c) There were outbreaks or there were increases in numbers that suggested the possibility of outbreaks. (d) The increase in numbers always seemed to be due to an alleviation of the shortage of food. (e) The shortage of food was sometimes alleviated when the plant was "stressed" as in the classic example of *Cardiaspina*. But there were a number of examples in which obviously "non stressful" components in the plant's environment seemed to produce the same result; not all of them were shown to have increased the concentration of soluble nitrogen in the tissues of the plant, but this remained the most plausible explanation.

## 10.32   Relaxation of Extrinsic Predation-Pressure

Any variation in the environment of a predator may influence the predation-pressure on the primary animal (sec. 5.2). Perhaps the most interesting if not the most usual condition for extrinsic predation-pressure is the facultative predator with an alternative source of food in its environment. The classic example comes from the introduction of *Ptychomyia* into Fiji, which was described by Tothill, Taylor and Payne (1930) and summarized in Andrewartha and Birch (1954, 486). The tachinid fly *Ptychomyia* was introduced into Fiji to control the moth *Levuana*, which was a pest of the coconut. *Ptychomyia* preferred *Levuana*, but it could maintain its numbers when *Levuana* was scarce by eating *Eublemma* and *Plutella*. On one island where the alternative food was scarce *Ptychomyia* died out and there was an outbreak of *Levuana*. The phenomenon is commonplace. We describe a number of examples in this book. On Isle Royale wolves prey on moose, but beavers are an alternative source of food for the wolves (sec. 5.1). The alternative food for *Comperiella* is *Aonidiella aurantii* (sec. 10.22). The alternative food for *Danaus* in Barbados is *Calotropis*; the preferred food of *Danaus* is *Asclepias*, which is the sole food of *Oncopeltus*. The consequent severe predation of *Danaus* on *Asclepias* seemed likely to cause a fatal extrinsic shortage of food for *Oncopeltus* (sec. 8.432). And the vertebrate predators of the rabbit in Australia may be kept scarce by a dearth of alternative food during the summer when there are few juvenile rabbits (sec. 12.432).

In general, when the primary animal is preferred by the predator the predation-pressure on the primary animal will be positively related to the supply of alternative food for the predator; when the primary animal is not preferred by the predator, the predation-pressure on the primary animal is likely to be negatively related to the supply of alternative food for the predator.

## 10.33   Relaxation of an Intrinsic Shortage of Food or Intrinsic Predation-Pressure

The idea of intrinsic shortage of food is familiar (sec. 3.132). The idea of intrinsic predation-pressure implies that the activity (abundance) of a predator depends on the supply of food in the previous generation (fig. 1.03A). This is the numerical response of some authors.

Depending on which member of a predator prey couple is the primary animal, the other member will be either food or predator in the centrum of the primary animal (secs 5.0, 5.1). This coupling is a fact of nature and has to be incorporated into our explanation of outbreaks. Paradoxically, the first, but very persistent, attempt to recognize this coupling with strict logic has proved a dismal failure. It is a well-known attempt, so it can be dismissed briefly in section 10.331. The lesson to be learned from this failure is that we must be more realistic (sec. 10.332).

### 10.331   An Attempt to Abstract the Predator-Prey Interaction from the Rest of the Environment

In the logistic model (Lotka 1932; Volterra 1931), the rate of increase of a population that lives in a limited space with a constant supply of food may be expressed by the equation (Andrewartha and Birch 1954, 349):

$$r = r_m (1 - N) N/K$$

K is the upper limit to the population.
N is the size of the population at any moment of time.
$r_m$ is the rate of increase of an individual that is reproducing free from any restraint from too many other animals of its own kind.

The rate of increase in the population is fully determined by N, K, and $r_m$: all other components have been abstracted from the model.

The logistic formula may be extended to cover the abstract predator-prey interaction by writing a pair of equations (Andrewartha and Birch 1954, 413):

$$r' = N'r'_m - C'N'',$$

$$r'' = N''d'' + C''N'.$$

$d''$ is a death-rate of predators that is related to the number of animals in the population.
$C'$ and $C''$ are constants that depend only on the niches of the predator and prey.

When these equations are solved with arbitrary values for $r'_m$, $d''$, $C'$ and $C''$ they generate graphs as in figure 10.08A,B. The curve in figure 10.08B is the orbit of a point that expresses N' relative to N''. At specific values of N' and N'' the orbit condenses to a point—the "singular point", which has also been called the "steady density". In figure 10.08A oscillations decline and finally disappear as the curves approach the "steady-density". The curves in figure 10.08A are "coupled oscillations". The coupling is explicit in the mathematics: the exponential growth of N' is corrected by subtracting a quantity that is a function of N''; and the exponential decline in N'' is corrected by adding a quantity that is a function of N'. In layman's language this mathematics means that the curves in figure 10.08A are mutually determined by intrinsic shortage of food for the predator and intrinsic predation-pressure on the prey. All extrinsic influences, indeed all other components of environment, are excluded from the equation. This end is achieved by making $r'_m$, $d''$, $C'$, and $C''$ constants. In nature $r_m$, $d''$, $C'$, and $C''$ are variables that are determined by the environment. To assume constancy for these four terms is to imply that the animals have no measurable environment except the densities of their own populations. This assumption measures the extreme lack of realism in this model. Of course it is not possible to impose such a condition on real animals either in natural populations or in laboratory models. A few laboratory models have produced persistent oscillations in the numbers of predator and prey, but only by dint of carefully manipulating the environment to match the particular life-styles of the animals (Andrewartha and Birch 1954, 438; Utida 1957; Huffaker 1958). None of the empirical oscillations has been shown to be critically like the coupled oscillations that are predicted by the model. Whether they conform or not, it seems certain that they are not to be related to the predictions of a model that specifically denies influence to all components of environment except the densities of the populations of predator and prey.

This criticism also holds for attempts to fit this model to natural populations and for any model, no matter how sophisticated it might be, that embraces the fundamental assumptions of logistic theory, especially the assumption that $r'_m$, $d''$, $C'$, and $C''$ are constant. That is why we doubt the relevance of Caughley's (1977, 130) statement about a prediction he got when he attributed arbitrary values to all the constants of one such model. He wrote: "Most wildlife managers will recognize [its] trajectories as a blue-print for an ungulate eruption". For the same reason we doubt the relevance of many similar models (sec. 9.02).

### 10.332   An Attempt to Set the Predator-Prey Interaction in the Perspective of the Theory of Environment

Perhaps the best place to look for the predator-prey interaction at its strongest would be in cell 4 of table 3.02 or in the bottom right corner of figure 3.01. These places house the successful agents of biological control such as *Cactoblastis* or *Asolcus* and certain pests such as *Hemitragus* (Himalayan thar). Such species arrive in the bottom right corner of figure 3.01 by traversing a pathway similar to the one indicated on figure 3.01 for *Cactoblastis*. It is customary, when speaking of large mammals like thar, to refer to their movement across the top of the figure as an "eruption". But the other species also "erupt" in similar fashion. All such species, after they have settled in the bottom right corner, fluctuate in response to any component of environment that happens to be influential and variable. To explain the fluctuations one must delve into the web of predator or prey as far as may be necessary.

Caughley (1970) reconstructed the passage of thar across and around figure 3.01 by the neat device of making distance speak for time. The Himalayan thar, a goatlike ungulate, was introduced into New Zealand in 1904. The original immigrants were all released together in the same place, and they have slowly extended their distribution through a large area of rugged country. Caughley (1970) sampled extensively (chiefly by shooting) in four localities. In (a) the original center of release; in (d), near the present margin of the distribution; and in (b) and (c), intermediate between (a) and (d) but (b) was closer to (a) and (c) was closer to (d). He held the hypothesis, taken over from Riney (1964) that: (1) At the locality (a), i.e., at the original center of invasion, the population would have become adjusted to the new condition of the food, that is, to the changes in the distribution and abundance of the vegetation that had been caused by the original eruptive outbreak in the numbers of thar. (2) At the margin of the distribution (on the assumption that the distribution was still expanding), the population would be geared for rapid increase in response to abundant food that had not been depleted by the feeding of previous generations. (3) At locality (b) the population would be geared for zero rate of increase (r = 0). (4) At locality (c) the population would be declining (r < 0). This hypothesis rested on the assumption that a transect across an expanding distribution would recapitulate the history (in time) of the original population at its place of origin.

Caughley constructed life tables and calculated exponential rates of increase; he considered rates of twinning; he measured fat as an index of r; and he inspected the conditions of the herbage that was the staple diet of the thar. Not surprisingly considering the technological difficulties and the number of assumptions, the empirical results did not mirror the hypothesis very closely, but large and highly significant differences between the estimates of r at critical localities (the difference between locality b and locality c was most important) were in keeping with the hypothesis. Certain other criteria pointed in the same direction, and none of the observations contradicted the hypothesis.

Caughley reconstructed the history of the original colony. It is likely that the local population of thar, at the place where they were released, quickly erupted into a massive outbreak in response to the abundance of food. They would overshoot the food supply. The increase would first be halted, then converted into a catastrophic decline by a severe intrinsic shortage of food. The nature of the vegetation would change, the distribution of the food would become more patchy, and its abundance would be reduced. Some of it might be replaced by inedible plants. Ultimately the population would become adjusted to the new food regime, and the animals would respond in the usual way to an intrinsic relative shortage of food (sec. 3.131). The distribution and abundance of the food would fluctuate with vagaries in the weather or in some other component of the environment of the plants, and the rate of increase in the population of thar would follow that in the food more or less faithfully. Variations in the age structure of the population and in other components in the environment of the thar might modify the simple response to food. Occasionally an unusual run of favorable events in the environment of the food might spark off an outbreak of thar, but the outbreak would be temporary and probably local. As the thar expanded its distribution, pristine localities would be colonized, and history would be repeated in an ever-expanding wave spreading away from the original center of invasion until all the suitable habitat would have been occupied. The occasional outbreaks that might occur in the more settled areas behind the frontal wave might differ quantitatively from the eruptive outbreaks that occur with the colonization of a pristine locality, but they do not differ qualitatively. So Caughley could see no fundamental difference between the ecology of a colonist and that of an animal in an area that had long been occupied.

Caughley reviewed the literature about other ungulates that had been released in New Zealand and some that had been released in the Northern Hemisphere, mostly on islands. He thought their ecologies might be essentially the same as that of the thar. There seems to be nothing in any of these observations to suggest that the populations are responding to density-governing mechanisms of the sort that generate cyclical coupled oscillations as in the Lotka-Volterra theory (secs. 7.21, 10.331).

The history of a success in the biological control of an insect pest or a weed reads remarkably like Caughley's understanding of the history of the thar in New Zealand (Osmond and Monro 1981). At first the colonists find themselves in the midst of abundant food (top left corner of fig. 3.01), and they breed exceedingly. As a consequence, they move to the top right corner of the figure. This position connotes the temporary condition that is defined by cell 4 of table 3.02; it is the condition that precedes the crash. The crash, and recovery from it, takes the predator to the bottom right corner of fig. 3.01. This position connotes an intrinsic relative shortage of food. As with thar, a favorable change in any component in the food's environment would probably allow the predator to increase. But in the realm of pest control the spotlight is on the pest, so we emphasize that any unfavorable change in the environment of the predator would probably allow the food (i.e., the pest) to increase.

In southern California local outbreaks of *Icerya* have been observed in places where citrus trees have been inadvertently sprayed with an insecticide that kills many *Rodolia* and *Cryptochaetum* but few *Icerya*, and in areas where Quezada and De Bach deliberately sprayed citrus trees with similar insecticides (sec. 10.24). Such outbreaks are obviously caused by a relaxation of predation-pressure. A number of similar examples are mentioned in section 5.3.

The poisonous insecticide is a malentity, but in nature any component of environment might be influential. A number of diverse examples are mentioned in section 5.22. Certain aspects of the weather might influence both predator and prey, for example, as in *Encarsia* and *Trialeurodes* or *Winthemia* and *Protoparce* (sec. 5.225). When the component in the web of the prey is an alternative source of food for the predator (secs. 5.221, 10.22), the distal cause of the outbreak might be found in the environment of the alternative food.

All the causes for outbreaks in the prey or the predator that have been mentioned in this section can be traced back into the web of the primary animal, whichever one it may be. This is the feature that distinguishes them from the causes of the hypothetical "coupled oscillations" that are the subject of section 10.331.

## 10.4   Conclusion

Controversy is most fruitful when the facts have been strictly ascertained and there is agreement about the limitations of scientific knowledge (sec. 10.0). Controversy is usually about how to explain certain facts, the truth of which is agreed upon. Theory feeds on controversy, and controversy leads to better explanations. There is an analogy with genetic evolution. Theories "struggle" to explain the facts and thus to absorb them. Success in the struggle is judged by the plausibility of the explanation. With each success the theory grows more general and more powerful. The losers fade away.

In section 10.2 the theory of environment opened the way to the question: How is it that certain sorts of animals can live together, professing the same niche in the same habitat? And to the answer: Because they are alike—not identical, but sufficiently alike with respect to the niche that they have in common (relative to the heterogeneity of the habitat) to be able to coexist. Conversely, when two species fail to coexist it is usually because they differ with respect to their success in professing a niche that they have in common. This explanation is more general and more plausible than the best explanation that is consistent with competition theory. For a long time ecologists with nothing better than competition theory to help them have had to be satisfied to ask: Why do animals fail to coexist? And to accept the strangely implausible answer: Because they are alike.

In section 10.3 we showed that competition theory and the theory of environment lead to different explanations for outbreaks or, indeed, for fluctuations of any kind. At its worst, competition theory leads to a futile cul-de-sac: sophisticated mathematical models based on unrealistic assumptions make predictions that cannot be tested because they cannot be related to our experience of nature. At its best, competition theory tends to overemphasize intrinsic predation-pressure and intrinsic shortage of food at the expense of extrinsic predation-pressure and extrinsic shortage of food. The theory of environment leads to more realistic explanations because it looks at the full environment. Fruitlessly abstract models are less likely to be involved. But the risk of fruitlessly complex ones still has to be avoided (sec. 2.32).

# 11

# Spreading the Risk in Evolution

## 11.0 Introduction

The theory of the distribution and abundance of animals based on spreading the risk in the multipartite population in a multipartite habitat is discussed in chapter 9. In that chapter we refer briefly to spreading the risk through phenotypic variation which may or may not be genetic. Variation in environment and variation between individuals were two ways in which den Boer (1968) conceived spreading of risk in his original account of this concept. He wrote (p. 167), "spreading of risk is the statistical outcome of selection varying from generation to generation within a heterogeneous population".

When the local populations of a natural population are exposed to different environments, it is to the advantage of the local population to have genes that adapt it to its particular environment. When that environment changes it is to the advantage of the local population to have a reserve of genetic variability enabling it to adapt to the new environment as rapidly as it changes. In a homogeneous environment that changes little there may be one or a few genetic constitutions that best adapt the individual to that environment. It would be to the immediate advantage of the species to be genetically specialized to match the specialized environment. But the cost of specialization is the risk of extinction when environment changes. So, again, even for the specialist it may be an advantage to possess some reserve of genetic variability to spread the risk. Or there may be combinations of strategies, as, for example, when an aphid reproduces parthenogenetically and multiplies its own kind of genotype when the environment is favorable but reproduces sexually with the onset of autumn when the environment is becoming harsher and this risk needs to be spread. Alternation of sexual and asexual generations is common in plants and invertebrates but not in vertebrates. Here we may suppose that "population structure", that is to say the genetically multipartite population, achieves similar ends (sec. 11.1).

Spreading the risk genetically is a concept that has largely been ignored by ecological theorists, yet it is of fundamental importance in determining the chance of a local or a natural population to survive and reproduce. This idea should therefore be incorporated into any general theory of population ecology. We attempt to do so in this chapter.

## 11.1 Genetic Diversity among Local Populations of a Species

Unless special efforts are made, the boundaries of the local populations that constitute a natural population may not be identified. We discuss first some examples in which the boundaries were recognized.

The local population of the black pineleaf scale *Nuculaspis californica* is the population on a single ponderosa pine tree *Pinus ponderosa*. This was established by the careful study by Edmunds and Alstad (1978, 1981), who have shown that even in a dense forest of ponderosa pine individual trees support distinct local populations of scale insects. Dispersal of the crawlers is largely restricted to the tree on which they were born. They tend to fall from higher branches to those below, their lateral movement being quite restricted.

Individual trees differ in the toxins that occur in their leaves. These toxins are the tree's defense. They are a variety of terpenes, terpene acids, polyphenols, and tannin-like substances, all of which are under strong genetic control. Almost every tree in the forest has a unique constitution in terms of these defensive compounds.

A striking characteristic of infestations of the black pineleaf scale is the high variation in density of scale populations from tree to tree. Trees that are free of scale frequently stand for years beside trees infested with as many as ten scales per centimetre of needle, even though branches of neighboring trees may intertwine. When twigs were reciprocally grafted from adjacent infested and uninfested trees, the grafted twigs retained the level of infestation characteristic of the tree from which they came, not the tree onto which they were grafted. Nor did scale from the grafted twig infest the tree onto which it was grafted. Twigs infested with mature scales about to reproduce were hung on limbs of trees that were uninfested. Of eighty-one such transfers twenty-eight (36%) failed to produce young. Of those that produced young, in only two cases did the number of offspring exceed the total number of mature potential parents. Within-tree transfers resulted in sixteen times as many larvae surviving to second instar than in transfers between trees. These results are consistent with the hypothesis that the population of scale insects on an individual tree are selected for resistance to the particular combination of toxic compounds on that tree. However, an alternative hypothesis is not excluded—that scales transferred from infested to uninfested trees may fail because all uninfested trees are unsuitable for all scale insects. The critical experiment that was not done (no doubt because it would be difficult to carry out) was to transfer scale insects to different infested trees.

A single pine tree survives on the average for about two hundred generations of the scale insect. So it is theoretically possible for a local population of scale insects and their descendants to be subject to the same selection for resistance to the tree for many generations. This saves the local population from having to evolve a large number of detoxifying mechanisms. Edmunds and Alstad (1978) argued that the chance of gene exchange between scales on different trees is least when the density of scales is highest, because the chance of a female's mating with a male from the same tree is greater then. Females are sessile. Most of the newly hatched larvae crawl only a few centimeters to needles on the same twig. They could, of course, be blown from tree to tree, but the chances of recolonization are small because of the differences between trees. That is an advantage to the tree. It would be disadvantageous for the insect to become exclusively adapted to one tree, since that would result in the extinction of the local population with the death of the tree. No doubt many local populations become extinct that way, but presumably not all do. Evolution involves a compromise between doing the best in today's world and keeping some reserves for an uncertain long-term future.

For vertebrates other than man, probably the most extensive analysis of genetic differentiation between local populations has been done on the house mouse *Mus musculus*. Experimental and field studies suggest that populations are subdivided into family groups composed of a dominant male and several females and subordinate males. Intergroup dispersal is said to be rare. The genetically effective size of the groups is less than ten. Given this social system, genetic heterogeneity is inevitable. Populations inhabiting different barns on the same farm often have a different frequency of alleles, apparently because they were established by a few founders that multiplied rapidly without receiving much in the way of immigrants from elsewhere. In large barns there is genetic differentiation between populations within the barn. The work of Anderson (1970) and Selander (1970) suggests that for small populations stochastic processes may play a dominant part in determining genotype or gene frequency. However, this may not be a general phenomenon. Baker (1981) found that an allele introduced into a mouse population on a poultry farm spread rapidly throughout the mice on the farm, suggesting that the social structure of the population in this case did not prevent the spread of the allele.

The social structure of populations of the European rabbit in Australia is well known (sec. 12.2). Subdivided as they are into independent breeding groups with mating restricted to members of the same group and certain individuals in that group, one might anticipate that local populations might show genetic differences. However, Daly (1981) showed that this was not the case in her studies of the European rabbit in Australia. Genetic analysis by means of electrophoresis showed that genotypes were randomly distributed among warren groups and that genetic differences between groups could be ascribed to chance events. Daly (1981) gives reasons to suggest that the same probably applies to other mammals.

Wright (1978) proposed a statistic $F_ST$ for measuring the amount of genetic differentiation among local populations of a species. The value of the statistic can range from zero to one. For thirty-five species studied, the value of $F_ST$ varied from a low of 0.005 for the fish *Notropis stramineus* to as high as 0.94 for species of the lizard *Anolis*. In all cases this statistic of genetic diversity had a value greater than zero (p. 288).

These examples illustrate the principle that the multipartite population in a multipartite environment (chap. 9) is also genetically multipartite. Genetic diversity aids survival in a multipartite environment. This concept is elaborated mathematically by Wright (1977, 1978) in his "shifting balance" model of randomly differentiated local populations. According to Wright, this population structure promotes rapid evolution.

The obverse of the principle that genetic diversity aids survival in a multipartite environment is the principle that the multipartite environment helps to maintain genetic diversity. Powell (1971) maintained thirteen experimental populations of *Drosophila willistoni* in the laboratory. All populations were initiated from five

**Table 11.01**  Average Genetic Variablity of *Drosophila will-istoni* Kept in Environments in Which Zero, One, or Three Components Were Varied

| Variables | % Heterozygosity per Individual (± SE) | Alleles per Locus (± SE) |
|---|---|---|
| 0 | 7.81 ± 0.31 | 1.67 ± 0.02 |
| 1 | 9.62 ± 0.84 | 1.92 ± 0.13 |
| 3 | 13.36 ± 0.76 | 2.12 ± 0.17 |

*Source:* After Powell (1971).

hundred flies that were descendants of five hundred single females collected in the wild. Some were kept in a homogeneous environment; in one set of experiments one of the following components was varied, in another set three of these components were varied. The variables were two sorts of yeast, two sorts of larval medium, and two temperatures (alternate weeks). After forty-five weeks (fifteen generations) samples of adults were assayed by electrophoresis for genetic variability at twenty-two gene loci. There were two measures of genetic variability: percentage of heterogeneity per individual and average number of alleles per locus. Table 11.01 shows the average genetic variability for populations that were kept in environments in which none, one, or three of these components were varied. More variable environments maintained more genetic heterogeneity by either measure. King (1972) argued that, since *D. willistoni* flies are polymorphic for a large number of chromosomal inversions, the inversion polymorphisms and not the single gene loci have been the direct targets of selection in Powell's experiments.

McDonald and Ayala (1974) repeated these experiments using *Drosophila pseudoobscura*, which has inversions only on the third chromosome. They studied seventeen gene loci on other chromosomes. Table 11.02 shows that the average heterozygosity increased as the number of heterogeneous components of environment increased from zero to three but decreased slightly from three to four. McDonald and Ayala (1974) concluded that their experiments demonstrated that genetic variability increased with environmental heterogeneity, thus supporting Powell's (1971) findings with *D. willistoni*.

**Table 11.02**  Percentage of Heterozygous Loci per Individual of *Drosophila pseudoobscura* Kept in Environments in Which Zero, One, Two, Three, or Four Components Were Varied

| Variables | % Heterozygosity per Individual (± SE) |
|---|---|
| 0 | 0.159 ± 0.032 |
| 1 | 0.193 ± 0.044 |
| 2 | 0.217 ± 0.046 |
| 3 | 0.224 ± 0.052 |
| 4 | 0.204 ± 0.051 |

*Source:* After McDonald and Ayala (1974).

Powell and Wistrand (1978) used *D. pseudoobscura* that had no chromosomal inversions and obtained an increase in proportion of heterozygous loci per individual as environmental variables were increased from zero, one, or two to three. All three sets of experiments support the thesis that genetic diversity is maintained by environmental heterogeneity. Powell and Taylor (1979) tested this thesis in the field. They predicted that in a multipartite habitat one would expect to find different frequencies of genotypes in the different localities. They collected *Drosophila persimilis* from several distinct localities that were close to one another in Mather, California. They determined the frequencies of genotypes based on chromosome inversion polymorphism and protein polymorphisms. Even over short distances gene and inversion frequencies differed from locality to locality. Because *D. persimilis* is an active disperser, they did not consider that differential selection between the different localities could alone account for their observations. Movement of flies would quickly swamp out any differences that might arise from any but the strongest selection. They predicted, and were able to show, that flies removed from one sort of locality and released in another tended to return to the same sort of locality where they originated. So selection, such as may occur, is reinforced by "habitat choice".

McLeod et al. (1981) studied the alleles at ten gene loci in the freshwater clam *Musculium partumeium* in an ephemeral pond and a permanent pond. The ephemeral pond was a more variable environment for the clams than the permanent pond. In samples of about fifty clams the percentage of loci that were polymorphic was about 20% in clams in the ephemeral pond, whereas in the permanent pond it was 0. In clams in the ephemeral pond 6% of the loci were heterozygous but none were in the clams in the permanent pond.

Schaffer (1974) argued that populations that live in heterogeneous environments should be polymorphic for reproductive characteristics. Some evidence supporting this hypothesis comes from the Atlantic salmon, which spawns in freshwater streams. Since salmon tend to spawn in the streams of their origin, there is opportunity for local populations to become differentiated from each other through natural selection. This seems to have led to great variation in the age of first spawning of salmon in different streams (Schaffer and Elson 1975). Giesel (1976) found that local populations of *Drosophila melanogaster* that were widely separated from one another had different age schedules of fecundity. Those from the north of New York State and other northern localities had their peak fecundity early in adult life and their egg laying phased out earlier, while those from Alabama and other southern states had their peak fecundity later and continued laying eggs longer. These differences were correlated with shorter growing seasons and greater extremes of weather in the North than in the South.

In addition to mutation, recombination of genes, and immigration of genes into a local population, there is a fourth source of genetic diversity between local populations—the random loss of genes in small populations. If a previously unoccupied locality is colonized by one or a few individuals, the genes in these individuals are unlikely to be representative of the gene pool of the natural population as a whole. All the individuals that spring from the original colonists will carry the genes of the founder population, which may be unrepresentative. Moreover, local populations that differ because of the differences in their founders will, under the same regime of selection, continue to give rise to different populations. This is known as the "founder effect" or the "founder principle". The principle applies also to populations that have been reduced to very low numbers. This bottleneck effect is illustrated in the northern elephant seal *Mirounga angustirostris* (sec. 11.2). This species was reduced by hunting to about twenty animals in the 1890s. Since then the population has grown to more than thirty thousand. Bonnell and Selander (1974) found no genetic variation in a sample of twenty-four gene loci examined electrophoretically. Genetic variation is known to exist in the southern elephant seal *M. leonina*, which was never reduced so severely in its numbers. Reference is made earlier in this section to genetic differences between local colonies of the house mouse that can be attributed to the small number of colonists.

Dobzhansky and Pavlovsky (1957) made one of the few experimental studies of the founder effect. They set up twenty populations of *Drosophila pseudoobscura*. The foundation stock of all the populations were $F_2$ hybrids between flies from Texas which had the Pikes Peak (PP) inversion in the third chromosome and flies from California which had the Arrowhead (AR) inversion in the same chromosome. In ten of the populations there were four thousand founders. In the other ten populations there were only 20 founders. The frequencies of PP and AR chromosomes in all initial populations were 50%. The flies were kept as ongoing populations in population cages at 25°C. Eighteen months later the frequencies of PP varied from 20% to 35% in populations descended from the large number of founders, and from 16% to 47% in those descended from small numbers of founders (fig. 11.01). The heterogeneity of the frequencies for the populations that started with the small number of founders was greater than that of those that were

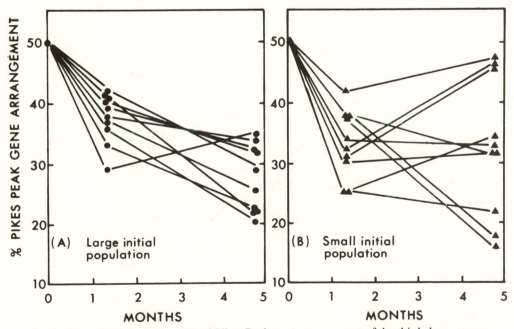

**Fig. 11.01** The frequencies (%) of Pikes Peak gene arrangement of the third chromosome of *Drosophila pseudoobscura* in (A) ten replicate populations initiated from four thousand founders, and (B) ten replicate populations initiated from twenty founders. (After Dobzhansky and Pavlovsky 1957)

descended from the large number of founders. The frequencies of PP and AR are controlled by natural selection, since the three genotypes have different adaptive values (fitness). In experimental populations of the same geographic origin, selection moves the frequencies of the gene arrangements to the same equilibrium frequency (sec. 11.311). But in this experiment with wild populations of mixed geographic origin the selective fates of the chromosomal inversions depended upon the rest of the genotype, which is highly variable. Owing to chance the background genotype of the populations that had started from the small number of founders must have varied more than that of those that had started from the large number.

Jones and Yamazaki (1974) established a number of lines of *Drosophila pseudoobscura*, each homozygous for one of two alleles at the esterase-5 locus and each initiated from a single fertilized female. A number of populations were founded in population cages with a chosen initial frequency of the two alleles. These founding populations were initiated from either two lines, twenty-two lines, or forty-four lines. The frequencies of the alleles were then monitored in the population cages for seventeen months. Populations founded with only two lines showed marked temporal changes in frequencies of the alleles and in the final frequency reached. By contrast, populations founded with twenty-two and forty-four lines showed no temporal change in frequency of the alleles. Jones and Yamazaki (1974) attribute the difference to the different genetic background of the founding population, that is, to genes other than those at the esterase locus. Linked loci appear to be important in influencing gene frequency at the esterase locus when a small sample of the natural population is used as the founding population. A large sample does not appear to lead to major deviations from apparent selective neutrality of the alleles at the esterase locus.

## 11.2   Genetic Diversity and the Rate of Evolutionary Change

According to Fisher's fundamental theorem of natural selection, "the rate of increase in fitness of an organism at any time is equal to its genetic variance in fitness at that time" (Fisher 1958b, 37). In bacteria and other organisms, such as some insects that exist in huge numbers and have a high rate of increase, recurrent mutation may provide much of the genetic variability that is immediately available for evolutionary change. According to Dickson (1960), the parthenogenetically reproducing spotted alfalfa aphid *Therioaphis maculata* was introduced into North America in 1953. Within three years populations arose that were resistant to their hosts and to organophosphorus insecticides. In the Imperial Valley in California alone there were 68,000 ha of alfalfa and some 50 million aphids per hectare. They produced twenty-five to thirty-five generations each year. Therefore the number of aphids living there before the discovery of resistance must have exceeded $1.7 \times 10^{11}$. An assumed rate of mutation of $10^{-5}$ would still give 1.7 million occurrences of a given mutation during the available time.

Although mutation is the ultimate source of genetic variability, the rate of mutation may be too slow for most organisms, other than those like bacteria and some insects, to provide the necessary variability in times when the environment is changing rapidly. A secondary source of genetic variability that then becomes important is stored genetic variability, which was first demonstrated in field popu-

**Table 11.03** Percentage of Wild Chromosomes of *Drosophila pseudoobscura* (from Stanislaus National Forest, California) with Different Effects on the Survival of Homozygous Flies

| | | Chromosome | | |
|---|---|---|---|---|
| Effect on Survival | Survival Relative to Heterozygous Flies | Second | Third | Fourth |
| Lethal and semilethal | Less than 50% | 33.0 | 25.0 | 25.9 |
| Subvitals | More than 50% but less than two standard deviations from the mean | 62.6 | 58.7 | 51.8 |
| Quasi-normals | Between two standard deviations below and two standard deviations above the mean | 4.3 | 16.3 | 22.3 |
| Supervitals | More than two standard deviations from the mean | < 0.1 | < 0.1 | < 0.1 |

*Source:* After Dobzhansky and Spassky (1953).

lations of *Drosophila melanogaster* by Chetverikov (1926). The flies he collected in the Russian steppes all looked much the same on the surface, but by making appropriate crosses he was able to reveal hidden variability in the form of recessive genes. Since then many populations in the field have been studied in a similar way. Notable among these studies are those of Dobzhansky and his colleagues; for example, Dobzhansky and Spassky (1953, 1963) on *Drosophila pseudoobscura* and those of Mourão, Ayala, and Anderson (1972) on *D. willistoni*. Their technique consisted of crossing flies from field populations to obtain many flies homozygous for a single chromosome. Recessive alleles present in the wild chromosome are expressed in the homozygous flies. Some chromosomes when homozygous are lethal. Others have less drastic effects compared with flies that are heterozygous for random combinations of wild chromosomes. Table 11.03 shows typical results for *D. pseudoobscura*, in which the effect of the chromosomes on survival of the flies from egg to adult was measured. Over 25% of second, third, and fourth chromosomes from the wild populations are lethal or semilethal when homozygous. Even a higher proportion of flies have chromosomes that when homozygous have less drastic effects. The point of the exercise is that it demonstrates an enormous quantity of stored genetic variability, much of which is deleterious when homozygous.

Evidence of the sort shown in table 11.03 demonstrates the large store of genetic variability that exists in natural populations. But it does not tell us much about the number and diversity of genes that are involved. For that we need to know the extent to which allelic variants exist at each gene locus—in other words, how polymorphic the genes at a locus are. The number of polymorphic genes at a locus can be measured by the technique of electrophoresis as developed by Lewontin and Hubby (1966) for *Drosophila*. In most species studied, over 30% of the gene loci have more than two alleles (i.e., are polymorphic), and individuals are heterozygous at some 10% of their testable loci (Sammeta and Levins 1970; Nevo 1978). In *Drosophila willistoni*, for example, the proportion of polymorphic loci is 53%, and 18% of the loci per individual are heterozygous. Table 11.04 summarizes the

**Table 11.04**   Genetic Variation in Natural Populations of Major Groups of Animals and Plants Based on 242 Species

| | Number of Species | Proportion of Polymorphic Loci per Population | | Proportion of Heterozygous Loci per Population | |
|---|---|---|---|---|---|
| | | Mean | SD | Mean | SD |
| Plants | 15 | 0.259 | 0.166 | 0.0706 | 0.0706 |
| Invertebrata (except Insecta) | 27 | 0.399 | 0.275 | 0.1001 | 0.0740 |
| Insecta (except *Drosophila*) | 23 | 0.329 | 0.203 | 0.0743 | 0.0810 |
| *Drosophila* | 43 | 0.431 | 0.130 | 0.1402 | 0.0534 |
| Osteichthyes | 51 | 0.152 | 0.098 | 0.0513 | 0.0338 |
| Amphibia | 13 | 0.269 | 0.134 | 0.0788 | 0.0420 |
| Reptilia | 17 | 0.219 | 0.129 | 0.0471 | 0.0228 |
| Aves | 7 | 0.150 | 0.111 | 0.0473 | 0.0360 |
| Mammalia | 46 | 0.147 | 0.098 | 0.0359 | 0.0245 |
|    Total | 242 | 0.263 | 0.153 | 0.0741 | 0.0510 |
| *Mean values* | | | | | |
| Plants (1) | 15 | 0.259 | 0.166 | 0.0706 | 0.0706 |
| Invertebrates (2–4) | 93 | 0.397 | 0.201 | 0.1123 | 0.0720 |
| Vertebrates (5–9) | 134 | 0.173 | 0.119 | 0.0494 | 0.0365 |

*Source:* After Nevo (1978).

results of such studies on 242 species of plants and animals in which fourteen or more loci were studied. The number of loci that have been studied in the species from which table 11.04 was compiled ranges from fourteen in the barnacle *Balanus* to seventy-one in man. These numbers probably represent less than one per thousand of all loci in these organisms. There is no way of knowing whether the gene loci studied represent a random sample. It should also be pointed out that some species show little or no gene polymorphism at their gene loci. The reasons are not understood, though various suggestions have been made. For example, Bonnell and Selander (1974) found no polymorphism in the gene loci that they studied in the northern elephant seal *Mirounga angustirostris*. They suggested this could be due to loss of alleles caused by heavy mortality inflicted by sealing in the nineteenth century, which reduced local populations to very low numbers. Nevo (1978) claims that fossorial rodents are likewise virtually monomorphic and suggests this may be related to their life-style.

Lewontin and Birch (1966) tested the hypothesis that the infusion of genes from *Dacus neohumeralis* into *Dacus tryoni* would provide additional genetic variability that would promote the evolution of populations better adapted to extreme environments. The reason for posing the hypothesis in the first place was that during the course of the past one hundred years *D. tryoni* in Australia had spread from its northern tropical home into the cool temperate South. Bateman (1967) had demonstrated that the populations had changed genetically by measuring the innate capacity for increase $r_m$ at different temperatures (sec. 11.4). He showed that tolerance to extremes of temperature increases toward the south. *D. neohumeralis* is sympatric with *D. tryoni* over much of the range of the latter. There is evidence that some hybrids are produced in the field and they are readily produced in the laboratory. Population cages were established with flies from Brisbane at 20°C, 25°C, and 31.5°C at 70% R.H. The middle temperature is near the optimum for both species, and 31.5°C is the highest temperature at which *D. tryoni* can be kept in continuous culture. The low temperature is the lowest giving a reasonable generation time. At each temperature two replicate populations of each of *D. tryoni* and the $F_1$ hybrids were established in population cages. The object of the experiment was to allow both the single species and the hybrid population to evolve at the optimal and extreme temperatures and to measure the adaptation of the populations to these temperatures. The measure of adaptation used was the innate capacity for increase in numbers of the population $r_m$ after two years in population cages (sec. 11.4). In addition, the total production of pupae was determined each week. At less frequent intervals the populations were scored for morphological characters that indicated the extent to which *D. neohumeralis* genes had become incorporated in the populations.

Table 11.05 shows that after two years at 20°C populations of *D. tryoni* were barely superior to hybrid populations. At 25°C there is no difference between *D. tryoni* and hybrid populations. However, at 31.5°C the hybrid populations had a much higher innate capacity for increase than the *D. tryoni* populations. Hybridization per se did not produce better-adapted populations at any temperature. The evidence for this is that the production of pupae for hybrids was lower than for *D. tryoni* at the beginning of the experiment. After a year the hybrid productivity had reached the same level as that of *D. tryoni*, and after that the hybrid population at the high stress temperature was superior. Scoring of morphological characters showed that in the hybrid population there was a rapid change from the fifty-fifty composition of the hybrids in the $F_1$ to dominantly *D. tryoni* characters after the $F_2$. The hybrids as such seemed to be at a disadvantage, but the ingression of some genes of *D. neohumeralis* into *D. tryoni* was highly advantageous at the stress temperature of 31.5°C. There is no doubt that introgression of genes from *D. neohumeralis* into *D. tryoni* accelerated the genetic adaptation of the population to a high stress temperature. It was not possible to experiment at really low temperatures because of the great length of the generation.

Birch and Vogt (1970) found evidence for much genetic variability in local populations of *D. tryoni*. Strong selection evidently influenced the genetic constitution of local populations. Within any local population there is a constant produc-

**Table 11.05**   Innate Capacity for Increase $r_m$ (and Standard Errors) for Populations of *Dacus tryoni* and Hybrid Populations of *D. tryoni* and *D. neohumeralis* after Two Years at the Temperatures Shown

|  | 20°C | 25°C | 31.5°C |
|---|---|---|---|
| *D. tryoni* mean | 0.0752 ± 0.00090 | 0.1337 ± 0.00401 | 0.0944 ± 0.00352 |
| Hybrid mean | 0.0710 ± 0.00176 | 0.1295 ± 0.00462 | 0.1178 ± 0.00352 |
| *D. tryoni* vs. hybrids |  |  |  |
| Diff. | 0.0042 | 0.0042 | −0.023 |
| SE | 0.00198 | 0.00612 | 0.00498 |

*Source:* After Lewontin and Birch (1966).

**Table 11.06**   Productivity and Size of the Adult Population of Flies from Widely Separated Local Populations of Two Related Species of *Drosophila* Compared with "Hybrid" Populations Made by Crosses between Flies from Different Local Populations

| Population | Temperature (°C) | Productivity (Number/Food Unit) | Population Size |
|---|---|---|---|
| *Drosophila serrata* | | | |
| Sydney | 25 | 550 ± 17 | 1,782 ± 76 |
| Cooktown | 25 | 568 ± 20 | 2,221 ± 80 |
| Popondetta | 25 | 477 ± 13 | 1,828 ± 90 |
| Sydney | 19 | 483 ± 13 | 1,803 ± 87 |
| Cooktown | 19 | 486 ± 12 | 2,017 ± 84 |
| Popondetta | 19 | 357 ± 8 | 1,580 ± 52 |
| Hybrid population | | | |
| Sydney × Cooktown | 25 | 593 ± 16 | 2,360 ± 74 |
| Sydney × Popondetta | 25 | 622 ± 18 | 2,541 ± 117 |
| Cooktown × Popondetta | 25 | 540 ± 18 | 2,419 ± 76 |
| Sydney × Cooktown | 19 | 554 ± 19 | 2,418 ± 171 |
| Sydney × Popondetta | 19 | 572 ± 14 | 2,448 ± 86 |
| Cooktown × Popondetta | 19 | 479 ± 12 | 2,227 ± 172 |
| *Drosophila birchii* | | | |
| Cairns | 25 | 351 ± 16 | 1,262 ± 83 |
| Popondetta | 25 | 152 ± 9 | 469 ± 49 |
| Cairns | 19 | 324 ± 11 | 1,091 ± 66 |
| Popondetta | 19 | 121 ± 5 | 428 ± 33 |
| Cairns × Popondetta | 25 | 342 ± 26 | 1,331 ± 123 |
| Cairns × Popondetta | 19 | 303 ± 10 | 1,203 ± 55 |

*Source:* After Ayala (1965b).

tion of variants that can be identified by their morphological characters. But there is strong selection against most variants, so that the variance after selection is small. By selection in the laboratory the mean of each of the characters can be moved quite rapidly.

Ayala (1965a,b) crossed *Drosophila serrata* from widely separated localities and compared the "adaptation" of "hybrid" progeny with each of the parent populations. He did the same for *D. birchii*. His measure of "adaptation" was the mean productivity of flies (numbers per unit of food) and the size of the adult population in the experimental populations after a year. Comparing the means of the hybrid populations with the means of the two parental populations (table 11.06) shows that the productivity and size of the hybrid population were always significantly greater.

The influence of genetic variability on the rate of evolutionary change is illustrated by experiments by Ayala (1965a,b, 1968) on *Drosophila serrata*, a species endemic to eastern Australia and to Papua New Guinea and New Britain. Flies were collected from local populations in Popondetta (Papua New Guinea) and Sydney (Australia). Two populations were established in the laboratory with flies from Popondetta, one at 25°C and the other at 19°C. Likewise, two populations were established with the Sydney population. Two other populations were established with $F_1$ progenies of mass crosses between flies from Popondetta and flies from Sydney, one was kept at 15°C and the other at 19°C. Each experiment was replicated. The flies were allowed to become very crowded in culture bottles in which

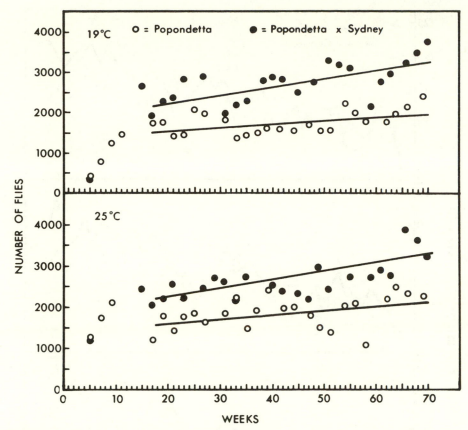

**Fig. 11.02**   Change in size of population of *Drosophila serrata* at two temperatures owing to natural selection. Popondetta is a strain from the locality of this name in New Guinea. Popondetta × Sydney is a hybrid population of strains from these two places. The Sydney strain is not shown, since it gave about the same results as the Popondetta strain. (After Ayala 1965a)

food was regularly renewed. Adaptation of the flies to the experimental environment was measured by the number of adult flies in the population during the seventy weeks of the experiment. Figure 11.02 shows that the size of the hybrid population (Popondetta × Sydney) was significantly larger than that of the nonhybrid population from Popondetta. (The nonhybrid population from Sydney was about the same as that from Popondetta). All populations increased their numbers with time. At 25°C the Popondetta population increased at a rate of 10.5 flies per week, the Sydney population increased at about the same rate. The hybrid population increased at a rate of 19.5 flies per week.

In a similar series of experiments Ayala (1966, 1969) increased the genetic variation of populations of *Drosophila serrata* and its sibling species *D. birchii* by exposing them for three generations to moderate doses of X rays. Initially the number of flies in the irradiated populations was less than in the control populations (fig. 11.03). Ayala attributed this to the deleterious mutations induced by radiation, which were eliminated in early generations. Thereafter the irradiated populations increased in numbers at a faster rate throughout the two and a half years of the

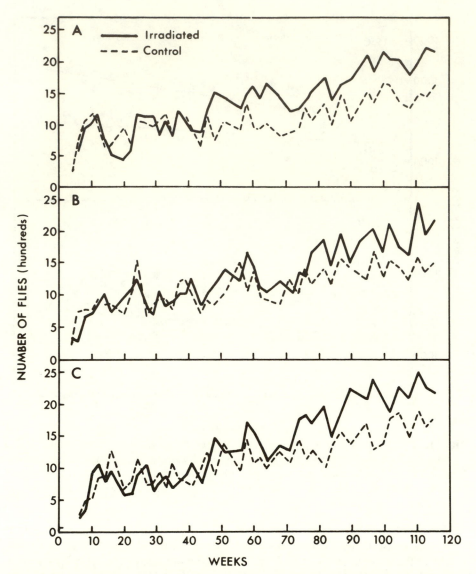

**Fig. 11.03** Change in size of populations of irradiated and nonirradiated populations of *Drosophila birchii* owing to natural selection. A, B, and C are replicate populations. (After Ayala 1969)

experiment. Ayala attributed this to the minority of mutations that proved to be favorable and enabled the population to grow faster.

A nice example of the practical utility of the theory of the multipartite population, especially of spreading the risk by genetic differences between local populations, comes from an attempt to eradicate the screwworm fly *Cochliomyia hominivorax* from the southern United States. The project began about 1950. It was very effective until 1968, but since then outbreaks have occurred despite the continued release of sterile flies in enormous numbers. This failure seems to be due to the mismatch between the sterile flies and the wild flies among which they are liberated. Instead of the wild flies being a homogeneous group as initially thought, it now appears that there are at least nine "races". The races differ in the genes in their populations, in the structure of the chromosomes in some cases, and in such qualities as the time of mating. The races come from different localities in Mexico. They have migrated into the United States, where they have become sympatric, thus presenting a formidable problem in the eradication program. It now seems that effective control of screwworm in the United States will depend upon matching sterile flies with the particular races that exist in the places where eradication is being carried out. This task may be less formidable than it seems if the release of much greater numbers than is normally used happens to overcome disinclination to mate and other problems (Richardson, Ellison, and Averhoff 1982).

## 11.3   Components of Natural Selection

Mutation of genes, recombination of genes, immigration of new genes into local populations from other local populations, and random loss of genes in small populations are the sources of variability between local populations on which natural selection acts to mold the population (sec. 11.2). The components of selection are the components of environment which were discussed in chapter 1. Anything that influences the chance to survive and to reproduce is a component of environment and ipso facto a component of selection. This rule includes directly acting components in the centrum and indirectly acting components in the web.

### 11.31   Resources

Resources are defined in table 1.01. We choose to discuss heat and food because something is known about the influence of these two resources as agents of natural selection.

### 11.311   Heat

Many populations of *Drosophila* spp. are polymorphic for a particular sort of mutation, which is called a chromosomal inversion because a certain genetic sequence along the chromosome seems to have been rotated through 180 degrees. Chromosomal inversions can be identified under a microscope. Dobzhansky (1943) and Wright and Dobzhansky (1946) demonstrated that different chromosomal inversions confer different qualities on the flies that carry them. They were able to demonstrate that these qualities were adaptive. Furthermore, Dobzhansky (1970) postulated that each chromosomal inversion carried particular adaptive combinations of alleles at its various loci that were "coadapted" to function well together. His predictions were consistent with the finding of Prakash and Lewontin (1968) for *D. pseudoobscura* and *D. persimilis*. "Closely related" inversions shared alleles that were different from those less "closely related". Closeness of relationship is defined by proximity on the phylogenetic chart of the chromosomal inversions (fig. 11.04).

Chromosomal inversions are given names from the places where they were first found. An arrow (e.g., from Standard to Arrowhead) indicates that the inversion at the head of the arrow can be derived by a single inversion within the chromosome

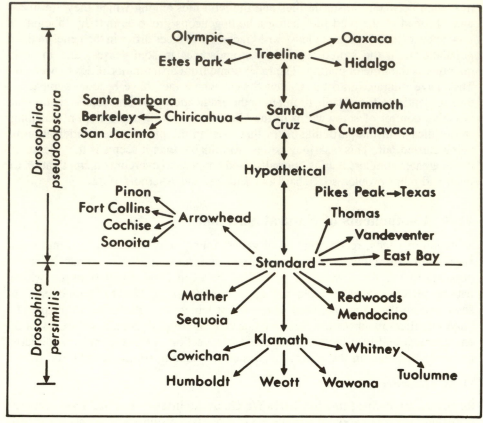

**Fig. 11.04**  Phylogenetic chart of the gene arrangements (chromosome inversions) in the third chromosome of *Drosophila pseudoobscura* and *Drosophila persimilis*. (After Anderson et al. 1975)

at the base of the arrow. Presumably this is how the chromosomal inversion Arrowhead was evolved. Olympic at the top left can be derived by a succession of four inversions from the original standard chromosome. Prakash and Lewontin (1968) found that "related inversions" in the group Santa Cruz, Chiricahua, and Treeline shared alleles that were different from those prevalent in the more distantly related group, Standard, Pikes Peak, and Arrowhead. Hence genetic diversity in this case is provided not only by the changed sequence of genes but also by changed blocks of genes being held together in the chromosomal inversion. Indeed, the chromosomal inversion may be regarded as a way of keeping combinations of genes intact, because recombination of genes is almost completely suppressed over the entire chromosome in inversion heterozygotes. There is virtually no gene exchange between differently inverted chromosomes within a population, but there is free mixing of the gene contents among chromosomes of the same arrangement (Dobzhansky and Epling 1948).

Dobzhansky (1943) measured the relative frequencies of the three chromosomal types—Chiricahua (CH), Arrowhead (AR), and Standard (ST)—at different seasons of the year for four years in local populations of *Drosophila pseudoobscura* living in three localities in California. The relative frequency of ST decreased and

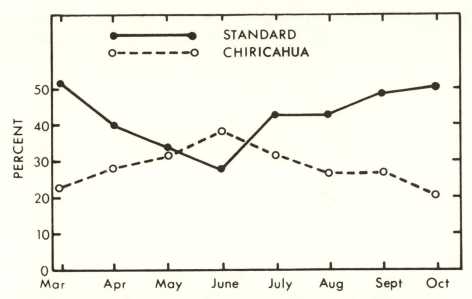

**Fig. 11.05**   Changes in the frequency of the two gene arrangements (chromosome inversions) Standard and Chiricahua in a population of *Drosophila pseudoobscura* at Piñon Flat, San Jacinto Mountains, California. (After Dobzhanksy 1943)

that of CH increased from March to June. The trend was in the opposite direction during the hot season, from June to September (fig. 11.05). The changes in the relative frequencies of AR were less regular, but on the whole they seemed to follow a trend somewhat similar to that of CH. There was a remarkable similarity in the seasonal trends in the three localities. The seasonal trend was repeated each year but with minor variations, which were no doubt associated with the vagaries of the weather.

Wright and Dobzhansky (1946) showed that trends in the relative frequences of the different chromosomal inversions in *Drosophila pseudoobscura*, rather like the seasonal trends that had been observed in nature, could be produced in experimental populations by manipulating the temperature in the cage where the flies were breeding. A number of flies was placed in a population cage with an abundance of food, which was replenished at frequent intervals. The cage soon became crowded, and the flies were allowed to continue breeding in these circumstances for a number of generations. Many more eggs were laid than emerged as adults, indicating that "the struggle for existence" was intense. In each experiment the proportion of flies carrying each chromosomal inversion was known for the initial population, and it was measured at intervals during the experiment by the following method. Some eggs were taken from the cage and reared uncrowded and with plenty of food. The distributions of the chromosome inversions in these samples gave reliable estimates of the composition of the population of adults in the cage at the time when the eggs were laid.

Figure 11.06 shows what happened in one such experiment conducted at 25°C with flies from Piñon Flat, California, of which 11% carried ST and 89% CH at the beginning of the experiment. Within about four months the proportion of ST chromosomes in the population had quadrupled. It then rose more slowly to about

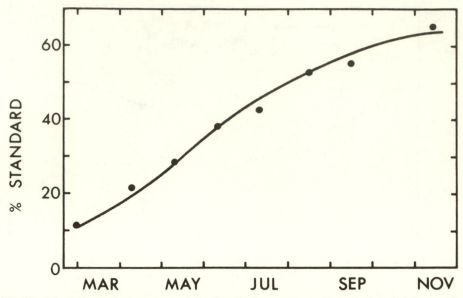

**Fig. 11.06**   Change in frequency of the Standard gene arrangement in *Drosophila pseudo-obscura* in a "population cage" at 25°C. The population was initiated with 10.7% Standard and 89.3% Chiricahua chromosomes from Piñon Flat. (After Dobzhansky 1947)

70%, after which no further change occurred. In other experiments with flies from the same source the same equilibrium was approached, irrespective of whether more or fewer than 70% of them carried ST at the beginning. The increase of ST from 11% to 70% was due to natural selection.

When the same experiment was repeated at 16.5°C, the proportions of ST and CH in the population at the end of the experiment were the same as they had been at the beginning. The advantage that ST had possessed over CH at 25°C disappeared at the lower temperature.

The increase in the frequency of ST in the experiment (fig. 11.06) simulates the increase of ST in the field in summer (fig. 11.05). But what could account for the fall in ST in the cooler months of the year before June? Birch (1955) repeated these experiments at 25°C with one difference. The flies in his experiments were kept uncrowded both as larvae and as adults. When the population was commenced with 80% ST and 20% CH, the frequency of ST fell. When it was started with 20% ST, the frequency of ST rose, but only to 35% (fig. 11.07). Further experiments showed that the direction of selection was a function of the density of the population of larvae. CH was favored by low density; Standard was favored by high density. Selection in these experiments is analogous to the decrease in frequency of ST in the spring. One might postulate that at that time of the year the larval population is more likely to be at a low rather than a high density. The failure of selection to proceed beyond the equilibrium of 70% ST in the experiment that is illustrated in figure 11.06 was explained by Wright and Dobzhansky (1946) and Dobzhansky (1951) as follows:

The heterozygotes (individuals having one ST and one CH chromosome) are superior (in crowded population cages at 25°C) to both homozygotes, CH/CH and

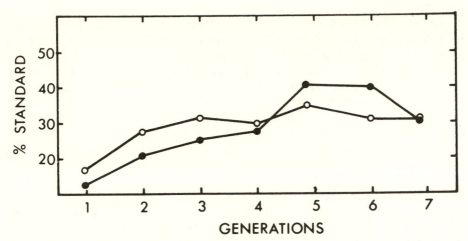

**Fig. 11.07**   Change in frequency of the standard gene arrangement in *Drosophila pseudo-obscura* at 25°C in two replicate populations when both larvae and adults are uncrowded. (After Birch 1955)

ST/ST. In other words, the rate of increase of the heterozygotes in these experiments was greater than the rate of increase of either homozygote. Dobzhansky referred to the relative rates of increase as the adaptive values of the different polymorphic forms. From the genetic point of view, the adaptive value is "the relative capacity of carriers of a given genotype to transmit their genes to the gene pool of the following generations" (Dobzhansky 1951, 78). The extent to which the genes are transmitted to the gene pool is related to the rate of increase of the genotypes in the population. Dobzhansky went on to say:

The adaptive value is, then, a statistical concept which epitomizes the reproductive efficiency of a genotype in a certain environment. Now, the adaptive value is obviously influenced by the ability of a type to survive. The adaptive value of a homozygote for a lethal gene is evidently zero. But the individual's somatic vigour, its viability, is only one of the variables which determine the adaptive value. The duration of the reproductive period, the number of eggs produced (fecundity), the intensity of the sexual drive in animals, the efficiency of the mechanisms which conduce to successful pollination in plants, and many other variables are likewise important.

Dobzhansky's adaptive value is like our concept of r, the actual rate of increase. It has meaning only in relation to the components of environment that help to determine it. A difference is that r measures an absolute rate of increase, whereas adaptive values are stated relative to one another and have no meaning in terms of absolute rate of increase. An example will make this clear. In the experiment described above, Dobzhansky arbitrarily gave the heterozygotes an adaptive value of 1 and then worked out the adaptive values of the homozygotes ST/ST and CH/CH as 0.7 and 0.4, respectively. The smooth theoretical curve in figure 11.06 is derived from these ratios. Wright and Dobzhansky (1946) referred to the differences between the adaptive values of the homozygotes and the heterozygotes as the selective coefficients of the homozygotes. The selective coefficients of ST/ST and CH/CH were thus 0.3 and 0.6, respectively. The selective coefficient of the heterozygote was, by definition, zero. Further experiments by Dobzhansky (1947) showed that the superiority of the heterozygotes was in their capacity to survive better from the egg to the adult stage. He showed this in the following way.

If, in a population of flies living in a cage, the frequencies of the chromosomal types ST and CH are represented by q and $1 - q$, respectively, and if the flies mate at random with respect to chromosomal type and the number of eggs laid by individuals of the different chromosomal types are the same, then the proportions of heterozygotes and homozygotes among the eggs laid will be given by the expansion of the binomial $\{ q + (1 - q)\}^2$, which gives:

$$ST/ST:ST/CH:CH/CH = q^2:2q(1\text{-}q):(1\text{-}q)^2. \tag{1}$$

But if the different classes of zygotes survive at different rates, the observed proportions of adults that survive will depart from those predicted by the foregoing expression. If we suppose the survival rates of ST/ST, ST/CH, and CH/CH to be, respectively, $W_1$, $W_2$, and $W_3$, then we have:

$$ST/ST:ST/CH:CH/CH = q^2W_1:2q(1 - q)W_2:(1 - q)^2W_3. \tag{2}$$

Dobzhansky determined the relative frequencies of the three genotypes in samples of larvae hatched from eggs that had been deposited in population cages but were reared in culture bottles, where nearly all of them survived. The observed proportions deviated only slightly from those predicted by equation (1). This result indicated that the flies were mating at random with respect to genotype and that each genotype was laying about the same number of eggs. Samples of adults were then taken from the same crowded cages and classified. There were too many heterozygotes and too few homozygotes to fit the frequencies predicted by equation (1). These figures indicated a differential elimination of the homozygotes CH/CH and ST/ST at some stage between egg and adult. Further calculations showed that the change was on a scale that was more than sufficient to account for the change in the ratios of the three genotypes that occurred in the population cages at 25°C (fig. 11.06).

Further evidence of components of weather as probably influencing the frequency of chromosomal types in *D. pseudoobscura* is shown in figure 11.08. Standard (ST) decreases in frequency across North America from east to west. Pikes Peak (PP) increases from west to east. Arrowhead (AR) has low frequencies to the east and west but is high in the center. Chiricahua (CH) and Tree Line (TL) show less conspicuous changes in their frequencies. Ward et al. (1974) showed a latitudinal gradient in the frequency of a single chromosomal inversion in *D. pachea*.

It is not necessary to look to large distances for changes in the frequencies of chromosomal types in *Drosophila pseudoobscura*. The changes that occur within a few kilometers at different elevations on a mountainside may be dramatically large. Table 11.07 shows the frequencies of four chromosomal types in local populations living at different elevations in the Sierra Nevada, California (Dobzhansky 1948). Note how the distributions of ST and AR seem to be complementary, the former decreasing and the latter increasing with altitude. The seasonal trends within these local populations may be summarized by saying that during the summer the upland populations become more like the lowland ones and during the winter the lowland populations become more like the upland ones. So

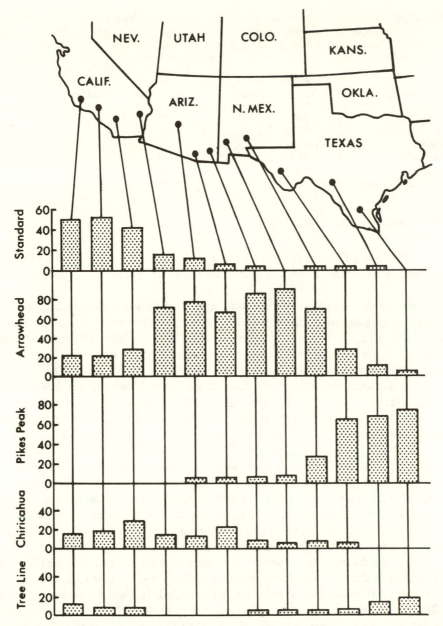

**Fig. 11.08** Change in frequency of five gene arrangements (chromosome inversions) in the third chromosome of *Drosophila pseudoobscura* in the southwestern United States. (After Powell, Levene, and Dobzhansky 1973)

the evidence seems to suggest that the physiological difference between the carriers of ST and AR is closely related to temperature; those carrying ST may be better adapted to higher temperatures, while those carrying AR may have the advantage at lower temperatures. Spiess (1950) described similar gradients in the frequencies of chromosomal types in *D. persimilis* associated with altitude. He also confirmed experimentally that there were differences in the adaptive values of the different

**Table 11.07**    Frequency (%) of Four Common Chromosomal Types of the Third Chromosome of *Drosophila pseudoobscura* along a Transect at Different Elevations
in the Sierra Nevada of California

| Locality | Elevation (Meters) | Chromosomal Inversions | | | |
|---|---|---|---|---|---|
| | | Standard | Arrowhead | Chiricahua | Treeline |
| Jacksonville | 260 | 46 | 25 | 16 | 8 |
| Lost Claim | 900 | 41 | 35 | 14 | 6 |
| Mather | 1,400 | 32 | 37 | 19 | 9 |
| Aspen Valley | 1,850 | 26 | 44 | 16 | 11 |
| Porcupine Flat | 2,400 | 14 | 45 | 27 | 9 |
| Tuolumne Meadows | 2,600 | 11 | 55 | 22 | 9 |
| Timber Line | 3,000 | 10 | 50 | 20 | 10 |

*Source:* After Dobzhansky (1948).

types. Powell and Taylor (1979) found that local populations of *D. persimilis* that
were only a few hundred meters apart had different inversion frequencies (sec.
11.1).

Stalker (1980) was able to demonstrate a relation between chromosome type in
*Drosophila melanogaster* and the ability to fly at different temperatures. In the
midwestern and eastern United States populations of *D. melanogaster* have a higher
frequency of the Standard third chromosome (and a lower frequency of chromosomes with inversions) in the North than in the South. In Missouri the frequency
of Standard chromosomes regularly rises during the cold season and drops during
the warm season, thus paralleling the north-south change in frequency. Populations
of flies in the field with a high frequency of Standard chromosomes tend to fly at
low temperatures (13–15°C). Those with a low frequency of Standard chromosomes tend to fly at high temperatures (16–28°C). Furthermore, flies with the high
frequency of Standard chromosomes (which fly at low temperatures) have a low
wing-load index, whereas those with a low frequency of Standard chromosomes
(which fly at high temperatures) have a high wing-load index. The wing-load index
is the ratio of thorax volume to wing length, so flies with a small wing-load index
have relatively large wings and a slow wing-beat, which enables them to fly better
at low temperatures than flies with relatively short wings.

Much information on the genetic constitution at different gene loci has become
available since the use of electrophoresis to identify enzymes produced by genes.
While some studies of enzyme polymorphism, such as in United States populations
of *Drosophila pseudoobscura*, reveal a uniformity of allele frequency among local
populations over both space and time (Lewontin 1974, chap. 3), there are some loci
in a variety of animals that show a diversity between local populations over space
and time that correlates with differences in the temperature of the environment of
the local population.

Mitton and Koehn (1975) found that the frequency of an allele at a locus for
malate dehydrogenase in the marine fish *Fundulus heteroclitus* was 0.720 and
0.739 when measured a year apart in the warm water of the pond of a power plant.
In adjacent Long Island Sound it was 0.95. In warmer New Jersey waters it was
0.019.

Taylor and Mitton (1974) made a careful analysis of gene frequencies of local populations of the ant *Pogonomyrmex barbatus* in a variety of localities in Texas that differed in temperature and rainfall. The frequency of particular alleles at the esterase locus Esr varied as localities changed from cold and wet to hot and dry. The frequency of particular alleles at the esterase locus Esr and a malic dehydrogenase (Mdh) locus were related to the severity of extreme temperatures in winter and summer. For example, local populations that lived in places with an extremely cold winter and an extremely hot summer had higher frequencies of the allele Esr-2 and Mdh-2 compared with places with less extreme temperatures.

The crested blenny *Anoplorchus purpurescens* lives in the rocky intertidal zone of the seashore. The A allele of a lactate dehydrogenase locus changes in frequency from 0.02 to 0.30 from north to south over three degrees of latitude in Puget Sound. At any particular site it is also higher in frequency in the spring than in the winter. Allele frequency had a correlation of 0.69 with surface temperature of the water in August. The relation to temperature was confirmed in laboratory experiments with larvae. Genotype $A^1A^1$ had higher survival at high temperature (16°C) than AA and a lower survival at lower temperature (4°C) than AA (Johnson 1971).

While there is much information on the influence of temperature or latitude on the genetic constitution of local populations, there are relatively few cases in which we know the way a particular genetic constitution acts in adapting a population to a particular temperature range. In the freshwater fish *Catostomus clarki* there is a latitudinal cline in the Colorado River basin for the alleles at an esterase locus. An allele that we shall call *a* varied in frequency from 1.0 in southern Arizona to 0.18 in northern Nevada. Its frequency was linearly correlated with latitude. Esterases produced by the three genotypes we shall call *aa*, *ab*, and *bb* were measured for their activity at different temperatures. The homozygote most common in the warmer part of the range *aa* had the highest activity at the highest temperature used in the study (37°C) and was ten times that of *bb*. The homozygote most common in the coldest part of the range *bb* had the highest activity at the lowest temperature used (0°C) and was ten times that of *aa*. The heterozygote had the highest activity at an intermediate temperature. The different activities at different temperatures were considered to be adaptive (Koehn and Rasmussen 1967; Koehn 1969).

Similarly, in the flathead minnow *Pimephales promelas* there is a north-south cline in frequency of two alleles at the lactase dehydrogenase locus. The enzyme produced by the $A^1A^1$ homozygote differed from that produced by the AA and the $AA^1$ genotypes. The frequency of the $A^1$ allele increases from zero in the South to 1.0 in the North. The mean water temperature in the hottest month of the year is 26.7°C in the South and 23.9°C in the North. A number of biochemical properties of the three genotypes were measured and showed that the genotype $A^1A^1$ produced an enzyme with properties that would be advantageous at a lower rather than a higher temperature (Merritt 1972).

Another enzyme whose biochemical properties vary with temperature is $\alpha$-glycerol dehydrogenase ($\alpha$-GDH) in *Drosophila melanogaster*. The electrophoretically slow allele of $\alpha$-GDH is higher in frequency in places with a low mean annual temperature and in other places during the cool autumn months (Berger 1971; Johnson and Schaffer 1973). The property of the enzyme associated with the slow allele correlates with its predominance in cold places (Miller, Pearcy, and

Berger 1975). The enzyme has an important function in the metabolism of insect flight-muscle. Mutants in *D. melanogaster* that lack the enzyme altogether cannot fly. The allele of $\alpha$-GDH became predominant in screwworm populations under the culture conditions used by the United States Department of Agriculture for mass rearing for eradication by means of release of sterile flies. Apparently the high temperature and lack of opportunity to fly (to reduce damage) favored flies with the $\alpha$-GDH allele. When released into the field these flies remained inactive until early afternoon. Mating of the wild flies was therefore completed before the sterile flies had become sexually active (Bush, Neck, and Kitto 1976). Further examples of the way heat influences the genetic constitution of local populations are given in section 11.4. In *Drosophila melanogaster* two alleles control the nature and production of alcohol dehydrogenase (ADH). The allele $ADH^6$ decreases in frequency from 0.9 in Florida to 0.5 in Maine, whereas the other allele $ADH^4$ correspondingly increases. Adult flies of the three genotypes were together held at extreme temperatures until 90% died. There was significantly higher survival of flies with the $ADH^6$ allele at the high temperature of 41°C and low survival of such flies at 0°C and vice versa with flies having the alternative allele. These experimental results corresponded with the greater frequency of the $ADH^6$ allele in the warmer southern regions (Johnson and Powell 1974).

## 11.312   Food

The appearance of species of native insects as pests of plants of economic importance is a well-known phenomenon. Related to this is the existence of local populations or "races" of insects that have different host preferences, though without any recognizable morphological difference (Andrewartha and Birch 1954, sec. 15.12). The so-called host races have in some cases been shown to be previously unrecognized sibling species that do not exchange genes. Others appear to retain their distinctive host preferences in the absence of any barriers to gene flow between the "host races". The codling moth *Laspeyresia pomonella* was introduced into North America from Europe in 1750 and reached California in 1873. Some twenty-six years later it became a pest of walnuts in California (Bush 1975). In 1930 it became a pest of plums in a walnut-growing area in California. There is now substantial evidence that there are genetic differences at least between the apple and the walnut populations. Phillips and Barnes (1975) provided the three sorts of codling moth from the same region with a choice of apples on apple trees, walnuts on walnut trees, and plums on plum trees in a glass house. Table 11.08 shows that the apple race had a strong preference for laying its eggs on apples, with only 1% each laid on walnuts and plums. The walnut race showed a preference for walnuts (61%), but nothing like as strong as the preference of the apple race for apples. The plum race, on the other hand, preferred walnuts.

Further evidence for genetic differences between these host races is indicated by the preferences of the $F_1$ hybrids of the different races (table 11.08), which are different from either parental population. Furthermore, when bred for thirteen generations on apples the walnut race and the plum race increased their preference for apples. But this might be an example of "host conditioning"; that is, the adult prefers to lay its eggs on fruits in which the larvae have developed, or in fruits that have the smell of the place where the adults emerged from their pupae.

**Table 11.08**    Percentage of Eggs of the Codling Moth *Laspeyresia pomonella* Laid by Three "Host" Races on Three Sorts of Fruits

|                      | Host |        |      |
| -------------------- | ---- | ------ | ---- |
| Host Race            | Apple | Walnut | Plum |
| Apple                | 98   | 1      | 1    |
| Walnut               | 39   | 61     | 0    |
| Plum                 | 28   | 71     | 1    |
| Walnut × Apple       | 49   | 18     | 33   |
| Plum × Apple         | 39   | 50     | 11   |
| Walnut × Plum        | 6    | 74     | 20   |
| *After thirteen generations on apples in the laboratory* | | | |
| Walnut               | 45   | 30     | 25   |
| Plum                 | 53   | 10     | 37   |

*Source:* After Phillips and Barnes (1975).

The data in the top section of table 11.08 suggest that the population on walnuts is genetically different from that on apples. Such differences must have come about through natural selection, with or without an initial step of host conditioning. The moths tend to mate on trees in the orchard from which they have emerged, which would help keep the host races genetically distinct despite their presence in the same district.

Another much-cited example of host races is the apple maggot *Rhagoletis pomonella*, whose native host is hawthorn *(Crataegus)*. In 1864 it had become a pest of apples *(Malus)* in the Hudson River valley, and in 1960 it had become a pest of cherries *(Prunus)* in Wisconsin (Bush 1975). According to Bush (1975), it spread rapidly as a pest from the Hudson River valley. The cherry population remained confined to Door County in Wisconsin. In this region cherries are the first fruits to become available for oviposition, then come apples and last of all hawthorn. There is some overlap in the emergence of flies from apples and cherries and apples and hawthorn, but there is very little, if any at all, between cherries and hawthorn. Are the flies on apples, cherries and hawthorn three different host races that have differentiated genetically on three hosts? There are no known morphological or genetic differences between the populations on cherries and hawthorn. Bush (1974) says there are slight differences in size and length of the ovipositor between sympatric populations on apples and on hawthorn. However, in the laboratory wild flies reared from hawthorn and apples both preferred to lay their eggs in freshly picked apples rather than on fruits of hawthorn until the last week in August. Thereafter both preferred to lay their eggs in fruits of hawthorn (Reissig and Smith 1978). In the field, flies associated with either hawthorn or apple readily laid their eggs in fruits of other species hung in the trees (Reissig and Smith 1978). These results suggest that host conditioning does not occur in *Rhagoletis pomonella*; they provide no evidence of genetic difference between the supposed host races (see also Futuyma and Mayer 1980).

It seems that the evidence is not yet sufficient to establish that host races exist in *Rhagoletis pomonella*. How then does one explain the fact that for most of its history in North America this fly has had only one host, the hawthorn, but more recently has extended its host range? The possibility of change in crop husbandry cannot be excluded as a contributing cause. Indeed, Bush (1974) points out that the move to cherries was associated with a change in agricultural practice. In Door County Wisconsin cherries are frequently not picked because of a declining market. Orchards are abandoned and left unsprayed. The fruit thus remains on the tree much longer than in earlier times, when it was sprayed and picked. It thus provides a new resource for those *Rhagoletis* that emerge early in the season.

The western cherry fly *Rhagoletis indifferens* in North America has as its host the native bitter cherry *Prunus emarginata*. Some eighty-nine years after domestic cherries were introduced to Oregon, Washington, and British Columbia, the fly established permanent populations on domestic cherries. Farther south in California domestic cherries are rarely infested with the fly. Domestic cherries in California are grown for the most part at much lower altitudes than the native cherry, and this seems to provide a considerable degree of isolation of the domestic cherry from the flies in the wild cherry (Bush 1969). Although Bush (1969) refers to two races of western cherry fly, one in domestic cherries and the other in wild cherries, this supposition has yet to be confirmed. The existence of host races seems to be better established for the sawfly *Neodiprion abietis*, which has different populations or races on each of balsam fir, white spruce, and black spruce. Each race prefers the "correct" tree for egg laying, and the larvae show high mortality on the "wrong" tree (Knerer and Atwood 1973).

According to Bush (1974) the only direct evidence for genetic differences influencing host selection between sibling species of tephritid fruit flies is the study of Huettel and Bush (1972) on two sibling species of the gall-forming genus *Procecidochares*. Each species lays its eggs into a different genus of Compositae. Experiments in hybridizing the two species of flies led Huettel and Bush (1972) to conclude that host selection was determined by a single pair of alleles, each species of fly differing in one allele. Futuyma and Mayer (1980) challenge this interpretation of the experiments. On the other hand, there is good evidence that gene differences at a single locus can influence survival of the Hessian fly on different hosts. Hatchett and Gallun (1970) showed that the ability of race A to survive on the resistant wheat cultivar Seneca and of race E to survive on Monon are controlled by pairs of recessive genes at different loci. In both races there is a gene for gene relationship between the insect and the host plant, with each resistant gene in wheat having a complementary gene for survival in the insect (sec. 11.323). Similar gene-for-gene relations have been found for cultivars and their aphid and nematode predators (Bush 1974).

Three mechanisms that could initiate a variation in oviposition preference within a species are larval conditioning, adult conditioning, and genetic variation. As we have indicated, direct evidence for genetic variation as the cause is difficult to establish, though this is certainly the case in the Hessian fly. And it is difficult to interpret the results of experiments on the walnut and apple races of codling moth in any other way. Adult conditioning refers to the formation of a search pattern for hosts based on previous experience. Hershberger and Smith (1967) showed that adult *Drosophila melanogaster* that had been kept on peppermint-scented food subsequently selected the peppermint scented arm of an insect olfactometer more often than did control flies raised on a nonscented medium. Thorpe (1939) investigated larval conditioning in *D. melanogaster* in a series of carefully controlled experiments. He reared larvae on a medium containing 0.5% peppermint oil. When the larvae had emerged and pupated, he removed the pupae and washed them thoroughly with distilled water three times to remove any smell of peppermint, then let the adults emerge into a peppermint-free environment. In this way he hoped to separate the effect of larval conditioning from any effect of adult conditioning. These adults were given a choice of a peppermint-scented and a nonscented arm of an olfactometer. Unlike the controls (in which only 35% chose the scented arm) 53% flies whose larvae had been reared on the medium that smelled of peppermint chose the scented arm of the olfactometer. This difference was significant. Thorpe (1939) quite rightly interpreted his results as demonstrating larval conditioning. He also found that if the pupae were not washed and the adults thus emerged in a peppermint-smelling environment, they selected the scented arm of the olfactometer even more strongly. Jaenike (1982) did somewhat similar experiments with *D. melanogaster* and with other species of *Drosophila*. He could not demonstrate larval conditioning, but he did demonstrate adult conditioning. However, his experiments are hardly comparable with Thorpe's, since he used a different larval medium and a different concentration (0.05%) of peppermint oil. Furthermore, instead of testing the adults in an olfactometer he counted the number of eggs laid on slices of medium containing peppermint and slices free from peppermint. Manning (1967) obtained results similar to those of Thorpe using geraniol instead of peppermint. Whereas 88% of the control flies showed aversion to geraniol, only 53% of the flies that had been reared on it as larvae showed aversion to it. Moreover, he selected for reduced aversion to geraniol and after two generations only 40% of the flies showed this aversion. In any particular case in nature it is difficult to sort out the causes determining the choice of host by any species. The larvae of the butterfly *Colias eurytheme* have alfalfa as their main host and vetch as their second host in California. Before the cultivation of alfalfa 125 years ago, they fed on a variety of native legumes. Tabashnik et al., (1981) showed that the butterflies prefer to oviposit on plants other than those that they fed on as larvae. While not conclusive, the evidence suggests that some of the present variation is genetically controlled. Possibly different hosts are best at different times, and this ecological variation may help to maintain genetic variability, a subject that is discussed in section 11.1.

In section 11.311 we discuss genetic variability in *Drosophila pseudoobscura* in the form of chromosomal inversions. Different inversions enable the flies to survive and reproduce at different temperatures. Likewise, different sorts of inversions confer adaptations to different sorts of foods. Da Cunha (1951) isolated nine species of yeasts and two species of bacteria from the crops of *Drosophila pseudoobscura* collected in California. These microorganisms, in separate cultures, were fed to populations of flies in which the Chiricahua (CH) and Standard (ST) chromosomal types were represented; the experiments were done in the usual population cages. The adaptive values of the homozygotes ST/ST and CH/CH and the heterozygotes ST/CH were estimated in the way we have already described, the adaptive value of the heterozygote being taken as unity. The results in table 11.09 show that the adaptive values of each genotype depended upon the sort of food provided in the

**Table 11.09**   Estimates of Adaptive Values and Frequencies at Equilibrium of Standard (ST) and Chiricahua (CH) Chromosomal Types in Experimental Populations of *Drosophila pseudoobscura* at 25°C That Were Fed with Different Microorganisms

| Food | "Adaptive Value" | | | Frequency at Equilibrium | |
|---|---|---|---|---|---|
| | ST/ST | ST/CH | CH/CH | ST | CH |
| *Saccharomyces cerevisiae* | 0.71 | 1 | 0.32 | 0.70 | 0.30 |
| *Rhodotorula mucilaginosa* | 0.83 | 1 | 0.49 | 0.75 | .25 |
| *Candida parapsilosis* | 0.89 | 1 | 0.54 | 0.80 | .20 |
| *Zygosaccharomyces dobzhanskii* | 1.06 | 1 | 0.35 | 1.00 | .00 |
| *Z. drosophilae* | < 1 | 1 | < 1 | ±0.73 | ±.27 |
| *Kloeckeraspora apiculatus* | 1.55 | 1 | 1.00 | 1.00 | .00 |
| *Candida guilliermondii* | 0.80 | 1 | 0.32 | 0.77 | .23 |
| *C. krusei* | < 1 | 1 | < 1 | ± 0.75 | ±.25 |
| Bacteria strain no. 3 | 1.14 | 1 | 0.37 | 1.00 | .00 |
| Bacteria strain no. 9 | 1.15 | 1 | 0.57 | 1.00 | 0.00 |

*Source:* After da Cunha (1951).

cage. The heterozygote was superior to the homozygote (heterosis) in most of the experiments, but at least two species of yeasts and two species of bacteria caused the adaptive value of the heterozygotes to fall below that of the ST homozygote. El-Tabey, Shihata, and Mrak (1952) reported that most flies collected in nature contained only one species of yeast in the crop at any one locality at any one time. This might be due to the dominance of one yeast in the field or to the selection of food by the flies. In nature the relative abundance of the different species of yeasts shows seasonal fluctuations. Hence it seems likely that, in nature, food may be one of the more important components of environment determining the fluctuations in relative frequency of the different chromosomal types. Nothing is known about whether larvae of different genotypes within the species can select different yeasts. However, Fogleman Starmer, and Heed (1981) were able to show that the larvae of *Drosophila mojavensis*, which feed on yeasts in rotting parts of cactus, distinguish between different patches of different species of yeasts in the substrate. They spend more time in patches of preferred yeasts. Typically the substrate may contain about five of a possible twenty or more species of yeast found in rotting cactus. Larvae did not have random samples of these yeasts in their guts. For example, one widely distributed yeast was typically in greater frequency in the larvae than in the substrate. Another yeast showed the opposite relationship. The difference was due not to differential digestion of yeast but to selective ingestion of yeasts.

   *Drosophila willistoni* is a widely distributed tropical species in the Americas. Associated with the wide distribution is the high number of forty-four chromosomal inversions, which contrasts with the small number in less widely distributed species. One obvious variable in their environment is the great variety of tropical fruits in which the larvae live. It is possible that different genotypes are favored in different fruits.

   Birch and Battaglia (1957) reared *D. willistoni* in two different fruits, *Psidium araca* and *Philodendron* sp. When larvae developed in the fruits of *Philodendron* sp. there was selection between flies, depending upon which of the nine chromosomal inversions they carried. When larvae developed in *Psidium* fruits there was no selection between flies carrying different inversions. In other words, in *Philodendron* sp. some genotypes were favored over others. For example, in *Philodendron* heterozygotes of one particular inversion survived better than the homozygotes of this inversion. In *Psidium* the difference was less pronounced or absent. Although these differences as a whole are small, they nevertheless show that food is an agent of selection for the immature stages of *D. willistoni*.

## 11.32 Predators

As discussed in chapters 1 and 5, predators include parasites, parasitoids, pathogens, herbivores, and carnivores. We discuss the coevolution of plants and their predators and animals and their predators in section 5.21. In this section we are primarily concerned with genetic variation within a species that results in differences between local populations of prey and local populations of the predator.

### 11.321 Moths and Their Predators in Industrial and Nonindustrial Areas

The caterpillars of the peppered moth *Biston betularia* feed on the foliage of a wide range of deciduous trees. They pupate in the soil where they spend the winter. The adult moth flies at night and rests by day on the bark of tree trunks, where it would be ready prey for birds were it not well camouflaged to look like the lichen-covered trunk of a tree. In rural areas of Cornwall, Wales, and Scotland, the peppered moth is, as its name implies, light in color with small dark markings. Since the middle of the nineteenth century a coal black form of the moth (the *carbonaria* form) has been found in industrial areas of England (Bishop and Cook 1980). By 1890 these moths had become the predominant form in industrial areas in England, where some 90% of the moths are black. Kettlewell (1955, 1956, 1973) showed in a series of experiments that the background on which the moth rests during the day provides camouflage from its bird predators. The light-colored moths are well camouflaged in rural areas, where tree trunks are covered with lichens and other epiphytes. The dark forms are well camouflaged in industrial areas, where air pollution kills the lichens and other epiphytes and the bark may become covered with soot.

The directly acting component of environment that causes a differential death rate in the two sorts of moths is predators. Pollutants in the air constitute part of the web in the environment of the moth, as indicated in this segment of the envirogram of the moth:

Industries $\longrightarrow$ pollutants $\longrightarrow$ tree trunks $\longrightarrow$ predators $\longrightarrow$ the peppered moth
in the air      and their
epiphytes

For more than a century natural selection has favored the black form in industrial areas and the light form in rural areas. But in recent years, with the introduction of controls on air pollution, the trend has begun to be reversed. For example, between 1961 and 1964 the frequency of light forms increased from 5.2% to 8.9% at a locality on the Wirral peninsula. By 1974 the frequency of the light form had risen to 10.5% (Clarke and Sheppard 1966; Bishop, Cook, and Muggleton 1978). In Manchester there were no light forms of the peppered moth in samples taken in the late 1950s. They now represent about 2.5% of the sample. In regions where backgrounds of resting moths vary there has been an evolution of behavior such that a moth tends to choose a tree trunk on which it is well hidden (Bishop and Cook 1980).

A single dominant gene is responsible for the difference between the completely black moths and the original light form. There are thus two sorts of dark moths: homozygotes and heterozygotes for the gene at the *carbonaria* locus. There are many intermediate forms, which are apparently due to different alleles at the *carbonaria* locus. But in any one locality the range of forms is very restricted. This is apparently due to differences in tree trunks between localities which tend to favor one form over another. In no locality, even in highly industrial areas, are all the moths of the dark form. This may be because the heterozygote has the greater chance to survive and to reproduce compared with the black homozygote in industrial areas, so some light forms will always be produced. Or there may be some diversity of localities in industrial areas that favor the light form. Or there may be other unknown causes (Bishop and Cook 1980).

Figure 11.09 shows the change in frequency of the dark form in a stretch of country from the industrial areas of Liverpool to the Welsh countryside. In the industrial northeast of the transect (top left corner) the dark form predominates, with over 90% of the population being dark. In the southwest of the transect (bottom right) there are no dark moths. The "peaks" represent local populations that have a higher proportion of dark forms, while the "valleys" represent local popu-

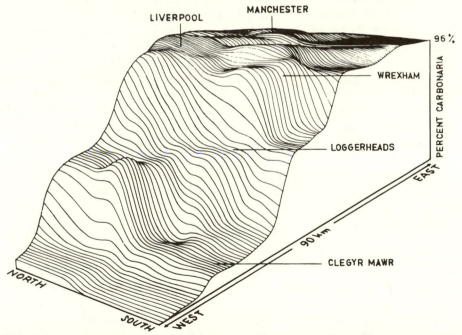

**Fig. 11.09**   The frequency of the dark (*carbonaria*) form of the peppered moth *Biston betularia* in a stretch of country from the industrial areas of Manchester and Liverpool in the north to Wales in the south. The diagram is based on data from 145 sites. (After Bishop, Cook, and Muggleton 1978; Bishop and Cook 1980)

lations with a lower proportion of dark forms than neighboring localities. The decrease in proportion of dark forms along this transect from north to south is correlated with increased reflectance of light from tree trunks and increase in epiphytes on the trunks. Experiments have also shown that death from predation is greater for the dark form the farther south one goes on this transect (Bishop 1972; Bishop et al. 1975).

The distribution of light and dark forms of the scalloped hazel moth *Gonodontis bidentata* between Manchester and Merseyside is shown in figure 11.10. It is a part of the area represented by the plateau in figure 11.09. The hills and valleys are much more marked. This is readily explained as a consequence of a difference in dispersal between the two species of moths. The peppered moth, perhaps because it is sparsely distributed, is likely to have to travel much farther to find a mate than the scalloped hazel moth. In capture, mark, recapture experiments the scalloped hazel moths seldom flew more than 150 m. On the other hand, many peppered moths flew farther than 1 km (Bishop and Cook 1975). The scalloped hazel moths may therefore not disperse their genes as widely as the peppered moth. This gives rise to much greater differentiation between localities and the rugged appearance of the diagram in figure 11.10. This interpretation is supported by the data on the yellow underwing moth *Triphaena pronuba*, which moves farther than the peppered moth. It too is polymorphic. The proportion of different morphs in different localities is almost invariant, resulting in a "smooth landscape" compared with figures 11.09 and 11.10 (Bishop and Cook 1975). These moths are only a few of

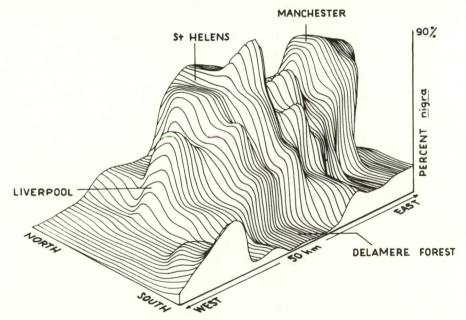

**Fig. 11.10** The frequency of the dark (*nigra*) form of the scalloped hazel moth *Gonodontis bidentata* between Manchester and Merseyside. The area lies entirely within the plateau section of figure 11.09. The diagram is based on data from 112 sites. (After Bishop, Cook, and Muggleton 1978; Bishop and Cook 1980)

more than two hundred species that have evolved populations with predominantly dark forms in industrial areas (Kettlewell 1973; Bishop, Cook, and Muggleton 1976).

### 11.322   The Land Snail *Cepaea nemoralis* and Its Bird Predators

*Cepaea nemoralis* is a land snail that lives in a variety of habitats in western Europe that vary from coastal sand dunes through cultivated lands to altitudes as high as 2,100 m in the Pyrenees. The snail is highly polymorphic for color and for the presence, number, and appearance of up to five dark bands on the whorl of the shell. The main color classes are yellow, pink, and brown, with several shades in between. The bands may be fully pigmented, or the pigmentation may be interrupted. It is also polymorphic for the color of the lip of the shell and for a range of colors of the body. Genes controlling polymorphisms for shell color, presence or absence of bands, lip color and type of band pigmentation are borne together as a supergene. The number of bands on banded shells is determined by unlinked genes (Jones, Leith, and Rawlings 1977). The snails live in small isolated local populations. There are striking differences in the frequency of the morphs in neighboring populations that are only a few hundred meters apart. Although there is much dispute about the cause of the differences between local populations, at least some of them can be attributed to differences in natural selection.

Goodhart (1962) sampled populations of *Cepaea nemoralis* along a three-kilometer stretch of bank of the New Bedford River and the verge of the adjacent road along the riverbank (fig. 11.11). He classified local populations according to the proportion of effectively banded shells and the proportion of yellow shells. Effectively banded shells are those with one or the other or both of the two uppermost bands on the whorls. The proportion of effectively banded shells ranged from 26% to 76% on the bank and from 26% to 87% on the verge of the road. Yellow shells ranged from 24% to 90% on the bank and 20% to 89% on the verge. He measured the proportion of several other morphs as well. They all showed comparable variety in abundance, and none remained constant in proportion along the transect. The samples varied from one hundred to six hundred snails, so the differences can be regarded as real. Goodhart could not recognize any differences in environment along the transect, despite the big difference between the local populations of snails, nor could he suggest any influences of natural selection that might be different between local populations.

In other places there is a predominance of a few morphs over large areas, embracing many local populations of snails in a variety of habitats. Between such areas the frequency of the morphs may change dramatically over a couple of hundred meters or less. Again there is no obvious difference in environment to account for these sudden transitions (Cain and Currey 1963; Carter 1968). Cain and Currey (1963) argued that unidentified selective influences must change over these short distances.

Various influences of selection have been postulated as determining differences in morph frequencies in *Cepaea nemoralis*. The most conclusive evidence points to predation. Figure 11.12 shows the frequency of yellow morphs and effectively banded morphs in three sorts of adjacent habitats: in woodlands, in hedgerows and rough herbage, and in short turf. In woodlands, brown unbanded snails which

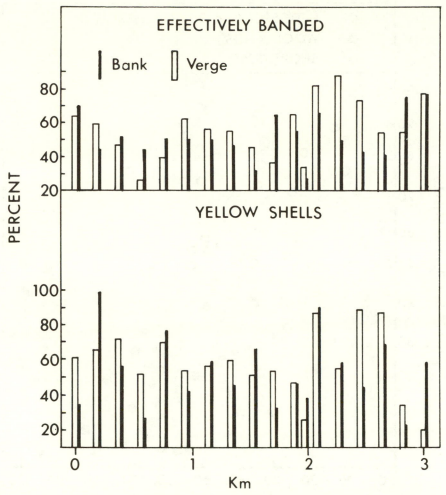

**Fig. 11.11** Variation in the proportion of banded shells (*above*) and yellow shells (*below*) of the snail *Cepaea nemoralis* between local populations along three kilometers of bank of the New Bedford River and the verge along the road running by the bank. (After Goodhart 1962)

match the predominant background of uniform brown leaf litter predominate. In hedgerows and rough herbage there was often a greater proportion of yellow-banded forms, which more nearly matched the greenish background with linear shapes. These differences seem to be due to differential predation by the song thrush *Turdus ericetorum*. According to Sheppard (1951), *Turdus* has the habit of picking up snails and breaking open the shells on stones. Because the birds do not carry the snails far, it is possible to compare the frequencies of types in the selected shells with those that escaped predation. The thrushes discriminate against conspicuous shells. Sheppard marked more than a thousand snails and released them at random in the vicinity of stones on which thrushes were in the habit of crushing snails. The snails were marked in different colors so that local populations from

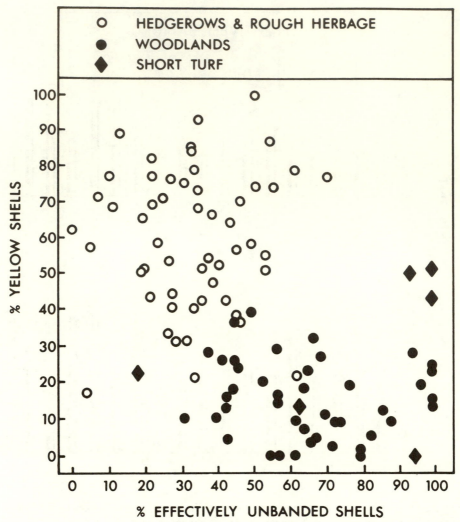

**Fig. 11.12**   Relation between percentage of yellow shells and percentage of banded shells of the snail *Cepaea nemoralis* in local populations in different habitats. In woodlands yellow shells are less abundant and unbanded shells more abundant than in hedgerows and rough herbage. (After Cain and Sheppard 1954)

which the thrushes collected the snails could be identified. Sheppard counted the number of shells of the yellow, brown, and pink varieties found around the stones and compared them with the numbers found in random collections made in the same area that the thrushes hunted over. In the collections made from the stones, the yellow variety dropped from 43% at the beginning of April to 14% at the end of May (fig. 11.13), but in the random collections the proportion remained nearly constant at 24% and 28%, respectively. So the birds were collecting relatively fewer yellow snails as the season advanced, despite the fact that there was no change in the relative abundance of the different varieties in the area where the birds

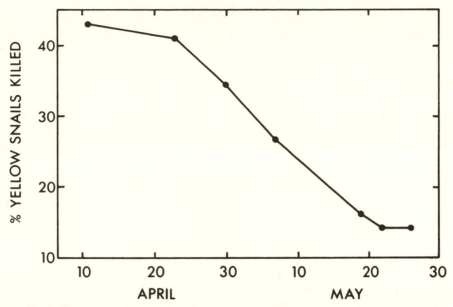

**Fig. 11.13** Seasonal changes in the proportion of yellow shells of the snail *Cepaea nemoralis* killed by thrushes in Marley Wood. (After Sheppard 1951)

were hunting. This was attributed to the fact that the background had become greener as the season advanced. Sheppard reached the conclusion that the morphs selected by thrushes changed markedly from season to season as a result of slight changes in the background colour. Similar experiments later in the year showed a reversal of this trend (Carter 1968). Further evidence pointing to the same conclusion was provided by Clarke and Murray (1962), who compared the frequency of morphs on sand dunes in Somerset in 1926 and in 1959–60. The overall proportion of single-banded snails had doubled over this period, and that of dark brown ones had halved. The most obvious changes to the dunes in this period was that they had stabilized and were overgrown with sea buckthorn. This vegetation provided not only a suitable place for thrushes to hunt, but also changed the background color, in a way that was consistent with the change in frequency of morphs between the two times when populations were sampled.

There are probably many causes for the persistence of many morphs in a region. During one season one is favored over another, but that may be reversed in another season. Predation is not the only influence in selection; there is some evidence that yellow shells are favored in warmer places and darker shells are favored in colder places (Jones, Leith, and Rawlings 1977). Predators form a "search image" for the first morph encountered and subsequently tend to hunt for it to the exclusion of other morphs. So a rare morph will be protected from predation because predators are unlikely to encounter it in their first sample (Jones, Leith, and Rawlings 1977). Also, migration between local populations will tend to maintain diversity of morphs. The persistence of a variety of genotypes in different local populations is a subject we return to in section 11.4.

11.323    Some Insect Predators and Their Prey

The rapid evolution of the wheat rust pathogens to overcome resistance in the wheat plant involves a continuing battle between the wheat breeder and the pathogen. An analogous situation occurs with wheat and the Hessian fly *Mayetiola destructor*. Dominant alleles at each of five loci in wheat confer resistance to the Hessian fly. Corresponding to each of these genes is a recessive allele in the fly that confers counter-resistance (Hatchett and Gallun 1970). Similarly, different strains of both the pea aphid *Acyrthosiphon pisum* and the spotted alfalfa aphid *Therioaphis maculata* are adapted to survive on different strains of alfalfa, each aphid overcoming the resistance of each strain of alfalfa (Cartier et al., 1965).

Pimentel and Stone (1968) reared the housefly *Musca domestica* and the predatory wasp *Nasonia vitripennis* in culture together for two years. They then compared the rate of increase of the wasp on the population of flies it had lived with against its rate of increase on a control population of flies with which it had had no previous contact. The rate of increase of wasps preying on the population that they had been reared with was 75% less than that of those on a population with which they had had no previous contact. Pimentel and Stone (1968) argued that the prey had become resistant to the wasp over the two years of predation. One might argue that the result could have gone in the opposite direction, the wasp evolving to become more effective in preying on the fly. Indeed, people working in biological control generally consider that insect predators control more of their prey as time goes on (e.g., Messenger and van den Bosch 1971).

11.324    The European Rabbit and Its Predator the *Myxoma* Virus in Australia

The history of the development of resistance of the European rabbit to the *Myxoma* virus in Australia is given in section 12.431. The rabbit became more resistant to the virus in time, and the virus became less virulent. This is because it is transmitted by a mosquito that feeds only on living rabbits, so virulent genotypes of the virus will be less likely to be spread than benign genotypes. Avirulence benefits the virus. In Brazil, where *Myxoma* is endemic, rabbits of the genus *Sylvilagus* that are infected with the virus develop a localized tumor that does not seriously influence their health. But the rabbit remains infectious to the mosquitoes for many months, so it is an effective reservoir of infection. Whether a similar relationship will evolve between the European rabbit and *Myxoma* in Australia remains to be seen (Fenner 1965).

## 11.33    Malentities

11.331    Insecticides

For the target animal a pesticide is a malentity, albeit one invented and applied by man. In principle a pesticide in the environment of an animal is similar to a poison or deterrent that an herbivore might find in the leaves or fruit that it eats. And just as animals have overcome the risk of poisons in their food by evolving resistances (sec. 5.211), so too they have evolved resistance to pesticides. The first evidence of inherited resistance was reported by Melander (1914) when he called attention to the fact that it was becoming increasingly difficult to control the San Jose scale,

*Aspidiotus perniciosus*, with lime sulfur sprays. Melander suggested that the scale had become more resistant to the insecticide.

This suggestion met with a good deal of skepticism at the time, though it has not been disproved by any subsequent investigation (Babers 1949). However, a second example of apparent resistance to an insecticide that followed closely on this one captured the attention of entomologists; careful investigations soon substantiated it fully. This was Quayle's claim in 1916 that the red scale *Aonidiella aurantii* on citrus at Corona, California, had become resistant to cyanide. Smith (1941) gathered together this and other evidence that had accumulated since 1916 and showed clearly that a number of insects had developed resistance to insecticides; besides the red scale, both the black scale *Saissetia oleae* and the citricola scale *Coccus pseudomagnoliarum* had become harder to kill than they once were.

The evolution of resistance to pesticides had been documented for 225 species of arthropods by 1972 (Georghiou 1972). The number had grown to 374 by 1976 (Georghiou and Taylor 1976) and to more than 400 by 1982 (Georghiou and Mellon 1982). In many cases a single gene is responsible for conferring resistance: for example, resistance of some houseflies to DDT (e.g., Lichtwardt 1964), resistance to dieldrin in several Diptera (e.g., Busvine, Bell, and Guneidy 1963), and a two-thousand-fold increase in resistance to the organophosphate parathion, that is conferred on spider mites of the genus *Tetranychus* by a single gene (Ballantyne and Harrison 1967). In other cases many genes are involved, as for example in some forms of resistance of houseflies to DDT (Sawicki and Farnham 1967), and DDT resistance in *Drosophila* (Crow 1957).

Genes that confer resistance do so by changing the physiology or behavior of the animal, which in turn influences survival, birthrate, and rate of development (Georghiou 1972). For example, three physiological changes can result in increased resistance of houseflies to DDT. A semidominant gene results in production of the enzyme dehydrochlorinase, which converts DDT to the relatively nontoxic DDE; this is probably the primary cause of resistance to DDT in houseflies (Oppenoorth and Welling 1976). In some cases microsomal detoxification is brought about by the action of a second sort of semidominant gene. And, a recessive gene has been identified that causes reduction in permeability of the integument to DDT; the specific change in the integument has not been identified (Georghiou 1972). The latter, combined with genes that control detoxification, greatly enhances resistance.

An example of change in behavior that amounts to an increase in resistance is increased irritability and tendency to escape from the pesticide. Individuals of *Anopheles atroparvus* were selected for capacity to escape rapidly from a tube lined with paper impregnated with DDT. After ten generations of selection the resultant strains escaped much more rapidly than the unselected strains. By the thirty-second generation the presence of DDT was not necessary for the escape reaction, though it did enhance the response (Gerold and Laarman 1967).

Genes conferring resistance to pesticides are generally rare in environments lacking the pesticide. They are presumably kept in the population by recurrent mutation or by a genetic mechanism such as heterosis. There must be some disadvantage in having such genes in pesticide-free environments, otherwise they would not be so rare. If there is a disadvantage conferred by a particular gene, it is to be expected that such a gene would become scarce again when the pesticide was withdrawn from the environment. The retrogression would happen even in a population in which selection had proceeded so far that, in the presence of the pesticide, the frequency of the gene conferring resistance was high. Such retrogressions have been observed in many instances (Crow 1957). Ferrari and Georghiou (1981) compared the innate capacity for increase $r_m$ of a strain of the mosquito *Culex quinquefasciatus* that was highly resistant to the organophosphorus insecticide temephos with that of a nonresistant strain in an environment free of the insecticide. The nonresistant strain had an $r_m$ of 0.294 compared with an $r_m$ of 0.233 for the resistant strain, so the resistant strain was poorer off in the insecticide-free environment. Its fecundity and survival rate were lower, and its development was slower. Early workers assumed that genes conferring resistance would be extremely rare in environments lacking the pesticide. However, as Crow (1957) has remarked, when selection for resistance is accompanied by a great deal of natural selection for general fitness, the only kind of resistance genes that would become frequent in the population would be those that cause little reduction in fitness, so presumably one should find them in pesticide-free populations. Such indeed seems to be the case with two species of mosquitoes. A high frequency of genes conferring resistance to dieldrin in *Anopheles gambiae* and *A. funestus* has been found in an area in northern Nigeria that had never been sprayed. The nearest sprayed area was 300 km away, and the populations there had different genes for resistance. So migration of flies from this area is not an explanation. Some 74% of the population of *A. gambiae* were resistant to dieldrin; 45% were homozygous for the resistant gene. The conclusion in these examples was that genes conferring resistance to the insecticide were not deleterious to populations in localities lacking the insecticide (Service 1964; Service and Davidson 1964).

Resistance of the red scale *Aonidiella aurantii* to cyanide used as a fumigant is widespread in California. Writing in 1941, Smith commented on the fact that the resistant race was not then present in the San Gabriel Valley, although fumigation had been practiced there longer than in any other part of California. Smith put forward a tentative explanation which may be summarized as follows: There is experimental evidence that, in the absence of cyanide, the nonresistant race has a higher capacity for increase than the resistant race. If this were not so, it would be difficult to explain why the nonresistant race persists anywhere. It was possible to obtain resistant individuals by collecting in "nonresistant" areas, though they were not abundant there. Whether one race or another becomes predominant in an area must depend upon their relative rates of increase between fumigations and their survival rates during fumigation. In an area where the resistant race is numerous, fumigation may be practiced annually or even more frequently, whereas in an area where the nonresistant race is the more numerous fumigations may be separated by several years. Evidently, infrequent fumigations enable the nonresistant race to increase to such proportions between fumigations that it becomes the dominant race; but with frequent fumigations its numbers are kept low relative to the race that breeds more slowly but has a better chance of being alive after a fumigation. This was put forward by Smith as a particular explanation for a specific phenomenon, but the reader will recognize it as an example of the general theory of distribution and abundance which we discuss in chapter 9.

There seems every reason to expect that the contest between insects and men will continue, one producing resistant races while the other produces new insecticides, as it has with bacteria that become resistant to antibiotics and with fungi that are capable of attacking hitherto resistant plants. These are all striking examples of genetic plasticity leading to evolutionary changes in a species as its circumstances of life change.

## 11.34 Mates

According to Darwin (1859, 88) sexual selection "depends. . . . on a struggle between males for possession of females; the result is not death to the unsuccessful competitor, but few or no offspring." Sexual selection or deviation from random mating may occur in local populations of a species, but it is difficult to demonstrate irrefutably. Ayala (1972) and Ayala and Campbell (1974) give examples in *Drosophila* and *Tribolium*, in which they claim that departure from random mating is a function of the frequency of the genotypes involved (frequency-dependent sexual selection). In some cases the mating-success of a genotype was inversely related to its frequency.

Even though a mate may be selected at random, the particular mate found influences the individual's chance to reproduce. An extreme example is the recessive allele at the t locus in house mice which causes homozygous males to be sterile. However, a male +t heterozygote produces 80–95% t-bearing sperm. This distortion of segregation, or meiotic drive, results in the t-allele having a selective advantage. Lewontin (1962) simulated a population on a computer and found that a large population would have an equilibrium gene frequency of 0.37 (as found in the wild) only when it was subdivided into small inbreeding groups, with some loss of genes from and some extinctions of the small populations. His model fits nicely into the concept of the multipartite population in a multipartite environment.

In some species the male mates with a number of females. In others the female mates with a number of males. Still others are monogamous. Any deviations from random mating will of course influence the gene frequency in the progeny. The consequences of "mating systems" seem to have been studied less than their evolutionary origin (see e.g., Wittenberger and Tilson 1980).

## 11.35 Components of the Web

Directly acting components of environment are not the only agents of selection; any component of the web may act in this way. An individual's chance to survive and reproduce may be influenced by the number of other animals of the same kind sharing resources with it (chap. 7). Further, the influence may depend on the particular genotypes that share these resources—that is, the frequency of genotypes in the population. This is known as frequency-dependent selection. Levene, Pavlovsky, and Dobzhansky (1954) studied experimental populations of *Drosophila pseudoobscura* containing three different chromosomal inversions (and thus six genotypes). The adaptive value (fitness) of a genotype depended on which other genotypes were present and on their frequency. The adaptive value of some of the genotypes was inversely related to their own frequency. This relationship would help to maintain polymorphism in the population. Many experiments with *Drosophila*, *Tribolium*, and *Musca* have shown that survival of larvae is influenced by the frequency of different genotypes sharing the larval resources (Ayala and Campbell 1974).

Likewise, an individual's chance to survive and reproduce may depend not only upon the number of other organisms of a different species that share resources with it, but upon the genotype of those organisms and the frequencies of these genotypes. In a tube of flour, *Tribolium confusum* influences the "activity" of the resource for *T. castaneum* and vice versa. Lerner and Ho (1961), using inbred strains of both species, showed that different strains in each significantly differed in their influence on the other's chance to survive and reproduce. Similar results were obtained by Park, Leslie, and Mertz (1964) with *Tribolium* and by Barker (1963) with *Drosophila* spp. Until this aspect of the ecology of the two species of *Tribolium* was appreciated, it was not possible to properly interpret experiments on the influence of one species on the other.

## 11.4   Diversity in Life-History Patterns

A natural population consists of not only a diversity of genotypes but also a number of different developmental stages, or age-classes, generally showing differences in tolerance, preference, and behavior and consequently having different chances to survive and reproduce. Some stages are dispersive, others may be resistant to drought or cold, and some may have a diapause. And within each of these stages and states there may be genetically determined variability. If there is a high variation in the rate of development or the time of reproduction or both, individuals of the same developmental stage will be exposed to different environments at different times of the year. Hence, within the same stage the chance to survive and also to reproduce may be spread over time. This aspect of spreading the risk was pointed out by den Boer (1968) when he first wrote on this subject. He said at the time (p. 169), "the quantitative influences of these ways of spreading the risk which are dependent on the heterogeneity of the population will be intricate." Subsequent studies, most of which have been theoretical, have vindicated this judgment (e.g., the extensive review by Stearns 1976, 1977, 1980). This is not the place to discuss differences in life histories between species and how these differences may have evolved. That has been a main concern of many students of this subject. We are primarily interested in differences within the natural population of the species and how these differences contribute to the spreading of risk.

One approach to this subject is to consider the variation in the components of the innate capacity for increase $r_m$ (Andrewartha and Birch 1954, chap. 3). This statistic is, of course, an abstraction from nature, since it measures the rate of increase of a population of stable age distribution growing in an environment of unlimited resources. The components of $r_m$ are the age schedule of survival from egg to the end of reproductive life (designated $l_x$) and the age schedule of female births (designated $m_x$). These include within themselves the rate of development from birth to sexual maturity. Figures 11.14 and 11.15 show these two sets of components of $r_m$ for *Drosophila serrata* from two different local populations and the sibling species *D. birchii* from three local populations. The measurements were

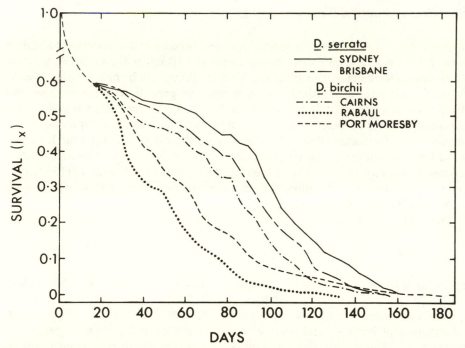

**Fig. 11.14**  The survival from egg to the end of adult life ($l_x$) of females of two geographically separated local populations of *Drosophila serrata* and three geographically separated local populations of the sibling species *D. birchii* at 20°C. The dotted line at top left represents survival of the immature stages, which is the same for all populations. (After Birch et al. 1963)

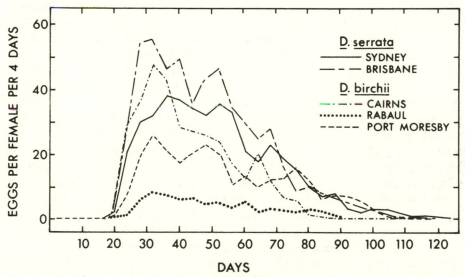

**Fig. 11.15**  The birthrate (eggs laid per female per four-day period) during the lifetime of two geographically separated local populations of *Drosophila serrata* and three geographically separated local populations of the sibling species *D. birchii* at 20°C. Zero day is the same as in figure 11.14. (After Birch et al. 1963)

made on flies collected from these five localities, bred up in the laboratory, and then kept at 20°C and 25°C on a standard and ample diet (Birch et al., 1963). At the time these studies were made the sibling species were thought to be races of one species. Ayala (1970) later established that there was no gene flow between them and designated them species. It is clear from figures 11.14 and 11.15 that the local populations differ in the chance to survive ($1_x$) and the chance to reproduce ($m_x$). An analysis of variance showed that locality significantly influenced the length of adult life but neither the survival nor the length of life of the immature stages. Likewise, locality significantly influenced the birthrate. Birch et al. (1963) proceeded a step further and estimated from these data the innate capacity for increase $r_m$ of the different local populations.

The Lotka equation

$$\sum_{o}^{t} e^{-r_m x} l_x m_x = 1$$

gives the interrelations between age-specific survival ($1_x$), age-specific fecundity ($m_x$), age (x), and rate of population growth $r_m$ for an exponentially growing population, where 0 to t is the reproductive life-span. Changes in any one of the components of birthrate and survival rate will influence the innate capacity for increase, $r_m$. And insofar as $r_m$ is a measure of fitness these components will be subject to selection. An increase in $r_m$ that has been brought about by altering one component may lead to changes in others that may tend in the opposite direction. Hence the notion of trade-offs between life-history characteristics; for example, higher birthrates may be compensated for by shorter life. We pursue some of these ideas below.

Given the values of $1_x$ and $m_x$ we can estimate $r_m$ using the Lotka equation. Figure 11.16 shows the results of such calculations for the five local populations of *D. serrata* and *D. birchii* at two temperatures, 20°C and 25°C. The rate of increase is expressed as the finite rate of increase $\lambda$ where $\lambda = \text{antilog}_e \, r_m$. The reason for expressing rates of increase this way is that the infinitesimal rate ($r_m$) is less readily visualized than the finite rate which we use in everyday speech. Figure 11.16 shows a trend in the value of $\lambda$ at two temperatures. Considered as a sequence from south to north, $\lambda$ for *D. serrata* rose from Sydney to Brisbane. For *D. birchii* $\lambda$ fell from Cairns to Rabaul. Differences in birthrates had an overriding influence in determining these differences in $\lambda$. Not enough is known about the ecology of the two species to speculate on the possible adaptive significance of this pattern.

In a similar study of the Queensland fruit fly *Dacus tryoni*, Bateman (1967) showed adaptive differences along a cline from Cairns in the north of Australia to Gippsland in the south (see map of area, fig. 14.05). Figure 11.17 shows the finite rate of increase $\lambda$ of four geographically distinct local populations from Cairns in the tropical north, Brisbane in the subtropics, Sydney in the temperate region, and Gippsland in a region with a Mediterranean climate. The differences correlated well with the climatic differences. At the lowest temperature (20°C) Sydney had the highest $\lambda$ and Cairns had the lowest. At the highest temperature (30°C) Cairns had the highest $\lambda$ and Sydney had the lowest. At the intermediate temperature (25°C) there was no significant difference between strains, suggesting that the populations are selected by extremes of temperature, not optimum temperatures. Local popu-

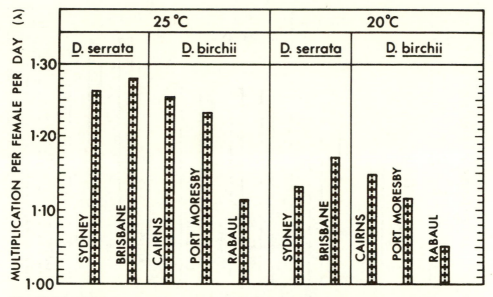

**Fig. 11.16** The finite rate of increase $\lambda$ (multiplication per female per day), where $\lambda = e^{r_m}$ for two geographically separated local populations of *Drosophila serrata* and three geographically separated local populations of the sibling species *D. birchii*. (After Birch et al. 1963)

lations from the north are better adapted to survive and reproduce at higher temperatures. Those from the south are adapted to survive and reproduce at low temperatures. The southernmost population (Gippsland) lives in a climate of both extreme low temperatures (relative to those in the rest of the distribution) in the winter and extreme high temperatures in the summer. It therefore makes sense to find this population having a higher $\lambda$ at low and high temperatures than most of the others. As with *Drosophila serrata* and *D. birchii*, the major differences in values of $\lambda$ (shown in fig. 11.17) are primarily due to differences in birthrate and secondly to differences in survival of adults and immature stages but not in speed of development of immature stages.

We wish to illustrate with these examples that there are differences in characteristics of the life history between local populations. We carry this analysis a step further with data on the influence of different chromosomal inversions in *Drosophila pseudoobscura* on the rate of increase ($r_m$) of the populations that carry them. Dobzhansky, Lewontin, and Pavlovsky (1963) found that populations of *D. pseudoobscura* that were polymorphic for the inversions Arrowhead, Chiricahua, and Pikes Peak had a higher $r_m$ at two temperatures than monomorphic populations. The superiority of the polymorphic populations was presumably dependent upon their greater genetic variability and the superiority of the heterozygotes.

In *Drosophila* spp. all components of the life cycle that contribute to $r_m$ are under genetic control (Ayala 1968). Presumably natural selection molds the local population to fit the environment of its locality. If the outcome is to increase $r_m$, this can be explained in a number of ways. A glance at figures 11.14 and 11.15 indicates that when $l_x m_x$ is plotted against time the result is an inverted v-shaped curve. A

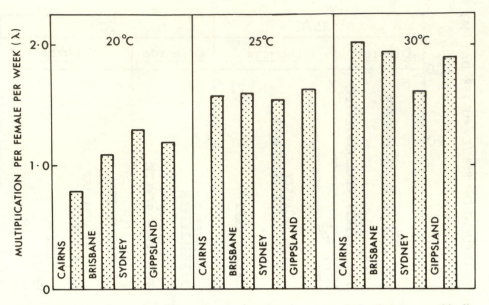

**Fig. 11.17**  The finite rate of increase (λ) per female per week for four geographically separated populations of *Dacus tryoni* from Cairns in the north of Australia to Gippsland in the south. (After Bateman 1967)

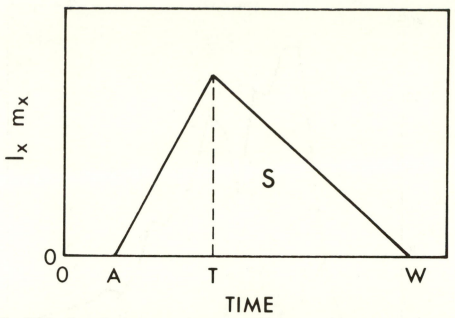

**Fig. 11.18** Generalized triangular function of $l_x m_x$ plotted against time: A, age at first birth; W, age at last birth; T, age at maximum $l_x m_x$; S, area under triangle, which is total number of births.

generalized shape is shown in figure 11.18. Any alteration of the shape of the curve and its position on the time axis will influence $r_m$. A heightening of the curve will increase $r_m$; so will a move to the left. What then would amount to an equivalent move in either of these ways? Lewontin (1965) took the case where the age of first births A = 12 units of time; the age of last births W = 55 units of time; the age of peak production T = 23 units of time; the total number of eggs per female, which is the area under the curve S = 780, and $r_m$ = 0.30. He found that any one of the following would increase $r_m$ from 0.30 to 0.33:

an increase of total births per female from 780 to 1,350
a decrease in age of first births by 2.20 units of time
a decrease in age of last births by 21 units of time
a decrease in age of maximum birthrate by 5.5 units of time
a decrease of each of age of first birth, age of last birth, and age of maximum
    birthrate by 1.55 units of time.

He also found that the number of units of time equivalent to a given increase in total births was smallest when total births were high and the time was short and was largest when total births were low and the time was long.

The overriding importance of early birth in determining the value of $r_m$ is illustrated in figure 11.19, which compares the $l_x m_x$ values of two populations of *Drosophila birchii*: a Rabaul population at 25°C and a Cairns population at 20°C, both of which have the same $r_m$ = 0.135. The low values of $l_x m_x$ for the Rabaul

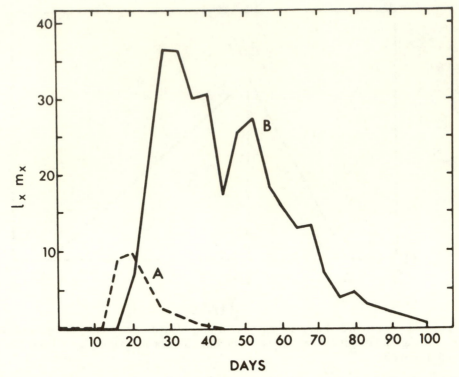

**Fig. 11.19**   The $l_x m_x$ values of two local populations of *Drosophila birchii* that have quite different characteristics but have the same intrinsic rate of natural increase $r_m$. Curve A is a local population from Rabaul at 25°C; curve B is a local population from Cairns at 20°C. (After Lewontin 1965; Birch et al. 1963)

population (due primarily to low $m_x$ values) are compensated for by the displacement of its curve to the left compared with that of the Cairns population at 20°C.

Meats (1971) explored a broad range of possibilities and showed that the mean age of reproduction was by far the most important variable of the reproductive schedule, the age of first births came next, and the age of final reproduction was the least important.

These models apply to the abstract situation of a population with a stable age distribution growing at the rate $r_m$ in an environment that does not change. These are circumstances that may not often apply in nature. However, they do establish the principle that any particular value of $r_m$ can be achieved in a variety of ways— which way presumably depending on the selective pressures that the population has experienced.

Considerations of this sort have led MacArthur and Wilson (1967, 147) to postulate that early maturation, larger broods, and shorter life-span should characterize populations that live in fluctuating environments subject to frequent episodes of colonization (so-called r-selected species). Populations in stable environments in which numbers hover around the "carrying capacity" of the environment should evolve toward a combination of shorter life-spans and a spread of reproductive effort (k-selected species). While some examples seem to fulfill these predictions,

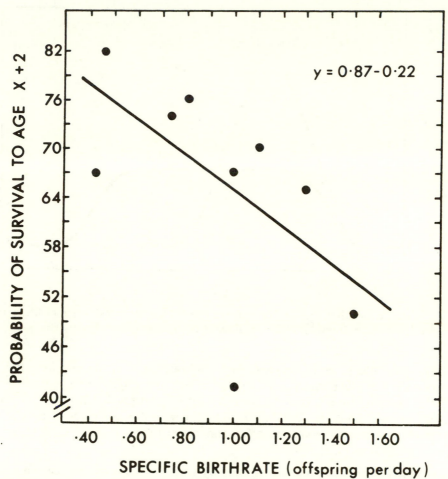

**Fig. 11.20** Regression of probability of survival of the rotifer *Asplancha brightwelli* to age class x + 2 (days) and age-specific birthrate (offspring per day, at 25°C). (After Snell and King 1977)

many do not (Stearns 1980). Nature seems to abhor these simple models; indeed, the work of Tallamy and Denno (1981) on four species of lace bugs suggests that different combinations of life-history traits may be equally adaptive in particular environments. However, it is instructive to examine variations between local populations of a species to see if differences in life histories could contribute to spreading of risk. The rotifer *Asplanchna brightwelli* reproduces parthenogenetically. Snell and King (1977) established twenty-one clones in the laboratory and studied differences in life-history characteristics between clones at different temperatures. They found an inverse relation between birthrate and lifespan. Figure 11.20 shows the negative regression of age-specific birthrates for a particular age class and probability of survival to age-class x+2, that is, $l_{x+2}/l_x$. Age-specific birthrate accounts for 38% of the variance in the probability of survival to age-class x + 2. In other words, the higher the birthrate of a particular age-class, the lower the survival rate. The general relation is shown in figure 11.21,

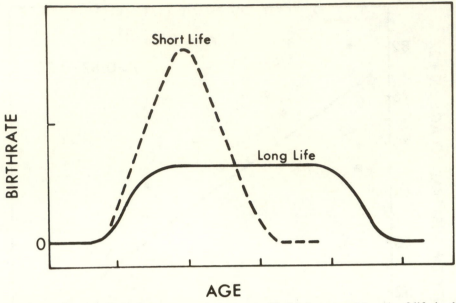

**Fig. 11.21**   Generalized relation between age-specific birthrates and length of life in the rotifer *Asplancha brightwelli*. (After Snell and King 1977)

where short life is associated with high birthrate early in life and long life is associated with low birthrate early and late in life. One might postulate that in one sort of environment one of these patterns might be best and in another the other might be best. Snell and King (1977) did not pursue this prediction. This idea of a trade-off between reproductive effort and longevity was implicit in the question posed by Fisher (1958b, 47):

It would be instructive to know not only by what physiological mechanism a just apportionment is made between the nutriment devoted to the gonads and that devoted to the rest of the parental organism, but also what circumstances in the life-history and environment would render profitable the diversion of a greater or lesser share of the available resources towards reproduction.

The answer to the second question may be none. Moeur and Istock (1980) showed that there was no "trade-off" between reproductive effort and adult longevity for the pitcher-plant mosquito *Wyeomyia smithii*. They fed diets that differed in nutritive value to different groups of larvae. There were differences in fecundity and adult longevity between the adults from these groups, but there was no correlation between their fecundity and length of life. Flies with high fecundity did not live shorter lives. Nutrients that accumulated in the fourth instar larvae were used only in oogenesis and virtually not at all for maintenance of the adult mosquito.

Life histories of some insects are regulated by the interpolation of diapause at some stage of the life cycle. Several aspects of diapause are known to be genetically determined: the inception of diapause (by photoperiod or temperature), the duration of diapause, development during diapause, and cold-hardiness exhibited during diapause. All these aspects of diapause have responded to artificial selection in a number of species that have been studied. Furthermore, differences in these aspects of diapause have been shown to vary adaptively with latitude within certain species. The details of such studies go beyond the scope of this chapter and have been reviewed by various authors in Dingle (1978, 53–190). We refer to just two examples. The distribution of the codling moth *Laspeyresia pomonella* closely matches that of its principal host, the apple tree. Hence its distribution covers a wide range of latitudes in places with quite different climates. Despite the wide range of climates, the codling moth is able to synchronize its seasonal activity with the presence of developing apples to exploit to the full the period when fruits are available. This is achieved by diapause in the mature larvae. The mature larvae recognize approaching adverse seasons by the shortening length of day and falling temperature. The period of dormancy is usually terminated by exposure to cold, after which development is resumed in warm weather, resulting in moths emerging when the new crop is being set. Local populations in different latitudes are genetically different in their response to shorter lengths of day. A latitudinal shift of 10 degrees to the north in North America corresponds to an increase of 1.25 hours in the critical photoperiod that induces diapause. The corresponding increase in photoperiod in Europe is 1.5 hours (Riedl and Croft 1978). Within a local population there is a fine tuning to changes in photoperiod from year to year. The temperatures to which the larvae are exposed during pre-diapause modify their reaction to photoperiod in an adaptive way.

In the pitcher-plant mosquito *Wyeomyia smithii* there is polygenic variation in both rate of development of the immature stages and incidence of diapause (Istock, Zisfein, and Vavra 1976; Istock 1981). This provides a genetic means for continual adaptation to a changing environment. A regression of developmental time of parents against that of the offspring showed that 20–40% of the variation in rate of development is genetically determined. Selection for fast rate of development resulted in a decrease in the proportion of larvae going into diapause in the third instar from 56% in the unselected population to 1–2% after five generations of selection. Selection for slow rate of development also reduced the percentage going into diapause. Selection for diapause quickly produced a population in which nearly every member went into diapause. According to Istock, Zisfein, and Vavra (1976), the influence of rate of development and diapause on the innate capacity for increase $r_m$ was shown by calculating $r_m$ for fast developers, slow developers, and diapause individuals, giving each an egg production of 50. The values of $r_m$ that were so estimated were 0.075, 0.058, and 0.004 respectively. These results suggest, other things being equal, that natural selection would lead to fast developers. Why then is there so much variation in these characteristics in the natural population? Istock, Zisfein, and Vavra (1976) suggest that heterogeneity in time (seasonal changes) and in space may account for why such variability is maintained in the natural population as a whole.

The examples given in this section all illustrate the principle of spreading of risk through genetically controlled variation in the life history of different local populations of the species. In some examples (e.g., *Drosophila serrata*, *D. birchii*, and *Dacus tryoni*) the local populations form a cline over long distances from tropical to temperate climates. In other examples (e.g., the pitcher-plant mosquito) the genetic diversity in life-history characteristics is evident within local populations and between local populations not far apart. In these cases the genetic variability presumably reflects a variable environment in time or in space.

In this chapter we have shown that there is an evolutionary component to the distribution and abundance of animals. To the extent that theory neglects this component, it lacks realism. The multipartite population is gentically multipartite in both space and time. It evolves. This quality in species has enabled them to survive, both in the short term and in the long term, in the face of hostile and changing environments.

PART 3

# *Population Ecology as It Is Practiced*

# Introduction to Part 3

Parts 1 and 2 are largely theoretical. Part 3 is largely practical. Part 3 should be read with the theory of the first two parts in mind. The theory may be judged plausible insofar as it helps to explain the practical detail.

The theory rests on three concepts—about the meaning of "environment", the meaning of "population" and the importance of chance (luck) in the operation of environment. All three concepts are essential to the theory, but perhaps environment is the most fundamental, because without a rational analysis of environment the other two concepts would lose much of their force. The concept of the dendriform environment is established verbally in chapter 1 and illustrated visually in the envirograms of chapter 2. The concept of the multipartite population is established verbally in chapter 8 and illustrated visually by demographs in section 9.2. Spreading the risk, which emphasizes both the importance of chance in the operation of environment and the multipartite nature of environment, is discussed in section 9.3.

The plausibility of our general theory is demonstrated in the eight ecologies that constitute part 3. Envirograms have been drawn for six of them and might readily be filled in for the other two. To draw an envirogram is to enhance one's insight into the particular ecology that is under review. More particularly, the envirogram reduces the risk that an important component of environment might be overlooked.

When it comes to apply the principle of the multipartite population to examples discussed in part 3, it may be helpful to keep figures 9.05 and 9.06 in mind as examples of the sort of model that can be constructed for any natural population. At the same time we emphasize that such diagrams are abstractions and are no more than visual guides to thinking about the multipartite population.

The principle of the multipartite population can be seen operating on a vast scale in the ecology of the rabbit (chap. 12, sec. 12.6). The distribution of the grey teal in Australia is even wider than that of the rabbit—it covers the whole continent. Because of asynchrony in weather and variability in topography, the birds in local populations experience a wide range of rapidly changing environments (sec. 13.2). The concept is also relevant to the other ecologies that are discussed in part 3.

Once the principle of the "multipartite population" has been accepted, the principle of "spreading the risk" indubitably follows—at least for anyone who has learned the habit of thinking in probabilities, which should be part of the mental equipment of every ecologist.

# 12

# The Ecology of the European Rabbit (*Oryctolagus cuniculus*) in Southeastern Australia

## 12.0 Introduction

According to Ratcliffe (1959), domestic rabbits escaped or were released from captivity on many occasions before 1859, but they invariably failed to establish themselves in the wild. The first lot of wild-type rabbits, about twenty-five, were released near Geelong in 1859; they established themselves immediately and thrived. During the next fifty years the rabbits in southeastern Australia increased from twenty-five individuals to many hundreds of millions, and their distribution expanded from a few hectares to more than $3.2 \times 10^6 \, km^2$. Much of this dispersal must have been artificial. In 1950 about 100 million carcasses and skins were exported, and many more were consumed locally without causing a noticeable reduction in the population. Fenner and Ratcliffe (1965) referred to 4,000 rabbits on a study area of 200 ha in 1950 as a dense population, and Myers (1962) estimated that about 30% of the farms that he surveyed in the extensive Riverina district carried infestations that he ranked as severe or medium.

During the period of the original eruption, say from 1860 to 1900, the rabbit in Australia became a legend which was handed down by the pioneers who were themselves colonizing the country. There are tales of great hordes, their supplies of food or water exhausted, wandering along netting fences until, trapped in a corner, they died. We are told that their bodies formed a ramp that the latecomers would climb and so breast the fence. And there are tales of how rabbits ousted the colonial burrowing marsupials *Bettongia leseuri*, the boodies, and usurped their burrows (sec. 8.432). It is true that boodies were once abundant in inland eastern Australia and are now extinct; and certainly today there are rabbit warrens that look to naturalists as if they had been originally made by boodies. But while the rabbits were colonizing the country, so were sheep. Farmers of those days held exaggerated opinions of the carrying capacity of inland pastures; by 1891 the Western Division of New South Wales carried $15 \times 10^6$ sheep, with disastrous consequences to the pristine vegetation and the soil. Today the same area is considered to be fully stocked with $7 \times 10^6$ sheep. So who can tell whether the rabbits ousted the boodies or merely occupied burrows already vacant because their original occupants had been unable to live with so many sheep?

After allowing for exaggeration, it still seems clear that these legends describe an eruption such as is implied for *Cactoblastis* or for thar by their temporary occupation of a position near the top right corner of figure 3.01 and by the rising part of the curve in figure 9.03. When the introductory eruption subsided, the rabbit had moved not to the bottom right corner (like *Cactoblastis* or *Asolcus*); it was much nearer to the bottom left corner; perhaps below and to the left of thar would be the best position to assign it. In other words, in terms of its interaction with resources the rabbit, for most of the time over most of its distribution, might be modeled by figures 9.03 and 9.05. That is, for the rabbit shortages of resources are usually, although not universally, extrinsic (sec. 12.42).

## 12.1   Natural History

Most of the rabbits in eastern Australia live in a region with a Mediterranean climate—that is, the winter is cool and moist and the summer is hot and dry (fig. 12.04). Toward the south the Mediterranean region extends to the sea; inland it merges into arid rangeland where rainfall is low relative to evaporation, unreliable, and not seasonal; toward the eastern coast the pattern is complicated by altitude, but there is a tendency for rain to be more evenly distributed through the year in the south and to be concentrated in the summer in the north (sec. 12.421).

Throughout their entire distribution rabbits characteristically live in burrows in the ground which provide shelter and a place to make nests for the young. The burrows occur in clusters called warrens. The warrens themselves are also distributed patchily depending on soil and topography. Depending on the size and number of social groups that occupy the warren, it may often have only one entrance in sandy soil (Parer 1977) or, according to Myers and Parker (1965), as few as three or four entrances or as many as seventy, though usually not more than twenty, in hard soil.

During their heyday, say from 1890 to 1950, rabbits established warrens throughout their entire distribution in virtually all the places that were habitable. There are few extensive areas in southeastern Australia except on the black soil plains, in the high mountains, or in the dense forests where climate and variability in soil or topography do not provide the rabbit at least locally with suitable places to make burrows. And there are thousands of square kilometers of gently undulating plain where suitable places are widespread and food is often plentiful. In the Mediterranean climate that prevails over all but the extreme ranges of the distribution, the rabbits come into breeding condition when the drought-breaking rains in the autumn bring a flush of green to the pastures as the seeds of the annuals germinate and the perennials put out fresh shoots (sec. 12.421). Not every female starts breeding at once—some delay until the season is well advanced—but once a female has come into breeding condition she usually continues to produce litters of three to eight kittens at intervals of about thirty days until the season closes. The season ends, usually abruptly, as the herbage ripens in the spring and there are no longer enough lush, growing shoots to maintain the rabbits in breeding condition.

The young weigh about 200 g at birth and, if food is plentiful, grow to about 1 kg at 3 months. A female is ready to breed at about 5 months; a well-grown one may weigh 1.5 kg at this age, but a poorly grown one may breed at 5 months though weighing no more than 1 kg. So, in a season that is prolonged, if the food has been good and abundant, the new season's young may produce a litter or two before the end of the season. But this is unusual in most of the drier parts of the distribution. Most of the new generation must survive a summer before they get a chance to breed.

The scientific study of the ecology of the rabbit in eastern Australia got under way during the late 1940s. The following account is largely based on the publications of a group of ecologists led by K. Myers of the Commonwealth Scientific and Industrial Research Organization Division of Wildlife Research and B. D. Cooke of the Department of Agriculture of South Australia, and on many discussions with them.

## 12.2  Behavior

The rabbit is a social animal; an individual has little chance to survive or to reproduce unless it can sustain membership in a social group. So it is not surprising that the rabbit has evolved deeply ingrained patterns of behavior that bind it strongly to its social group and to the particular cluster of burrows in a warren that is the core of the group's territory.

The social and territorial behavior of rabbits was first studied by confining rabbits in small fenced "enclosures", usually about 1 ha of natural pasture (Mykytowycz 1958, 1959, 1960, 1961; Myers and Poole 1959, 1961, 1962, 1963a,b). The whole area was marked out with a rectangular grid of numbered pegs which allowed a rabbit's position to be fixed precisely, wherever observed. A tall "hide" which commanded the whole enclosure allowed an observer to see without being seen or heard. The lights that illuminated the area at night did not seem to disturb the rabbits. The populations in the enclosures were started by releasing from five to twenty rabbits. The rabbits were individually marked with ear tags and dye; the young were also marked as soon as they were big enough. The ventral fur of females was dyed because the rabbit plucks this fur to line her nest; so the nest and the nestlings could be identified. The rabbits were weighed at intervals and the females were palpated to determine their breeding condition. Mykytowycz dug artificial warrens, which the rabbits used without hesitation. He made special provision for inspecting the nests. Myers and Poole provided hollow logs for shelter but always intervened to prevent the rabbits from establishing deep permanent warrens, which would have been more natural but would have made the nests inaccessible. The experiments in enclosures were artificial, but they permitted the researchers to record the intimate details of social behavior, breeding condition, fecundity, and growth rates for individual rabbits much more precisely than could have been done for rabbits living more naturally. The observations made in the enclosures have been largely confirmed by watching rabbits living naturally—not restrained by fences, not disturbed by lights at night, and only infrequently handled to be marked and weighed. The information obtained from the enclosures formed the base from which the field experiments were planned. In particular the subtle feedback between the social group and its territory could hardly have been inferred so clearly without the precise observations that the enclosures permitted.

## 12.21   The Influence of Reproductive Condition on Behavior

The populations in the enclosures were allowed to breed for two or three years. Each year as the herbage matured in the spring lactating females dried up, litters were neglected, fetuses were resorbed, females became anestrous, and males became infertile. As the breeding condition disappeared, social pressures that had been rampant during the breeding season slackened; social groups and social status within the groups persisted during summer but were not paraded. With the onset of winter the pastures suddenly became green with germinating annuals and sprouting perennials. Almost as suddenly, usually within a few days of the appearance of green food, a dramatic change occurred in the behavior of the rabbits as they came back into breeding condition.

It is not known how the change to green food brings the rabbits back into breeding condition. With females, an early consequence of feeding on young green herbage is the secretion of gonadotrophic hormones which stimulate the growth of the ovarian follicles, but development of the follicles stops short of ovulation unless the female receives an additional stimulus. In nature the female in this condition usually copulates, and ovulation follows copulation, but ovulation may also be induced without copulation by the courting of a sexually stimulated male.

Implantation occurs on the seventh day and parturition on the thirtieth day. There is a post partum estrus, and the female usually copulates again on the first or second day after parturition. Females held separate from males show a persistent estrus with a seven-day cycle that is correlated with a cycle of growth in the ovarian follicles.

During pregnancy (and pseudopregnancy) the female shows signs of heightened "maternal" behavior at intervals of seven days: she digs to make or enlarge burrows; she is attracted by a burrow or by other females digging, but at the same time she is aggressive toward other females; she builds grass nests and lines them with fur. While she is in this condition the female attracts more attention from the male, but this attention always stops short of copulation. This cycle of "maternal" behavior occurs despite a continuous noncyclic development of the ovarian follicles, the uterus, and the corpus luteum (Hughes and Myers 1966).

## 12.22   The Social Hierarchy

The characteristic behavior of the pregnant females at the beginning of the breeding season sparks off a round of social behavior that quickly re-establishes and reinforces the social grouping that had been latent during summer. Four steps in the process can be recognized: (a) The number of females in each group and their social status in the group is established. Most of last year's adults and some of last year's juveniles are likely to be competing for places. Normally the older ones are likely to achieve higher status than the young ones; among the old ones last year's hierarchy is likely to be preserved, but no standing is impregnable. Challenges may be issued by residents or by intruders, young or old, and though usually lost they may be won. As the struggle resolves itself, one female emerges as clearly dominant, and the others seem to be arranged in a "dominance hierarchy". The group rarely contains more than five females. Usually some will be driven out; the expatriates will include many of last year's young. There is little information that might explain the departure of the unsuccessful ones. Whatever drives them out is powerful, because it is opposed by two strong counter-attractions: they are strongly attached to the home where they grew up, and they are strongly attracted to any warren, especially one that is well worked and obviously a going concern. Perhaps most of those that leave are naturally subordinate yet not so submissive that they can tolerate more than four or five superiors in the social hierarchy. There is, however, evidence that at least a few of them are by nature dominants that leave home because they cannot win first place there and will not tolerate a lesser place. Several such rabbits have been observed to challenge a dominant in another group and win. (b) The group of burrows that can be claimed and defended by the social group has been clearly demarcated. (c) The dominant female has established her right to take precedence over any other female in the group and especially to have first choice of the best nesting sites in the territory. (d) The new season's flush of green food that sparked off this chain of behavior in the females also brought the males into breeding condition and made them socially active. But whereas the females were primarily interested in making and establishing ownership of burrows (where nests would subsequently be built), the males were primarily interested in the females. They became intensely aggressive toward other males and, in the course of much chasing and fighting, soon established a social hierarchy with one male clearly dominating the one or two others in the group. Usually the new season's round of aggression confirmed last year's dominant in his old position, but occasionally he was successfully challenged by a stranger or by a young male of the new generation. After the matter had been settled the group usually contained two or three males. But Parer (1977), in a three-year study of a free-living population that was less crowded than those in the enclosures, found that many groups contained only one male.

## 12.23   The Feedback between Membership in a Group and Burrowing

During this seasonal establishment or reestablishment of the social groups, their social hierarchies, and their territories, the males also do a lot of digging, but the stimulus seems always to come from a female that is or has been digging: the males join the females in this activity but do little to initiate it.

In marked contrast to the persistent, efficient, orderly home building undertaken by a socially organized group is the remarkable failure of rabbits to dig shelters for themselves in other circumstances. Rabbits that are forced from home may renovate an old warren if they find one ready made, but usually they seek improvised shelter—a hollow log or a heap of brush. This is true whether they are adults or juveniles, whether they have been driven from home by shortage of food or by social pressures at the beginning of the breeding season. If a new burrow is to be built it must start as a "stop" (sec. 12.26), dug by a pregnant female to house a nest. The stop may be enlarged if it is used again, and in due course it may grow into a capacious warren providing good shelter for the socially organized groups that live in it; but it owed its beginnings and its development to the female's habit of making an underground nest for her young. The rabbit, famous as a burrowing

**Plate 4**   The European rabbit *Oryctolagus cuniculus* in Australia, "chinning" a post in an experimental enclosure. (Photo by Ederic Slater)

animal, is strictly limited with respect to the circumstances in which it will burrow.

The social group, which is also the breeding group, cannot exist without its territory of burrows, but the burrows cannot be made except by animals in breeding condition. This subtle reaction between the social group and its home was discovered by studying rabbits in artificial enclosures.

## 12.24   The Maintenance of the Group

The social behavior of the members within a group, and especially of the dominant male, strongly suggests that the members of the group recognize each other and are bound together by the common odor of the group, which derives largely from the dominant male.

As a pregnant female approaches her term, the dominant male consorts with her, driving away all other males; during the postpartum estrus he copulates with her. During courtship and after copulation he urinates over her and smears her with the secretion from his chin gland. At other times he "chins" the females, the kittens, and the juveniles in the group (plate 4). It seems that by chinning the kittens and juveniles he protects them from the aggression of the females, because it has been shown that females will accept strange juveniles that have been chinned by the dominant male of their group, whereas they will drive out others that have not been so marked (Mykytowycz, pers. comm.).

## 12.25 The Defense of Territory

The territory that the group defends is also marked and defined by pheromones. Urine is important, and so are the pheromones secreted by the anal glands and the submandibular gland. The anal glands lie on the sides of the rectum close to the anus; their secretion is smeared over the fecal pellets as they are excreted. The rabbits deposit small mounds of fecal pellets on the main pathways leading from the burrows. Elsewhere on the territory fecal pellets that are not dropped in mounds seem to carry less of the secretion from the anal glands. The secretion from the submandibular (chin) gland is smeared on a variety of landmarks in the territory—posts, sticks, and almost any object that is not likely to be marked with feces or urine. As the fecal pellets become old and weathered, they too will be smeared with secretion from the chin gland. If a foreign rabbit drops feces in the territory they will be chinned vigorously.

Once marked, the territory is defended strongly, especially while good food remains plentiful and the rabbits are breeding freely. All members of the group will join in defending of the territory against intruders, but individuals react more strongly to intruders of their own sex. The dominants of both sexes are most forceful and consistent in defense. The burrows and the immediate approaches to the burrows are policed most energetically; as might be expected, the strictness and the vigor of the defense declines as distance from the burrow increases. It is likely that the characteristic odor of a group and its territory gives confidence to the members of the group and, conversely, saps the confidence of an intruder. Certainly this impression is confirmed by watching the behavior of an intruder: the demeanor which has been confident changes to one of marked and obvious wariness and diffidence as soon as the intruder is in a position to pick up the odors of a strange, well-marked territory. Because the odors are stronger near the burrows, the defenders become fiercer and more persistent and the intruder becomes less antagonistic the closer they are to the burrows; but as the fight or the chase approaches the boundary of the marked territory the difference between the antagonists tends to disappear or might be reversed.

## 12.26 The Powerful Attraction of an Established Warren

Males intrude because they are attracted to females in breeding condition in the established group. Females intrude because a pregnant female is strongly attracted to a deep, well-worked warren. Myers and Schneider (1964, 142) described how one expatriate in a natural population contrived to drop two litters in a well defended warren.

Warren A possessed as satellites a strong male, a young sub-adult male and a female all of which lived on the surface in squats beneath tussocks of grass. . . It was a common occurrence each night to observe the active satellite male and the dominant male from the warren running along a line around the warren, like two dogs along a fence, scratching the ground and threatening each other and behaving exactly as had been described for confined animals (Myers and Poole 1961). In the face of violent opposition the single satellite female separately attempted to gain entrance to the warren (79 observations) but had to wait until the other females were out feeding, in the evening, before she could enter. In this way, although living outside the warren, and not accepted by the other females, she managed to drop two litters within the warren and came in each evening to feed them.

Mykytowycz and Gambale (1965) watched a free-living population of rabbits on 18 ha near Canberra and reported similar evidence of the strong attraction that the warrens have for a rabbit that has been evicted from one of the warren's established social groups. Kittens that had been reared in stops on the fringe of the marked territory persistently sought to enter the warrens.

Despite the attraction of the occupied warrens, rabbits that fail to find a place in a social group are likely to be completely excluded. Expatriates may be numerous, especially if, at the beginning of the new breeding season, old warrens are scarce and existing ones crowded as a result of rapid breeding and high survival during the preceding winter and summer. Expatriates may be found living in temporary or makeshift shelters in a no-man's-land between the warrens. Bushes, heaps of brush, or hollow logs are used. The most such expatriates will do to make shelter for themselves is to dig a "squat" which is a primitive open trench in which they crouch. Males are usually more numerous than females among expatriates. Females, when they have a litter to drop, invariably dig a burrow and make a nest at the end of it. An expatriate, surrounded only by virgin soil, digs a primitive burrow called a "stop"; it slopes gently down for about 2 m, and the nest at the end is usually about 30 cm below the surface of the ground; the entrance is plugged with soil and marked with feces and urine, which prevents other rabbits from molesting it (Mykytowycz 1968). But ever since the outbreak of myxomatosis in 1950–52 there have been extensive areas where derelict warrens abound, especially in regions of durable soil and low rainfall. In loose, sandy soil the warrens weather quickly if they are not maintained; in the better-watered and therefore more valuable grazing country many of the old warrens have been destroyed by "ripping". Where derelict warrens are present, an expatriate female is more likely to clean out a burrow in an old warren than to build a stop in virgin ground. Not only is it easier to clean out the old burrow, but the rabbit has a strongly ingrained tendency to prefer a large, well-worked warren. This is not surprising, because both the adult and her litter have a much better chance to survive in such a shelter (secs. 12.412, 12.413). A female that successfully pioneers such a burrow is likely to be joined by other expatriate females and males to form a social group.

## 12.3   Physiology

The environmental physiology of the rabbit has not been studied as deeply as its behavior. There are some experiments in enclosures and some observations in the field to show that lactating females and growing juveniles need a richer diet than adults in a nonbreeding condition and that water is chiefly important because a ready supply allows a rabbit to choose a diet that is rich in other nutrients.

### 12.31   Diet for Breeding Females and Their Young

The influence of green food on fecundity and lactation was measured by Stoddart and Myers (1966). Groups of rabbits consisting of five or six females and several males were kept in pens of 0.15 ha. In one pen the natural pasture was sprayed with a herbicide to prevent its growth, and the rabbits were given oaten hay and oats. In the other two pens the natural pasture was left unsprayed, and in one of them freshly cut green lucerne was supplied daily. The experiments began on 27 June and ended on 16 January. The pastures matured during November. Drinking water was available in all the pens. The results of the experiment are summarized in tables 12.01–12.03. They show the influence of diet on the number of litters born (fecundity), the survival rate of kittens to weaning (lactation), and the growth rate and survival rate of kittens after weaning.

The variances were not given, but the differences between means are large and almost certainly significant. The opportunity to eat green food as the natural pasture matured prolonged the breeding season, enhanced fecundity and lactation, and greatly improved the young rabbits' chance to survive, especially those born late in the season. On the other hand, the adults, though failing to breed on dry food, maintained their weight just as well as those on natural pasture or pasture supplemented with green lucerne. The green food contains some component which is specifically required by the breeding adult and the growing juvenile (sec. 10.311).

The differential survival rate of juveniles born early or late in the season and of juveniles and adults when the pasture begins to dry off has been confirmed on a number of occasions with rabbits living in enclosures and in the wild. Myers and Poole (1963b) measured the survival rate of juvenile and adult rabbits in 0.8 ha enclosures of a pasture that had originally been dominated by rye grass and subterranean clover (sec. 12.427). At the beginning of the breeding season there were 55 adults in the enclosure; the population increased to 360 by the beginning of summer. Food became very scarce, and before the summer was out most of the rabbits had died of starvation. But before food became absolutely scarce it dried up so that the diet consisted of seeds and straw. During this period scarcely any adults died; but many juveniles died; and the younger they were the more likely they were to die (table 12.04). But the hardiest of all were those born early enough to have become adult before November when the pasture dried off, especially the cohort that was born during May–June.

Similar but more forceful results were obtained by Parer (pers. comm.) with a sparse population of rabbits at Urana in 1967—a year of drought when the pasture grew poorly and dried off early (table 12.05; sec. 12.421). At the beginning of the breeding season in 1967 the study area of 280 ha carried 153 adult rabbits. About three-quarters of them were living in a sandy ridge in one corner of the area; the figures in table 12.05 refer to the population on the sandy ridge.

Throughout the breeding season young rabbits were trapped and marked individually. Subsequent records were made either by retrapping them or by watching them from a hide. The numbers of new season's young and old season's adults that were known to be alive on specified dates during the spring and summer are shown in table 12.05. "Survival" means known to be living on the home warren. The advantage of being born early in the season is clearly brought out by these figures. In order to explain the disappearance of so many juveniles, Parer searched the rest of the study area away from the sandy ridge.

Although most of the rabbits lived on or near the sandy ridge, there were burrows scattered thinly through the rest of the area, some of which might afford shelter for a wanderer. Parer set traps near all these warrens. Among the rabbits that he caught

**Table 12.01**   Influence of Green Food on Fecundity and Lactation

| Diet | Number of Adult Females | Litters per Female | Kittens per Litter | Mean Length of Breeding Season (days) | Mean Number of Days Female Not Pregnant | Kittens per Female | Survival Rate to Weaning (%) |
|---|---|---|---|---|---|---|---|
| Pasture | 6 | 2.8 | 4.4 | 106 | 21 | 12.3 | 82 |
| Pasture and lucerne | 5 | 4.6 | 4.9 | 141 | 3 | 22.6 | 93 |
| Hay and oats | 5 | 1.8 | 2.6 | 77 | 23 | 4.6 | 77 |

*Source:* After Stoddart and Myers (1966).

**Table 12.02**    Influence of Green Food on the Growth Rate and Survival Rate of Juveniles after Weaning

| Diet | Number of Rabbits | Growth Rate (g/day) | Percentage Alive on Eightieth Day of Those Alive at Weaning | |
|---|---|---|---|---|
| | | | Born Early: July–August | Born Late: September–October |
| Pasture | 14 | 9.8 | 56 | 17 |
| Pasture and lucerne | 38 | 11.5 | 73 | 79 |
| Hay and oats | 3 | 9.6 | — | 0 |

*Source:* After Stoddart and Myers (1966).

**Table 12.03**    Weights of Adult Rabbits in Same Pens as the Young Rabbits Reported in Table 12.02

| Diet | Weight (kg) | | | |
|---|---|---|---|---|
| | Beginning of Experiment 2 July | | End of Experiment 15 January | |
| | Male | Female | Male | Female |
| Pasture | 1.55 | 1.44 | 1.44 | 1.48 |
| Pasture and lucerne | 1.35 | 1.46 | 1.50 | 1.45 |
| Hay and oats | 1.30 | 1.43 | 1.48 | 1.59 |

*Source:* After Stoddart and Myers (1966).

**Table 12.04**    Differential Survival Rate of Mature and Juvenile Rabbits in an Enclosure as the Food Matured and Green Food Became Scarce

| Date of Birth | Survival Rate (%) | | |
|---|---|---|---|
| | 1 October to 31 December | 1 October to 31 January | 30 November to 31 January |
| May–June | 100 | 69 | 69 |
| August | 72 | 31 | 39 |
| October | — | — | 15 |
| Adults | 100 | 50 | 50 |

*Source:* After Myers and Poole (1963b).

there were 44 (all juveniles) that had been marked on warrens in the sandy ridge and were more than 450 m from their birthplace. This distance was arbitrarily chosen as sufficient evidence that they had made a clear break from their birthplace. Trapping on sparsely distributed warrens over 200 ha did not constitute an intensive search, so it is likely that there were more than 44 wandering at this time. Parer continued to trap in the same places; 16 rabbits were not seen again; 20 were seen more than 25 days later; 11 were seen more than 50 days later; and 5 survived the

**Table 12.05** Survival Rate of New Season's Juveniles and Old Season's Adults Subsequent to the Breeding Season of 1967 at Urana

| Date of Birth | Number Alive on Home Warren on Specified Date | | | | | "Survival" Rate on 6 March as Percentage of Those Alive on Earlier Date $100\dfrac{N_5}{N_1}$ |
|---|---|---|---|---|---|---|
| | 6 September (N 1) | 4 October (N 2) | 31 October (N 3) | 30 December (N 4) | 6 March (N 5) | |
| June–July | 234 | 197 | 117 | 14 | 8 | 3.4 |
| August | — | 134 | 83 | 2 | 1 | 0.8 |
| September | — | — | 90 | 0 | 0 | 0.0 |
| Old season's adults | 67 | 66 | 63 | 51 | 22 | 32.9 |

*Source:* I. Parer, pers. comm.

*Note:* The pastures dried off about the end of October.

summer and were still alive at the end of the 1968 breeding season. All 5 were born early, before the end of July. Four of the 5 returned to the sandy ridge for the breeding season.

From these figures it seems likely that most of the young rabbits that left home after the 1967 breeding season died before the summer was far advanced because (a) most of the juveniles that were living on the warrens at the beginning of October had disappeared from the home warrens by early December; (b) although there was a range of four months in the birth dates, there was a range of only two months in the dates when the first and the last wanderers were first seen, and a range of only eleven days in the mean date of wandering for the three cohorts; and (c) the dates when the juveniles were wandering coincided with the period when the pastures were maturing. So it seems that the young rabbits were forced from home because they were starving on a diet that was sufficient for the nonbreeding adults. The physiological experiments that might have explained these striking results seem not to have been done. Cooke (1974), after reading the literature on diet for herbivorous mammals, especially small ones, suggested that green growing herbage might promote breeding in the adult and survival and growth in the juvenile by virtue of the amino acids that it contains.

For the nonbreeding adult green growing herbage is not essential for survival, but a proper balance between water and nutrients is (sec. 12.32).

## 12.32 Diet for Nonbreeding Adults: The Interaction between Water and Nutrients

In order to measure the influence of water-shortage on the intake and metabolism of nutrients, Cooke (1974) placed fifteen wild rabbits in metabolism cages. For food they were offered an excess of air-dry stock pellets. Water was rationed; five rabbits were offered 20 ml daily, five received 50 ml, and the third group of five were allowed to drink as much as they would. The ambient temperature and relative humidity were about 13°C and 65%. The rabbits that were on a short ration of water drank the whole ration each day as soon as it was offered. The amount of water the third group drank was measured each day, and the daily consumption of food was measured for the rabbits in all three groups. Urine and feces were collected under oil, measured, and analyzed. When the rabbits were killed at the end of the experiment the water content was measured in samples of digesta from the hindgut, muscle from the leg, and skin from the back. It was known that the caged rabbits were likely to lose weight quickly for the first five days, so they were weighed at the beginning, after the sixth day, and after the eleventh day, when they were killed. In addition to the statistics that are set out in table 12.06, it was found that there was no significant difference between treatments in the water content of feces, the digesta in the hindgut, or the muscle, or in the osmolarity of the urine. And from table 12.06 it can be seen that the differences in the water content of the skin, though significant, were small. Cooke (1974, 41) commented on the loss of weight:

Since the loss of water from the skin was small i.e., less than 10% of the total water-content, and there was no evidence of dehydration of the muscles it seems that dehydration did not contribute appreciably to the weight losses of the rabbits.

**Table 12.06**   Responses of Wild Male Rabbits to Shortage of Water in the Diet When Kept for Eleven Days in Metabolism Cages

| Statistic | Daily Ration of Water | | | $p$ (var. ratio) |
|---|---|---|---|---|
| | 20 ml | 50 ml | Ad Lib. | |
| Ratio, water $\times$ (water + dry matter)$^{-1}$ (%) | 55.8 | 57.0 | 64.3 | < 0.01 |
| Food intake, g $\times$ kg$^{-0.75}$ $\times$ day$^{-1}$ | 12.9 | 24.2 | 56.3 | < 0.001 |
| Nitrogen in urine, g $\times$ (100 ml)$^{-1}$ | 5.2 | 3.7 | 2.7 | < 0.001 |
| Volume of urine, ml $\times$ kg$^{-0.75}$ $\times$ day$^{-1}$ | 15.3 | 23.5 | 40.7 | < 0.001 |
| Weight of gut, g $\times$ kg$^{-0.75}$ | 104.7 | 129.1 | 138.6 | < 0.001 |
| Water in skin, % wet weight | 54.1 | 54.3 | 59.0 | < 0.01 |
| Loss of weight, g $\times$ kg$^{-0.75}$ $\times$ day$^{-1}$ | 12.9 | 7.5 | 7.0 | < 0.01 |
| Digestibility of dry matter (%) | 71.9 | 68.0 | 61.4 | < 0.01 |

*Source:* After Cooke (1974).

*Note:* Loss in liveweight, intake of food, weight of gut, and volume of urine are related to corrected initial liveweight ($W^{0.75}$).

This meant that after changes in the gut-fill, catabolism of the tissues must have accounted for the remaining weight loss. In fact the observed losses agree well with published data on the energy requirements of rabbits. For example one rabbit, which initially weighed 1.965 kg, lost 20.2 g daily during the final 5 days of the experiment. Because food intake and presumably gut-weight were constant during this time, it can be assumed that the loss represented catabolized tissue. This would yield 77.4 cal. daily because Hellberg (1949) showed that 3.83 cal. are produced for every gram of tissue used. In addition 67.3 cal. were derived daily from the food that the rabbit ate because the rabbit's intake was 22.2 g pellets daily and the pellets had a dry-matter digestibility of 73% and yielded about 4.15 cal. per gram digestible matter. The calculated metabolic rate of the rabbit was therefore 144.7 cal. per day, quite close to the expected maintenance metabolic rate of rabbits of this size e.g., 153 cal. (Crampton and Lloyd, 1959) or 164 cal. (Krasnianski and Nessonowa, 1934).

The rabbits with the small intake of water produced less urine, but the concentration of nitrogen increased from 2.7 to 5.2 g $\times$ 100 ml$^{-1}$; because they ate less, the rabbits produced less fecal matter, but the feces were no drier. Despite the substantial reduction in the excretion of water through urine and feces the rabbits on the low rations of water seemed to maintain normal metabolic rates. So the normal amount of water was probably lost through transpiration and respiration, but it was not measured.

A shortage of water limited the intake of dry food. When, as in this experiment, the food was stock pellets, the ratio of dry matter to water was 43:57 for a daily ration of 50 ml of water and 44:56 for a daily ration of 20 ml of water. That is, the intake of dry matter per g of water was 0.75 g with the daily ration of 50 ml and 0.79 g with the daily ration of 20 ml. These figures are to be compared with 0.56 g of dry matter per gram of water when drinking water was freely available. The difference between 0.75 g and 0.79 g is small, which suggests that the rabbit might not have been able to do better than 0.79 g of dry matter per milliliter of water if the daily ration had been reduced below 20 ml. In nature the figures will vary with the quality of the food, but the principle will hold. During drought the rabbit will die of starvation (even in the presence of ample food) unless it can consume enough water to allow the intake of enough calories to maintain weight or at least to prevent a fatal loss of weight. The water ration may come either as drinking water or, more usually, in succulent plant-tissues. If an adequate ration of water can be got only by consuming large quantities of fibrous, energy-poor succulent material (e.g., by gnawing bark), there may not be time enough in the day or room in the stomach for an adequate ration of energy-rich food even though it may be accessible. Death, when it comes, may seem to be due to a shortage of water, which is indeed the distal cause of death, but the proximate cause is nevertheless starvation: even though there is nutritious food easily accessible, it is too dry to eat (sec. 12.424). Compared with *Dipodomys merriami* (sec. 3.22), the rabbit seems poorly adapted for life in dry places; yet it seems to do fairly well. Perhaps its extreme opportunism counterbalances some other shortcomings (sec. 12.5).

Cooke (1974) followed a population of rabbits on a study area, Witchitie, in the rangeland of South Australia, about 320 km southwest from Myers' study area at Tero Creek. During a severe drought in the summer of 1969/70 the population crashed from a relative density of about 400 to about 7 (fig. 12.01). Toward the end

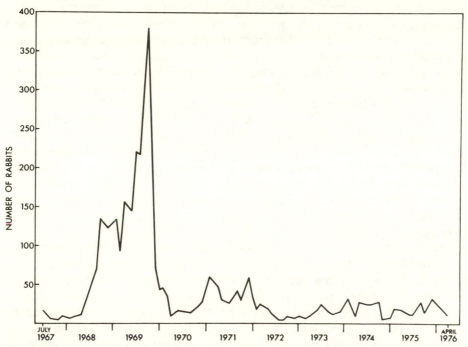

**Fig. 12.01**   The number of rabbits seen by spotlight on a routine transect at Witchitie. The large number at the end of the breeding season of 1969 reflected favorable weather during the breeding and survival seasons of 1968 and 1969. From the weather, as measured by the last column of table 12.16, and from Cooke's survey of vegetation (fig. 12.05), large numbers were also expected at the beginning of the breeding seasons in 1974 and 1976. But the rabbits were not unusually abundant at these times: *Myxoma* was endemic, and fleas, which are vectors, were abundant. (After Cooke 1974)

**Table 12.07**   Weights of Organs from Adult Male Rabbits Shot before and after a Population Crash That Occurred during a Severe Drought

| Statistic | October 1969 | March 1979 | Difference % | p |
|---|---|---|---|---|
| Eye lens (mg) | 172 | 186 | + 8.5 | n.s. |
| Liveweight (g) | 1,647 | 1,280 | −22.3 | < .001 |
| Stomach, with contents (g) | 86 | 76 | −11.2 | .1 > p > .05 |
| Whole gut, with contents (g) | 359 | 335 | − 6.7 | n.s. |
| Kidney (g) | 13 | 10 | −19.3 | < .001 |
| Spleen (mg) | 427 | 181 | −57.6 | < .001 |
| Liver (g) | 43 | 27 | −37.4 | < .001 |

*Source:* After Cooke (1974).

of the decline Cooke shot a sample of adult male rabbits and compared certain physiological measurements with a similar set of measurements from a similar sample that had been shot about six months earlier—before the drought was obvious and before the decline had begun. The measurements are summarized in table 12.07. From the weights of the eye lenses it seems that the second sample might properly be regarded as having been taken from the same population, allowing for the individuals to have aged by six months. There was no immature rabbit in either sample. None of the rabbits in these samples had access to drinking water. The rabbits in the second sample had lost, on the average, about 22% of their liveweight compared with those in the first sample. Cooke thought that a rabbit that had lost about 30% of its initial weight was likely to die soon. The survivors were consistent in that liveweight, liver, kidney, spleen, testes, and kidney fat had all been significantly reduced in weight, indicating starvation. The gut weighed less but not significantly so. After six months' drought the rabbits had been severely starved, but they were still finding enough food to fill their guts. The inference was that the rabbits had been forced to seek out succulent nonnutritious foods to keep up their ration of water, but this had left too little room for more nutritious items of diet—hence the starvation. This inference was confirmed by an experiment that Cooke did on an adjacent area while the rabbits on the main study area were starving as indicated by the data in table 12.07.

Cooke established two small troughs of water on an area where he had marked all the rabbits individually. A fence was built around each trough, enclosing an area just large enough to entrap the rabbits that came to drink. The fence contained trapdoors. He set the traps for four consecutive nights once a month; otherwise the trapdoors were propped open. He also trapped, in cage traps, all the rabbits on the area and mapped the home ranges of many of them. He found that no rabbit traveled more than 500 m to water. He also found that there were a number of rabbits living much closer than this to water which did not once visit the water. At the end of the experiment he trapped, in cage traps, most of the rabbits that were living within 590 m of the water. He found that he had fifteen drinkers and sixteen non-drinkers. Their weights are given in table 12.08.

Rabbits that drank, and hence were free to choose a more nutritious diet, lost proportionately less weight than the nondrinkers. Even at the end of a long dry summer, on an area where a large population had been reduced, apparently by

**Table 12.08** Loss of Weight in Rabbits: Drinkers Compared with Nondrinkers

| Mean Weight (g) | Drinkers (*n* = 15) | Nondrinkers (*n* = 16) |
|---|---|---|
| At beginning (November 1969) | 1,575 | 1,733 |
| At end (March 1970) | 1,480 | 1,481 |

*Source:* After Cooke (1974).

*Note:* The experiment lasted from December to March, through the end of a hot, dry summer. A dense population had already crashed to low numbers on this area.

starvation, to 2% of its initial density, a rabbit could still find enough good food if it was not hampered by a severe inaccessibility of water. It seems that starvation was induced by a relative shortage of water. These results confirm that the distal cause of the very high death rate was a shortage of water, but the proximate cause was a shortage of food (sec. 12.422).

## 12.4 Environment

The discussion in this section follows the pattern that is set by the envirogram of figure 2.01 except that it begins with the warren. To appreciate the rabbit's ecology, it is first necessary to appreciate the uses of the warren.

### 12.41 The Warren

Wherever the rabbit goes it makes warrens. It has not penetrated north of the tropic of Capricorn or into certain areas where the soils, being heavy black clays or clay loams, are too hard for digging, or into unspoiled forest where food is scarce. Elsewhere warrens, or the signs of them, are ubiquitous except where they have been destroyed by farmers to control the pest (sec. 12.44). In all but the most sandy soils, natural erosion by water and wind takes a long time to remove all signs, and a warren or its remains may still be visible, from the air at 100 m if not from the ground, for many years after it has been abandoned. Light friable soils, especially consolidated sand dunes or low sandy ridges, are the most favored places for warrens, but the rabbit will make a warren almost anywhere it can dig; in hard or stony soils, it favors the faces of breakaway cliffs or the banks of channels and other such places.

The warren provides shelter from the weather and a refuge from predators and occupies several other minor places in the web. In the centrum, the warren has an important place as a token.

### 12.411 The Warren as a Token

The rabbit's response to the warren was described in sections 12.22–12.26. The warren is clearly a token, because without the warren the social group is not formed (or at any rate not consummated); without the social group a territory is not marked out and defended; and an expatriate rabbit without a territory is a poor rudderless beast compared to one that has been fulfilled by membership in a social group in a territory. In a word, although the rabbit makes the warren, it is the warren that makes the rabbit.

**Table 12.09**    Predation by Foxes on Rabbits Making Their Burrows in Sandy, Loamy, or Stony Soils

| Region of Study Area | Soil | Warrens | | | Number of Nests Destroyed | |
|---|---|---|---|---|---|---|
| | | Total | Attacked | | | |
| | | | Total | % | Total | Per Warren |
| Snowy Plains | Stony banks | 829 | 0 | 0 | 0 | 0 |
| Mitchell | Loam | 871 | 66 | 7.6 | 78 | 1.2 |
| Tero Creek | Stony hills | 806 | 11 | 1.4 | 12 | 1.1 |
| Tero Creek | Stony pediments | 371 | 46 | 12.4 | 58 | 1.3 |
| Tero Creek | Desert loam | 324 | 148 | 45.7 | 255 | 1.7 |
| Tero Creek | Sand dunes | 3,309 | 2,339 | 70.1 | 2,520 | 1.1 |

*Source:* After Myers and Parker (1965).

*Note:* For location and climate of study areas see figure 12.04 and table 12.12.

### 12.412    The Warren as a Refuge from Predators

No warren would serve as a refuge from *Myxoma* or the goanna, *Varanus* spp. (secs. 12.431, 12.432); any warren will serve as a refuge from raptorial birds (sec. 12.432); against the fox and the feral cat, which are the only vertebrate predators that matter, some warrens afford better protection than others.

The cat, which depends on an ambush (sec. 12.432), is more likely to make a kill when many rabbits share the same exit; it will have less success when few rabbits share many exits. In a relatively sparse population in a 300 ha study area, where most of the warrens had always been built on a low sandy ridge, Parer (1977) found that during three breeding seasons the mean number of exits per rabbit varied between one and five. In the sand, exits remain open only if they are consistently renovated, but in hard soil or in a stony face a burrow, once made, is likely to remain open for many years (sec. 12.414).

The fox relies on excavating a nestful of kittens; it digs down from the surface and is likely to be successful in soft soil, especially if the nest is not too deep, but it rarely succeeds in hard soil. Table 12.09 (from Myers and Parker, 1965) shows that predation by foxes was most severe in sand dunes and loams and was negligible in hard, stony soils. The 6,510 warrens recorded in this table were counted during extensive aerial surveys in rangeland and in subtropical and subalpine regions.

### 12.413    The Warren as a Shelter from Heat and Dryness

Hayward (1961) found that the temperature in a deep burrow was likely to be as much as 15°C lower than the air outside on a hot day, and the absolute humidity might be higher by as much as 15 mg per liter. Remarkably little work has been done on the microclimate of the rabbit warren, but presumably the rabbit gains from its burrow the same sort of protection from extremes of heat and dryness as does *Dipodomys* (sec. 3.221).

In flat country warrens tend to be built on rises and ridges, where the soils are lighter and easier to dig. However, it is not known whether instinct tells the rabbit that such places will also be better drained. Perhaps they are chosen simply because experience teaches that the digging is easier. Or perhaps the warrens that we see today are there because they have survivied extremes of wet weather whereas others that were built in lower places did not. The matter has not been investigated with these questions in mind.

### 12.414   The Warren as a Heritage

The rabbits in southeastern Australia probably reached their greatest penetration, if not their greatest numbers, during the outbreak that came to an end during 1950–52. The rabbits that died in such large numbers at this time from myxomatosis left behind them warrens that could be inherited by future generations. Warrens that had been built in sand were not durable, but in hard soil they would last a long time.

However, the prevailing low numbers of rabbits after the panzootic encouraged farmers, especially those that lived in good farming country where the husbandry was intensive, to step up the destruction of warrens by ripping and digging (sec. 12.44).

The outcome, in the more humid parts of the Mediterranean region, was that few warrens were left for the survivors to use or for their descendants to inherit. For the most part there remained only those warrens that had been built on land that was not good for farming or in places that were inaccessible to the ripping-machines. A few years later when Myers (1962) sampled the area he found that the rabbits were still largely restricted to the soil types and the localities where their warrens had not been destroyed. Myers estimated (chiefly by talking with farmers) the distribution and abundance of active warrens in three habitats at Urana before and after the panzootic of myxomatosis, which reached Urana in 1952. The first survey was made in 1951, the last in 1960. The results are summarized in figure 12.02. The curves reflect not only the impact of myxomatosis but also the effort that farmers put into chemical and mechanical control measures. Myers conducted surveys with similar results at two other places in the Mediterranean region, Albury and Corowa, where the climate is more humid and the farming more intensive than at Urana. A later survey in 1976 showed little change (Myers, pers. comm.). Myers' extensive surveys showed that the rabbits had not been able to recolonize the good farming areas where their old warrens had been destroyed. Parer's (1977) intensive three-year study of a sparse population on a 700 acre fenced study area confirmed Myers' findings and suggested at least a partial explanation for them.

Parer described the population that he studied as "a low density population in an area where the rabbits had previously been abundant." There were many signs that ripping and digging had been actively pursued in the past. When the project began in 1968 there were 153 rabbits, all adults, on the study area; a drought during the first year reduced them to 90; they rose to 182 at the beginning of the third breeding season, and there were 180 at the beginning of the fourth breeding season when the study was terminated. There was no evidence that the population had been much larger than this since the panzootic of myxomatosis in 1952, nor any indication of a trend during the three years of the study.

The warrens were mostly on slightly elevated sandy ridges. By 1965 (thirteen years after the panzootic) most of the warrens that had not recently been renovated had collapsed owing to rain and wind and were choked, if not obliterated. Parer thought the number of warrens had reached some sort of equilibrium with the population, which, though sparse, seemed to have settled down to a "steady state".

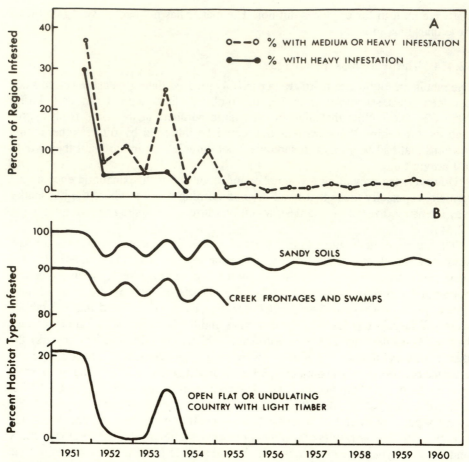

**Fig. 12.02** (A) Outbreaks of *Myxoma* in 1952–54 greatly reduced the number of rabbits around Urana. (B) Rabbits did manage to reestablish moderate to dense populations on some of the poor sandy soils. They failed to recolonize good farming land, where warrens had been destroyed by digging or ripping. (After Myers 1962)

Parer counted the number of warrens, the number of "open" entrances, and the number of "active" entrances. He recorded the number of entrances per warren and the number of active entrances per rabbit. An "open" entrance was one that was not visibly choked with soil or litter so far as could be seen from the surface. An active entrance was one that carried visible signs of being used by rabbits. Some of these records are summarized in table 12.10 and in figure 12.03. The number of active entrances per rabbit varied from 1.2 to 5.2. The number of rabbits varied between ninety and four hundred. The increase in the number of warrens during the breeding season was mostly due to the restoration of old (inherited) warrens, just as the increase in the number of active entrances per warren was mostly due to the restoration of old (inherited) burrows in warrens that were already in use. It seemed, from these data, that the rabbits were unwilling or unable to build new burrows or even to open up old ones to meet the demand of increasing numbers of juveniles. Even the demands of pregnant females were met almost entirely by

**Table 12.10** Number of Warrens and Open Entrances at the Beginning and End of the Breeding Season on Parer's Study Area at Urana

| Number of Warrens | | | | | | Active Entrances | | | | |
| Total | | With Three or Fewer Active Entrances | | With Four or More Active Entrances | | Total | | Per Warren | |
| Beginning | End | Beginning | End | Beginning | End | Beginning | End | Beginning | End |
|---|---|---|---|---|---|---|---|---|---|
| 94 | 112 | 65 | 73 | 29 | 39 | 292 | 390 | 3.0 | 3.5 |

*Source:* After Parer (1977).

*Note:* The figures are the mean of four breeding seasons (1967–70 inclusive). The increase in the number of small warrens (one to three) active entrances) is also compared with the increase in the number of large warrens (four to twenty active entrances).

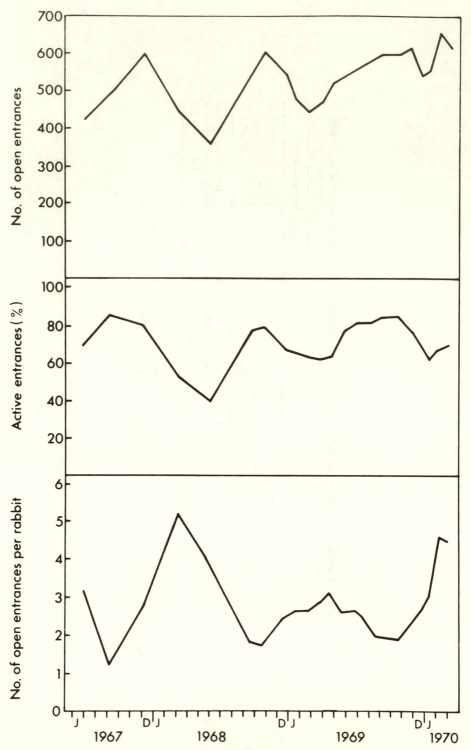

**Fig. 12.03** Changes in the number of open entrances, the percentage of open entrances that were active, and the number of open entrances per rabbit. (After Parer 1977)

restoring inherited burrows. Juveniles that had left home sometimes opened a choked burrow for themselves. Parer (1977, 201) commented on the apparent reluctance of rabbits to make new warrens or to enlarge old ones by adding new burrows:

it is surprising that an animal with the rabbit's capacity for digging did not establish more warrens on the study site, and that existing warrens were not enlarged during the breeding season to accommodate the increased numbers. Myers *et al.* (1975) found that although warren space was very limited in at least one of the three years of their study, only one warren was added to the initial 43.

It is proposed that some control operating through social organization is imposed on the number of entrances to a warren and on the number of warrens in an area. Mykytowycz (1960) suggested that some inhibition on the digging of isolated breeding-stops by subordinate does operated early in the breeding season, and it was only after conditions became crowded and pasture began to deteriorate that such digging occurred. Interference by dominant does was thought to be important in mediating this inhibition.

It seems that the upper limit to the number and size of warrens (as measured by active entrances) tends to be set by the females on the approach of the breeding season (sec. 12.21) and modified upward, late in the breeding season, only under extreme pressure from an excess of pregnant females seeking a good place to make a nest. It follows that: (a) In a place where the soil is hard and the warrens are durable, the number of entrances (the sum of derelict entrances plus those in use) gives a measure of historical populations at their most dense. (b) In a place where the soil is soft and the warrens are not durable, the heritage may be less valuable than in a place where the warrens are durable.

In Parer's study area the soil was sandy. Erosion by wind and water was severe and rapid. Moreover, farmland in the vicinity of Urana was productive enough to repay the cost of artficial destruction of warrens by ripping and digging (sec. 12.44).

In contrast, Myers and Parker (1965) studied the size of warrens in three soil types in rangeland near Tero Creek (table 12.11, fig. 12.04). The soils were all hard and the warrens durable. The land was not productive enough to repay the cost of destroying the warrens artificially. A small sample of Myers and Parker's results is shown in table 12.11. That the mode is about 3 and the median is less than the mean confirms Parer's conclusion that the rabbits strive to keep the warrens small. For Parer's data the mean was 3.5, the mode 1, and the median less than 2. By any

**Table 12.11**  Size of Warrens (Number of Open Entrances) in Three Hard Soils in the Rangeland Climatic Region

| Region and Soil | Number of Warrens | Size of Warrens (Number of Entrances) | | | |
|---|---|---|---|---|---|
| | | Mean | Mode | Median | Range |
| Rangeland: stony hills | 806 | 9.0 | 2 | 6 | 1–52 |
| Rangeland: stony pediments | 371 | 9.1 | 3 | 6 | 1–57 |
| Rangeland: desert loam | 96 | 6.7 | 5 | 5 | 1–22 |

*Source:* After Myers and Parker (1965).

*Note:* Study area 5 in table 12.12 is in the rangeland climatic region.

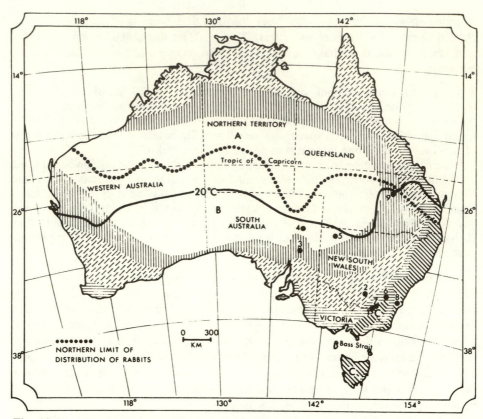

**Fig. 12.04**   The distribution of the European rabbit in Australia and study areas 1–9 listed in table 12.12 in relation to climatic zones. The climatic zones are based on the monthly ratio of rainfall to evaporation. The hatched areas indicate the number of consecutive months in a year during which this ratio exceeds 0.5. Note the concentric arrangement of the climatic zones and how aridity increases away from the coast. The zonation of the climate is reflected in the distribution of vegetation that supplies food for the herbivorous animals of the country. The northern limit of the distribution of the rabbit cuts across the pattern of climate made by rainfall and evaporation but runs roughly parallel to the 20°C isotherm for mean annual temperature. (Climatic zones after Davidson, unpublished; distribution of rabbits after Myers and Parker 1965).

| CLIMATIC ZONES | | | | | | | |
|---|---|---|---|---|---|---|---|
| TEMPERATURE | | | MOISTURE | | | | |
| ZONE | | MEAN °C | DESERT | ARID | SEMI ARID | SEMI HUMID | HUMID |
| | | | NO. OF MONTHS $P/E > 0.5$ | | | | |
| | | | 0 | 1-3 | 4-6 | 7-9 | 10-12 |
| A | HOT | ANNUAL >20°C | | | | | |
| B | WARM TEMP | COLDEST Mo >6°C | | | | | |
| C | COOL TEMP | COLDEST Mo <6°C | NIL | | | | |

of these criteria the warrens on Parer's study area were smaller than those in the hard soil of the rangeland. In the rangeland many burrows and warrens were derelict, but they still persisted. Some of them looked as if they had originally belonged to the extinct marsupial *Bettongia leseuri*. It is well known, in popular opinion, that rabbits were very numerous in this country about 1890. Perhaps the warrens we see today reflect the density of those historical populations.

One consequence of the durability of these old warrens is mentioned in section 12.423. A rabbit population that had been reduced to a very low number by a drought recovered when good rains broke the drought. The crash had been just as severe in the hard-soil country as in the adjacent sand dune country (where the pre-drought population had been denser), but the recovery came more quickly in the hard-soil country, especially where the inherited warrens abutted drainage channels. The advantage of inheriting ready-made warrens that needed only to be renovated was pronounced. The value of the heritage depends on the quality of the soil.

### 12.415   The Warren as a Self-Made Trap

Goannas, *Varanus* spp., readily go down a rabbit's burrow and will eat the nestlings. Sometimes the goanna will bring the nestlings to the surface to eat or cache them. But *Varanus*, though widely distributed, is rarely numerous enough to be important. The flea *Spilopsyllus cuniculi*, a vector of *Myxoma*, breeds in the warrens. Certain mosquito vectors of *Myxoma* also rest in the warren during the day. With respect to these predators and these vectors of a predator, the warren is not a refuge but a trap, because the rabbits in the warren are held there by their behavior; they are at the mercy of the predator that is adapted to seek them there.

## 12.42   Food

Because the supply of food depends largely on the weather and because climate varies widely across the distribution of the rabbit in Australia, this section begins with the arbitrary subdivision of the distribution into six climatic regions (sec. 12.421). The supply of food is discussed under two major headings, food for the breeding season and food for the holding operation that fills the time between breeding seasons. We call this the survival season (sec. 12.422). Water as a component of diet, including drinking water, is discussed in section 12.424. Sodium in the diet is discussed in section 12.426. The short-term consequences of grazing of pastures by rabbits, sheep, and cattle are discussed in section 12.426, and changes in the floristics of the pasture, the long-term consequence of such grazing, in section 12.428. In all these contexts the Mediterranean region is regarded as the norm because that is where most of the rabbits live. It is generally agreed that the species probably originated near the Mediterranean Sea in Europe. An Australian ecologist would feel at home looking at rabbits in the Spanish Coto Donana. The other regions are brought in for comparison when appropriate.

### 12.421   The Growing Season for Food in Different Climatic Regions

Davidson (1936) showed that the number of successive months in which the monthly rainfall was equal to or greater than half the evaporation of water from a standard Australian evaporimeter tank was a good measure of the duration of the growing season for crops and pastures in temperate Australia. Figure 12.04 is a simplified version of Davidson's (1936) map of "bioclimatic zones".

In southeastern Australia the Mediterranean region reaches the coast in the south and merges into rangeland along any transect that heads inland. Any transect that heads for the southeastern coast first crosses a relatively narrow region of eastern tableland and then a still narrower belt of coastal country. For about 500 km north of latitude 38 degrees south the eastern tablelands run up to highland which rises to about 2,100 m at its highest point. Rabbits have colonized certain habitats in these hills, which Myers calls the subalpine region. It is the only part of the rabbit's distribution in Australia where snow lies during winter. Toward the north the eastern tablelands and eastern part of the Mediterranean region merge into the subtropical region. The subtropical pattern of summer rainfall pushes south along a narrow belt of coastal country which Myers calls the south coastal region.

Table 12.12 gives the names of the study areas mentioned in this chapter, the key by which they can be located on the map (fig. 12.04), the regions to which they belong, and certain climatic characteristics of each region (or part of the region) to which a study area belongs. Note that in the Mediterranean region most of the rain falls during the winter. The effectiveness of the winter rain is enhanced by the low rate of evaporation at this time. The winter pattern is so reliable that most of the indigenous plants and many of the successful exotics are adapted to exploit the winter rainfall, so there is little response to rain that falls during summer. In the subtropical region most of the rain falls during summer. In the south coastal region, the eastern highlands, and the rangeland the rain tends to be more evenly distributed through the year. There is a tendency toward two modes. "Rangeland", along with the other regional names in table 12.12, is a climatic classification. The climate is too arid to support arable farming or even intensive grazing. Flocks of cattle or sheep are grazed extensively (about fifteen sheep per 100 ha) on the drought-hardy indigenous vegetation. Rabbits and other herbivores that cannot exist on an exclusive diet of sclerophyllous foliage from shrubs or trees are largely restricted to the vicinity of places that, by virtue of their topography, are better watered than the surrounding countryside (Andrewartha, Davidson and Swan 1938; Andrewartha 1940). In the western rangeland the winter pattern of the south seems still to be slightly dominant, but in the eastern part of the rangeland, the eastern tableland, and the south coastal region slightly more rain falls during summer. This may be offset by high evaporation during summer. The rabbits come into breeding condition whenever rain after drought or, in the subalpine region, warmth after cold produces a crop of green growing herbage. In the Mediterranean region the breeding season is closely linked to winter and the survival season to summer. Because evaporation is everywhere lower in the winter and rainfall therefore more effective then, this correlation tends to linger in all the regions; but it lingers only weakly in places where the rainfall is strongly biased toward summer, as in the subtropical, or highly unreliable, as in the rangeland, or where the winter is cold, as in the subalpine region. In all regions except the south coastal and subalpine the survival season tends to be dry.

In the Mediterranean region, where most of the rabbits live, the pattern of winter rainfall virtually restricts the breeding season to the winter and early spring

**Table 12.12** Climatic Regions in the Distribution of the Rabbit in Eastern Australia

| Study Area | | | Rainfall | | | | Mean Annual Temperature (°C) |
|---|---|---|---|---|---|---|---|
| Number | Name | Region | Mean Annual (mm) | Percentage in May-October | Number of Consecutive Months R/E > .5 | Wettest Four Months Begin | |
| 1 | Albury | Mediterranean | 630 | 60 | 7 | May | 17 |
| 2 | Urana | Mediterranean | 383 | 57 | 5 | May | 17 |
| 3 | Carrieton | Mediterranean | 315 | 63 | 4 | May | 16 |
| 4 | Witchitie | Rangeland | 212 | 49 | 3 | — | 19 |
| 5 | Tero Creek | Rangeland | 180 | 44 | 2 | December | 21 |
| 6 | Canberra | Eastern tableland | 495 | 51 | 8 | May (March) | 14 |
| 7 | Snowy Plains | Subalpine | 743 | 45 | 9 | June | 13 |
| 8 | Mogo | East coastal | 833 | 46 | 7 | January | 15 |
| 9 | Mitchell | Subtropical | 518 | 34 | 2 | November | 19 |

**Table 12.13**   Variability in Annual Rainfall: Comparison between Adjacent Areas
of Dry Mediterranean and Rangeland Climates

| Region | Value of $\dfrac{100(a - b)}{c}$ to Be Expected Once in | | |
|---|---|---|---|
| | Five Years | Ten Years | Twenty Years |
| Mediterranean | 62 | 96 | 127 |
| Rangeland | 75 | 119 | 159 |

*Source:* After Andrewartha (1943).

*Note:* a − b is the difference between the largest and the smallest amount of rain to be expected
once in five, ten, or twenty years; c is the mean rainfall.

(May–October). Rain that falls during summer in sufficient quantity to produce a
green flush may bring the rabbits temporarily into breeding condition, but it comes
to nothing. More fruitfully, unseasonal rain may start the herbage growing as early
as March or prolong its growth into November—but this is unusual. In most of the
other regions the tendency to breed during winter persists but is not so strongly
manifest.

Table 12.12 gives the expected (i.e., the mean) values for certain statistics for
a number of regions and districts. These statistics were chosen because they seemed
relevant to the fecundity or the life expectancy of the rabbit. But the rabbit is an
opportunist, so the variability of the rainfall is also important. In general, for
southeastern Australia the variability of the rainfall tends to increase as the amount
decreases, and the aridity increases inland away from the coast. For example, table
12.13 compares the variability of the annual rainfall for a number of stations in the
vicinity of study area 4, Witchitie, with that for a number of stations in the vicinity
of study area 3, Carrieton. Carrieton, near the dry edge of the Mediterranean
region, is about 240 km south of Witchitie, which is just over the boundary into the
rangeland region. As the mean annual rainfall decreases from 315 to 212 mm, its
variability increases as in table 12.13. We draw on two studies, one at Urana in the
heart of the Mediterranean region and the other at Witchitie in typical rangeland
climate, to illustrate the large influence of weather, especially rainfall, on the
supply of food for the rabbit.

## 12.422   Food for Breeding and Food for Surviving

So long as the supply of green growing herbage lasts, the breeding rabbit will
produce litters at intervals of about thirty days. But the last three or four cohorts
are usually wasted because an immature rabbit (less than 4 months old or weighing
less than 1.5 kg) has little chance to survive a long dry spell on nongrowing food
(sec. 12.31). The minimum effective breeding season is about four months, but
above this minimum the fecundity increases at the rate of one litter (three to eight
kittens) per month. For the nonbreeding adults that have to survive the arid summer, the water content of their food may be of chief importance.

Parer's (1977) study of a free-living population of rabbits at Urana recorded the
changes in the population through a severe drought followed by a year with about
normal rainfall. The results are summarized in table 12.14. At the beginning of the

**Table 12.14**  Influence of Winter Rain on Breeding and Summer Rain on Survival of Rabbits at Urana, in the Mediterranean Region

| Years | Number of Adults at Beginning of Breeding Season (1) | Births (2) | Death Rates (%) | | Rainfall (R) | | Probability of R | | |
|---|---|---|---|---|---|---|---|---|---|
| | | | Juveniles during May–December (3) | Adults during January–April (4) | Breeding Season (x mm) (5) | Survival Season (y mm) (6) | $R \geqq x$ (7) | $R \leqq 0.7x$ (8) | $R \leqq y$ (9) |
| 1967–68 | 153 | 1,000 | > 80 | 74 | 156 | 41 | 0.9 | 0.05 | 0.1 |
| 1968–69 | 90 | 500 | 60 | 38 | 257 | 107 | 0.5 | — | 0.4 |
| 1969 | 182 | — | — | — | — | — | — | — | — |

*Note:* In this and subsequent tables rainfall is the mean of a number of stations in the meteorological district that includes the study area. For columns 5 and 6 the breeding season is arbitrarily assumed to run from May through October and the survival season from January through April. Column 7 gives the probability of at least as much rain during the breeding season as in 1967 or 1968. Column 8 gives the probability of a breeding season so short that no juvenile matures before the food dries up. Column 9 gives the probability of rain during the survival season equal to or less than that in 1968 or 1969.

1967 breeding season there were 153 rabbits (all adults) in the study area; the breeding season was very dry—only one in ten might have been drier (table 12.14). More than 1,000 young were born, but more than 800 of them had died by the end of December. The drought that separated this breeding season from the next was unusually severe—only one year in ten might have been worse; only 26% of the adults that entered the summer were alive at the end of it, and the breeding season of 1968 began with only 90 adults present. Despite the smaller nucleus and the consequent fewer births (about 500) in 1968, the breeding population (182) was twice as numerous as it had been in 1968. The increase was largely due to the higher survival rate of adults during the milder drought of 1968/69. To put the empirical observations in perspective with the climate of this part of the Mediterranean region, table 12.14 also shows the probablility of getting a breeding season (May–October) with at least as much rain as in 1967 or 1968. The probabilities are read from tables of deciles that have been calculated by the Australian Bureau of Meteorology for every rainfall district in Australia. Urana is one of ten recording stations in its district. Similarly, the table shows the probability of getting no more rain than that which fell at Urana during the relevant survival seasons (January through April) 1968 and 1969. Also, because a breeding season of less than four months is unlikely to contribute recruits to the next breeding season and on the arbitrary assumption that a rainfall of no more than 0.7 of that which fell in 1967 would generate a breeding season of four months or less, the probability of such low rainfall is included in table 12.14 (column 8). Five times in one hundred years, on these assumptions, the population might have to face two consecutive survival seasons without recruits from an intermediate breeding season.

A first estimate of the extremes of scarcity and plenty in food that might be experienced by rabbits in this part of the Mediterranean region as a result of fluctuations in the rainfall, may be made by taking the products of the appropriate probabilities. At one extreme the chance of a fruitless breeding season followed by a drought at least as severe as 1967 is $0.05 \times 0.1 = 0.005$. On the other hand, the probability of getting about the median rainfall in both the breeding season and the survival season (as in 1968/69) would be about 0.25.

Water in the form of rain is undoubtedly an important component in the environment of the rabbit in the Mediterranean region. It is less important in the more humid parts, such as around Albury (fig. 12.04, locality 1): as the amount of rain increases so does its reliability; but more important than this, with higher rainfall the financial return from a hectare of farmland increases, and so mechanical and chemical control measures against rabbits become more practicable and more remunerative. Food tends to be outweighed by a malentity which becomes the most important component (sec. 12.44).

The harmful influence of shortage of rain has an ambiguous explanation (sec. 12.32); starvation is likely to be the proximate cause of death, but the shortage of nutrients arises because the rabbits have to spend too much of their time looking for and filling their stomachs with bulkly, nonnutritious "food" for the sake of the water it contains. This risk is great in the Mediterranean region, where the flora is dominated by species that are strongly adapted to growing during the winter and are likely to present only sclerophyllous tissues during summer. In the more arid climate of the rangeland the rainfall is less seasonal, and the flora is made up of opportunistic species, including some that store water in succulent leaves.

Cooke (1974, and pers. comm.) estimated the relative density of the population of rabbits on his rangeland study area at Witchitie (fig. 12.04) from 1967 through 1976; at intervals of about six weeks he ran transects and counted the rabbits that were seen in a spotlight beam (fig. 12.01). At the beginning of the breeding season in 1967 the relative density (all adults) was 15. It had risen to 100 by the beginning of the 1969 season and to 400 by November 1969. This last count may have included some subadults (less than 1.5 kg), which might have been doomed to die young through lack of green growing herbage in the diet. Cooke thought that at this density the rabbits had cleaned out and renovated all the traditional warrens and were making full use of them. The soil was "hard", so it seems likely that in 1969 the rabbits were as numerous as they had ever been during the one hundred years or so of their history in this area.

Table 12.15 shows the number of rabbits counted in routine samples on study areas at Witchitie and Belton from the beginning of the breeding season in 1968 through the survival season of 1976 (i.e., to the beginning of the breeding season in 1976). To keep the rainfall records in table 12.15 comparable, the breeding season is arbitrarily assumed to last from May through October and the survival season from January through April. The real breeding season begins when rain breaks the summer drought and causes the vegetation to germinate or sprout new growth. The real survival season begins when all the immatures that will die for lack of suitable food have died. So the figures in columns 1–4 in table 12.15 have been read from a graph drawn through the empirical counts.

The figures in the last column are the product of columns 6 and 8. They may be taken as an index of favorable weather, taking into account the actual wetness, relative to the average, of each breeding season and of the survival season that followed it. For example, the expectation is that the favorableness of the combined breeding and survival seasons (based on rainfall) would exceed that of 1973/74 only once in eleven years, whereas only once in two hundred years would we expect the combination to be as unfavorable as in 1977/78. Only twice in the ten years since 1968 has the combination of breeding and survival season been more favorable than in 1968/69. In 1969/70 rain during both breeding and survival seasons was close to the mode, giving the "index of favourableness" a value of 0.25, which is inconsistent with the decline in the number of rabbits from 154 in May 1969 to 12 in May 1970. An alternative explanation for this "crash" in numbers is offered in section 12.424.

Belton was about 20 km from Witchitie, in similar country. It might, for the purpose of this survey, be regarded as replicating the study area at Witchitie from 1968 through 1973. The rabbit flea, which had been present in low numbers at Witchitie since 1969, suddenly became widespread and abundant toward the end of 1973. The flea was absent from Belton until late in 1974, but within a year it had become widespread on this site also—in late 1975 80% of all the rabbits that were shot carried at least a few fleas. The "climate index" in the last column of table 12.15 suggests that rabbits should have been numerous at the end of the survival season (i.e., at the beginning of the breeding season) in 1974. Columns 1 and 2

**Table 12.15**  Variability in Rainfall during Breeding and Survival Seasons at Witchitie

| Year (1 May to 30 April) | Number of Adult Rabbits at Beginning of Season | | | | Rainfall (R) | | | | |
| | Breeding Season | | Survival Season | | May–October (Breeding Season) | | January–April (Survival Season) | | Combined Probabilities |
| | Witchitie (1) | Belton (2) | Witchitie (3) | Belton (4) | Rain (x mm) (5) | Probability (R ≤ x) (6) | Rain (y mm) (7) | Probability (R ≤ y) (8) | [P, R ≤ x]x [P, R ≤ y] (9) |
|---|---|---|---|---|---|---|---|---|---|
| 1968/69 | 27 | 20 | 127 | 72 | 118 | 0.65 | 126 | 0.80 | 0.52 |
| 1969/70 | 154[a] | 103 | 43 | 44 | 99 | 0.52 | 49 | 0.49 | 0.25 |
| 1970/71 | 12 | 55 | 27 | 55 | 90 | 0.41 | 129 | 0.82 | 0.33 |
| 1971/72 | 8 | 38 | 29 | 52 | 103 | 0.58 | 125 | 0.80 | 0.46 |
| 1972/73 | 14 | 76 | 7 | 35 | 55 | 0.20 | 189 | 0.95 | 0.19 |
| 1973/74 | 14 | 47 | 22 | 90 | 183 | 0.92 | 283 | 0.99 | 0.91 |
| 1974/75 | 17 | 73 | 13 | 36 | 228 | 0.99 | 51 | 0.51 | 0.50 |
| 1975/76 | 14 | 32 | 32 | — | 170 | 0.90 | 145 | 0.88 | 0.79 |
| 1976/77 | 11 | — | — | — | 115[b] | 0.63 | 12 | 0.05 | 0.03 |
| 1977/78 | — | — | — | — | 41 | 0.07 | 14 | 0.07 | 0.005 |

*Source:* Data from Cooke (1974 and pers. comm.).

*Note:* Column 1 shows the number of rabbits counted along a routine transect in May, that is, at the beginning of the breeding season (or at the end of the previous survival season). Column 3 shows the number of rabbits seen along the same transect in January, that is, at the beginning of the ensuing survival season. Column 5 shows the amount of rain that fell from May through October, and the relative favorableness of the breeding season may be judged by the probability of this much rain at this time of year (column 6). Similarly with the survival season (columns 7, 8). Column 9 (the product of columns 6 and 8, shows one way to combine the estimates of favorableness for a breeding season and its ensuing survival season. See sections 12.32, 12.424 for a fuller explanation of the decline in numbers between May 1969 and May 1970.

[a] In November 1969 the numbers reached a maximum of 400.

[b] An isolated record of 83 mm in October distorts the record.

show that rabbits were numerous at this time at Belton but not at Witchitie. Cooke (pers. comm.) pointed out that for the years 1967–72 there was a significant correlation (r=0.883) between the counts made at Witchitie and Belton, but for 1973 and 1974 the correlation was not significant (r=0.326). Cooke suggested that difference might be due to *Myxoma* transmitted by fleas at Witchitie during 1973 and 1974. If this explanation holds up we may find that, in the rangeland region, the overriding importance of rain, in the pathway through food, may be challenged by *Myxoma* transmitted by fleas.

Several points emerge in this section. The rabbit is well adapted to live in a Mediterranean climate. With fail-safe methods to ensure maximum fecundity, it exploits the opportunity to breed during the rainy winter. It will breed continuously so long as there is enough rain to maintain a green flush of growth on the grasses, forbs, and shrubs that are its food. The duration of the breeding season is the most important variable. Anything less than four months is likely to be fruitless (i.e., fail to add new recruits to the adult population) because a juvenile needs to reach maturity before the green flush fades if it is to have a chance to survive the summer. At Urana, on the oversimple criterion of total rain from May through October, the chance of a year's passing without a fruitful breeding season seems to be about 0.05. On the other hand, the chance of a good breeding season (at least good enough to make a full recovery in one season) seems to be at least 0.5.

Rain is also of chief importance in determining an adult rabbit's chance to remain alive between breeding seasons. Every summer is a drought, but it is the severity of the drought that matters. On the oversimple criterion of total rainfall from January through April, it seems that a drought severe enough to kill three rabbits out of every four might be expected at Urana about once in ten years. There are places in the Mediterranean region that are wetter and others that are drier than Urana, but, taking into account all the other components of environment as well, Myers thought that Urana might be one of the best places in Australia for rabbits to live.

Cooke's study of the population at Witchitie showed that the rabbit was well able to cope with the variations in rainfall that occur in the rangeland. However, not all credit goes to the adaptability of the rabbit: in the more arid climate variations in topography and vegetation tend to compensate for shortage of rain. We continue this discussion in section 12.423.

### 12.423   Interaction between Region, Habitat and Weather

A comparison of x and y in tables 12.14 and 12.15 suggests that less rain was needed for a good breeding season at Witchitie than at Urana. This happens largely because in the soft soils at Urana the rain tends to run into the soil where it falls—as is usual in good farmland. In the hard soils at Witchitie a larger proportion of the water runs from the minor elevations into the minor depressions before it sinks into the soil. The lower areas acquire more than their share from even a light fall of rain and thus come to have a moister microclimate than the bold meteorological records indicate for the countryside at large. Such "oases" may be quite small and in-timately related to their own catchment areas. This relation between topography, local drainage, and microclimate is characteristic of arid habitats (Andrewartha,

Davidon, and Swan 1938; Andrewartha 1940). At Witchitie and elsewhere in the rangeland, rabbits live in or are associated with such better-watered habitats.

During a drought, local populations are likely to be extinguished from the habitats that get only the rain that falls on them. Numbers may be reduced everywhere, but some will persist in places that get not only their own share of rain but also some runoff from adjacent hard upland. The most secure sort of place would be an alluvial "flat" at the base of a small hillock of hard, impermeable soil. The more intimate the better: light showers are the rule during drought, and even the smallest runoff is valuable. The flat will support tussocks of hard perennial grasses and other food for rabbits; the hillock will house durable warrens. Oases on a larger scale are also possible. A large "swamp", the terminus of a substantial drainage system, once filled, might take several years to dry out completely. As the water recedes the banks grow food for rabbits. If the adjacent upland is sandy the burrows will not be durable, but new ones will be readily made.

From 1963 through 1969 Myers and Parker (1975a,b) estimated the relative densities of populations of rabbits in two substantial study areas near Tero Creek (fig. 12.04, locality 5). A succession of wet years had preceded 1963, and rabbits were numerous when the study began. A severe drought set in during 1964 and lasted through the first part of 1966. Rabbits became sparse during the drought. There was a dry spell during the latter part of 1967, then abundant rain fell during 1968 and through the first part of 1969, and rabbits began to increase again. Over the whole study area, all habitats combined, the relative density of the population, as measured by active entrances, fell by 98% between 1963 and 1966; it increased by a factor of ninety between 1966 and 1969.

One area covered 110 km², the other 165 km². By day Myers and Parker counted the open entrances from a light airplane flying at about 100 m; by night they counted the rabbits that were seen by spotlight from a car driven over a routine transect. An intensive survey was done once a year from 1963 through 1969. One study area comprised three and the other eight habitats ("landforms"). Each habitat characteristically contained either no refuge or few or many refuges—and the nature and quality of the refuges tended to be characteristic of the habitat. There was a large swamp set in sand dunes. If the weather had been all that mattered, this might have been the best place for rabbits. But there were many foxes and feral cats, not only because the rabbits were vulnerable in the sand, but because the sand dunes also supported large supplies of alternative food for the predators. So the most secure places for rabbits during the drought were found in the hard country—some variation of the intimate system of the "irrigated" flat adjacent to the hillock of hard country with its durable warrens. Much of the country contained no refuge at all. It is not necessary to go into detail. The important point for our present purpose was made by Myers and Parker (1975a, 11) in their summary.

This paper describes a dramatic fall in rabbit populations in a large area of semi-arid northwestern New South Wales, due to a severe drought, and the increase in numbers which followed [the breaking of the drought]. The reduction in numbers differed markedly in different land-systems. The areas which supported rabbit populations throughout the study were limited to the proximity of swamps in sandy habitats, and close to drainage channels in stony habitats; populations became extinct over large areas of sandy habitats.

In the stony habitats populations increased very rapidly in the 12 months following the breaking of the drought, whereas in the sandy habitats it took about 3 years before a noticeable increase. The difference was probably due . . . to predation and to the availability of open warrens.

For further comment see section 8.412.

## 12.424 Water in the Diet

During the breeding season water presents no problem because the same plants that provide nutrients also provide plenty of water. During the drought it is different because the chief supply of nutrients may be locked up in dry food—seeds and straw of annual plants.

On Cooke's study area at Witchitie the potential rabbit food had the characteristic facies of rangeland ephemerals. The plants were mostly prostrate forbs or low grasses, making a low and incomplete cover for the soil. Cooke wanted to know whether such vegetation might supply an adequate ration of water and nutrients during a run of dry weather. The first step was to measure the relative proportions of the ground that were covered by green growing or, green mature plants, dry stubble, or litter. He related this measurement to an index for drought (Slatyer 1962; Newsome 1966). Starting with a substantial, drought-breaking rain, he calculated a running total by adding daily rainfall and subtracting daily evaporation. The rainfall was the ordinary meteorological record; the evaporation was an arbitrary estimate of the water that might be lost from the soil based on the atmospheric temperature and humidity. When the running total reached zero the soil was considered "dry". But the calculation was continued; as the total became increasingly negative the drought was considered to become increasingly severe. Figure 12.05 summarizes the result for nine consecutive years, 1968–76.

The vegetation, being adapted to the rangeland climate, will respond vigorously to rain at any season of the year. While there is green growing vegetation present the rabbits might start a breeding season and will not go short of water. In the green mature condition the vegetation may be sclerophyllous and the price that the rabbits pay for getting water from it may be to fill their stomachs with fiber.

This problem may be less serious in the rangeland than in the Mediterranean region, because the vegetation in the rangeland is more likely to include xerophytes with succulent leaves that can maintain a lot of water in their leaves even after evaporation from the soil has dried it out well below the wilting point for ordinary plants. Moreover, some of these xerophytes can absorb water vapor into their leaves from the air provided the relative humidity is not below 85%; they may wilt during the day and regain turgidity during the night; no change in absolute humidity is needed for this operation, merely the usual diurnal drop in temperature and the concomitant increase in relative humidity. At Witchitie *Scleretaena (Bassia) lanicuspis* is such a plant; it is fairly abundant and is much sought by rabbits, which eat it during hot, dry weather. Figure 12.05 relates the moisture content of the leaves of *S. lanicuspis* to the "drought index". The samples of leaves were collected about dawn to ensure maximum water content.

From the data in figures 12.05 and 12.06 and other observations Cooke concluded that during the summers of 1970/71 and 1971/72 the rabbits would have

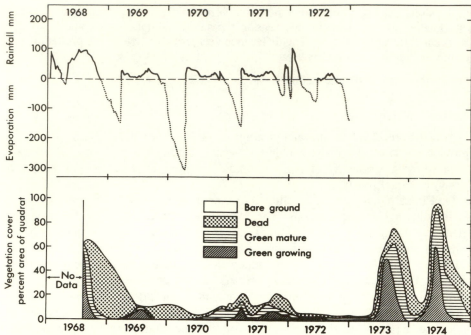

**Fig. 12.05** Drought index and food supply for the rabbit as measured in Cooke's (1974) study area of the rabbit at Witchitie in South Australia. *Above:* Drought index is measured in millimeters of rain or evaporation. The graphs reflect the cumulative sum of rain (positive) and evaporation (negative). *Below:* The vegetation cover is expressed as the percentage of the area covered by green growing vegetation, green mature vegetation, and dead vegetation. The percentages are summed for each entry, so the unshaded part below the 100% line represents bare ground.

been able to get plenty of water from the leaves of *Scleretaena* and other such plants without spending too much time on it or filling their stomachs with fiber to the exclusion of nutritious food.

Table 12.16 shows that there was more rain during summer (December through April) in 1970/71 and 1971/72 than in 1969/70. The summer rainfall was well

**Table 12.16**   Rainfall at Witchitie during Summer for the Years 1968/69, 1970/71 and 1971/72, and the Probability That This Much Rain or More Might Fall during the Period December through April.

| Year | Rain (mm) (x) | Probability (R $\geqq$ x) |
|---|---|---|
| 1969/70 | 50 | 0.67 |
| 1970/71 | 135 | 0.20 |
| 1971/72 | 140 | 0.17 |

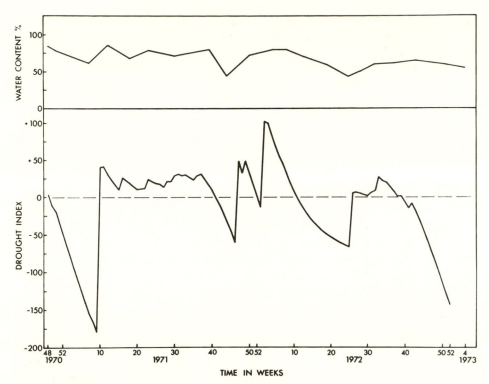

**Fig. 12.06**   The moisture content of the leaves and stems of *Scleretaena lanicuspis* (*above*) in relation to the drought index (*below*). Because the leaves of *S. lanicuspis* can absorb water vapor from air in which the relative humidity is above 85%, the plants retain substantial amounts of water in their tissues even though the drought index remains below zero for a considerable time. (After Cook 1974)

above the median in 1970/71 and 1971/72: the expectation is that for only one year in five would the summer be wetter. On the other hand, the summer rainfall in 1969/70 was below the median but not excessively so: the expectation is that the summer would be drier one year in three. The difference in rainfall does not seem sufficient to explain the high death rate during the summer of 1968/69. That is, the direct action of the drought on the succulent plants seems not to be sufficient to explain the extreme shortage of food (water) for rabbits that is documented in section 12.32. Cooke inferred that the rabbits had themselves caused the shortage by stripping the *Scleretaena* and other plants of all their leaves. The plants had not been able to replace themselves while under such heavy grazing (figs. 12.01, 12.05). If this is the correct explanation, we should ask what is the chance of getting weather as favorable as that in 1968 and 1969 which led to the high numbers at the end of the breeding season in 1969. Consulting the tables of deciles for an answer that considers only the weather, we get the results in table 12.16.

Since domestic sheep or cattle and rabbits share the same pastures almost everywhere, it might seem that water should be no problem for the rabbit, because it might share the watering points that farmers make for domestic stock. Not so. The rabbit is not specially adapted to seek out drinking water. When it has been deprived of water it will drink if it happens across a supply. In Cooke's experiment (sec. 12.32) the rabbits were seriously deprived of water, but none whose home range was more than 550 m from a watering point came to water during an experiment that lasted through the summer; and some that lived no more than 180 m away failed to find the water. The artificial watering points that farmers make for sheep are spaced more widely than this. In the rangeland it is generally accepted that a sheep might make use of grazing that is 8 km from water. Moreover, in the dry country sheep, and especially cattle, tend to denude the approaches to a watering point. Artificial watering points must be regarded as unimportant in the ecology of the rabbit.

### 12.425   Sodium in the Diet

The breeding season in the subalpine region begins and ends quite independently of soil moisture. It begins in the spring, usually about the end of August, when the herbage first begins to sprout new growth in response to the warmer weather. It comes to a puzzling close about the end of December while the soil is still moist and the herbage is still green and growing. Dunsmore (1974, 15) commented on the seemingly premature close to the breeding season:

It is less easy to suggest why breeding ceases in December each year. Water is always abundant and temperature does not reach the levels at which reproduction ceases in other parts of Australia. The pasture is still green and growing and according to Myers (1971) contains a fairly high level of protein at all times of the year. Blair-West *et al.* (1968) have shown that the Snowy Plains area is sodium-deficient and Myers (1971) has pointed out that the "stress on sodium homeostasis . . . undoubtedly affects reproduction significantly." Dunsmore (1966a, 1966b, 1966c) and Dunsmore and Dudzinski (1968) have shown that all three species of nematode parasites of the rabbit in Australia reach very high numbers in female rabbits at Snowy Plains late in the breeding season. The combination of low sodium intake together with large numbers of nematodes in the intestinal tract probably imposes a negative sodium balance of such magnitude that reproduction, with its very great drain on the body sodium, cannot be maintained for longer than is observed.

Blair-West et al. (1968) found that:

1. Herbage collected from subalpine grassland contained less sodium than herbage collected from rabbit habitats in the south coastal and rangeland regions (table 12.17).
2. In the subalpine region the rabbits sought out and ate softwood pegs that had been soaked in NaCl or $NaHCO_3$, but not pegs that had been soaked in KCl, $MgCl_2$, or $H_2O$. In the rangeland the rabbits ignored all the pegs, and in the eastern tableland the response was almost negative but they did eat a little bit of a few pegs.
3. The urine was analyzed for sodium and potassium, the peripheral blood for aldosterone, the kidney for renin, and the feces for sodium and potassium. The results for the subalpine region were what would be expected from rabbits that had been severely deprived of sodium (table 12.18). The table also shows the normal concentrations of these substances in rabbits from other regions.

**Table 12.17**   Concentration of Sodium (Mg Equivalent per Kg Dry Weight) in Herbage Collected from the Feeding Places of Rabbits in Four Climatic Regions

| Region | Season | Herbage | Sodium |
|---|---|---|---|
| Subalpine | Spring | Mixed | < 0.1–8.5 |
| Subalpine | Summer | Mixed | 0.9–1.7 |
| Subalpine | Autumn | Mixed | 1.6–2.0 |
| Subalpine | Autumn | Stylidium stems | 83 |
| Subalpine | Autumn | Stylidium seeds | 20 |
| Subalpine | Autumn | Stylidium seed husks | 175 |
| Subalpine | Winter | Mixed | 0.6–3.8 |
| Eastern tableland | Breeding season | Mixed | 5–10 |
| South coastal | Breeding season | Mixed | 68–309 |
| Rangeland | Breeding season | Mixed | 96–200 |

*Source:* After Blair-West et al. (1968).

**Table 12.18**   Physiological Measurements That Indicate a Shortage of Sodium in the Diet of Rabbits in the Subalpine Region

| Source of Material | Urine (mg equivalent per liter) | | Peripheral Blood Aldosterone (ng/100 ml) | Kidney Renin (U/g cortex) | Feces (mg equivalent/kg dry weight) | | | |
|---|---|---|---|---|---|---|---|---|
| | | | | | Hard | | Soft | |
| | Na | K | | | Na | K | Na | K |
| Subalpine, spring | 0.59 | 208 | 130 | 162 | 2 | 155 | 5 | 518 |
| Subalpine, summer | 0.53 | 269 | 69 | 103 | — | — | — | — |
| Subalpine, autumn | 2.6 | 466 | — | 18 | — | — | — | — |
| Subalpine, winter | 6.4 | 390 | 74 | — | — | — | — | — |
| Eastern tableland | 18.0 | 219 | 21 | 39 | — | — | — | — |
| Rangeland | 139.0 | 319 | 9 | 30 | 28 | 90 | 42 | 343 |

*Source:* After Blair-West et al. (1968).

The simplest explanation for these facts may be, as Dunsmore suggested, that the rabbits are able to build up a reserve of sodium in the body during the nonbreeding season and that as the breeding season draws to a close this reserve is exhausted. But the figures in table 12.17, especially with the analyses of *Stylidium* among them, give rise to the speculation that the rabbits, with their great demand for sodium during the breeding season, are selectively grazing certain sodium-rich components of the pasture and thereby making an intrinsic shortage of sodium, which might not be obvious to an observer. Such speculation might be relevant beyond the ecology of the rabbit. Some of the crashes in the populations of microtine rodents are known to occur while to the human eye there is still a good supply of green food.

### 12.426   Other Rabbits, Sheep and Other Species of Herbivores

Whatever the verdict on sodium may be (when the necessary investigation has been made), there is already evidence that, in the Mediterranean and rangelands regions, local populations may sometimes cause an intrinsic shortage of the green growing herbage that is needed to sustain their breeding or of the water that has to be garnered from succulent food during the dry survival season when air-dry seeds and straw are the only important sources of nitrogen and energy.

In 1967, at Urana, Parer documented a very high death rate among juveniles and subadults toward the end of the breeding season. It was striking that at least three cohorts of youngsters were driven to move away from their home ranges at the same time quite independent of their age. It was not social pressure that drove them out synchronously: it was hunger. They left home because they were starving on a diet of dry food that was adequate for the adults (sec. 12.31). They were starving because green growing herbage was absent from their diet. The weather was chiefly but not entirely responsible.

Parer had three substantially separate local populations under close observation. In September there were three hundred rabbits in group 1 and seventy in group 2. By the end of December 95% of the juveniles and subadults from group 1 were dead, as were 90% of those from group 2. The dry weather had caused the grass to stop growing and dry up, thus causing an extrinsic shortage of green growing grass. But the rabbits had made the extrinsic shortage into an intrinsic one that was even more severe. Parer pointed out that the intrinsic shortage was most severe in area 1 because the rabbits were most numerous there.

During the survival season at Witchitie in 1969/70 a dense population of rabbits 'crashed' to low numbers (secs. 12.32, 12.424). Most of the deaths were attributed to starvation, not because nutritious dry food (the staple diet for this season) was scarce or inaccessible, but because the rabbits were forced to spend too much time, energy, or stomach space seeking water in "succulent" food that was not truly succulent; many of them were observed gnawing bark. This catastrophe occurred on the same study area where a smaller population of rabbits remained in good condition through the summers of 1970/71 and 1971/72 because they were able to get plenty of water from the succulent leaves of *Scleretaena lanicuspis* and other such plants.

Table 12.15 shows the relative densities of the populations at the beginning of the survival period for the years 1969/70, 1970/71, and 1971/72, the rainfall during the four-month period January through April for each year, and the probability of getting no more rain than this during this four-month period. The survival period of 1969/70 was drier than either of the subsequent ones, and doubtless there would have been less water stored in succulent leaves (and so available to rabbits) during the more severe drought. Cooke monitored the disappearance of green vegetation from the study area during 1969/70 (fig. 12.05) and concluded that a large part of the shortage of water and the concomitant shortage of nutrients in that year was probably an intrinsic shortage owing to the relative abundance of the rabbits at the end of the 1969 breeding season (sec. 12.32).

Since sheep share the pastures with rabbits almost everywhere the rabbit occurs, it is appropriate to consider the chance that the sheep's grazing might cause an extrinsic shortage of food for the rabbit. There are no experimental data bearing on this question, but a general knowledge of sheep husbandry in Australia allows reasonable speculation about it. In most parts of the Mediterranean and rangeland regions, especially in those areas where the rabbit has managed to remain a pest despite myxomatosis (sec. 12.431) and control measures (sec. 12.44), it is good husbandry to conserve ample stocks of "dry feed" to last the sheep through the summer and into the next growing season. The only way to achieve this goal is to ensure that the green pasture is not overgrazed during its growing season by either sheep or rabbits.

Once the summer has set in, the sheep and the rabbits depend on essentially the same stocks of "dry feed". The sheep have the advantage because they have access to drinking water, but the rabbits must seek water in less accessible places. They are better off in the rangeland, where they may find water in the succulent leaves of such xerophytes as *Scleretaena lanicuspis*, than in the Mediterranean region where they rely on the fibrous leaves of closely grazed tussocks of perennial grasses such as *Stipa* spp. or the fleshy roots of such perennial herbs as *Hypochoeris*. They may even be reduced to gnawing bark. The sheep seek these resources too, but not so urgently or so intimately. It is unlikely that the sheep appreciably reduce the supply of them for the rabbits.

The advantage that the sheep have over the rabbits in exploiting dry food during the summer is not likely to be pressed home either, because once again it is good husbandry to graze the "dry feed" conservatively, not only as insurance against an unusual extension of the summer drought, but also to protect the soil from erosion and loss of structure.

Reid (1953) estimated that between 1950 and 1953 the yield of wool per sheep increased from 3.90 to 4.17 kg, and the number of sheep increased by $4 \times 10^6$; he attributed 66% of the increased yield and 50% of the increase in numbers to myxomatosis. Given that there were about $8.2 \times 10^7$ sheep in southeastern Australia in 1950, it follows that the yield of wool for 1950 would have been $320 \times 10^6$ kg; and the increase in yield from 1950 to 1953 that might be attributed to myxomatosis was $30.4 \times 10^6$ kg, about 10%. Assuming that the production of wool is linearly related to the consumption of food, we conclude that in 1950 the rabbits got one tenth and sheep nine-tenths of the food in the pastures that they shared. Of course such a widely sweeping estimate ignores, among other things, the extreme patchiness of the distribution of both rabbits and sheep.

There is only anecdotal evidence about patchiness. Fenner and Ratcliffe (1965, 27) reported that, after the original panzootic of myxomatosis, "Ranges of hills that had been kept in a semi-denuded state by the pressure of rabbit grazing became grass-covered and green almost overnight". They went on to speak of a grazing property of 4,000 ha in eastern central New South Wales (the same climatic region as Albury):

The owner took the extreme step of eradicating rabbits from his land and keeping it rabbit-free. He was able to compare his production before and after the clearance, against that of his neighbours who took what might be regarded as normal measures to reduce rabbit numbers. He found that complete freedom from rabbits doubled the number of sheep that he could safely carry.

Elsewhere Fenner and Ratcliffe refer to a study area near Albury that was used in the early field experiments with myxomatosis. An area of 200 hectares supported four thousand rabbits. They said this was a severe infestation. Myers (1962), reporting on an extensive survey he made near Urana in 1951, said that about 30% of the properties carried a medium to severe infestation.

There must have been many medium to severe infestations where the rabbits got more than 10% of the pasture that was eaten. On the other hand, there must have been many places, like the grazing property that was mentioned by Fenner and Ratcliffe, where the sheep had access to virtually all of the pasture. In neither instance is it likely that the husbandry would be so relaxed that the sheep would cause a substantial shortage of food for the rabbits.

### 12.427    Long-Term Changes in Vegetation under Heavy Grazing by Exotics

During the two hundred years since Europeans colonized Australia, great changes have occurred in the vegetation (sec. 8.221). Pristine communities of sclerophyll forests, savanna woodland, scrub, and shrub steppes have been destroyed and replaced by farmland and rangeland and other rural economies. Grazing has always been prominent in the Australian rural economy, and this has suited the rabbit. Many woody sclerophyllous plants were replaced by grasses, herbs, and forbs that were more edible and nutritious. The newcomers included many exotics that had been brought in intentionally or accidentally, usefully or uselessly, as well as many opportunistic indigenous ones that took advantage of the disturbance of the old order. So in a good season the rabbits were offered a feast of good food. There would be a great many rabbits because even in the good farming areas control measures were still primitive. Neither the modern machinery for ripping nor the modern poison sodium fluoroacetate, was available until after the Second World War. Common lore and the large numbers of old warrens that can still be seen in areas of hard soil suggest that rabbits had been even more numerous in the past than they were in 1950. This is not surprising, because the new pastures in the early stages of colonization would have contained many plant species that were not adapted to live with rabbits. It is likely that natural selection reduced the more palatable species and that the modern stages in the succession are less nutritious than the early ones were.

## 12.43    Predators

Predators of the rabbit in Australia include the virus *Myxoma*, a number of Sporozoa, *Eimeria* spp., a number of nematodes, a number of arthropod ectoparasites, a lizard, a number of raptors and two mammals. The most important predator is *Myxoma*; the vertebrates are also important locally in certain circumstances.

### 12.431    *Myxoma*

Myxomatosis, the disease that is caused by the virus *Myxoma*, was first noticed in *Oryctolagus* in 1896 when a highly lethal disease broke out among captive rabbits in the Institute of Hygiene in Montevideo, Uruguay. *Myxoma* and *Fibroma* are two closely related viruses of the pox group which are widely distributed in three species of *Sylvilagus* indigenous to North and South America.

The strain of *Myxoma* that began the great panzootic of myxomatosis in Australia in 1950/51 was collected from a naturally infected European rabbit in Brazil in 1910. The virus had been maintained in culture for forty years by serial passage through rabbits first in New York and then in London (Fenner and Ratcliffe 1965, 216). In Australia it came to be known as "the standard laboratory strain". In tests using standardized procedures this strain can be expected to kill 99% of rabbits that are as free from genetic or acquired immunity as were the rabbits in Australia in 1950. Death usually occurs about the tenth day after infection with such a virulent strain. After the fifth day the virus can be recovered from almost any tissue, but even at the time of death the virus has not as a rule reached a high titer in such essential organs as the adrenals, kidneys, spleen, liver, lungs, or brain, so the proximate cause of death is obscure (Fenner and Ratcliffe 1965, 102).

The original virulence was not sustained in the field. As early as 1953 strains of *Myxoma* were recovered from the field that were killing a smaller proportion of the population and taking longer to kill the individuals (Mykytowicz 1953; Fenner 1953). By 1959 672 samples had been tested and five categories of virulence had been arbitrarily erected (table 12.19); by 1959 grade IIIA was modal; and by 1964 the mode had moved to grade IIIB (table 12.20). Seventeen years later (i.e., twenty-seven years after *Myxoma* became widely established in Australia) Myers and Calaby (1977, 162) wrote:

Myxomatosis is now an endemic infection in Australian wild rabbits, flaring up into outbreaks annually or less regularly according to local conditions. The initial case mortality rate of 99.9% has now fallen to 40% in those areas where annual epizootics occur. In more marginal populations case mortality rates are still high. The decrease in mortality is due to widespread replacement of virulent strains by less virulent field strains which are more readily transmitted, the development of genetic resistance and various factors related to physiological adaptation in the host.

**Table 12.19**  Classification of Strains of *Myxoma* Virus into Grades of Virulence

| Virulence Grade | Mean Survival Time (days) | Estimated Case Mortality Rate (%) |
|---|---|---|
| I | < 13 | > 99 |
| II | 13–16 | 95–99 |
| IIIA | 17–22 | 90–95 |
| IIIB | 23–28 | 70–90 |
| IV | 29–50 | 50–70 |
| V | — | < 50 |

*Source:* After Fenner and Ratcliffe (1965, 216).

**Table 12.20**  Frequency of Different Strains of *Myxoma* Virus (Defined by Their Virulence) in Samples Collected in the Field from Naturally Infected Rabbits

| Date Sample Collected | Percentage of Samples in Specified Virulence Grade[a] | | | | | |
|---|---|---|---|---|---|---|
| | I | II | IIIA | IIIB | IV | V |
| 1950–51 | 100 | 0 | 0 | 0 | 0 | 0 |
| 1958–59 | 0 | 25.0 | 29.0 | 27.0 | 14.0 | 5.0 |
| 1963–64 | 0 | 0.3 | 26.0 | 34.0 | 31.3 | 8.3 |

*Source:* Fenner and Ratcliffe (1965, 342).

[a] As defined in table 12.19.

The genetic changes in the virus are obvious and easily explained. A rabbit becomes infectious about six to eight days after infection and remains so until it dies or recovers. Table 12.19 shows that a rabbit infected with grade III virus remains infectious for eleven to twenty days compared with less than five days for a grade I virus.

Genetic changes in the rabbit were also observed from the very beginning. Myxomatosis reached the country around Urana in 1952. For the next eight years at least, there was an epizootic every summer. Each year from 1953 through 1958 young rabbits that had been born between epizootics and therefore had not been exposed to *Myxoma* were reared to maturity in quarantine. They were tested to ensure that they were free from antibody. Then they were inoculated with a standard dose of a standardized grade III virus. The results are set out in table 12.21. It seems that the rabbits living around Urana were rapidly evolving a good measure of genetic immunity to *Myxoma*.

Genetic changes in the rabbit as seen in the field are obscured by genetic changes in the virulence of the virus and by interactions between virulence, weather, and vectors. For example, during Parer's three-year (1968–71) study of a population at Urana, virtually every rabbit was individually marked and closely watched. In particular a complete and detailed record was kept of their exposure and response to *Myxoma* (Williams and Parer 1972). Because Parer had a full knowledge of their births and their deaths—not only when but also how—Williams and Parer (1972, 403) were able to infer:

It is clear that the situation at Urana from 1968 to 1972 approached that which Rendel (1971) predicted: namely that, due to a build-up in genetic resistance, field strains of myxomatosis would be killing only 3–4% of animals in areas with annual epizootics.

It seems to be true that *Myxoma* was the proximate cause of death for relatively few rabbits at Urana in 1968–71, but this is not the measure that we need to compare with the data in table 12.20. The death rate from *Myxoma* was low because many rabbits died from starvation or were eaten by vertebrate predators before they were killed by *Myxoma*. If they had been given good food and protected from other predators, as were the rabbits in table 12.20, more might have died from *Myxoma*. Even so, we would still have required to know about the virulence of the virus that was operating at Urana between 1968 and 1971. Although Williams and Parer's observations do not provide critical evidence of a continuation of the trend that was apparent in table 12.20, the data certainly are consistent with such a trend.

The results at Urana are not typical of events in other regions where epizootics do not recur every year or where they are more likely to occur during the breeding season than after it. Williams and Parer (1972, 403) commented:

The present low mortalities from myxomatosis at Urana are in sharp contrast to the very high mortalities in winter epizootics elsewhere in Australia (Dunsmore, Williams and Price, 1971; Williams et al. 1972, 1973). The high mortalities in winter probably reflect the slower build-up in genetic resistance, since winter outbreaks do not occur annually, combined with the low ambient temperatures prevailing, since Marshall (1959) showed that mortality from myxomatosis increases as temperature decreases.

**Table 12.21**   Severity of Myxomatosis in Samples of Wild Rabbits Collected at Urana 1953–58

| Origin of Sample | Number of Epizootics to Which Ancestors Had Been Exposed | Severity of Diseases (% in each category) | | |
| --- | --- | --- | --- | --- |
| | | Severe, Including Fatal | Moderate | Mild |
| Wild rabbits before myxomatosis | 0 | 93 | 5 | 2 |
| Lake Urana, 1953 | 2 | 95 | 5 | 0 |
| Lake Urana, 1954 | 3 | 93 | 5 | 2 |
| Lake Urana, 1955 | 4 | 61 | 26 | 13 |
| Lake Urana, 1956 | 5 | 75 | 14 | 11 |
| Lake Urana, 1958 | 7 | 54 | 16 | 30 |

*Source:* After Fenner and Ratcliffe (1965, 246).

Obviously the evolution of virulence in the virus and resistance in the rabbit may still have further to go. A new vector, the flea *Spilopsyllus cuniculi*, was released in Australia in 1969. If this flea becomes widespread and abundant, it may give a new direction to the evolution of virulence in the virus or of resistance in the rabbit.

Being carried by a "vector" seems to be the only way *Myxoma* can disperse. All the known vectors are bloodsucking arthropods. The infectious particles, clusters of molecules called "virions", are so large, about $286 \times 227 \times 80 \, \mu$, that they can be seen with a light microscope. They are too large to pass up the lumen of a mosquito's proboscis; they are carried from a sick rabbit to a healthy one because they stick to the outside of the proboscis and some of them rub off when the mosquito pierces the skin of its next host. Hit and miss as this method might seem, in one experiment it was found that a mosquito infected the first nineteen healthy rabbits it attacked after feeding once on one infectious rabbit. The most important ecological consequence of the external transmission of the virus is that any blood-sucking arthropod will do for a vector—artificially an ordinary needle makes quite a good vector. So when looking for vectors in nature any species that feeds readily on rabbits and disperses actively between meals is a likely candidate.

The virus *Myxoma* was established in Australia on the second attempt. The first attempt failed for lack of vectors; the second came to the brink of failure for the same reason before it was revived by a lucky accident with the weather. The full story is a striking demonstration of the essential importance of mosquito vectors and of their dependence on rain.

The first experiments with *Myxoma* virus for the "biological control" of rabbits in Australia were reported by Bull and Mules (1944). In a number of experiments, done over a period of half a year in rangeland near Witchitie, they inoculated several hundred rabbits. They observed a number of diseased rabbits and recorded that when the disease was carried into a warren by a diseased rabbit the whole population of the warren often died. Nevertheless, in every experiment the disease died out after a month or two. Stickfast fleas *Echidnophaga myrmicobii* were abundant, but mosquitoes were scarce.

It had been known since Aragão's (1920) work that a variety of arthropod vectors might distribute *Myxoma* virus; and Bull and Mules speculated that an abundance of winged vectors might be necessary for the virus to spread well. So when the next experiments were undertaken in 1950 a study area was chosen near the Murray River, where there was a reasonable expectation of numerous mosquitoes. This expectation was not realized, because an unusual drought had caused a dearth of mosquitoes; but there was a dense rabbit population: four thousand rabbits on 80 ha. So the exercise began.

Gin traps that had been modified to puncture the skin and inject an inoculum were put out in May 1950. The inoculation was successful because within a few weeks seventy-seven diseased rabbits were seen, well distributed over the area, but the disease did not spread; by July no diseased rabbit could be found, though healthy ones were numerous. So the research team recorded a negative result and moved about 240 km upstream to set up four new study areas near Albury. But there had been no worthwhile rain; mosquitoes were still scarce, and myxomatosis failed to become established in any of the new areas even though in one of them, Rutherglen, very close attention was paid to every detail. Fenner and Ratcliffe (1965, 274) described the course of events at Rutherglen:

Counts of diseased rabbits were made daily, and carcasses were counted and removed. Population counts were made at intervals. These were carried out in a carefully standardised way, and although they did not provide a measure of the total number of rabbits on the site, they can be accepted as a reliable index of population density and its changes.

Twenty-seven rabbits were inoculated with virus on 7 September when the population count was about 700. The disease failed to become established, and a further inoculation of 66 rabbits was made on 27 October, when the population count had risen to 900 as a result of breeding. Between then and 16 December, when observations ceased, four generations of natural infection could be distinguished each one smaller than the preceding one . . . The ratio of infection from one generation to the next was only 0.6

Fenner and Ratcliffe (1965, 276) summarized the results of the season's work:

By the beginning of December, 1950, myxomatosis had apparently died out everywhere except at Rutherglen, where the infection was fading towards extinction. The Murray Valley trials had been thorough: they had been carried out at a number of sites where the rabbit population was dense, and sometimes very dense indeed. Aedine mosquitoes had been reasonably abundant at Coreen; and at Balldale, where they were common, one of the three species collected-*Aedes alboannulatus*—had earlier been shown by Bull and Mules (1944) to be capable of transmitting myxomatosis under laboratory conditions. [But, during the intensive studies of vectors that was undertaken subsequently, *Myxoma* was recorded from this species on only one occasion.]. At that juncture, therefore, the 1950 trials provided what seemed a clear confirmation of the conclusion reached by Bull and Mules on the results of the South Australian trials carried out seven years earlier, i.e. that *Myxoma* virus held little promise of being of practical value in rabbit control.

But even as that conclusion was being reached there was beginning a massive and spectacular panzootic, of which Fenner and Ratcliffe (1965, 276) said: "for scale and speed of spread [it] must be almost without parallel in the history of infections." The extent of the panzootic in 1950/51 is shown in figure 12.07. Fenner and Ratcliffe (1965, 277) commented on this map:

Two things stand out. The first is the sheer size of the event recorded. The area over which the virus was dispersed—albeit patchily in some regions-measured nearly 1000 miles [1,600 km] from south to north and 1100 miles [1,770 km] from east to west. [Apart from a few outlying areas where inoculation had been artifical] the map is a record of the natural spread of the infection. The second thing that stands out is the dominating importance of the Murray-Darling river system in the distribution of myxomatosis activity. As we realized when we had learned something of the insect vectors, this was due to the fact that epizootic transmission had depended on summer mosquitoes that breed in persistent water, notably *Culex annulirostris*.

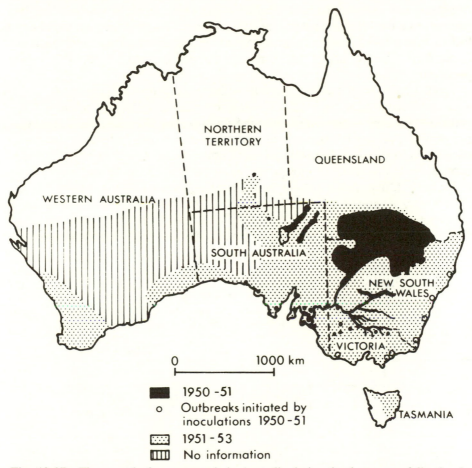

**Fig. 12.07** The spread of myxomatosis in Australia during the three years following its escape in 1950 from a test site near Albury in southeastern New South Wales. The isolated occurrences in 1950–51 (*solid black circles*) ranged from single cases to outbreaks covering a few km². The two in central Australia, north and west of Lake Eyre, are included in the first season's dispersal, although they were not reported until August and September 1951.

The unusual rainfall distribution during 1950 reflected in the great difference in the extent of disease activity in the southern and northern portions of the river system. In Victoria and New South Wales, 1950 was a year of sub-average rainfall, and by mid-summer the countryside in general was very dry. In marked contrast, the northern parts of New South Wales and most of Queensland received record-breaking precipitations, resulting in flooding on an almost unprecedented scale in the northern part of the epizootic area. The last flood rains had occurred in November 1950; but their effects in the form of abundant surface water lingered for at least a couple of months.

The spectacular panzootic of 1950/51 was caused by the extraordinary abundance of *Culex annulirostris* and the favorable turbulences and winds that dispersed them so widely. As Fenner and Ratcliffe pointed out in the passage quoted above, this was an accident of the weather at that time. With more normal weather *Anopheles annulipes* might have been abundant earlier in the season and the first attempts to establish myxomatosis might not have failed—but then we might have had to wait a long time for an unequivocal demonstration of the overwhelming importance of vectors and of the relative unimportance of the density of the rabbit population.

Table 12.22 shows that the rainfall in the vicinity of the study areas where *Myxoma* failed to establish itself was below average throughout the whole of 1949 and also during the winter (April through September) of 1950. But the drought was not severe. Two successive years of equivalent dryness might be expected about once in six years. One suspects that some other component of environment also contributed to the dearth of mosquitoes in all the study areas during 1950.

Table 12.22 also shows that the record-breaking wetness of the northern districts in 1950 was in strong contrast to the drought in the south. The extreme wetness of extensive areas in northern New South Wales and southern Queensland seems to explain beyond reasonable doubt the plenitude of the vector *Culex annulirostris* that

**Table 12.22**    Rainfall for Four Districts Representing the Summer Rainfall Areas of Northern New South Wales and Southern Queensland and Four Districts Representing Areas of Winter Rainfall in Southern New South Wales and Victoria That Embrace Albury and the Other Study Areas Where *Myxoma* Was Liberated on a Number of Occasions between May and December 1950.

| Northern Area | | | | Southern Area | | | | |
|---|---|---|---|---|---|---|---|---|
| | Rain (mm) | | | | | Rain (mm) | | |
| District Number | Mean (67 years) | 1950 | Maximum Ever Recorded | District Number | Mean Full Year | 1949 Full Year | Mean Winter | 1950 Winter |
| 43 | 549 | 1,075 | 1,075 | 72 | 807 | 784 | 454 | 405 |
| 51 | 477 | 1,130 | 1,130 | 74 | 477 | 440 | 253 | 221 |
| 54 | 680 | 1,125 | 1,125 | 79 | 501 | 447 | 305 | 295 |
| 55 | 663 | 1,075 | 1,075 | 82 | 778 | 755 | 451 | 325 |

*Note:* Albury is close to the boundary between districts 72 and 82. Figures give annual total, January through December, except for 1950 in the southern area, which is total winter rain, April through September.

spread *Myxoma* so widely and so quickly across the countryside during 1950/51 (fig. 12.07). Ratcliffe and his colleagues who were trying to get *Myxoma* established were lucky: their project happened to coincide with weather that might be expected to occur about once in one hundred years.

Although a variety of species of mosquitoes, gnats, fleas, and other bloodsucking arthropods may transmit myxomatosis, the two vectors of outstanding importance in eastern Australia are *Culex annulirostris* and *Anopheles annulipes*. They owe their importance to (a) their ability to breed in the sorts of waters that are likely to be abundant after rain; (b) their behavior in feeding at dusk and during the night; (c) their behavior in seeking shelter in cavernous places, which leads them into rabbit burrows; (d) their preference for feeding as adults on rabbits, coupled with their ability to support themselves, in the absence of rabbits, on a variety of other hosts that are abundant—notably cattle, sheep, and men.

According to Fenner and Ratcliffe (1965, 182):

*Culex annulirostris* breeds in permanent or persistent water, preferably where there is aquatic or emergent vegetation. Its main stronghold within the rabbits' range is provided by the river-frontages, swamps and lagoons of the Murray-Darling system—from the plains of Queensland in the north and the foothills of the eastern ranges, down to the Murray mouth in South Australia. In the southern part of its range, in Victoria, in addition to the breeding grounds associated with the tributary streams of the Murray, the stockwater dams of the northwest (Mallee) and the extensive system of channels which feed them produce large numbers by February.

To the north, towards the Queensland border *C. annulirostris* may be found at any time of the year where conditions are favourable. Elsewhere it is typically active in mid and late summer when, in southern Australia, the open country is normally dry. Under these conditions it rarely ranges beyond the vegetation fringing the breeding grounds, being sensitive to low atmospheric humidity; and those myxomatosis outbreaks that are based on its activity reflect this very obviously. However when the exceptional season turns up with water lying and persisting over the countryside in summer, it is quick to take advantage of the situation. It then not only breeds in almost any persistent ground water, but becomes widely ranging in its flight behaviour. . .

*Anopheles annulipes* will be found breeding in many, if not most, of the situations favoured by *C. annulirostris* and, like the latter, is produced in abundance by the billabongs and swamps associated with the Murray and its tributaries. However it exploits a number of situations not normally colonized by *Culex* such as extensive surface waters present in the Riverina plains after heavy winter rain, the slowly moving water on the margins of rivers and creeks (particularly when and where a substantial growth of the filamentous alga, *Spirogyra* has developed), and pools formed in the course of the drying up of small hill-fed streams, which are such an important feature of the extensive belt of undulating pastoral country in eastern Australia. These little streams may run for a few days in the warmer parts of the year; but it is by no means unusual for rainfall distribution to be such that pools persist at various points along their courses well into the spring and early summer, or that such *A. annulipes* breeding places are created by autumn falls.

Although the adults of *A. annulipes* and *C. annulirostris* prefer to feed on rabbits, it seems quite unlikely that they will go short of food when rabbits are scarce. On the other hand, there is a persistent shortage of suitable waters for the larval stages, which is ameliorated by rain falling seasonally each year; but because the incidence of rain is variable the shortage is ameliorated more in some years than in others. So the distribution and abundance of the vectors depends (apart from the initial pattern laid down by accidents of topography) very largely on the seasonal incidence of rain and only slightly on the abundance of the rabbits (providing food for adults).

Fenner and Ratcliffe thought it most likely that *Myxoma* persisted in an active condition between outbreaks. More recent evidence suggests that it may also persist in a latent condition in "carriers" that show no sign of the disease. The suggestion is that the disease is reactivated when the carrier is exposed to "stress". This explanation is gaining support from a number of experienced workers in the field. Perhaps the critical experiment has not yet been done, but the circumstantial evidence is accumulating (Williams and Parer 1972; Williams et al. 1972, 1973; see also sec. 11.324).

However myxomatosis may persist between outbreaks, it is at a low density; but however small the nucleus, it seems to be able to multiply and become widespread whenever the vectors are abundant. When vectors are few the disease cannot increase no matter how numerous the rabbits may be. Extrapolating from the theory of the epidemiology of malaria (MacDonald 1957), there may be a critical low density of hosts below which the disease is slow to spread even with plenty of vectors; but this question has not been investigated empirically for myxomatosis. With this one proviso, it seems that the multiplication and spread of myxomatosis depends largely on the abundance of vectors and only slightly on the abundance of rabbits.

In all but the driest parts of the Mediterranean region an outbreak of myxomatosis is likely to occur during the first half of the summer; the vectors are always mosquitoes. In one such place, Urana, which is probably fairly typical, myxomatosis seems to be the proximate cause of fewer than 10% of the deaths (see above). This is not to say that if the other restraints were lifted temporarily myxomatosis might not account for a larger proportion of the deaths. But it seems to indicate that there must be some explanation other than the current activity of *Myxoma* for the currently low numbers of rabbits relative to the large numbers that prevailed during the years that led up to 1950 (sec. 12.5).

In the rangeland and the drier parts of the Mediterranean regions rainfall is not reliable enough to ensure a crop of mosquito vectors every summer. Moreover, there is a greater chance of a breeding season so dry or so short that no rabbit will be recruited into the oversummering population of adult, and largely immune, rabbits; thus the prospect of an outbreak is postponed. In such places, where outbreaks are not likely to occur every year, myxomatosis seems still to be killing a high proportion of the population.

If the newly introduced vector *Spilopsyllus* becomes widespread and abundant in any region, a different condition may develop. If, as seems likely, outbreaks that depend on the flea occur chiefly during the breeding season (winter in the Mediterranean region), the exposure to infection in each generation may become more complete, with consequences that have not yet been fully examined.

12.432   Vertebrates

The following predators have been mentioned in the literature, but none has been deeply studied.

Fox, *Vulpes vulpes* (Myers and Parker 1965)
Feral cat, *Felis cattus* (Myers and Schneider 1964)
Australian goshawk, *Accipiter fasciatus* (Myers and Schneider 1964)
Little eagle, *Hieraaetus morphnoides* (Calaby 1951)
Wedgetail eagle, *Aquila audax* (Myers and Poole 1963b)
Peregrine falcon, *Falco peregrinus* (Myers and Poole 1963b)
Barn owl, *Tyto alba* (Myers and Schneider 1964)
Barking owl, *Ninox connivens* (Mykytowycz 1959)
Raven, *Corvus coconoides* (Myers and Poole 1963b)
Goanna, *Varanus varius* (Myers and Parker 1965)

Only the wedgetail eagle and the adult cat can take an adult rabbit; the others, being restricted to kittens or young juveniles, must find other food during the summer when the rabbits are not breeding. Each one has its own style of hunting, and all of them are limited by their own disabilities. The hawks do better if they have a suitable "perch" near a warren. The fox, because it is clumsy, rarely catches a rabbit on the surface, even a small one, unless it is decrepit from disease. The fox's style is to dig for nestlings. It can, with great skill, locate the nest from the surface; then it digs vertically downward. Myers and Parker (1965) recorded the number of nests that had been dug out by foxes in different sorts of soil (table 12.09). The fox's method of hunting is successful only in the softer soils. The feral cat favors the ambush; it commonly crouches on the roof of the burrow and pounces on a young rabbit as it emerges.

Although the mature cat has been observed to prey on adult rabbits, it takes mostly juveniles; certainly the immature cat cannot take any but immature rabbits. The feral cat was the most important vertebrate predator on Parer's (1977) study area at Urana. Nevertheless, most of the cats left the area during the summer, probably in search of food; the mouse *Mus musculus* was the most likely alternative, but in most summers it would also be scarce. Myers and Parker (1975a) in their survey of rangeland habitats, found cats, and evidence of their predation, only in sandy habitats that housed an alternative source of food for them. So perhaps cats, like all other vertebrate predators of the rabbit except the wedgetail eagle, must rely on finding alternative sources of food to tide them over the summer when the rabbits are not breeding. But the eagle does not specialize in rabbits either.

Of course a facultative predator that can find plenty of alternative food may become the most effective of all (sec. 12.22). But it is clear that the vertebrate predators of rabbits in 1950 were not likely to find alternative supplies of food for the summer that would match the abundance of rabbits during winter. For example, on a study area of about 165 ha that carried about 4,500 adult rabbits Calaby (1951), after watching for five hundred hours on seventy-one consecutive days, concluded that the area supported 13 cats, 8 little eagles, 2 goshawks, 2 peregrine falcons, 2 barking owls, and 2 wedgetail eagles—about 1 predator for every 155 adult rabbits.

The females would have been producing litters at about monthly intervals. Parer (1977, 202) compared this ratio with his own observations at Urana seventeen years later:

In contrast, at Urana, during September and October of 1968 and 1969, there were about one of the predators listed above to 15 adult rabbits. If each adult female produced four kittens a month and if each predator killed one kitten each day, at the end of the month 90% of the kittens would remain uneaten in the situation described by Calaby, but none would remain uneaten in the Urana situation.

Parer thought the reduction in numbers that was brought about in the first place by the original panzootic of myxomatosis and later enhanced and sustained by farmers' control measures might have given the vertebrate predators a measure of importance which they had lacked in 1950.

## 12.433   Invertebrates

At least six species of Sporozoa, all belonging to the genus *Eimeria*, and three species of Nematoda are commonly found in *Oryctolagus* in Australia.

On a study area near Mogo in the south coastal region Dunsmore (1971) found that coccidia, *Eimeria* spp., especially *E. stiedae*, were an important component in the rabbit's environment. The chief breeding season for the rabbit in the south coastal region occurs in the spring, with a small outburst in the autumn. Having reviewed the evidence for food, myxomatosis, and vertebrate predators and found no suggestion that any one of them was of major importance, Dunsmore (p. 370) commented:

Because this region is unusually wet as a habitat for *Oryctolagus*, coccidiosis, especially the hepatic type caused by *E. stiedae*, is very prevalent and in normal years probably causes severe losses in kittens in the < 3-month-old category. The intensity of *E. stiedae* infection is likely to operate in a generally density-dependent manner, but in years with subnormal rainfall considerably fewer animals than usual may be affected to a degree that would cause death.

Three nematodes are commonly found in rabbits in southeastern Australia, *Graphidium strigosum*, *Trichostrongylus retortaeformis*, and *Passalurus ambiguus*. Not all of them have been found in every region, and none is of major importance anywhere with the possible exception of the subalpine region (Dunsmore and Dudzinski 1968). On a study area at Snowy Plains, in the subalpine region, all three species of nematodes were common; they usually increased greatly in female rabbits toward the end of the breeding season. Dunsmore (1974) thought that the heavy load of parasites in conjunction with a shortage of sodium in the diet might explain why the rabbit stops breeding while there is still, to the human eye at least, plenty of green food about.

## 12.44   Malentities

Of the possible malentities that come to mind, little is known about weather, in the form of excessive heat or drowning; perhaps it is unimportant except at the northern limit of the distribution, where high temperature may cause males to be sterile. Overcrowding by other animals of the same kind has been studied, because this idea has been so prominent in ecological theory about small mammals (Andrewartha 1970, 75–84), but it was shown not to be important for rabbits (Myers and Poole 1959; Myers 1964). The malentity of overriding importance in the ecology of the rabbit is the farmer who practices artificial control measures—ripping and poisoning.

## 12.441   Farmers

A hunter who kills rabbits for their flesh or their fur seems to fit into the envirogram as a predator; a farmer who catches rabbits and inoculates them with *Myxoma* is acting as an agent of dispersal for the virus; but a farmer who practices chemical or physical control measures seems to be an aggressive malentity (sec. 6.2).

A wide range of control measures have been used against rabbits. Farmers have coined their own names for them. To poison is to spread lethal bait along furrows or runways; to fumigate is to fill the warren with lethal gas—the exhaust from a tractor is often used; to dog is to use trained dogs to drive the rabbits into a warren before fumigation or into a corner of a netting fence so that they can be dispatched in some other way; to dig is to break up and fill in the warrens with a spade; to rip is to attain the same ends but more thoroughly by using deep tynes attached to a tractor. To shoot and to trap have also been used on occasion. After 1952 digging and ripping increased relative to the other methods. For example, at Urana before 1952 farmers who used digging or ripping as their chief method accounted for 20% of all who were regularly using control measures; after 1952 it was about 60%. In 1952, Myers (1962) estimated, for the three districts he surveyed, that at Corowa 92% of the farmers were regularly using control measures of one sort or another, at Albury 97%, and at Urana 98%. In 1959 the percentages were 27, 32, and 64. Referring back to figure 12.03 (sec. 12.414), we can see that before 1952 about 30% of the area was heavily infested with rabbits and a further 6–8% carried "medium" infestations. Even though virtually all (about 96%) of the farmers were regularly practicing control measures, the populations remained very dense. The effort that was put into control was not sufficient to make much impression on such a dense population. After 1952 even a reduced effort put into control measures served to prevent the population from increasing very much during the next seven years. There is an interesting comparison to be made with the vertebrate predators (sec. 12.432). In terms of figure 1.02, both the vertebrate predator and the malentity (a man with a ripper) have a negative influence on the rabbit's chance to survive and reproduce (fig. 1.02B), but with the predator limited by a shortage of alternative food during summer and the malentity limited perhaps by a shortage of money in the bank, neither could grow strongly enough to exert a noticeable influence on such a dense population of rabbits. The didactic solutions to both problems are essentially the same: the limit to the activity of the predator or the malentity might have been raised (by causes not related to the density of the rabbit population) until they matched or exceeded the rabbit's capacity to increase. In fact, the converse happened; the density of the rabbit population was reduced (by causes that were independent of the predation and the artificial "control measures") until the actual capacity for increase was matched or exceeded by the existing capacity of the predators or the malentity.

It so happened that after a while, in some areas, the effort exerted by the malentity was relaxed a little too far, and rabbits began to increase again (Myers, pers. comm.). Since an aggressive malentity usually gains some advantages for itself from its aggression and may be expected to relax its aggression when the need is not great, the behavior of man in this instance confirms his classification into the category of aggressive malentity.

But the farmer with his ripping machine is also a modifier of the first order, modifying "warren" (as a heritage) because the rabbits find it hard to replace warrens that have been destroyed (sec. 12.414).

## 12.5   Regional Variation in Certain Demographic Statistics

This section tells how Myers and his colleagues gathered such statistics as the age structure of the population and the age-specific fecundity and death rate, as well as a number of other physiological indicators of the condition of the individuals in the population, and used the differences between the regions to draw logical inferences about the environmental "control" of such statistics. Such inferences have the status of hypotheses (sec. 9.01). Figures 12.09 and 12.10 represent the sort of model that emerged from this five-year program. These models might have been more fruitful if they had been available at an earlier stage in the more than twenty-five years' project on the ecology of the rabbit.

For about five years between 1962 and 1967 Myers and his colleagues shot a sample of about fifty rabbits at about six-week intervals in each of the following regions: subtropical, rangeland, subalpine, and Mediterranean; they also had less intensive data from the south coastal region. All together there were records from about seven thousand rabbits. The rabbits were shot after dark with the aid of a spotlight. Immature rabbits (i.e., those less than 3 months old) were either not shot or not counted with the sample. Age was estimated from the weight of the dried eye lens (Myers and Gilbert 1968). Females in breeding condition were subdivided into those that were pregnant and lactating (the normal condition) and those that were lactating but not pregnant. For a rabbit to be lactating but not pregnant during a breeding season implies either that she did not conceive at the postpartum estrus or that she had lost her litter, probably through resorbtion. The litter was counted. For males the condition of the testes was recorded. Because the sampling was restricted to mature rabbits, the survival rate in immatures (birth to 3 months) was not estimated. This was an important omission, because the death rate among immatures may approach or even reach 100%, especially for those born during an unseasonable spell of breeding or toward the end of a normal breeding season. A preliminary account of the results of this massive program was published by Myers (1971). For the brief comment that follows we have drawn on this preliminary account and have also consulted some unpublished data (Myers, pers. comm.).

From the empirical data, supplemented by plausible assumptions about death rates for juveniles, age-specific natality was estimated and life tables were constructed, and from these statistics the "capacity for increase" was calculated (i.e., $r_c$ of Laughlin 1965). Alternatively it was possible to estimate, for any region, the combinations of natality and juvenile mortality that correspond to $r_c = 0$, that is, to a "steady state" in the density of the population. Such calculations lead to interesting comparisons between regions.

The age distributions of rabbits in five distinct regions are illustrated in figure 12.08. The age is measured by the weight of dried eye lenses. A weight of 204 mg indicates an age of 1 year, and a weight of 250 mg an age of just over 2 years. By

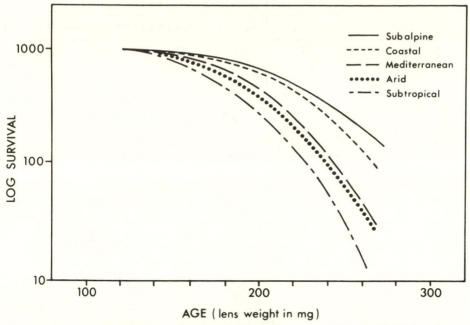

**Fig. 12.08**   Age distributions of rabbits plotted as log cumulative weight of dried eye lenses in different regions of Australia. (After Myers 1971)

the $\chi^2$ test the age distribution in each region is significantly different from that in every other region. The low expectation of life for a mature rabbit in the coastal region is to be attributed to the prevalence of parasites, especially coccidia, whereas in the subalpine region the risk is not only from an abundance of parasites (of a different sort) but also from a shortage of salt. In the Mediterranean region the population, on the average, was younger than in the subtropical region. Productivity was higher in the Mediterranean region because of a reliable breeding season in winter or spring, but the adults survived better in the subtropical region, perhaps because the supply of water and food was more reliable the year round.

The "productivity" (i.e., the potential fecundity based on evidence of conceptions) of the populations in the distinct regions is compared in table 12.23. Productivity is "lost" when the number of young that are born are fewer than the potential, as judged by the original conceptions or by likely conceptions. Losses would have been due chiefly to failure to conceive at the postpartum estrus or resorbtion of fetuses. Given the data from table 12.23 and figure 12.08, the actual rate of increase $r_c$ (Laughlin 1965) can be calculated, provided juvenile mortality (i.e., deaths up to the age of 3 months) and the rate at which litters are produced are also known. Alternatively, if the value of $r_c$ is arbitrarily fixed at zero (i.e., the density of the population is assumed to be in a steady state), it is possible to calculate the series of coordinate values of productivity and juvenile mortality that correspond to $r_c = 0$.

In figure 12.09, which is taken from Myers (1971), these coordinates are shown for values of juvenile mortality between zero and 95% and for productivity between 0.5 and 2.0 litters per three months. The empirical estimate of productivity during

**Table 12.23**  Mean Productivity in Distinct Regions

| | Region | | | | |
|---|---|---|---|---|---|
| | Subtropical | Subalpine | Rangeland | Mediterranean | Coastal |
| Number of females | 584 | 63 | 937 | 412 | 301 |
| Pregnant (%) | 32.1 | 24.1 | 24.9 | 43.4 | 26.6 |
| Lactating (%) | 46.4 | 33.5 | 24.9 | 41.0 | 36.5 |
| Size of litter | 4.80 | 4.53 | 4.49 | 5.65 | 5.23 |
| Mean productivity (per female per year) | 15.4 | 10.9 | 11.2 | 24.5 | 13.9 |
| Loss in productivity (%) | 46.0 | 44.7 | 34.5 | 21.2 | 41.1 |

*Source:* After Myers (1971).

*Note:* See text for definition of "productivity" and "loss in productivity."

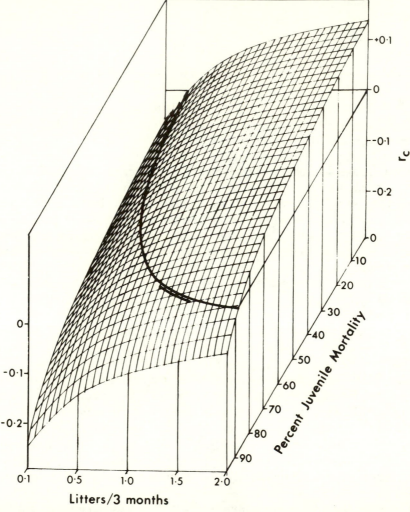

**Fig. 12.09**  The solid model shows how fecundity, juvenile mortality (birth to three months), and capacity for increase ($r_c$) were confounded in a population of rabbits living in the mediterranean region (near Urana). The curved line joins those points on the plane where $r_c = 0$. The other thick line represents the measured rate of reproduction (litters per three months). (After Myers 1971)

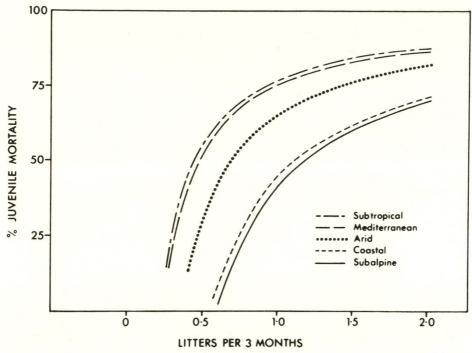

**Fig. 12.10**   Each curve is drawn through the coordinates of percentage juvenile (birth to three months) mortality and fecundity (litters per three months) that correspond to $r_c = 0$. That is, the curves define the interaction between these two statistics that is required to keep the population in a steady state, given the environments the rabbits experience in the regions shown.

the period 1962–1967 ranged from 1.0 to 1.5 litters per three months. For this range of productivity the corresponding range in juvenile mortality was 75–85%—figures that would not surprise a naturalist who was familiar with the rabbit in the Meditterranean region.

In figure 12.10 the regions are compared by the values that the two coordinates take as $r_c=0$. The confluence of the curves for the subtropical and mediterranean regions arises from the relatively high productivity in these regions, but the relatively high mortality of adults in the Mediterranean region compensates for the higher productivity in this region. The confluence of the curves for the subalpine and coastal regions reflects the low productivity in these regions and the relatively higher adult mortality in the coastal region, perhaps associated with the prevalence of coccidia.

These models are fruitful because, by rearranging the facts, they help to identify useful predictions about how environment influences the rabbit's chance to survive and reproduce. But, as the discussion above shows, experiments and observations based on environmental theory are needed to verify the predictions. Myers emphasized this point when he presented the models. He wrote: "Ecology is not a study of reproduction and survival but of those processes that affect them" (Myers 1971, 491).

## 12.6   Conclusion

The rabbits that live in southeastern Australia are contained by effective ecological barriers on all sides—the sea to the south and west, the tropics to the north, and inhospitable rangeland and desert to the west. There is no effective barrier inside these boundaries; only distance impedes the interaction of the local populations. So these rabbits constitute a natural population as we defined that idea in section 1.6. The studies of Myers and his colleagues encompassed the whole population over the whole distribution. This extensive but thorough study gives us an insight, which might be difficult to realize in any other way, of how vastly the "risk" (to pick up den Boer's idea again) might be spread in a large natural population (sec. 9.3).

The risk is spread widely because the population is distributed over five climatic regions: climate, soil, and vegetation vary widely between regions and less widely within them. Under the impact of man, sheep, rabbit and other exotic species, the soil, the vegetation, and other features of the environment have changed more rapidly and become more diverse than before. The evolution of important traits such as resistance to myxomatosis has gone further in some regions or subregions than in others. Because environment varies from place to place and from time to time, there are consonant variations in the age-structure of local populations (Myers 1971). Because the distribution is large and because tropical, continental, and polar air masses are interacting to cause its weather, the weather can be highly contradictory in distinct parts of the distribution at the same time; consider how, in 1950, unprecedented floods in the north coincided with much drier weather in the south (sec. 12.431). These points have been made elsewhere in this chapter, but to emphasize some of the spatial variability we recall how different components of environment may become chiefly important in distinct parts of the distribution. For the most part their influence causes an extrinsic shortage of resources as in figure 9.05.

In the Mediterranean region rain, in the pathway through food, is always important, but relatively more so in the drier parts of the region. In the more humid parts of the region food might still be the most important component if its importance were not overshadowed by farmers, vertebrate predators, and *Myxoma*. (Rain is also important in the pathway through *Myxoma*, but the two activities are distinct). In the rangeland region rain is most important in the pathway through food and less so in the pathway through *Myxoma*. In this region there is probably no other component to compare with rain, but inherited warrens make some sorts of places much more favorable than others for a recovery after a severe drought; and in sandy habitats vertebrate predators, chiefly foxes and cats, are likely to slow down the rate of recovery after severe drought. Also, if *Spilopsyllus* becomes well established the pathway through *Myxoma* may become more important. In the subtropical region the rabbit is approaching the northern limit of its distribution. The limit is probably set not by temperature acting alone but by a combination of temperature and a rainy season that comes in the summer. Green growing vegetation is present only during the summer, when the temperature is too high for healthy breeding. In the south coastal region, with a growing season that lasts seven months (table 12.13), there is also no shortage of food. There is no evidence that *Myxoma* is important. There is some evidence that coccidia, chiefly *Eimeria stiedae*, responding positively to wet weather and probably to the density of the rabbit population as well, is most important. In the subalpine region the breeding season seems to come to an end while, to the human eye at least, there is still plenty of green growing herbage to be had. It has been suggested that a shortage of sodium in the diet and an accumulation of nematodes in the gut of the female are sufficient to bring the breeding season to a premature end.

The density of the rabbit population has had to be taken into account to analyze certain reactions or interactions that have been observed. These analyses have been made within the framework of the theory of environment. For example, after the original panzootic of myxomatosis had reduced the rabbits to very low numbers, the disadvantages of *underpopulation* allowed vertebrate predators and a malentity (farmers) to prevent another outbreak. An *intrinsic* shortage of succulent leaves (of *Scleretaena lanicuspis*) caused the crash in numbers that occured at Witchitie during the summer of 1969/70, but the shortage was short-lived because other components in the environment of the rabbit kept the numbers of over-summering rabbits low during 1970/71 and 1972/72; so the *Scleretaena* recovered (secs. 12.422, 12.424).

But in general the most important components in the environment of the rabbit are all such that their activity is caused by events that are not responsive to the density of the rabbit population.

The rabbit has been in Australia for a mere 120 years, but who would dare to predict that it will soon be extinct. Why are we so sure? Myers and his half-dozen or so ecological colleagues, working as a team for rather more than twenty years, have produced an extensive and thorough account of the ecology of the rabbit. It reveals an environment, or rather a set of environments, that is immensely complex in the sense that there are vast opportunities for environmental reactions to be operating on different scales or in different directions in different places at the same time or at different times in the same place. Such an empirical demonstration of diversity makes it easier to realize how the elements of chance, when there are enough of them pushing in different directions, might reduce the likelihood of extinction so far that stability seems certain. And if this explanation holds for stability, might it not also explain the distribution and abundance of a species at a particular time? In other words, this deep study of the ecology of the rabbit appears to have strongly confirmed den Boer's theory of stability through spreading the risk. This theory seems likely to gain support as more ecologies come to be studied in similar depth (secs. 8.5, 9.3).

# 13

# Three Contrasting Case-Histories

## 13.0  Introduction

This chapter comprises three ecological case-histories, told briefly because the chief aim is to bring out the contrasts between the three ecologies. The black-backed magpie, *Gymnorhina tibicen*, is extremely territorial and sedentary. The grey teal, *Anas gibberifrons*, is extremely opportunistic and dispersive. The contrast between the two birds in sections 13.1 and 13.2 and the fish in section 13.3 is not only in their ecologies but also in the ecological theories that have shaped the investigations. Frith, who studied the grey teal, and Carrick, who studied the magpie, were guided by the theory of environment: they were alert for any component of environment that might be influential, and they found broadly satisfying explanations for the distribution and abundance of the species that they studied. In contrast, a substantial search of the literature on fish and fisheries revealed a wealth of material, especially theoretical models, based on predator-prey relations, especially on predation by man, but an abysmal neglect of the rest of the environment. With the popular hope that the oceans might help to feed the burgeoning populations of *Homo*, it was disappointing to find that the ecological background for fisheries' management is not more fundamental. The feature that distinguishes the ecology of *Homo* (chap. 15) is that man has, throughout his history and especially since the agricultural revolution, manipulated his own environment. As a consequence, he has maximized r (rate of increase), and his ecology in this respect has been like that of any other successful colonizing species. Now the task ahead, if he is to avoid the disastrous "crash" that characterizes the ecologies of other colonizing species, is to so manipulate his environment that his numbers remain within the carrying capacity of his habitat the earth. The magpie is a good example of how one species has achieved such a goal.

## 13.1  The Black-backed Magpie (*Gymnorhina tibicen*)

The following account of the ecology of *G. tibicen* is taken from Carrick (1972). Australian magpies are large (about 36–40 cm long) black-and-white birds (plate 5). They thrive in open woodland where groves of tall trees are interspersed through meadow, arable land, or scrubland where they can forage for insects and for other small animals and carrion (fig. 13.02). A grove of say twenty trees nicely situated alongside a feeding area of say 8 hectares might make an optimal territory which

NESTS
● ● ● eggs laid ● no eggs

**Plate 5** *Above*, the Australian black-backed magpie *Gymnorhina tibicen* singing in its territory. (Photo by Ederic Slater) *Below*, part of the intensive study area in the Australian Capital Territory in the breeding season in 1961 of the black-backed magpie studied by Carrick (1972). The illustration shows territories of the different sorts of groups and the flock. For further explanation, see text. (Sketch by June Jeffery from photograph in Carrick 1972)

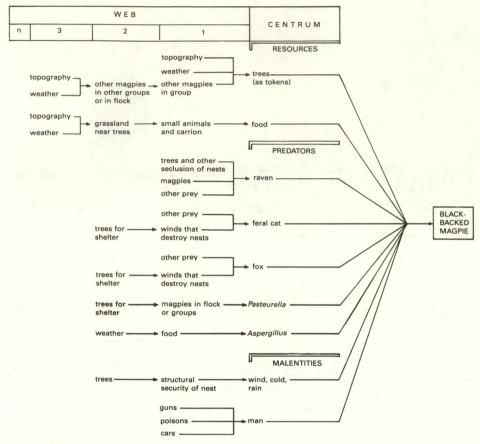

**Fig. 13.01**   The envirogram of the Australian black-backed magpie.

would support a group of five to ten birds for all their needs throughout the year and, with luck, for many years. There are records of groups that have held the same territory for upward of ten years. When an optimum territory is not available, territories that seem to the human eye to fall far short of the optimum will be occupied, but less successfully.

The black-backed magpie, *Gymnorhina tibicen*, occurs virtually throughout Australia, wherever the two prerequisites of nesting places and feeding places occur conjunctly, except for the southwest corner of the continent, where it is replaced ecologically by the white-backed *G. dorsalis*, and in the southeast corner, where it is replaced by another white-backed species, *G. hypoleuca*. Where *G. tibicen* and *G. hypoleuca* meet there is a hybrid zone. Carrick's study area was near the northern limit of the hybrid zone.

The ecology of the magpie is especially interesting because of the relatively dense, relatively steady populations that are maintained at least in the more typical habitats. Carrick has shown that the explanation lies largely in the social behavior of the birds in relation to food and trees. In the envirogram (fig. 13.01), food, trees and other animals of the same kind are important components. Trees are important as part of a token (sec. 3.31). The bird's response to the token is physiological, which is the only physiological point that is made in Carrick's explanations; for the most part the emphasis is on behavior. Behind the study of behavior there is the question, How does the magpie continue to maximize the proportion of "effective" food (sec. 3.121)? The answer is not a direct one, but it is all the more interesting for that. The only food that is used "ineffectively" (i.e., by individuals that are doomed to die before they reproduce) is food that is too far away from roosting and nesting places to be much good anyway. We follow Carrick as he established this knowledge.

In the vicinity of Canberra where Carrick had his study-area, all the suitable habitat had been claimed as territories. Each territory housed a small group of birds, several adults and a number of young, usually between four and ten all told. The adults defended the territory; the young were usually evicted from the territory shortly after they matured. Females were often allowed to stay until they were 2 years old, but males were rarely allowed to stay after they were one year old. Only the birds that were securely established in a good territory bred. There were many that did not (sec. 3.31).

In this country there was quite a lot of open pasture or arable land that was not incorporated into the territories because it did not encompass or abut suitable groves of trees. There were also trees that were not defended because they were surplus, sparsely dispersed, or poorly placed with respect to feeding places. Birds that could not maintain a place in a territory foraged in these undefended feeding places and roosted more or less communally in these undefended trees. Carrick (1972) called these expatriates "the flock". The flock consisted of birds that had been evicted from the territories and had not yet succeeded in fighting their way back into a territory. Many were young birds, 1 or 2 years old that had been driven out because they were approaching maturity. A few of the older birds had come to the flock singly, having been driven out of the territory by solitary invaders that had taken their places by force. But most of the older birds in the flock had come as the remnant of a group whose leader had died or been disabled. Because of such a weakness the group became vulnerable and was driven out by an invading group.

Carrick (1972) studied this population intensively for eleven years. He had an "intensive" study area (6 km²) which was part of the "main" study area (18 km²), which was surrounded by the "extensive" study area (500 km²). In the intensive study area every territory was mapped and every adult and nestling was banded (plate 5, fig. 13.02). In the main study area the territories were mapped, and most of the birds in the territories were banded. On three occasions 80% of the birds in the flock were banded; but turnover in the flock was rapid, so this proportion was not maintained for long. The extensive study area was periodically surveyed, sampled, and searched for birds that had been banded in the main study area. All together, 4,517 magpies were banded.

## 13.11 Behavior

The most striking feature of the behavior of an adult magpie is its relentless drive to attain membership in a group that defends a good permanent territory.

## 13.111  The Territories

Carrick recognized four categories of territory. Each sort conferred a specific status on the group of birds that lived in it. Carrick adopted four adjectives which he used to describe either the territory or its residents: "permanent", "marginal", "mobile" and "open". Each group of birds was given its own number, and for maps the convention was to use merely an arabic numeral for a "permanent" group or territory—for example, 79. A roman numeral was used for a "marginal" group or territory, such as XXIII. A circled arabic numeral with an arrow leaving the circle meant a "mobile" group (the arrow indicates commuting between an outer feeding ground and an inner group of trees), for example, ⑰➤, and an arabic numeral in parentheses meant an open group—(21). We follow this usage because we have quoted directly from Carrick's text and diagrams.

1. A permanent group was one that occupied a good territory and was satisfied with it. The group was most unlikely to leave such a territory voluntarily and almost equally unlikely to surrender to aggression from outside unless it had been weakened by disaster within, such as the death or disability of the leading defender. Satisfactory territories—that is, those that were permanently defended, varied from 1.2 to 2.4 ha; in a sample of 312 "permanent" territories, 67% were between 5 and 12 ha. The size of the group varied from four to ten birds, including adults and new season's young, but the area of the territory and the number of trees in it were independent of the size of the group. Carrick did not count the trees in all the territories, but it seemed that a minimum of about twenty might be required to ensure that a territory was defended permanently. There were several records of "permanent" groups with fewer than twenty trees that fought to enlarge their territory until it contained more than twenty trees. Water for drinking or bathing seemed not to be necessary.

2. A "marginal" group was one that was only marginally satisfied with its territory and was forever seeking a better one. Marginal territories were "inner marginal" or "outer marginal". Inner marginal territories were usually surrounded by other territories that were mostly permanent and therefore well and aggressively defended. So an inner marginal territory might or might not have sufficient trees, but it was unlikely to have enough feeding area. "Outer marginal" territories were usually on the fringe of the wooded country. They usually had plenty of feeding area but were short on trees. An aggressive, well-led group might seek to enlarge such a territory instead of seeking to evict the owners from a distinct permanent territory. There is a record of one group that started with an outer marginal territory with only one tree. With great persistence, they fought to extend their boundaries until the territory ultimately embraced twenty-six trees.

3. A mobile group defended a split territory. They would defend a tree or a small grove, usually in the midst of the permanent territories, that entailed virtually no feeding area that they could claim. Instead, they would also defend a patch of feeding area elsewhere, perhaps several kilometers away.

4. An open territory was hardly a territory at all; it was more like a rallying point—a patch of feeding area that had been carved out of the general feeding grounds of the flock. It was defended by a group that had crystallized out of the amorphous flock. To the human eye at least, and probably to the magpies as well,

the "territory" was lacking in landmarks and ill defined. This mattered little because it served merely as a rallying point for the newly-formed group. The immediate purpose of the group was to fight its way into a permanent territory. Figure 13.02 is a map of the territories in the intensive study area in 1961.

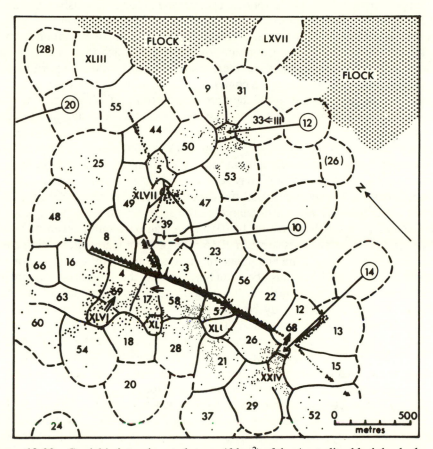

**Figure 13.02**  Carrick's intensive study area (6 km²) of the Australian black-backed magpie in the breeding season 1961, showing the distribution of the four sorts of territories or groups of birds and the flock. Solid lines indicate boundaries of territories; broken lines indicate less well defined boundaries where pressure from outside is minimal. Small dots indicate trees of *Eucalyptus;* triangles indicate pine trees. There were thirty-eight permanent territories (arabic numerals, e.g., 28). There were seven marginal territories (roman numerals, e.g., XXIV). There were four mobile groups (arabic numerals with circle and arrow, e.g., ⑩). The arrow indicates movement between an outer feeding area and a smaller inner group of trees. The mobile groups do not breed. There were four open groups (arabic numerals in parentheses, e.g. (28)). Open arrows trace the steps by which a group or individual moves from one territory to another as its life cycle progresses. For example, when the adult cock and one hen in 4 were killed on the road, XLVI at once took over and formed the new group 69. Similarly, when the adult female in 12 was collected after failing to breed for three years, mobile ⑭ moved in to form 68. (Modified after Carrick 1972)

### 13.112   The Territorial Groups

The "permanent" groups are permanent only in the sense that they retain control of their territories longer than any of the others. But, of course, all the other groups want to be rid of their territories and are always on the lookout for a chance to seize a satisfactory "permanent" territory. So the permanent groups are constantly under challenge from these others. As soon as a permanent group is seriously weakened—for example, by the death of its leader—it is likely to be driven out and its territory usurped by the victors. The victors are most likely to be an "open" group from the flock, but they may come from any of the other categories, even from a "permanent" group of neighbors that merely adds the newly acquired territory to its own. More often the neighbors are content to extend their boundaries a little at the expense of the new owners while they are still "settling in". The remnant of the defeated group, the victims of this aggression, usually disband and join the flock as individuals.

In addition to this process of total defeat and eviction, a permanent group may be changed, but not necessarily diminished, by attrition and exchange. Individuals from the flock or from a territory may seek and gain membership in an established group, often but not necessarily, at the expense of an established member. So a permanent group may remain in charge of the same territory for a long time by replenishing its strength gradually. In 1955 there had been thirty-seven permanent groups in the intensive study area; in 1966 nine of them were still holding essentially the same territory. The endurance of groups from the different categories is shown in table 13.01. It must be remembered that the tenure of a permanent territory can be ended only by the destruction of the group. The tenure of marginal, mobile, or open territories may be ended this way also, but more likely by the group's taking victorious possession of a more satisfactory territory.

The vicissitudes in the history of one group that formed rather loosely in the flock during 1957 and had achieved permanent status by 1959 is given in figure 13.03 and in the explanation that accompanies the figure. This group briefly defended a mobile and a marginal territory on its way to the top.

The drive that activated mobile group 5 on its way to becoming permanent group 49 is characteristic. The recycling between flock and territory and back to flock goes on incessantly, but through it all the number shape, and distribution of territories remains fairly steady. However, the figures in table 13.02 and Carrick's comment on them, which we include in the table, show that there was considerable

**Table 13.01**   Duration of Tenure of Territorial Groups Depending on the Quality of the Territory

| Kind of Territory | Number of Territories | Duration of Tenure (months) | | Less than Two Years (%) |
| --- | --- | --- | --- | --- |
| | | Maximum | Mean | |
| Permanent | 108 | > 183 | > 47 | 14 |
| Marginal | 54 | 52 | 11 | 85 |
| Mobile | 36 | 28 | 8 | 97 |
| Open | 47 | 38 | 8 | 92 |

*Source:* After Carrick (1972).

turnover of individuals. Figures 13.04 and 13.05 show the vicissitudes encountered by a solitary female (for seventeen years of her life) and by a solitary male (over his complete life of eleven years) as each one sought and ultimately achieved a place in a permanent group. These two life histories are typical of many that Carrick observed.

The impression of extreme sedentariness that these observations convey was confirmed by searching for marked birds farther afield. During the eleven years 1955 to 1966 Carrick had banded a great many birds in the flock in the main study area (19 km$^2$); on three occasions he succeeded in marking 80% of the flock birds. In 1965 he searched thoroughly for banded birds in an area of 500 km$^2$ around the study area. He found 70 banded birds in the banding area and 9 outside it. None was more than 21 km from where it had been banded. Carrick quoted further evidence of the sedentariness of the magpies. Of 178 territory-banded nestlings that moved through the flock into territories, 147 bred within 1.6 km of their birthplace. Of 43 territory-banded adults that returned to the flock, 23 later found territories within 1 km of their previous territory.

The size of the group and its composition varied according to the quality of the territory. The thirty-seven permanent territories in the intensive study-area averaged over eleven years 45 adult males and 59 adult females—that is, 2.8 adults per territory. New season's young less than 2 years old swelled the numbers by about 1.2 per territory, making the total about 4 birds per group.

The sex ratio of the adults varied according to the category of the group. There was a tendency for mobile and open groups to have fewer pairs, that is, to have larger groups than the permanent or marginal groups. In the mobile and open categories there was a tendency for males to outnumber females more frequently than in the permanent and marginal groups. Carrick observed on several occasions that a subordinate male was discarded shortly after an invading group had safely established itself in the captured territory. Some details of group size and sex ratios are given in table 13.03. Note that mobile and open groups produced no young, and marginal groups produced scarcely any. These groups waste no effort trying to breed in an inferior place but concentrate all their effort on establishing themselves in a good place to breed—even to the extent of keeping a supernumerary male or two in the group until this goal has been achieved.

### 13.113  Intraterritorial Behavior

When a group achieves a permanent territory it is time for the individuals to turn to their specialities. The special niche of the male—the dominant male if there is more than one—is to defend the territory. He will be assisted by every other adult bird, of both sexes, in the group, but the main responsibility is his. He may be kept busy. Greedy neighbors may seek to enlarge their territory at his expense. Land-hungry marauders from the flock or from an inferior territory may seek to evict him and his group. He may also have to oppose individuals that seek a place in his group. He is so preoccupied with defense that he has little time for any other activity, even for the prerequisites for breeding.

All that is left to the female. Alone, she chooses a place and builds the nest; when she is ready she solicits the male. She incubates the eggs and usually feeds the nestlings on her own. And she has to defend the eggs and nestlings from predators,

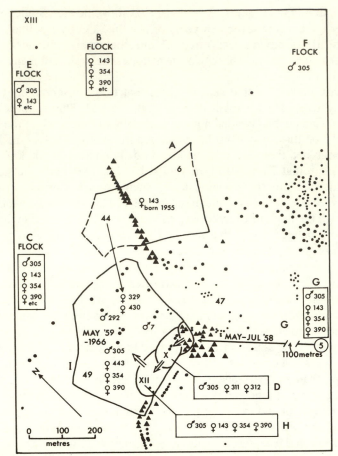

**Figure 13.03**  The history of ♀ 143 from birth in 1955 to her disappearance in 1959 and of the groups to which she belonged in sequence from A to I. Symbols have same meaning as in figure 13.02. This figure and the following account are slightly modified from Carrick's (1972) account.

(A) October 1955: ♀ 143 banded as nestling in 6; remained until August 1956. (B) March–May 1957: ♀ 143 in flock, with ♀ 354, ♀ 390, and others. (C) May–June 1957: during frosty weather, ♂ 305, ♀ 143, ♀ 354, and ♀ 390 were among many flock birds that invaded outer permanent territories to feed. (D) September–October 1957: ♂ 305, ♀ 311, and ♀ 312 claimed small four-acre inner marginal territory X; females built at the center and top of a conifer, but constant boundary fighting led to loss of eggs from the upper nest (probably taken by crows) and death of embryos in the other nest. ♂ 305 left X in December. (E, F) January–April 1958: ♂ 305 and ♀ 143 in same flock in January, then as ♂ 305 moved around in flock, ♀ 143 and other flock birds visited conifers held by ♀ 311 and ♀ 312 until mid-March. (G) May 1958: ♂ 305, ♀ 143, ♀ 354, and ♀ 390 formed mobile ⑤♂, commuting between conifers and pasture 1,100 meters away; they held less than 1 hectare in trees. (H) July 1958: ⑤♂ was enlarged to two hectares to become inner marginal XII, with small pasture area won from adjacent 7 (not shown); thus ♂ 305 had reoccupied its former X area. In September 1958, ♀ 143 bred, but the nestlings did not fledge, and another hen in XII built but did not lay; continual conflict with 7 inhibited breeding. In March 1959 XII increased to three hectares by extension around a dam. (I) May 1959: when adjacent 7

**Table 13.02** Number of Territories (Groups) Known in the Intensive Study Area 1955 to 1966

| Date | Number of Territories Based on Quarterly Counts | | | | | | |
|---|---|---|---|---|---|---|---|
| | Mean Annual | | | | Eleven Year Period | | |
| | Permanent | Marginal | Mobile | Open | Permanent | | |
| 1955 | 37 | 3 | 1 | 1 | Mean | Minimum | Maximum |
| 1961 | 38 | 7 | 4 | 4 | — | — | — |
| 1966 | 36 | 3 | | 2 | — | — | — |
| 1955–66 | — | — | | — | 37 | 34 | 39 |

*Source:* After Carrick (1972).

*Note:* Carrick's comment—Eight of the thirty-seven permanent groups present in 1954 survived to 1966 (though with gradual replacement of some or all members) as well as marginal III which became thirty-three in 1956. In 1966, twenty-one of these thirty-seven territories were essentially the same as their 1955 counterparts (though thirteen of them underwent complete replacement by new groups at some point).

chiefly the raven *Corvus coronoides*. Even so, she will take time from these duties to help defend the territory, if necessary, and to assert her dominance over any subordinate female in the group.

Dominant females differ widely in their aggressiveness toward a subordinate in their own group. Most dominants harry the subordinates so severely during nest building or incubation that the subordinates are unlikely to raise any young. The

---

lost ♂ 84, leaving two females and three first-year birds, XII at once evicted them and became 49 with ten hectares.

All four adults were replaced by 1961. When ♂ 305 was found freshly dead (septicemia) on 7 August 1959, five-year-old ♂ 7, from adjacent 47 (leaving its dominant brother ♂ 76 with two adult hens) had already settled into XII with the three females. During 23–29 August ♂ 7 reverted to 47 when challenged by ♂ 292, who deserted ♀ 380 in outer marginal XIII 900 meters away. In September 1959, the more aggressive hens caused ♀ 143 to leave. ♀ 390 died of injuries in August 1960. ♀ 354 had nestlings in 1959, fledged young in 1960, and eggs in 1961, but died about September leaving ♂ 292 alone.

In adjacent 44, the very tolerant ♀ 339 (not shown) had not evicted her offspring, four-year-old ♀ 329 and three-year-old ♀ 430, who now joined ♂ 292 (see arrow), though ♀ 329, with nestlings in 44, used both territories until February 1962. During 1962–66, ♀ 329 and ♀ 430 each produced free-flying young in 49 on three occasions, and they nested in all years. Their tolerance enabled their one-year-old offspring ♀ 662 to fledge young and ♀ 663 to lay eggs in 1963, and all four hens nested in the group of eucalyptus trees in the center of 49.

From September 1960, group 49 surrendered the same small group of conifers in which it started as (5)° to ♂ 7, who formed (13)° with two hens. The latter expanded as XLII (not shown) into the same area as X and XII by April 1961 and became permanent 70 (not shown) at the expense of 5 and 49 to regain the dam and some conifers. Both groups survived to 1966, with twelve-year-old ♂ 7 in 70 and nine-year-old ♀ 329 in 49, but on 12 September 1966, eight-year-old ♀ 430 drowned herself in a bathtub.

In summary, this history illustrates the limited home range of individuals throughout life, their tendency to return to familiar places with known companions, and the unremitting effort toward permanent ownership of the best resources with whatever partners offer most.

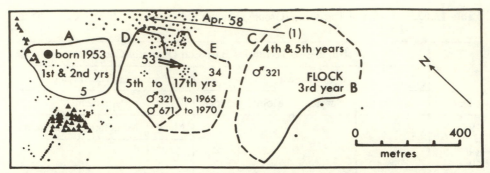

**Figure 13.04**    The life history of ♀ 29 from her birth in 1953 at A to 1970 at D. (After Carrick 1972)

(A) October 1953: ♀ 29 banded as nestling in 5; remained until August 1955. (B) April 1956: seen in flock where ♂ 321 was trapped in May 1956. (C) August 1957: formed open group (1) with ♂ 321; in April 1958, pair visited trees between 9 and 12 (fig. 13.02). (D) September 1958: ♂ 321 and ♀ 29 became 53 by displacing 47 (fig. 13.02) from former 14, into which it and 34 (not shown) had extended in May 1958 when 14 lost its hen. (E) April 1961: 53 displaced ♂ 219 and sick ♀ 218 from adjacent 34. July 1965: ♂ 321 presumably died, leaving ♀ 29 alone; 47 had two cocks and one hen, so subordinate ♂ 671 crossed into 53, which they still occupied in April 1970.

In summary, during sixteen and a half years, ♀ 29 had not been seen more than 1,250 meters from her birthplace and had spent the past eleven and a half years only 450 meters from it. She progressed through flock and open group to an average, then a large permanent territory. Her first opportunity to breed was in 1958 at five years old; though she collected nest material, constant conflict with new neighbors prevented it that year. In 1959 and 1960 she hatched young but failed to rear them. In 1961–66 (except 1962, not known), she fledged two or three young each year.

subordinate's condition may be ameliorated: in a large territory with many trees she may be able to "hide" from the dominant; or by postponing her own breeding until the dominant female is busy incubating eggs or feeding nestlings, she may escape serious molestation. In some groups the dominant female was more tolerant, and subordinates had a better chance to contribute to the next generation.

### 13.114    The Natural Selection for Territorial Behavior in the Magpie

A number of noticeable specializations in the behavior of the magpie suggest that selection for territorial behavior has been strong in this species. The members of the landless groups specialize in offense. An adaptation, which we discuss under the heading "tokens" in section 3.31, ensures that they will not be diverted from the main target by the distractions associated with breeding. Despite the appropriate length of day, maturation of the gonads does not proceed to fruition without the stimulus that comes from ownership of a proper territory. A group may enhance its striking power by the inclusion of one or more supernumerary males which are likely to be expelled when the group's chief objective changes from offense to defense. The pressure that can be exerted by the landless group is also maximized by a thorough and intimate knowledge of the territorial groups—their condition and the vicissitudes of their lives. This comes from the extreme sedentariness of the landless ones. In contrast to most other animals, the magpies seem to have had

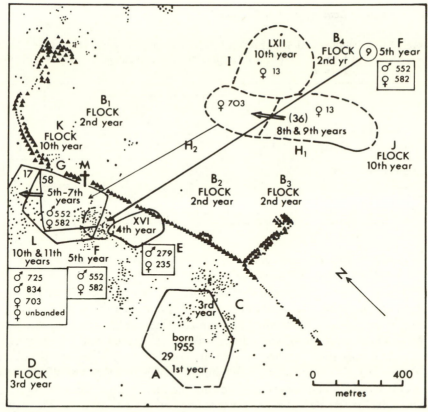

**Figure 13.05**   The life history of ♂ 176 from birth at A in 1955 to its death at M in 1966. (After Carrick 1972)

(A) October 1955: ♂ 176 banded as nestling in 29; remained until June 1956. (B₁–B₄) February–October 1957: mobile flock up to fifteen hundred meters from birthplace. (C) February 1958: revisited natal territory 29 (both parents gone and only one adult female of 1955 group left). (D) April–August 1958: in flock near 29. (E) September–November 1959: entered inner marginal XVI and evicted ♂ 279 but ♂ 176 and ♀ 235 were soon replaced by adjacent 26 (fig. 13.02). (F) April–July 1960: with ♂ 552 and ♀ 582 formed ⑨♂ at trees adjacent to former XVI, feeding at former B₄ flock location. (G) August 1960: expanded to become 58 by taking area of former 36 from 17. Evicted ♂ 552 about March 1961. First bred in 1961. Took rest of 17 in October 1961 when ♂ 62 was lost. In May 1962, ♂ 582 disappeared, and two hens from adjacent 67 joined ♂ 176. Between September 1962 and July 1963, 58 was replaced by 72. (H1) October 1963: in open group (36) with ten-year-old ♀ 13 (which had been there since October 1962). No breeding. (H2) July 1964: visited former 58 area, now 75, and aggressive to ♂ 594. September 1964: added two-year-old ♀ 703 and increased (36) area. No breeding. (I) February–May 1965: moved with ♀ 13 to LXII, with two small trees. (J,K) June–September 1965: in flock with ♀ 703 etc., invaded stock enclosures for food in drought year. (L) September 1965: when ♂ 707 disappeared from 80 (successor to 75 in former 58), ♂ 176, ♂ 725 (born in adjacent 51 in 1963), ♀ 703 and unbanded ♀ formed 91. ♂ 834 (born in adjacent 79 in 1964) joined 91 in February 1966 and was dead on road by July. About same time ♂ 725 and ♀ 703 disappeared. (M) September 1966: finally ♂ 176 was found dead on road.

In summary, the checkered career of ♂ 176 illustrates how magpies strive to attain permanent status by their own individual and group efforts, though the social bonds of the latter prove as fragile as necessity or a better opportunity dictate. The limited range of ♂ 176 is typical, but many life histories are simpler.

**Table 13.03**   Number and Sex of Adult Magpies (over Two Years Old) in the Four Sorts of Groups (Territories), Expressed as a Percentage of Groups with Various Sex Ratios

| Group | Number of Groups | Percentage of Groups with Specified Number of Male and Female | | | | Number of Young per Female |
|---|---|---|---|---|---|---|
| | | 1♂ 1♀ | 1♂ 2♀ | 2♂ 1♀ | Other | |
| Permanent | 381 | 47 | 35 | 5 | 13 | 0.56 |
| Marginal | 74 | 53 | 33 | 1 | 13 | 0.09 |
| Mobile | 36 | 20 | 30 | 25 | 25 | 0.0 |
| Open | 42 | 36 | 31 | 22 | 11 | 0.0 |

*Source:* After Carrick (1972).

dispersiveness virtually bred out of them. So they stay at home and keep a close watch on their prospective victims.

The groups in the permanent territories usually remain secure while their leadership is strong. Their defensive strength is based on their talent for specializing—the male specializes in defending the territory, the female in rearing the young.

Meanwhile, selection of individuals strong in territorial behavior goes on. The dominant female does her best to prevent her less-aggressive subordinates from rearing young, thus ensuring the predominance of her own genes. While the male seeks to extend the boundaries of his territory to include more trees, thus giving the second-ranking female a chance to rear young and his genes a chance to contribute to posterity through both females. Here is further support, if support is still needed, for the idea that natural selection for territorial behavior operates at the level of the individual. Territorial groups are preserved in the magpies' life-style because the fit individuals have achieved their fitness within this system. It is a positive feedback—a closed circle that seems most unlikely to be broken into.

### 13.12   The Environment and Some Statistics for the Life Table

Even in the territorial groups, not every female breeds every year (sec. 13.113). Those that do breed produce one clutch for the year. The mean number of eggs per clutch is 3.5. Apart from the reduction in fecundity of subordinate females owing to the aggression of the dominant female, eggs may be destroyed or nestlings killed by predators or malentities. The most important predator is the raven *Corvus coronoides*, but the feral cat and the fox also take some weaklings, especially young fledglings that have been blown prematurely from the nest. The most important malentities are high winds, which may blow the rather inadequate nest of sticks out of the tree, and cold, wet weather. Eggs or nestlings that are neglected during such weather are likely to die from exposure. Such risks are exacerbated by external aggression. All other duties are likely to be abandoned when the call to defend the territory becomes acute. The risks from predators and from exposure are greater for the eggs and nestlings of a subordinate female if she is likely to be harried by the dominant female.

From all these risks, the potential production of 3.5 young per female per year (see above) was reduced to an actual production of 0.56 per female per year (table 13.03). Apart from a few females that were allowed to stay in the territory where they were born, all the young were evicted as they approached maturity. Nearly all the males had gone before they were 1 year old, and nearly all the females by the time they were 2. During the period 1955–66 the territorial groups in the intensive study area produced 363 young adults; virtually all of them were driven out of the territories and entered the flock; about 35 regained a place in a territory. They had to wait, on the average, about 2 years in the flock. In the meantime many of their contemporaries were dying. The death rate in the flock was high; the expectation of life for a yearling joining the flock was about 18 months. The probability of any bird's living for five years in the flock was about 0.03 (tables 13.04, 13.05).

Carrick also estimated the same statistic for the territories—that is, the probability of being alive and still in a territory after five years was 0.42 for a male and 0.46 for a female. A number of individual records suggested that these estimates might have been conservative. Referring to these records Carrick commented: "These figures would be much higher in an environment free from human interference, especially the steady toll of road casualties" (Carrick 1972, 68).

In the main study area during 1956 through 1962, 74 adults from the territories died as follows: 17 died on the roads (hit by cars), 3 were electrocuted, 8 were deliberately killed by men with guns, poisons, and so forth, 7 died of disease (*Pasteurella, Staphyllococcus coccidiosis,* etc.), 6 were taken by vertebrate predators (fox, feral cat, raptors), 8 died from wounds (fighting or unidentified accident), 25 long-dead carcasses were found but the cause of death could not be identified.

Corresponding figures for the flock were less accurate but Carrick commented (p. 71): "In the wet cold winter of 1956 *Pasteurella pseudotuberculosis* killed so many healthy flock birds that dead and dying birds were picked up daily; contact-spread, it did not kill any birds in nearby territories, and it is as local and infrequent as the weather that favours it (Mykytowycz and Davies, 1959)." Carrick then went on to mention several other minor causes of death for the magpie, both in the flock and in the territorial groups, and he concluded (p. 71): "It seems that diseases, predators, human agencies, and the magpie's aggressive behaviour all contribute, directly or indirectly; it remains to be seen whether any individuals suffer nutritional inadequacy, and why, and if this exposes them to death from these known proximal causes."

**Table 13.04** Time Spent in the Flock before Dying or Regaining a Place in the Territories

| Sex and Age on Entry to Flock | Mean Time in Flock before Dying or Regaining a Territory | |
|---|---|---|
| | Died in Flock (months) | Regained a Place in a Territory (months) |
| Male, one year | 18 | 28 |
| Female, two years | 16 | 21 |
| Older birds from disbanded groups | — | 8.5 |

*Source:* After Carrick (1972).

*Note:* Young birds one to two years old that were in the flock for the first time are compared with older birds that rejoined the flock after their group had been evicted and disbanded.

**Table 13.05** Probability That a Bird First Banded in the Flock Would, after Five Years, Be Alive in the Flock, Alive in a Territory, or Dead without Regaining a Territory

| Fate | Probability (at end of five years) |
|---|---|
| Alive and still in the flock | 0.03 |
| Left the flock for a territory | 0.13 |
| Disappeared from flock; probably dead | 0.84 |

*Source:* After Carrick (1972).
*Note:* Age, sex, and history confounded.

Carrick's inquiry into food was indirect. He weighed and measured 2,163 birds from territories and 7,749 from the flock. The females in the territories tended to be underweight while they were feeding nestlings, but that could hardly be taken to indicate shortage of food. The subordinate birds in the flock tended to weigh a little less than the dominants, presumably because the subordinates were barred by "social pressure" from feeding in the best places. Perhaps the most interesting observation relates to the temporary shortage of food for the flock during the dry frosty winter of 1957. Carrick commented: "In the dry frosty winter of 1957, the open pasture [flock feeding ground] did not thaw during the day, and ground invertebrates were inactive, with the result that flock birds invaded some outer permanent territories and searched for insects and seeds around haystacks where they picked up spores of *Aspergillus* which thus became an important secondary cause of death that year."

It seems that the most important cause of the high death rate in the flock, relative to the territorial groups, is the shortage of trees for the flock birds. The shelter that is provided by trees seems to reduce the risk from certain predators and from certain weather-induced shortages of food.

## 13.13 The Distribution and Abundance of *Gymnorhina tibicen*

In the extensive study area during 1965 there were 350 km² of territories and 90 km² of flock habitat, and between them they supported about 7,000 birds. Carrick's estimates of the numbers in different categories based on age, sex, and breeding condition are given in table 13.06. The number of adults in the territories outnumbered the flock by 4,452 to 1,093, about four to one. Carrick thought that this density would be characteristic of populations in savanna woodland, which is about the optimal habitat for this magpie. Extrapolating from the empirical results from the intensive study area the population in the extensive study area might have been expected to produce 1,421 young recruits for the flock each year.

During the eleven years 1955–66 the weather varied widely: there was a hot drought and a cold, dry winter, both of which made it harder for all magpies to find an adequate ration of food, especially those in the flock; there was an excessively wet period which increased the incidence of certain diseases in the flock but not in the territories. The numbers in the flock varied from 1,090 to 1,770. The numbers in the territories varied scarcely at all: most groups were protected from the vagaries of the weather's influence on food by preempting a feeding area that had reserve

**Table 13.06**   Number of Magpies Estimated to Be Living in the Extensive Study Area in 1965, Number per Square Kilometer, Number of Recruits, and Proportion Successful in Rearing Young

|  | Territory | Flock | Total |
|---|---|---|---|
| Area (km$^2$) | 350 | 90 | 440 |
| Density (birds per km$^2$) | 16.8 | 15.7 | 16.6 |
| Number of adult males | 1,781.0 | 629.0 | 2,410.0 |
| Number of adult females | 2,671 | 414 | 3,085 |
| All birds | 5,873 | 1,416 | 7,289 |
| Proportion (%) of adult males not breeding | Few | 100 | 18+ |
| Proportion (%) of adult females not breeding | 19 | 100 | 33 |

*Source:* After Carrick (1972).
*Note:* All figures divided between territories and flock.

capacity for hard times and that was protected by trees. The trees also reduced the risk from *Pasteurella* during wet weather. Fluctuation in the size of the flock, by altering the social pressure on the groups, might have been expected to cause fluctuation in the numbers in the territories, but apparently it did not.

Carrick suggested that the size of the territory and the number of birds in the group depended largely on the innate aggressiveness of the leaders. If this is so, it seems that none of the fluctuations in environment during 1955–66 was big enough to offset the influence of territorial behavior. By practicing this niche the magpies reduce the risk that food might be used "ineffectively" during bad times (sec. 3.121). Presumably, changes in the distribution and abundance of tall trees and pastures would be influential. Only man's activity is likely to introduce such changes into the environment of *G. tibicen*, at least while we think in the time scale of ecology.

Carrick emphasized that man as a malentity (through cars, guns, and poison) was the chief cause of death of adult magpies. It is interesting to ask what would happen if this risk were reduced or removed. With leaders surviving longer, would the rate of turnover of territorial groups slacken? Would the overall death rate in territories and flock be reduced? Would the territories produce more young adults? Would the flock grow larger? If so, would increased social pressure from outside force the groups to accept smaller territories or larger groups? Or would the "policy" of the well-bred territorial birds prevail? If so, the only consequence of removing a malentity would be to increase the death rate in the flock from other causes. In the southwest corner of the continent a white-backed magpie, *G. dorsalis*, maintains larger groups than *G. tibicen* but we do not have an ecological explanation for this observation. Carrick's nice study, within the framework of the theory of environment, has brought us much closer to a full understanding of the ecology of *G. tibicen*.

It seems that the evolution of territorial behavior in the magpie has produced a species that is good at using its food supply effectively (sec. 3.121); that is, very little of the food is available to individuals that have a poor chance to contribute progeny to the next generation. An individual has to prove its quality (by the fruition of a long and arduous apprenticeship, sometimes lasting five years) before it is allowed to share the food and shelter that has been preempted for the breeders.

This is the sort of social organization whose origin might be explained by "group selection", and whose operation might be explained by "competition". But we think that the first is explained better by the conventional theory of selection of the individual and the second is explained better by the theory of environment.

The magpie provides an excellent example of intense intraspecific competition, which is obviously going on relentlessly in the flock and in the territories, but especially where flock and territory or territory and territory interact.

Obviously the idea of competition would be useful, indeed essential, for a student of evolution who has to explain the different kinds of magpies that constitute the population. But in ecology the primary question is always How many? For the ecologist who has to explain how many magpies are in the population, the idea of competition is irrelevant and may even be counterproductive, as we have seen in chapter 10 and elsewhere in this book.

## 13.2   The Ecology of the Grey Teal (*Anas gibberifrons*)

The following account of the ecology of the grey teal has been condensed from papers by Frith (1957, 1959a,b,c,d, 1962, 1963, 1967), Braithwaite and Frith (1969a,b), Frith, Braithwaite, and McKean (1969), and Braithwaite (1975).

The grey teal is the most widespread duck in Australia; none could be more so, because the teal is found throughout the length and breadth of the continent, breeding wherever and whenever the opportunity arises, but more often resting and feeding in nonbreeding places while regaining strength and condition to continue the search for somewhere to breed. For the grey teal is the most dispersive of all the Australian ducks. And there is no other duck, except perhaps the black, that is so numerous (fig. 13.06).

The grey teal has achieved this eminence among the waterfowl of the driest continent by specializing in certain niches which have proved extremely rewarding. It is well known that shallow temporary ponds or other waters that are filled by rising floodwater support extremely productive communities while the water lasts (Mozley 1955, 5). To exploit this rich supply of food it is necessary to resolve several problems. (a) Feeding habits must match the sort of food that is offered. (b) Since the opportunity is apt to be created abruptly and is likely to be short-lived, the feast should be discovered without delay. (c) Because time is short, the search for a new place must be readily resumed, and those that search must be sustained meanwhile.

The grey teal has resolved these problems by specializing in two niches which may be labeled "feeding in shallow temporary water" and "dispersing in search of temporary shallow water". Both niches are aimed at exploiting the rich supply of food that is found in temporary shallow water. Temporary shallow water is rich in the sort of food that the teal needs for breeding. The rising flood that leaves the shallow water behind it is a token that brings the teal into breeding condition. In the envirogram (fig. 13.07), the pathways that lead away from "food" and "token" (resources for breeding) lead to weather and topography, which are modifiers of outstanding importance. In the pathway that leads from food (for surviving), shallow water is still necessary in a "drought-refuge", but it need not be temporary.

Many drought-refuges are inlets from the sea; in them topography is still important, but they depend much less on the weather.

## 13.21 Breeding Places

The first requirement of a breeding place is the token that changes the teal from a nomad into a stay-at-home breeder. For the early arrivals that arrive while the water is still rising, it seems that the sight of rising water may be a sufficient token. But this does not explain why latecomers also join in the breeding without delay. Of course, by the time they come there is a rich carnivorous diet available, and many of their fellows are displaying and breeding all around them.

The second requirement of a breeding place is food. Mozley (1955) explained the wealth of aquatic life, both plant and animal, that usually fills a shallow temporary pond shortly after it has been filled by floodwater. Primary production is high because the water is well supplied with plant nutrients from the organic matter that was decomposed after the last flood had dried up. And the shallow pond is warm and full of light. Herbivores thrive because they find abundant food. So the ducklings find an abundance of small aquatic animals, and the adults find plenty of plant material as well.

The third requirement is an abundance of tree holes for nesting. We discuss these three requirements below, and we also discuss the causes and frequency of floods.

### 13.211 The Distribution and Abundance of Breeding Places

Grey teal are to be found throughout the length and breadth of Australia, persistently searching for a place where they can breed. Such places are likely to be found in any of the major drainage systems shown in figure 13.06, though not in much abundance in most of them. Frith estimated that more than half of Australia's grey teal are bred in the area that is heavily shaded in figure 13.06. This shaded area includes the Riverina, where the breeding is most concentrated. We follow Frith's description of the topography of the backwaters that are formed during the flooding of the rivers of the Riverina (Murray, Murrumbidgee, Lachlan).

The Murray and its great tributaries the Murrumbidgee and the Lachlan drain $1.07 \times 10^6$ km$^2$. In the Riverina these rivers and their tributaries meander through extensive flat plains. There are many old meanders that have been cut off from a river as it changed its course. Some are ancient, others are more recent. The old ones have weathered, some more, some less, so that they differ in the readiness with which they fill as the river floods and, once filled, in their depth and permanence. Local names for these topographic features reflect their quality.

A meander that is still connected by both ends to the main stream, at least at high flood, is called a billabong. A juvenile billabong usually has steep banks, and the water is likely to rise and fall in it as quickly as in the main stream; so it is not likely to be a good place for grey teal. A mature billabong is likely to have a wide channel with gently sloping banks and to provide good places for grey teal and their ducklings to feed. At a later stage of maturity a billabong may lose its boomerang shape and its well-defined connection with the river. It is then called a lagoon. A lagoon is less likely to fill than a billabong, but having filled it usually lasts longer. So the lagoon is likely to be colonized by aquatic plants, perhaps by water fern

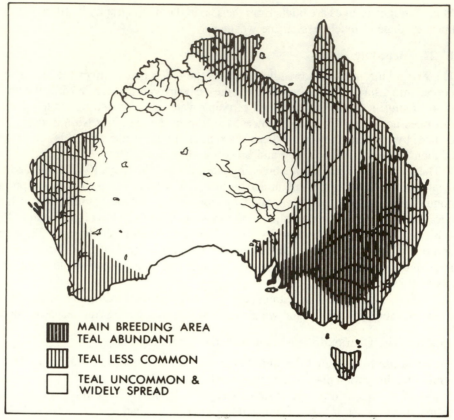

**MAIN BREEDING AREA**
**TEAL ABUNDANT**

**TEAL LESS COMMON**

**TEAL UNCOMMON &**
**WIDELY SPREAD**

**Fig. 13.06** The distribution of the grey teal in Australia. The main breeding area (*dark shading*) is associated with the floodplains of the Murray-Darling river system. There teal are abundant at all times. Teal are less common in the area of lighter shading but are still numerous. In the nonshaded area teal are widely spread but are uncommon and breed only in exceptionally favorable seasons. (After Frith 1962, 1967)

(*Azolla*) and nardoo (*Marsilea*) in the deeper parts and a wide variety of species in the shallows around the edges. Swamps are large depressions in the plains that may or may not have had their origins in ancient meanders. They fill during floods and have a reasonable chance of holding the water until the next flood. The vegetation in a swamp is a good indication of its depth and permanence. The deepest and most permanent swamps are dominated by cumbungi (*Typha angustifolia*); slightly shallower, less permanent ones may be dominated by cane-grass (*Glyceria ramigera*); lignum (*Muehlenbeckia cunninghamii*) grows in swamps that fill only rarely and go a long time without water. These plants give their names to the swamps that they dominate.

The water flowing away from the main channel and the billabongs during a flood usually flows in definite, though often broad and shallow, watercourses called effluent streams. The effluent streams are of great importance to the grey teal and other waterfowl. Frith (1959a, 100) described them as follows:

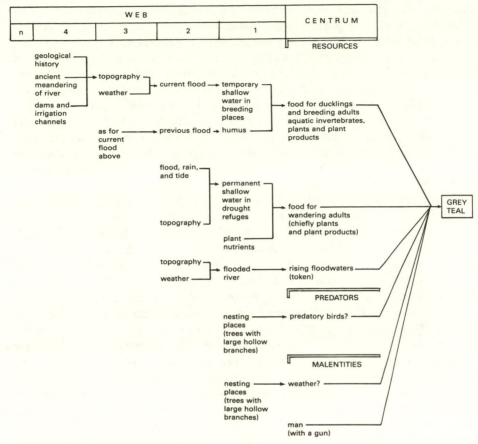

**Fig. 13.07** The envirogram of the grey teal.

They are normally dry, but when the river is in flood, water flows into them and is carried far into the arid plains. [One of] the principal effluents from the Lachlan is Willandra Creek [which] leaves the river near Lake Ballyrogan and flows westward for about 100 miles [160 km]; in times of high flood it may rejoin the Murray near Euston. . . The most important effluents from the Murrumbidgee are Gum Creek, Colleambally Creek, Eurolie Creek and Yanko Creek. These completely dissect the area between the Murrumbidgee and Murray Rivers, and in times of flood, water flows throughout the plains.

In addition to filling or replenishing the swamps, lagoons, and other such features, the effluent streams send floodwaters across the plains, covering vast areas with shallow water and filling depressions which, depending on soil and weather, may retain some water for as long as a year. Such residual floodwaters may make available to the grey teal an abundance of submerged dryland plants and also certain aquatic animals which, being adapted to exploit temporary water, often become extremely abundant in such places.

The region supports large areas of irrigated crops and pastures. The human artifacts—dams and irrigation channels—on which the irrigation depends—are extensive; they provide scarcely any favorable breeding places for grey teal; but by controlling floods they greatly reduce the activity of the natural breeding places.

The main streams of the Murray River and its tributaries have steep banks and the water level fluctuates widely; during severe drought flow virtually ceases, even in the Murray. But they are also subject to severe floods. The floodwaters originate in the headwaters of the rivers hundreds of kilometers from the plains of the Riverina. The rivers are fed by rain that falls chiefly in the winter (Murray and Murrumbidgee) or in the summer (Lachlan and Murrumbidgee) and by melting snow (Murray and Murrumbidgee). So floods do not occur at a regular season of the year. The grey teal is adapted to take advantage of flood waters whenever they occur. On the average, most parts of the region are likely to be inundated once every three or four years.

During the period when Frith was studying the ecology of the grey teal there were floods that promoted extensive breeding in 1952/53 and again in 1955/56, and there was a widespread drought that caused a catastrophic death rate in 1957. Floods occur sporadically, at irregular intervals, and vary widely in size. They may be localized on one tributary, or they may arise in all the tributaries at the same time; or the timing may be such that, in the lower reaches of the rivers successive flood-peaks may be experienced, that relate to floods arriving from distinct tributaries. Some of the smaller floods may not spill enough water to do much good for the teal. Substantial floods that might promote extensive breeding might be expected about thirty times in one hundred years. The vagaries of the system are magnified by the interactions of several weather systems. Water for the Murray comes from melting snow and winter rain, for the Murrumbidgee from melting snow and winter and summer rain and for the Lachlan and Darling from summer rain. Table 13.07 shows the number of occasions when there have been 0, 1, 2, 3 . . . flood peaks observed in one year. Table 13.08 shows the number of occasions when a run of 0, 1, 2, 3. . . consecutive years without a flood has been observed.

### 13.212   The Token That Indicates the Time for Breeding

Other species of the genus *Anas* which live in reliably watered regions in other continents and breed regularly at the same season every year are known to display and court during several months preceding the breeding season. This behavior is said to be adaptive because it cements the pair bond (Lorenz 1941; Hochbaum 1942; Höhn 1947). Sexual display in the grey teal is like that in other members of

**Table 13.07**   Number of Occasions, during about Seventy Years, When the Specified Number of Flood Peaks Have Been Observed in One Year on the Murray, Murrumbidgee, and Lachlan Rivers

| River | Years with Specified Number of Flood Peaks per Year | | | | | | |
|---|---|---|---|---|---|---|---|
| | 0 | 1 | 2 | 3 | 4 | 5 | > 5 |
| Murray (72 years) | 32 | 22 | 7 | 4 | 6 | 1 | 0 |
| Murrumbidgee (69 years) | 26 | 11 | 11 | 7 | 7 | 1 | 7 |
| Lachlan (78 years) | 52 | 7 | 8 | 4 | 1 | 3 | 2 |

*Source:* After Braithwaite (1975).

**Table 13.08**  Number of Occasions When the Specified Number of Consecutive Years without a Flood Were Observed, during about Seventy Years' Observations, on Murray, Murrumbidgee, and Lachlan Rivers

| River | Number of Consecutive Years without a Flood Peak | | | | | | |
|---|---|---|---|---|---|---|---|
| | 1 | 2 | 3 | 4 | 5 | 6 | > 6 |
| Murray (72 years) | 9 | 4 | 3 | 0 | 0 | 1 | 0 |
| Murrumbidgee (69 years) | 7 | 3 | 3 | 1 | 0 | 0 | 0 |
| Lachlan (78 years) | 4 | 4 | 0 | 1 | 3 | 2 | 1 |

*Source:* After Braithwaite (1975).

the genus in the Northern Hemisphere and a few unusually sedentary ones that live and breed on the cumbingi swamps of the Riverina or on permanent swamps in the swamplands of southeastern South Australia spend the same time on courtship; they begin displaying several months before they begin to breed. But the nomads that are already present on the billabongs or those flying in as the waters begin to rise seem to compress courtship and the formation of pairs into a week or two, after which the first eggs are laid. The advantage of such behaviour to a species that depends on floods that arrive erratically and cannot be relied on to endure is obvious.

It seems likely that the grey teal has carried adaptation so far that it relies on the token stimulus of rising water to trigger the processes that bring it into breeding condition, in anticipation of the abundant food that will become available to it and its ducklings as the floodwaters spread. Frith's evidence for this conclusion rests chiefly on three observations. (a) Ovulation usually occurred within two weeks of the water's starting to rise, which seems too short a period to be explained by a change in diet. (b) Grey teal on billabongs began to display within a day or two of the first sign of rising water, before water overflowed and therefore before it could have altered their diet (fig. 13.08). (c) On an occasion when a small increase in level was not sustained, the grey teal displayed briefly but did not ovulate.

On the other hand, the sight of rising water may not be the only stimulus that brings the grey teal into breeding condition. (a) During a long bout of breeding, as, for example, in 1956 when it lasted from March through December, fresh ducks flew in from time to time and started to breed, even though the water was no longer rising. (b) Braithwaite and Frith (1969b) reported one occasion when, on a swamp that was fed by local rain, the water rose 35 cm during June/July but the ducks, according to the condition of their gonads, did not respond strongly until August.

## 13.213  Nesting Places

Grey teal nest in trees in wet places. They prefer hollow branches well above the ground. The newly hatched ducklings fall, or are pushed, out and plunge into the shallow water below. Tall trees with dead branches are numerous in the Riverina, close to the breeding waters of the grey teal. The typical tree of the permanent streams is *Eucalyptus camaldulensis*, which fringes the streams and also sometimes forms dense forests on low-lying ground. Elsewhere *E. bicolor* is ubiquitous wherever water is likely to flow or lie, however temporarily. In this dry climate a

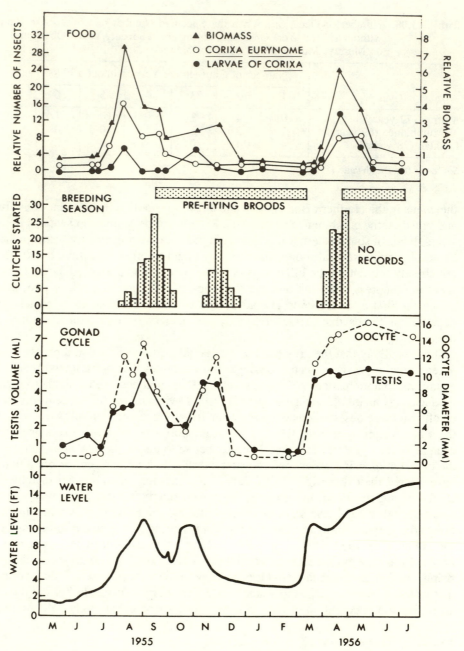

**Fig. 13.08** Breeding season, sexual cycle, and food supply of the ducklings of grey teal in relation to water level at Kooba Lagoon, New South Wales, in 1955 and 1956. (After Frith 1959d)

dead branch will persist on a tree for many years. So except for times when the teal are breeding in exceptional abundance there is a superfluity of nesting places, and only the most favorable are used. Usually a branch 9–12 m above the ground, horizontal, with an opening 15–18 cm in diameter and hollow extending back at least 2 m from the opening is chosen. When crowding is intense less favorable sites nearer the ground may be used, but even in 1956, when breeding was intense, rarely was a bird forced to use a site so unfavorable that it was not covered from above to make a narrow entrance. Elsewhere, in less favorable regions such as the breeding areas along the Lachlan River, grey teal have been observed to nest in reed beds, on the open ground, and in various other exposed places; but in most of the Riverina, at least in the better areas, it seems there is not likely to be a shortage of good nesting places.

### 13.214   Food in Breeding Places

As the water rises, especially as it pours over the flat plain, vast areas are covered by shallow water, which is a fertile habitat for many kinds of aquatic insects and other invertebrates. In the areas that Frith (1959a, 175) studied, water bugs *Corixa* spp. and water beetles *Berosus* spp. were dominant, but many other species were also abundant. The fertility of the habitat is assured by the wealth of decomposed organic matter that was left behind after the previous flood, and the shallow water is warm. As the flood expands, the animals in the second and higher trophic levels of the ecosystem, which are important in the diet of the teal, concentrate where their food is most abundant—not in the forefront of the flood, but not at the very rear either. In the big flood of 1955/56 they were so abundant that the teal, exceedingly numerous as they were at this time, seemed to make only a negligible impression on their numbers. As the flood subsides the edge of the shallow water retracts, and the aquatic animals are trapped in dense concentrations just behind the receding edge. The teal, especially the ducklings, concentrate their foraging here and never go short of food while the shallow water lasts. As the shallow water disappears— either draining into the channel or seeping or evaporating away—the aquatic animals die or disappear and the teal are confronted by a shortage of food. If the low water levels prevail, the shortage may become absolute over the whole of the erstwhile breeding area. The shortage is extrinsic because the influence of the changing water-level far outweighs the feeding of the teal.

The ducklings find all their food by dabbling around the edges of shallow fresh water. They seem to have no other way of feeding. Like the young of many other kinds of birds, the young ducklings need a high proportion of animal flesh in their diet. For the first week the diet, as indicated by stomach contents, was 100% animal; by the end of the fourth week it had fallen to 30%, the same as for adults that were feeding in the same place (table 13.09).

The adults thrive on a wider diet than the ducklings. When suitable animals are available in shallow water the ducks will eat them; when animals are scarce they will eat fruits, seeds, and other edible parts of plants if they can be taken from shallow water. When the necessity arises, an adult will make do on a purely vegetarian diet. The widening of the diet allows a concomitant widening of the sort of waters where it can be sought. For an adult duck the water need not be either fresh or temporary, because suitable vegetable food can be found in the shallow

**Table 13.09**  Food of Adult Grey Teal in Breeding Places in the Riverina

| Food | Proportion, as Percentage of Total Volume, of Specified Food in the Gizzard | | | |
| --- | --- | --- | --- | --- |
| | Murray | Lachlan | Murrumbidgee | Mean |
| Animals | 30.0 | 45.1 | 23.3 | 32.8 |
| Marsh plants | 35.9 | 29.0 | 59.1 | 41.3 |
| Land plants | 34.1 | 25.9 | 17.6 | 25.9 |
| Number of gizzards | 671 | 278 | 1,088 | 2,037 |

*Source:* After Frith (1959c).

reaches of permanent salt or brackish water. At the other extreme, the most recently formed or the most temporary of ponds will serve as a place where the adult may seek its food: terrestrial plants that have grown and matured edible parts on dry land will serve as food as soon as they are covered by a rising flood. Or such food may be taken from the most transient of pools after rain.

Frith studied the food of adults in the breeding-areas during four years, 1952–55, which included several floods when breeding was abundant. He shot 2,037 teal and recorded the contents of their gizzards. All three localities in table 13.09 are in the Riverina: one is fed from the Murray, one from the Murrumbidgee, and one from the Lachlan. The animals included representatives of the Insecta, Mollusca, Crustacea, and a few other groups. The insects accounted for 90% of all animals. The plants included representatives from Haloragidaceae, Chlorophyceae, Marsilaceae, Polygonaceae, Chenopodiaceae, Cyperaceae, Gramineae, Leguminaceae, Cucurbitaceae, and Compositae. The grey teal, although specializing strongly in the sort of place where it will feed, is apparently ready to eat almost anything edible that it can find in such a place. The relative abundance of the distinct items varies from place to place. In the Lachlan area animals predominated, and in the Murrumbidgee area it was marsh plants, but in the Murray area animals, marsh plants, and land plants were about equally represented. Animals are abundant and accessible in the shallows of rising and receding water. Dryland plants become available after flooding or heavy rain. The abundance of shallow claypans and lignum swamps in the Lachlan area and of large irrigated pastures in the Murray area helps explain the differences in the diets of the ducks in the two areas. Frith (1959c, 139) expanded these data and explained them fully by reference to the movements of the floodwaters.

For the adults, as for the ducklings, every change in water level, every movement in or out of the boundaries of shallow water, makes available a fresh supply of food. When the movements of the water's edge are intermittent with a reasonably short period, the outward movements are especially fruitful for the reasons mentioned above. By these particular circumstances of their food-gathering the adults are buffered from self-induced relative shortages of food even more effectively than the ducklings, because adults can use such a wide variety of food.

## 13.22  Dispersal from Breeding Areas

The grey teal does not, as a rule, fly away from a good breeding place until, with the drying up of the temporary water, food becomes scarce. Even with the massive breeding that was possible in the breeding places of the Riverina (floodplains of the Murray, Murrumbidgee, Lachlan river system) in 1956 there was no sign of an intrinsic shortage of food. The shortage, when it came, was an extrinsic shortage that was caused by the drying up of the shallow water where the teal had found their food.

Frith (1959b) studied the exodus of grey teal from two bodies of water in one important breeding area in the Riverina, Gum Creek and Kooba Lagoon. The data for figure 13.09 were gotten by the method of constant trapping effort; they measure relative densities (sec. 8.2). Gum Creek had been an extensive breeding area, but it dried up completely toward the end of summer (February/March). Kooba Lagoon retained a little water through the drought but not enough to support many teal. Lake Wyangan, a cumbungi swamp, was permanent; it served not only as a breeding area but as a drought refuge as well.

At Gum Creek, at the beginning, the grey teal congregated on the residual floodwater where they had bred. As the water dried up they left it, some to travel far afield but many to move locally. About this time teal began to accumulate on Kooba Lagoon, which also had been a breeding place but now became densely crowded with immigrants. As this water in turn receded, food became scarce and most of the teal flew away (about May/June; fig. 13.08). The great majority left the Riverina altogether, but a few moved locally to permanent waters in the region. About this time some teal began to arrive on Lake Wyangan. Such permanent cumbungi swamps are chiefly used by grey teal as a refuge during drought, but they may support a little breeding when they are filling or have recently filled. The condition of the water at Kooba Lagoon probably was modal for the region, because the great exodus from the whole Riverina coincided with the exodus from Kooba Lagoon.

The birds that were trapped to provide data for figure 13.09 were some of a number that had been banded in order to find out where they went. Of 1,613 that were banded at Gum Creek 189 were recovered, chiefly by hunters shooting for sport. Frith (1962) also studied the dispersal of teal away from a number of other spent breeding areas. For example, at Joanna (near Naracoorte, in the southeastern corner of South Australia) 1,346 birds were banded. The dispersals from Gum Creek and Joanna are illustrated in figure 13.10; the distances that had been covered by the 189 "recaptures" from Gum Creek are analyzed in table 13.10, and the direction of their flight is shown in table 13.11. From these and many other data Frith concluded (after allowing for known bias in the distribution of hunters) that the direction of at least the initial flight away from a spent breeding place was at random to the points of the compass. It is also clear that at least some of the teal covered great distances, over one 1,000 km.

In 1957 most of the continent was gripped by drought, and many of the teal that flew away from spent breeding places would have died of starvation. The lucky ones found water in which they could feed. Very few inland features still held water; the best bet lay in coastal features—lakes, lagoons, and creeks that were fed by the sea. That is why so many of the teal that left the Riverina flying in any direction between northeast and southwest had so far to go (table 13.10; fig. 13.10). Such places offered food for adult survival but not for breeding. Frith called them "drought-refuges".

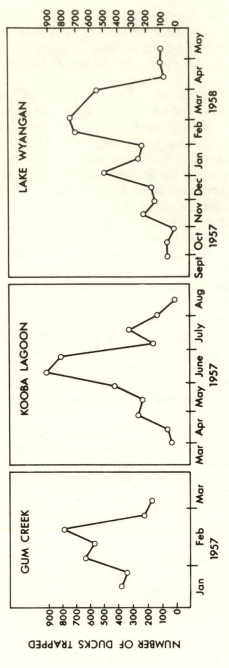

**Fig. 13.09** Changes in numbers of grey teal in three sites near Griffith, New South Wales during 1957, as shown by numbers trapped. (After Frith 1959b)

**Fig. 13.10** Dispersal of grey teal from the breeding places at Joanna, South Australia (*large triangle*) and Gum Creek, New South Wales (*large circle*). The small triangles show dispersal from Joanna, and the small circles show dispersal from Gum Creek. (After Frith 1959b)

**Table 13.10** Distances Traveled by Grey Teal That Emigrated from a Spent Breeding Place in 1957

| Distance from Banding Place (km) | Recoveries (% of 189) |
|---|---|
| 0–80 | 34.9 |
| 81–480 | 53.4 |
| 481–1,600 | 8.4 |
| More than 1,600 | 3.2 |

*Source:* After Frith (1962)

*Note:* Total banded, 1,613; total recovered, 189. Dispersal is measured by the percentage of recoveries at specified distance from where they started. Recoveries were made by hunters shooting for sport, so the distances do not record the maximum the birds might have traveled if left unmolested.

**Table 13.11**   Direction of Dispersal of Grey Teal Banded at Gum Creek (Riverina) and Recaptured More Than Eighty Kilometers away within Ninety Days

| Numbered Recaptured | Direction of Movement | | | |
|---|---|---|---|---|
| | North (%) | South (%) | East (%) | West (%) |
| 123 | 18.7 | 33.3 | 25.2 | 22.8 |

*Source:* After Frith (1962).

### 13.23   Drought-Refuges

It was quite a spectacle when the grey teal left their inland breeding places in 1957 (plate 6). Their numbers were large. Their departures were synchronized by the final widespread disappearance of shallow water during autumn (April/May) after a hot, dry summer (fig. 13.09). At about the same time or a little later, a number of observers in various parts of Australia reported grey teal arriving in large numbers on permanent waters remote from their breeding places (figs. 13.10, 13.11). A few of these drought-refuges were freshwater lakes, especially in the coastal highlands in the southeastern corner of the continent, but most of them were brackish or saltwater inlets from the sea. Reports of local congregations of grey teal

**Plate 6**   The grey teal *Anas gibberifrons* in flight. (Photo by Ederic Slater)

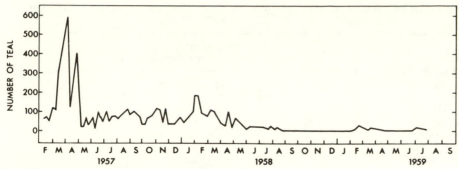

**Fig. 13.11** Numbers of grey teal counted on Lake Colac, an important drought refuge in Victoria. (After Frith 1962)

on such coastal refuges came in from places as far apart as Colac, Victoria; Sydney, New South Wales; Townsville, Queensland; Mandurah, Western Australia; and Humpty Doo, Northern Territory. These places span the whole continent from the east coast to the west coast and from the south coast to the north coast. Coastal inlets from the sea are the most typical and the most important drought-refuges for the grey teal. They are more frequent and more important in the southeastern corner of the continent. Nevertheless, permanent residues of floodwater in the breeding areas, such as Lake Wyangan (fig. 13.09), also serve as drought-refuges.

Information about what happens to grey teal on a coastal drought-refuge may be had from naturalists who shoot them for sport and therefore know them well. There is general agreement on two points. (a) Although a flock of grey teal on a particular body of water may persist large and dense for months, the individuals that make up the flock usually remain for a much shorter time. If the flock persists it is only because the rate at which birds depart is matched by the rate at which new ones arrive. (Compare fig. 13.09, Lake Wyangan, with fig. 13.11). (b) Even though a flock may persist for months, very few if any of the birds breed.

From what is known of the behavior of the grey teal in the breeding places, their failure to breed in the coastal drought-refuges is probably associated with the stability of the water in these places and the consequent lack of suitable food for breeding.

It is not known what stimulus causes a grey teal to fly away from a drought-refuge where it has been resting and feeding for a while. If the water dries up or becomes unproductive through ecological succession, shortage of food may hasten the evacuation of a particular refuge, but clearly shortage of water and therefore of food is not a necessary stimulus for emigration, nor does it seem to be the normal one except during severe drought. Considering the selective advantage attached to finding a place in which to breed, one might suggest that the teal tarry only long enough to restore their condition—which they may do more quickly with an abundance of food. Perhaps when they have rested enough and eaten enough they simply take off to resume the search for a breeding place.

## 13.24 Dispersal from Drought-Refuges

Frith banded a total of 7,510 grey teal in three drought-refuges—two coastal and one inland. They were Humpty Doo, near Darwin on the north coast, Sydney on

**Table 13.12** Distances Traveled by Grey Teal That Emigrated from Drought Refuges in 1957

| Distance from Banding Place (km) | Recoveries as Percentage of Total Recovery for Each Place | | |
| --- | --- | --- | --- |
| | Sydney | Lake Wyangan | Humpty Doo |
| 0–80 | 41 | 71 | 72 |
| 81–320 | 36 | 17 | 1 |
| 321–640 | 18 | 12 | 1 |
| 641–1,600 | 5 | — | 6 |
| > 1,600 | — | — | 19 |

*Source:* After Frith (1962).

*Note:* The totals banded in Sydney, Lake Wyangan, and Humpty Doo were 945, 1,216, and 5,349, and the totals recovered were 22, 94, and 168. Dispersal was measured as percentage recovered at specified distances from where they started. Recoveries were made by hunters shooting for sport.

the east coast, and Lake Wyangan in the Riverina. The distances covered by the teal that were recaptured are analyzed in table 13.12. From other evidence, Frith concluded that the initial dispersal from drought-refuges, as from the spent breeding places, was random with respect to direction.

On the other hand, there is evidence that grey teal can detect distant rain (either before or after it has fallen) and will fly toward it. Frith (1967, 192) gave the following examples:

The main movements of grey teal are drought-induced dispersions, but some movements strongly suggest that, in addition to being "pushed" by dry or otherwise unsuitable conditions, sometimes they can detect distant rain and move to it. Thus in 1952 grey teal were numerous in the Riverina district of New South Wales, and the rice growers were complaining bitterly. Heavy rain fell in central Queensland, and within two days the birds were rare in the Riverina but abundant in central Queensland. Other examples are the swamps on the lower Richmond River in north-east New South Wales. These are usually dry but whenever they fill flocks of grey teal arrive, almost overnight, some of them bearing bands that had been placed on them in inland New South Wales only a few days before. There are so many observations of grey teal quickly reaching newly flooded areas, in large numbers, that there seems little doubt that they do detect the distant rain, but in what manner is not known.

## 13.25 Conclusion

From 1954/55 through 1957/58 Frith (1963) banded a total of 13,331 grey teal—some in breeding places and some in drought-refuges. Of this total, 11,295 were banded in 1957/58 toward the end of the outbreak that was set off by the great floods of 1955/56. So the fate of these birds dominates the conclusions that might be drawn from the five years' banding and nine years' recapturing. The total recaptured was 1,405—mostly by hunters shooting for sport. The numbers that were recaptured during the first, second, . . . sixth year after they were banded were 845, 255, 161, 88, 39, and 12; 4 were recaptured from the seventh year through the ninth year. Converting the numbers that were recaptured during each of the second and subsequent years into a percentage of the number that were recaptured during the first year gives the series 30, 19, 10, 6, 1, 0.5%. On the reasonably safe assumption that the nonreturnable samples were not large enough to seriously bias the conclusions, and on the more doubtful assumption that the hunters maintained a constant "shooting-effort" from year to year, these percentage figures might be taken as estimates of survival rates. They suggest that 19% of the marked ducks were still alive three years after they had been marked, 10% after four years, and 6% after five years. These figures taken in conjunction with the probability of floods in the breeding areas suggest that the teal is well adapted to its nomadic way of life. The price that is paid for dispersiveness is the risk of death that is implicit in the above percentages. That is the price that is paid for the chance of finding a breeding place. There seems to be a wide margin for safety (table 13.08).

The margin is wide because in the grey teal the two niches "feeding in shallow temporary water" and "dispersing in search of shallow temporary water" are well developed. It seems reasonable to suppose that any change in the environment would provoke consonant evolutionary changes in these niches, in the direction of maintaining the margin of safety. Following this train of thought a little further, it seems reasonable to suggest that purely stochastic elements in the ecology of the grey teal are sufficient to explain not only its continuing existence as a species, but also its current distribution and abundance. There seems to be nothing in Frith's account of the ecology of the grey teal that suggests the need to invoke any aspect of competition theory to explain either of these events. This particular ecology is eminently consistent with the theory of spreading the risk.

The striking contrast in the ecologies of *G. tibicen* and *A. gibberifrons* is highlighted by comparing the pathways that come through resources. With respect to food: the magpies, by dint of their social behavior obviously belong with figure 9.08; the teal, by virtue of their opportunism, obviously belong with figure 9.05. Both species use a token to bring them into breeding conditions, but there are striking differences in their profession of this niche, as with the rest of their ecologies. Despite the extreme contrast between the two ecologies, both fit nicely into the theory of environment. The idea of "competition" is not needed to explain the distribution or the abundance of either species. The reader who wonders whether "competition" might help with the magpie is referred to the discussion of a hypothetical example in Andrewartha and Birch (1954, 23).

## 13.3   The Distribution and Abundance of Fish in the Sea

The study of the distribution and abundance of fish in the sea has primarily been oriented toward the fish as prey for man the predator, with scant attention to other components of the environment of the fish. The central question has been, What influence does human predation have on the abundance of fish? Masses of statistics have been gathered on the relation of size of catch to fishing effort—in other words, to the functional response of the fish to the predatory activities of fishermen. Some of these statistics have accumulated over almost a century, such as those for the Atlantic cod, *Gadus morhua* (Garrod 1977). From these statistics researchers infer how predation influences the numbers of prey. So it has become traditional that the mathematical models that have guided the management of fisheries are dominantly models of a predator and a single prey, with a move in recent years to models of predators that have more than one prey (May et al. 1979). Components of environment of fish, other than human predators, have rarely been taken into account. It is not surprising, therefore, that dependence upon these models has been an uncertain guide to management. Indeed, in a number of cases the abundance of fish has declined so drastically that fisheries have virtually ceased (Glantz and Thompson 1981). Figure 13.12 shows the fall in size of the populations of the four great pelagic fisheries of the world. The demise of these populations no doubt had a number of causes, but overfishing was one of them. Overfishing may be a consequence of poor management based on poor models, or it could be due to fishermen's failure to comply with proposals for good management. Whatever is the main cause of these declines, it is now being recognized by at least some students of fish populations that the time is overdue for a more realistic ecological approach to the management of fisheries and that the use of models that are restricted to predation alone has in many cases led to failure in practice (Robinson 1980, 20; Gulland 1976, 302; Cushing 1975, 216).

In temperate waters of the oceans fish tend to spawn at fixed seasons, in the same place each spawning season. However, the production of food for larvae does not have any fixed timing. Larvae drift away from spawning grounds and are carried by ocean currents to nursery grounds, which are often inshore. There the larvae metamorphose. With independent powers of movement, the mature fish swim away from the nursery grounds, usually to deeper waters (Cushing 1975, 220). Any individual fish that becomes a mature adult has an environment that changes from that in its spawning ground, through that in its pelagic life, to that of the free-swimming adult. Little is known of the details of the environments in these different habitats. Recruitment of adult Norwegian cod to the swimming cod population is correlated with surface salinity in the German Bight and with the temperature of a particular meridian called the Kila meridian in the Barents Sea. Both these parameters are functions of the degree of southerliness of the windstream across Europe during the spawning and drifting phases of the fish. Although the web of connections between southerly winds and the fish are not well known, there is evidence that the critical condition for high recruitment of adult fish is a match between time of production of larvae and of their food. Larvae that find themselves drifting in a current with a sparse supply of plankton may have a relative shortage of food (Cushing 1975, 223). Production of plankton depends upon the concentration of plant nutrients and the temperature of the surface water. Favorable levels of both seem dependent upon southerly winds.

But fish do not live alone with their food supply in the sea. They may be preyed upon by other fish as well as by man, and they may have to share their resources with other species. In the North Sea, larvae of the plaice *Pleuronectes platessa* are hatched in waters between the Thames River in England and the Rhine in the Netherlands. They drift toward Texel Island off the north coast of the Netherlands. At the same time, larvae of cod, sandeels, dabs, whiting, and herring drift in the same course. All the fish larvae feed on copepod nauplii, copepodites, and *Oikopleura*. Five species feed on three species of animals of the plankton. Little seems to be known about whether one species of fish influences another's chance to get adequate food (Cushing 1975, 239).

The multipartite nature of the population of fish in the sea is now being recognized for at least some species. To the extent that this idea is appreciated, the management of the fishery will change. For example, the Atlantic herring in the

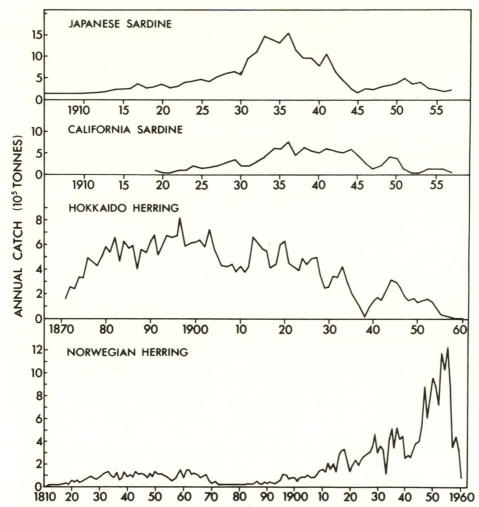

**Fig. 13.12** The fall in abundance of four pelagic fish stocks as indicated by the annual fishing catch over many years: the Japanese sardine *Sardinops melanostricta*, the California sardine *Sardinops caerula*, the Hokkaido herring *Clupea pallasi*, and the Norwegian herring *Clupea harengus*. (After Cushing 1975)

northwestern Atlantic Ocean consists of at least several dozen stocks (Iles and Sinclair 1982). Each stock is recognized by its own characteristic season for spawning and duration of spawning. Recognizably discrete stocks vary in size from hundreds to millions of tonnes. Small stocks are associated with small hydrographic features (e.g., the Gulf of Saint Lawrence), and large stocks are associated with large hydrographic features (e.g., Georges Bank). It seems that hydrographic features are important in defining both spawning areas and "retention areas" for the larvae and therefore to a large extent determine the size of the stock. So the habitat becomes very important in determining the size of the population. Recognizing the relatively discrete populations becomes important in the management of the fishery. For example, even a low fishing mortality for the stock as a whole could lead to the extinction of particular stocks. We see the notion of the multipartite population in a multipartite environment as becoming increasingly important in understanding the ecology of fish in the sea.

In some fisheries the recruitment of adult fish to the population depends upon upwelling of nutrient-rich waters at the time of spawning. However, in upwelling areas the spawning places are not fixed, and the time of spawning is highly variable (Cushing 1975, 235). The environment of the Peruvian anchovy *Engraulis ringens*, in all stages of its life cycle, depends on two ocean currents. The Peruvian Current, also known as the Humboldt Current, brings cold water from the south. It travels northward along the coast of Peru and deflects to the west just south of the equator. As the Peruvian Current is pushed northward by the prevailing southerly winds, it is continually deflected to the east by Coriolis force. As the deflected surface water moves offshore it is replaced by water welling up from below. The upwelling forms huge eddies and spirals and flows in the general direction of the equator. The extent of the upwelling varies seasonally with its location along the coast. Such waters have a high concentration of plant nutrients; on reaching the surface they support an abundant crop of phytoplankton. Some of the highest productivity in the sea has been recorded in the Peruvian Current. According to estimates made by Walsh (1981), the annual production of phytoplankton is a minimum of $1,000$ g C m$^{-2}$ yr$^{-1}$ in contrast to $200$–$500$ g C m$^{-2}$ yr$^{-1}$ for continental shelves in higher latitudes. Larvae of the anchovy feed on smaller copepods that have fed on the phytoplankton. Adult anchovies feed directly on the phytoplankton. The difference in timing and placing of upwellings associated with the Peruvian Current results in different environments for the anchovy (Anonymous 1975; Paulik 1971). Periodically a tongue of warm water of low nutrient concentration extends to the south over the Peruvian Current, with dire consequences both to the Peruvian fishing industry and to the population of guano birds which feed almost exclusively on anchovies. The phenomenon is known as El Niño (because it occurs around Christmastime). A severe El Niño seems to happen about every seven years, and less severe ones occur more frequently. Their influence on the numbers of anchovies taken by men or birds depends upon their duration and extent. The web of events that connect an El Niño to the anchovy is in dispute. Certainly fewer anchovies are caught during an El Niño. Figure 13.13 shows a small drop in the catch after the 1965–66 El Niño, a large drop at the time of the 1972 El Niño and a drop after the one in 1976. The 1957–58 El Niño is not reflected in the anchovy catch, perhaps because intensive fishing had hardly gotten under way until four years later. It has been widely assumed that the reduced catch is due to shortage of food for the fish in the warm waters. Walsh (1981) has made estimates of the production of phytoplankton for selected months in the years 1966–78 in the waters of the Peruvian shelf. They range from $1.3$ g C m$^{-2}$ day$^{-1}$ in June 1969 to $7.3$ g C m$^{-2}$ day$^{-1}$ in November 1977. In the vicinity of the last two El Niño years (1972 and 1976) his figures for carbon production are at the low end of the range. An alternative hypothesis for low catches of anchovies in El Niño years is proposed by Paulik (1971). He contends that the warm surface water of the El Niño forces the anchovy deeper, where it avoids the severe predation in the top 10 m from guano birds. Weak and immature birds are decimated by a severe El Niño because they are unable to dive deep enough to capture anchovies in their new position. Figure 13.13

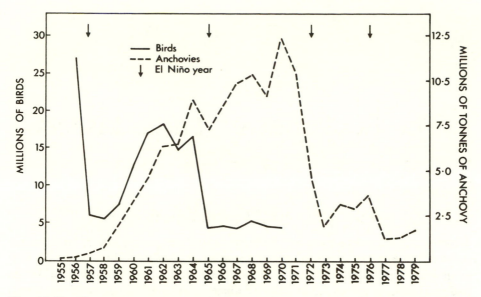

**Fig. 13.13** The annual catch of the Peruvian anchovy *Engraulis ringens* and the size of the population of guano birds off the coast of Peru. (After Paulik 1971; Idyll 1973; J. A. Gulland, pers. comm., for 1973–79)

shows that in 1956 the population of guano birds was estimated to be about twenty-five million. Between 1957 and 1958 it dropped to about six million. Similarly, in 1964 the population of guano birds was about seventeen million. Between 1965 and 1966 it fell to about four million. Each of these falls was associated with an El Niño.

The huge catches of anchovy from 1965 to 1971 are considered to reflect years of abnormal environmental conditions favoring the anchovy (FAO 1981, 15). Whatever the relation between abnormal environmental conditions, the El Niño, and the abundance of anchovy, it is certain that there is a connection, whether it be through pathways in the envirogram that lead through food or through changed predation by birds and man, or perhaps through these and other pathways as well. From the point of view of understanding how environment influences the anchovy's chance to survive and reproduce, it is of fundamental importance to know the links in the web that begin with changes in weather and ocean currents and eventually impinge on the anchovy.

We know of no single species of fish in the oceans for which information is available to construct an envirogram with any degree of completeness. This is in large part due to failure by students of fisheries to realize that the environment of a fish is more than its human predator. The examples we have given indicate that the distribution and abundance of fish depend on many components of environment, of which man is only one. It seems that a similar prejudice has restricted the study of freshwater fish, though perhaps to a lesser extent than for fish in the sea (see, e.g., Gerking 1978).

# 14

# The Ecologies of Three Insects

## 14.1   The Spruce Budworm (*Choristoneura fumiferana*)

Outbreaks of the spruce budworm, *Choristoneura fumiferana* (plate 7), in Canadian softwood forests may be severe and widespread. According to Prebble (1975, 81), some individual outbreaks in Ontario and Quebec have extended over $20 \times 10^6$ ha and have destroyed from 1.5 to $10 \times 10^9$ tonnes of timber, which is to be compared with $5 \times 10^9$ tonnes, the annual production of timber from all Canadian forests. The caterpillars defoliate the trees; after being severely defoliated for five successive summers, a tree is likely to die. When the 1949–57 outbreak had been going for about three years Morris (1963, 223) estimated, for one severely infested plot, that there were $4.5 \times 10^6$ mature or nearly mature larvae per hectare; and about 95% of the first- and second-year foliage had been destroyed.

Because an outbreak kills or impairs the growth of so many trees over such widespread areas, it has been possible, by studying tree-rings, to identify and date with considerable confidence outbreaks that occurred up to two hundred years ago. Outbreaks began in New Brunswick in 1770, 1806, 1878, 1912, and 1949 (Greenbank, in Morris 1963, 19). The 1949 outbreak lasted about eight years. Allowing about this time for the earlier ones, these figures suggest about thirty-two years for the interval between outbreaks.

According to Morris (1963, 7), *Choristoneura* is unusual among the insect pests of forests for its remarkable scarcity between outbreaks; he said that it may be rated as a rare species during what he called the "endemic phase" of its population cycle. It may also be remarkable for the speed with which it can change from very low to very high numbers. Between 1939 and 1944 in New Brunswick, routine samples were taken by beating low-growing fir and spruce with a 3-meter pole. Larvae of *Choristoneura* were found at the rate of ten per thousand samples. In 1945 a detailed investigation was set up, and routine sampling from a number of plots in New Brunswick failed to reveal any sign of an increasing population until 1947. By 1949 the outbreak was obvious, and by 1951 Morris (1963, 223) estimated more than $4 \times 10^6$ large larvae per hectare and 50% destruction of young foliage on some plots. Prebble (1975, 80) quoted estimates of increases from four thousand fold to ten thousand fold in six generations. Both the plots that Morris sampled and the

**Plate 7**   The spruce budworm *Choristoneura fumiferana: top left,* fifth-instar larva feeding on spruce; *top right,* severely forked and defoliated crown of spurce tree following an infestation; *bottom,* adult moth. (Photos by G. A. Van Sickle)

populations that Prebble cited were in New Brunswick, which is downwind, with respect to prevailing wind, from a number of local outbreaks that had already erupted west of New Brunswick. It is known that *Choristoneura*, as first- and second-instar larvae and as adults, may disperse on the wind from dense outbreaks in numbers comparable to those in a swarm of locusts. Because it is rarely possible to distinguish between immigrants and residents when sampling, conclusions about rates of increase must be qualified by this ignorance.

In New Brunswick the outbreak was well under way by 1951 (fig. 14.01); it probably peaked about 1953 or 1954, by which time virtually all of the northern part of the province was heavily infested. A decline in numbers was first noticed in 1956, and by 1960, throughout the whole of this area, *Choristoneura* was once again scarce. (Elsewhere, in central and southern New Brunswick, patches of forest still supported dense populations). There can be no doubt that seven or eight years' severe and repeated defoliation of young growth must have caused a marked change in the food available to *Choristoneura*; but the weather had also changed—in the opposite direction from what is thought to be necessary to get an outbreak going (sec. 14.1211, especially fig. 14.03).

The 1949 outbreak in New Brunswick provided striking confirmation of the theory that outbreaks are likely to originate in large continuous areas of mature or overmature balsam fir. Mature balsam fir produce staminate flowers in abundance, and the larvae of *Choristoneura* find good food plentiful in such trees (though not necessarily in the staminate buds themselves) and in the spruce that is associated with them in the same stand. The larvae and pupae of *Choristoneura* also find a microclimate that will suit them in the tall open crowns of mature and overmature trees (40 years old and older). In large continuous stands of such trees which, cover the ground with an unbroken supply of good food and good places to live, a dispersing *Choristoneura* has a good chance of making a landfall in a place where it is likely to survive. This theory was put forward by Swaine and Craighead (1924). It was used by Balch (1946) to predict the area in New Brunswick where outbreaks were most likely to originate. Figure 14.01 shows that the prediction was remarkably accurate. In 1951, when the outbreak was well under way, the distribution of the outbreak was still inside the boundaries of the "susceptible" stands that had been mapped by Balch five years earlier, with the trivial exception of two small areas just outside the eastern boundary of the predicted area. Balch's prediction was confirmed three or four years after it had been made, but these stands had been mature for much longer than this. Something more than the maturity of the forest seems to be needed to spark off an outbreak. According to Wellington, that something else is weather (Wellington et al. 1950; Wellington 1952).

The results of the detailed investigation that began in 1945 under the leadership of R. F. Morris and continued after 1956 under the leadership of C. A. Miller were summarized by Morris (1963). The findings of this investigation give us a remarkably detailed account of the ecology of *Choristoneura* during an outbreak; the routine sampling that was worked out at the beginning remained effective while the high numbers persisted. Our account is based largely on Morris' summary. The subsequent development of chemical and other artificial control measures may have introduced important new features into the ecology of *Choristoneura* between

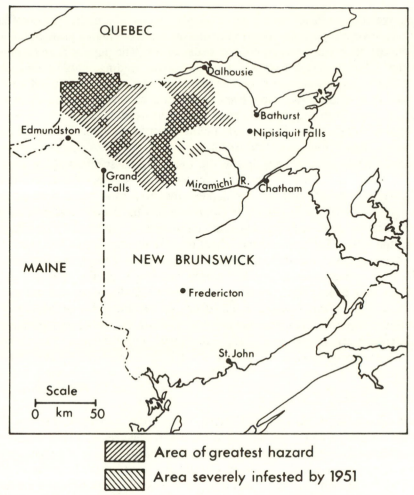

Area of greatest hazard

Area severely infested by 1951

**Fig. 14.01**  Map of New Brunswick showing (a) area of "greatest hazard" as mapped by Balch (1946), based on the preponderence of mature trees; (b) area that had been severely infested by 1951. Note that, with the exception of two small pockets, all of area b, comes inside area a, strongly confirming Balch's prediction. (After Greenbank in Morris 1963)

major outbreaks, so we confine this account to the results of the intensive study that was made of the 1949–57 outbreak in New Brunswick.

## 14.11  Life Cycle, Behavior, and Physiology

There is one generation a year. The eggs are laid late in July or in August. One female might lay 170–200 eggs, which she places among the foliage near the crown. The eggs hatch in about ten to twelve days. The first-instar larva, without feeding, spins a silken hibernaculum in some suitable crevice among the foliage and goes into diapause, a resistant condition in which it spends the winter. In the spring the larva molts into the second instar, which begins to feed by mining, for preference into the bud of a staminate flower of a balsam fir, but alternatively into a vegetative bud or a needle in the new season's vegetative growth. The older instars feed on the vegetative buds and needles that are not more than 2 years old, but they will eat the older leaves when nothing better offers. The pupa is found among the branches or on the trunk near where the larva was feeding. Large proportions of larvae in the first and second instars and of adults that have laid some of their eggs (at least among those that have been reared in a crowd, as during an outbreak) disperse away from the tree on which they were reared. All dispersal by small larvae and long-distance dispersal by moths depends on wind, but moths may of course fly short distances in fine weather.

Most animals have a stage in the life cycle that is specialized for dispersal; *Choristoneura* is unusual in having three: the first-instar larva, the second-instar larva, and the adult. Among Lepidoptera the first-instar larva or the adult is commonly adapted for dispersal. In *Choristoneura* the second-instar larva, being lighter than the first, is just as buoyant and dispersive, attached to its silken sail, as the first instar. This result comes about because the first instar does not eat, and there is a diapause interpolated between eclosion from the egg and the first larval molt. Diapause at this stage costs energy because it entails spinning a hibernaculum; such energy need not have been spent if the diapause were in the egg or the pupa. That *Choristoneura* could afford to pay the price of adding a third dispersive stage to a life cycle that already contains two indicates the importance of dispersal to *Choristoneura*. That the first instar disperses in the autumn and the second instar in the spring when the winds are different may be part of the profit from this deal.

### 14.111   Behavior in Relation to Dispersal

The first-instar larva, seeking to disperse, hangs from the tree on a silken thread. When the wind blows the thread snaps near the tree and serves as a sail to keep the larva afloat and traveling with the wind. When the larva hatches from the egg it will respond positively to diffuse light; on a bright day it is likely to be attracted toward the tip of the branch, whence it will drop on its silken thread. This is well-adapted behavior because good dispersal winds are more likely in this weather. On a humid, dull day the response to light may be largely replaced by a more complicated response to humidity (measured as evaporation) and perhaps temperature. The longer the larva stays away from the tip of the branch the more likely it is to meet a good place, at the base of the needles, to spend the winter and be stimulated to spin a hibernaculum. It will get another chance at dispersing as a second-instar larva in the spring.

The second-instar larva, on emerging from its hibernaculum in spring behaves in similar fashion. It may disperse, or it may stay and begin to feed. The issue is influenced by the weather, as before. Good dispersing weather is more likely in the spring than in the autumn.

During an outbreak, in the autumn while the first instars are active, and again in the spring after the second instars have emerged, any locality that supports a local population of *Choristoneura* is likely to lose small larvae by emigration and to gain some by immigration. In one large series of plots studied by Miller in Morris (1963, 12) the net loss was about 60% but this will depend on the relative densities of the population on the sample plot compared with those other localities that are contributing immigrants, because the probability that an emigrant will depart or an immigrant will settle is independent of density. Morris and Mott in Morris (1963, 180), choosing an occasion in the spring after a steady wind that would have been good for dispersal, counted the number of second-instar larvae that were established in stands known to be free from eggs in the autumn. Samples were counted in stands that were 1.6, 6.4, 16, and 35 km from the nearest outbreak. The numbers per square meter of foliage were respectively 46, 33, 9, and 9. Some of those counted may have arrived in the autumn as first instars, but many second instars had also been seen floating in the wind. Morris and Mott in Morris (1963) compared the number of hatched eggs with the number of established second-instar larvae in stands that supported populations of diverse densities. They inferred that, independent of density, about 96% of the original population of small larvae disappeared from the stand, chiefly through dispersal; and each stand, independent of the density of its population, received immigrants at the rate of about 37 per square meter. They commented that this experiment was done in weather that was extremely favorable to dispersal, so the results should not be regarded as modal.

The dispersal of adults has been studied using traps baited with pheromone or with light; the number of empty pupal cases has been compared with the number of eggs that were laid on the same stand; and dispersal has been inferred from diverse circumstantial observations. For example, large swarms of moths have been deposited, with the passing of a cold front, in towns that were 80 km or more from the nearest forest; and Greenbank in Morris (1963, 19) found it quite plausible to postulate that the 1949 outbreak in New Brunswick, might have been started by invasion: swarms of adults might have been blown in from outbreaks that were known to be in progress, but none closer than 40 km to the west.

A normal moth, well fed in the larval stage, might lay two hundred eggs. None of those that were caught dispersing carried a full quota of eggs; caged moths laid an average of thirty-eight eggs. It seems that most moths lay some of their eggs in the place where they were reared before dispersing, but the urge to disperse seems strong. Moths seem to prefer to lay their eggs among the needles in the crown. Defoliation may reduce the number of attractive sites for egg laying. Severe defoliation, by speeding the departure of locals and curtailing the stay of immigrants, may reduce the number of recruits in a way that is related to the density of the larval populations that had recently been there. Apart from this reaction, dispersal of the adults—like dispersal of the larvae, being largely involuntary—seems to be largely independent of density.

On the other hand, weather has a strong influence on the behavior of the moths: for most of the day they scarcely fly at all; virtually all flight activity is restricted to the brief period from just before sunset to a short time after sunset when the intensity of the light is changing rapidly, and even at this time they fly very little if the weather is humid and heavily overcast. They are strongly stimulated to fly in dry, clear weather provided the clouds are not too heavy or continuous. Some short flights, serving local dispersal, may be made within or below the canopy. But a moth that has laid some eggs locally and is ready to move farther afield will display a special flight behavior which clearly indicates that it is a volunteer to be picked up by a wind that might carry it much greater distances. Males also take part in these flights and often dominate them. The moths soar and spiral vertically up from the tops of the crown, then dart down again. After resting awhile they may repeat the flight. This behavior increases the probability that the moth will be caught up and carried by the wind for a short or long distance, depending on the nature of the wind.

Neither larva nor adult can make much voluntary contribution to the dispersal once it has been caught up in the wind. But both larva and adult volunteer most positively for dispersal by the wind.

### 14.112    Behavior in Relation to Feeding.

From observations in the field it was known that the larvae of *Choristoneura* eat less during wet, cloudy weather (Wellington et al. 1950; Greenbank, in Morris 1963, 174). And laboratory experiments suggest that the first cause for the reduced appetite lies in the behavior of the larvae, which move away from a place that is excessively moist (Wellington 1949a,b).

### 14.113    Diet and Nutrition

It is well known that outbreaks of *Choristoneura* are likely to develop in stands where the trees are mature, especially if they contain mature balsam firs in full flower. But during an outbreak in stands that had established a strong rhythm of biennial flowering, *Choristoneura* maintained high numbers through flowering and nonflowering years alike. It seems that staminate flowers may be a sufficient but not a necessary cause for a high rate of increase in *Choristoneura*. This conclusion was confirmed in the laboratory. Given a choice between flower buds and young needles, the young larvae strongly preferred flower buds and did well on this diet. In the absence of flower buds they readily mined into first- or second-year needles (that had come from a tree in a mature stand during an outbreak); although they grew more slowly on this diet, they survived well.

The physiological explanation for why outbreaks develop in mature stands after a spell of anticyclonic weather implies that such dry weather, while reducing the growth in the trees, causes both flowers and leaves to become highly nutritious for larval *Choristoneura*. In support of such an hypothesis Wellington et al. (1950) called attention to the series of narrow growth rings that often appear in trees for several years before an outbreak. This explanation for outbreaks of *Choristoneura* is a particular case of the general principle that we discuss at more length in section 10.311.

The summary in section 14.12 may be regarded as an enlargement of the envirogram in the upper part of figure 2.02.

## 14.12    The Environment of a Sedentary Resident

No matter what kind of an animal the primary animal may be, it is likely to have, in its life cycle, distinct stages which profess different niches and so require separate analyses. With the rabbit the essential difference was between the stage for survival and the stage for breeding (secs. 12.31, 12.32). With the budworm, not only is the adult quite different from the larva, but there is also a marked difference between the niches of the larva that remains a sedentary resident in the place where it was born or to which it has come and the larva that casts itself into the wind on a silken sail in search of a new place to live. It is convenient to consider the sedentary resident as a member of a local population and to look at its environment in terms of the habitat that the locality has to offer. On the other hand, the disperser, whether it be a larva of the first or second instar or an adult, is better regarded as a member of the natural population. So it is better to look at its environment largely in terms of the distribution and abundance of good places to live. The emphasis will be almost entirely on places that offer a rich supply of food, qualified perhaps by the presence of trees with open crowns that allow the penetration of sun and wind. The two things go together. Predators seem to be unimportant, perhaps because *Choristoneura* consistently and effectively outdistances them, but there is little empirical evidence to support such an inference.

It seems that more than half the young larvae are blown away from the place where they were born (one estimate went as high as 96%), but the others settle down and may be joined by immigrants from other localities (sec. 14.111). This section is about the environments that are experienced by the members of such a local population.

## 14.121   Resources

It is generally agreed that mature trees in a stand that is not too young (say 40 years or older) are necessary before a locality can give rise to or sustain an outbreak of *Choristoneura*. Such stands are recognized by the open crowns of the dominant trees and the abundance of staminate flowers, especially on the balsam fir (in the "on" years). According to Greenbank in Morris (1963, 202) the dominants in a stand may begin to produce a few flower clusters at about 20 years. He estimated $2 \times 10^5$ flower clusters on trees that were 30 years old and $3.5 \times 10^6$ on trees that were 75 years old. One explanation, which we follow in the next section, is that stands of such mature trees (if the weather is right) offer *Choristoneura* a good supply of nutritious food, whereas young, immature stands do not. See the pathway that converges on "food" in the upper envirogram of figure 2.02. An alternative explanation, which we discuss in section 14.1212, suggests that it is the open crown and the weather, not the food, that are of critical importance. See the pathway that leads to "heat" (as a resource) in the upper envirogram of figure 2.02.

### 14.1211   Food

In either case, if the limits of the locality for *Choristoneura* are to be set by the presence of trees that are mostly mature and all of the same age, then localities are likely to be large because in the Canadian softwood forests: (a) a new tree does not establish itself until an old one dies to make room for it; (b) the agents of death— severe storms, wildfire, clear-cut logging, and, most important of all, outbreaks of *Choristoneura*—usually kill all the trees in an area where they strike. The trees in the new stand that replaces the old will be of uniform age. The locality for *Choristoneura* will be large and, so far as the age of the trees goes, uniformly favorable. We return to this point in section 14.31.

But Wellington et al. (1950) and Wellington (1952) suggested that something more than large areas of susceptible trees was needed to spark off and sustain an outbreak of *Choristoneura*; the right sort of weather was also necessary.

In this part of Canada the weather seems to be dominated by the three kinds of air masses—polar continental, polar maritime, and tropical maritime. An air mass assumes a characteristic temperature and humidity from the region where it tarries before it moves in over the Canadian forest. The polar air masses give rise to anticyclones which bring clear skies and dry air: in winter cold, clear weather, in summer clear, dry weather, little rain, warm days, and plenty of radiant heat from the sun. The tropical air mass brings more cyclones, with warm, humid winters and cloudy, wet summers with plenty of rain. Once either polar or tropical air masses achieve control of the weather, they are likely to retain it for a number of years. By counting the cyclones recorded in the official meteorological records, it is possible to tell for any year, as far back as the records go, what kind of air mass was dominating any particular region of forest. Wellington counted the number of cycles in each year for fifteen years preceding and five years following each of eight outbreaks of *Choristoneura* that had been recorded in one part or another of the Canadian forests, mostly in Ontario and New Brunswick, between 1909 and 1940. The arrows in figure 14.02 indicate the first year of an outbreak. On at least six of the eight occasions it seems that the outbreak was preceded by a long period, at least several years, of anticyclonic weather—that is, weather that was dominated by dry polar air masses. Figure 14.03 suggests that the same was true for the outbreak that began in New Brunswick in 1949.

To explain this correlation in terms of food for *Choristoneura* is to suggest that exposure for several years to bright dry weather such as prevails when the system is dominated by polar air masses (i.e., anticylonic weather) induces physiological changes in mature balsam fir and spruce such that the trees come to offer an abundance of good food to the young larvae of *Choristoneura*. Mature trees, especially in such weather, produce many staminate flowers. In New Brunswick (and beyond) most trees flower every second year, and most stands are in step. Such concerted rhythm suggests that the timing was set by some extreme in the weather. Less extreme variations in weather may cause fluctuations in the productivity from one crop to another. For example, Greenbank in Morris (1963, 202) found that a sample of seventy-five-year-old trees carried 232 clusters of staminate flowers per square meter in 1950 compared with 92 in 1952; the comparable figures for trees that were 35 years old were 50 and 26. The crop was about twice as heavy in 1950 as in 1952. Figure 14.03 shows that the five-year running mean for rainfall had been below average for seven years before 1950 but was beginning to climb above average by 1952.

The outbreak continued through the biennial bearing of staminate flowers, and laboratory experiments showed that the flower buds were not a necessary additive to the diet of *Choristoneura* that already had access to leaf buds or young foliage from susceptible trees (sec. 14.113). It seems from this evidence that if the flower buds contain an essential nutrient that is lacking, or deficient in the foliage of immature trees, then the leaf buds and young foliage of mature trees that have been exposed to several years of anticyclonic weather must contain the same nutrient.

The observation by Wellington et al. (1950) that tree rings tend to be narrow for a number of years before an outbreak (sec. 14.113) suggests that the trees might have been "stressed", perhaps by shortage of water, during the bright, dry anticyclonic weather that usually precedes an outbreak. Conversely, in the 1949–57 outbreak that Morris and his colleagues studied in New Brunswick, the end came synchronously in stands that were severely defoliated and those that were not:

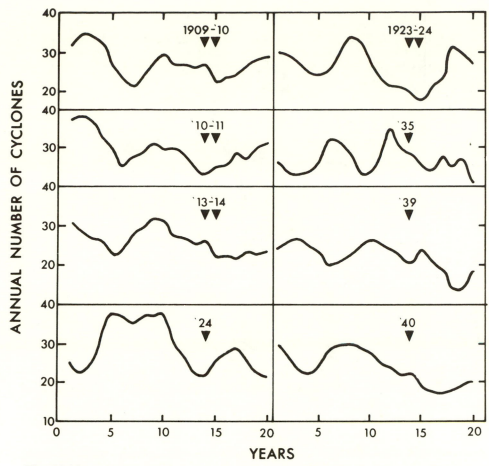

**Fig. 14.02** Each curve represents the annual count of cyclonic centers in a particular, but not necessarily the same, region during the fourteen years that preceded and the six years that succeeded an outbreak of *Choristoneura fumiferana* in that particular region. The arrows indicate the date on which the outbreak occurred. The figures alongside the pointers refer to the year when the outbreak began. (After Wellington et al. 1950)

intrinsic shortage of food seemed not to be a likely general explanation for the collapse. Predators were also ruled out as a general cause of the collapse (sec. 14.123). By a process of elimination weather seemed to be left as the most likely cause. According to figure 14.03 the spell of dry weather that had preceded the outbreak by five or six years came to a gradual end about 1950, and the outbreak began to decline about five or six years later. According to Wellington's explanation, the return to the above-average rainfall might be expected to reverse the nutritious condition of the trees. According to White (1974, 1978), it is commonplace for plants, especially woody plants, that have been "stressed", especially by shortage of water, to become highly nutritious for herbivorous insects and to return to their normal nonnutritious condition when the stress has been removed (sec. 10.311). The alleviation and reimposition of such extrinsic shortages of food seems to be important in promoting and terminating outbreaks of *Choristoneura* (White

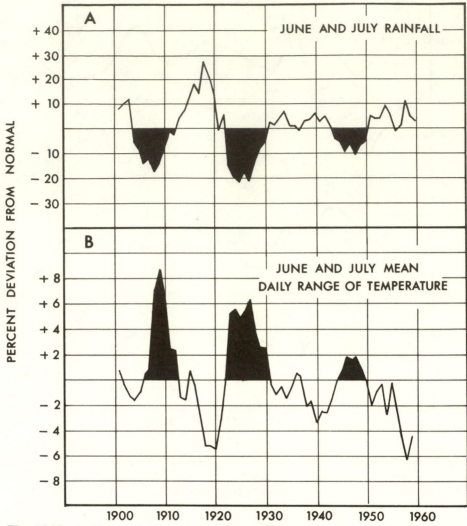

**Fig. 14.03**   Incidence of anticyclonic weather in New Brunswick, as indicated by deviations from the normal rainfall for June–July and deviations from daily range of temperature for June–July. Each curve is a five-year running average plotted against the fifth year. Records from nine meteorological stations were pooled. (After Greenback in Morris 1963)

1974, 296). However, during outbreaks intrinsic shortages of food also become important.

Severe defoliation is widespread over large areas of forest. The statistics cited in section 14.1 give some idea of the extent of the intrinsic shortages of food that develop during a widespread and prolonged outbreak. Miller in Morris (1963, 75) discussed one aspect of an intrinsic shortage of food. He wanted to relate the fecundity of the moth to the quantity and quality of the food that had been available to the larva. He was able to estimate fecundity from the size of the pupa. He estimated the relative density of the larval population by counting samples of larvae (third-instar and beyond) on 0.92 m² of foliage. He defined the "condition" of the tree in terms of the amount of new season's growth it put on in the spring. He quantified the condition by an index which was the reciprocal of the number of successive years that the young growth had been stripped from the tree by *Choristoneura*. The new growth was important because it was preferred and needed by the young second-instar larva when it began feeding for the first time in the spring. The new growth is also good food for the larger larvae, which prefer it. They move on to the old leaves when all the young foliage has been eaten. With successive years of complete removal of the young foliage, the old leaves got older and less nutritious. The index for condition took into account not only the decreasing amount of young foliage but also the decreasing quality of the old leaves as the years went by and each year all the young foliage was stripped from the tree. The tree was likely to die after about five years of such treatment.

Miller in Morris (1963) found that the mean fecundity decreased from about 195 eggs per moth when there were 9 larvae per square meter to about 105 when there were 232. The relation was approximately linear. Using the same data for fecundity, Miller found that the fecundity increased from 95 with the index for condition of the tree at 0.2 to about 180 with the index at 1.0. These figures relate only to the fecundity of survivors. The survival rate would have to be taken into account to explain the rate of increase of the population. Survival rate would be higher at lower densities. (See the discussion of "effective food" in sec. 3.121). Neither of the curves in figure 14.04 tells an unqualified story because the positive correlation between the two independent variates was significant. (Density of the population was at a minimum with the index at 0.2 and at a maximum with the index at 0.8). So a better indication of the association of fecundity with food is given by the partial regression which allows for this interaction:

$$Y = 100.72 - 0.16x_1 + 89.35x_2$$

$x_1$ is the number of larvae per 0.9 m²,
$x_2$ is the reciprocal of the number of years' defoliation,
$Y$ is the estimated number of eggs per moth.

This regression accounted for 85% of the variance of $Y$.

The independent variate $x_1$ reflects the influence of overcrowding in the current generation, and $x_2$ reflects the influence of overcrowding in past generations. There need be little doubt that changes in fecundity were caused by changes in the quantity and quality of food which were themselves caused by the feeding of the current and past generations of *Choristoneura*. It is generally true, when simple overcrowding can be demonstrated, that other animals of the same kind are modifying a critical component of environment, usually a resource in short supply and usually food. In the envirogram they appear in the web as in figure 2.02.

On trees where the consumption of new growth was less than total, Miller counted the number of young shoots per larva. But this attempt to get a direct measure of the amount of good food available to the local population failed because he could not get enough material for a critical analysis.

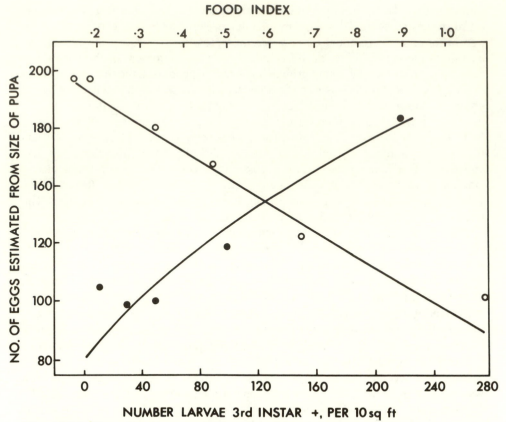

**Fig. 14.04**  Fecundity (as measured by the weight of the pupa) of *Choristoneura fumiferana* plotted against the density of the current population of larvae, third-instar and older (*open circles*) and against a "food index" which is the reciprocal of a qualitative "code of defoliation," based on the number of years of severe defoliation (*solid circles*). (After Miller in Morris 1963)

### 14.1212  Heat

There is another explanation for outbreaks that also begins with "old mature trees" and "anticyclonic weather" but does not invoke any change in the quality of food; it depends on the open crown of the mature tree and the immediate weather. According to this explanation, the tall, open crown of a mature tree allows drying winds and radiant heat to penetrate to the places where the eggs, larvae or pupae are living, keeping them warm and dry. The larvae are stimulated to eat more, and all stages grow and survive well. The anticyclonic weather provides a maximum of radiant heat and drying wind (Mott in Morris 1963, 189). This theory is represented in figure 2.02 by the pathway that leads to heat (as a resource).

There is circumstantial evidence to support both explanations. Doubtless both are right. The problem is to weigh them quantitatively against each other. This would require experiments that were aimed specifically at the critical differences between the two explanations.

## 14.122 Mates

The adults which emerge toward the end of summer copulate before they disperse. It seems that most of the females lay at least a few of their eggs in the place where they were born, but the matter has not been deeply studied. In one series of observations in which the number and size of empty female pupal cases were used to predict the number of eggs, the results over three years were extremely variable and seemed to be related to the weather. The number of eggs counted once exceeded the estimate threefold and on the other two occasions was 0.15 and 0.04 of the estimate. On the last occasion the weather was dull and rainy, and Greenbank in Morris (1963, 87) suggested that the very low number of eggs relative to empty female pupal cases might have been due to failure of the moths to copulate in such weather.

## 14.123 Predators

During the 1949–57 outbreak in New Brunswick records were kept of the numbers and activity of diverse predators and suspected predators. Birds were the only vertebrates of note, but there were many kinds of insects and spiders. Locally there were occasions when the predators seemed numerous enough to prevent the budworms from increasing very much in numbers, but it was thought that these local situations were atypical, perhaps to be explained by "edaphic factors" that favored the predator or disfavored the budworm. In general it seemed that the predators lacked the capacity for increase to match the rate of increase of the budworm. The abrupt ending of the outbreak seemed to be remarkably synchronized through localities that differed in shortage of food and in the presence of predators. This left weather as the most likely cause for the end of the outbreak. One hypothesis that was investigated with negative results was that the cloudy humid summer weather that prevailed during 1952–57 (fig. 14.03) may have fostered pathogens.

## 14.124 Malentities

Another suggestion from Morris' (1963) symposium was that the above-average rain that preceded and accompanied the end of the outbreak might have been directly harmful or lethal to the caterpillars of *Choristoneura*, but this suggestion was not confirmed by critical quantitative evidence.

The summary in section 14.13 may be regarded as an enlargement of the envirogram in the lower part of figure 2.02.

## 14.13  The Environment of a Dispersing Migrant

Every individual of *Choristoneura* is at considerable risk of becoming a footloose dispersing migrant three times during its lifetime. In every generation a high proportion of the adults, and larvae of the first and second instars, instinctively launch themselves into the wind to be carried wherever it may blow (sec. 14.111). Neither the adult nor the first-instar larva will feed, yet the dispersal of all three stages is essentially a search for food or, more particularly, for a new place to live where the food is good. If the wind happens to deposit a dispersing adult female in a good place, she will lay eggs which will promptly (i.e., before the autumn is spent) hatch into first-instar larvae. The newly hatched larva is likely (p > .5) to take off on a new voyage of dispersal of its own contriving. If the dispersing first-instar larva survives to make a landfall it will, without feeding, spin a silken cocoon (a hibernaculum) and enter diapause. Diapause disappears during winter, and in the spring the larva molts, still without feeding, into the second instar. If the original windfall happens to be in a place where the food is good, some of the newly molted second-instar larvae will stay to eat and grow; but many of them will set off once again on a voyage of dispersal. If the wind happens to blow one to a place where the food is good, it will at last settle down in the new place to eat and grow.

So in the environment of the dispersing migrant "food" is the only component in the centrum that is worth considering. There is only one pathway for the disperser in figure 2.02. The influential components in the envirogram will be found along the pathway that explains the distribution and abundance of stands of "susceptible" trees. If the individual areas cover a large part of the target area and are densely distributed, the dispersing *Choristoneura* will have a good chance to make a successful landfall at the end of its random voyaging.

At this stage the inquiry develops into a study of the ecology of the balsam fir, the spruce, and other trees that offer food for *Choristoneura* in Canadian forests.

### 14.131    The Age of the Trees and the Dispersion of Age-Classes

Canadian foresters recognize units which they call "stands". A stand is recognizable by the uniform age of its trees. If the complete history of a stand were known, it would be possible to point to an "agent of death"—a severe storm, a wildfire, a contract for clear-cut logging, an outbreak of *Choristoneura*—that had at some precise time in the past devastated the area that now houses a stand of uniformly aged trees. The precise conformity of the area that was devastated with the area of the stand happens because no new tree can grow in the forest until room has been made for it by the death of an old one. We do not have space to discuss the relative importance of the several agents of death. It will suffice to mention a few points about *Choristoneura* as an agent of death.

A severe outbreak of *Choristoneura*, lasting seven to eight years, inevitably leaves behind it large areas of forest in which all or most of the trees are dead. New stands arise in their place, all at about the same time. Forty or more years later, given a suitable period of anticylonic weather, they too will be at risk of being eaten. It is the size and contiguity of such susceptible stands that determine the chance that a dispersing *Choristoneura* will make a landfall in a good place with plenty of food.

In New Brunswick, records of recent outbreaks and data from growth rings suggest that, during the past two hundred years, the average period between the end of one outbreak and the beginning of the next was about thirty-two years which is not quite long enough for a stand to become mature and susceptible. So it may be that at least some outbreaks are fueled by a supply of food that was set in train not by the most recent outbreak, but by the one before.

### 14.132    Water

If anticyclonic weather really is changing the physiological condition of the food-trees as suggested in section 10.311 (White 1974), the water content of the soil is likely to be a proximate cause of the trees' condition. According to figure 14.03, in New Brunswick between 1900 and 1960 the rainfall for the two-month period, June–July, oscillated below and above the mean, with a period of twenty years. In figure 14.02 there is a suggestion of less regular oscillations with a shorter period in the frequency of anticylonic and cyclonic weather over broad areas of Canada. The data do not permit precise analysis, but it seems that neither a prolonged run of anticyclonic weather nor an abundance of mature stands of food trees is sufficient in itself to lift from *Choristoneura* the restraints that normally keep it rare. It seems that a concurrence of the two events might be necessary to spark an outbreak.

It would be a mistake to suggest too much uniformity among the stands of food-trees. Morris' (1963) symposium remarked on a number of stands that were out of step, and the phenomenon was usually explained by reference to "edaphic factors". If the critical component really is water in the soil, influencing the physiology of the tree, it would be expected that the activity of rain would be modified locally by topography and soil. Water runs off rises and into hollows, especially where the soil is hard. Clay soils store more water than sands if the rainfall is sufficient, but clay soils give up little water if the rainfall is not sufficient to wet them fully. Sandy soils store less water but readily give up most of what they have stored. Water-shortage may be much more severe in a clay soil than in a sandy soil if rainfall is light and infrequent. Depth of soil may also be important.

## 14.133   Wind

The small caterpillars and the moths that disperse launch themselves instinctively into the wind. Having done so, there is little they can do to influence the direction or distance of the dispersal or where they land. To understand the course of their dispersal one must to study the wind. Greenbank in Morris (1963, 87) studied the wind, especially in relation to the adults, but the comment also applies also to the dispersing larvae.

The nature of the wind and the weather depends on the origin and circulation of air masses and, more particularly, on the distribution and abundance of cyclones (low-pressure "cells" with air rotating counterclockwise) and anticyclones (high-pressure cells with air rotating clockwise) which are generated within and especially by the confluence of air masses. Clear, dry weather with long-distance horizontal winds is characteristic of anticyclones; humid, overcast weather, heavy clouds and persistent rain are characteristic of cyclones. Large-scale vertical turbulence develops at the confluence of the cyclone and the anticyclone. This turbulence is felt at ground level as the cyclone passes by. Meterologists call it "frontal circulation" and recognize several sorts of "fronts". For our purpose it is enough that frontal circulation may bring ahead of it weather that greatly stimulates the flight activity of the moths. Light changes rapidly, and so does air pressure, which generates air currents; the moths are stimulated and fly abundantly. Whole segments of large populations may be lifted high and carried far away before being dumped, perhaps in a barren place, but with luck where they can found a flourishing new colony. Greenbank (in Morris 1963) called this process "convectional transport". He said: "This type of transport is generally responsible for the tremendous flights of moths that have suddenly descended on towns as much as 80 km from the nearest infestation."

In anticyclonic weather the moths, engaging in their normal flight activity above the trees, are likely to be carried upward by local turbulence into a strong lateral wind blowing just above the treetops. Greenbank called this process "turbulent wind transport" and said of it: "In heavily populated areas, this type of dispersal is sometimes spectacular and can be likened only to the tremendous migrations of locusts".

Because the moths travel closer to the treetops in this sort of dispersal, the average distance traveled by a moth is likely to be less than with "convectional transport". "Turbulent wind transport" is very good for spreading an outbreak through adjacent areas where abundant good food is continuously distributed.

### 14.134   Intrinsic Shortage of Food

So far, in seeking to explain activity of food in the environment of the dispersing *Choristoneura*, we have explored only those pathways in the web that lead to an extrinsic shortage of food. So long as we consider only the initiation and early stages of an outbreak, there is no evidence of any widespread intrinsic shortage of food for the natural population. But as the outbreak runs its course, say after four or five years, the accumulated defoliation of vulnerable stands may be enormous (sec. 14.1). Even though there may still be vulnerable forest beyond the frontiers of the outbreak, a much smaller proportion of the total area will be available to the disperser for a favorable landfall. The chance of a favorable landfall has been reduced by an intrinsic shortage of food. Nevertheless the evidence is against the idea that such an intrinsic shortage of food contributed much to the ultimate ending of the 1949–57 outbreak in New Brunswick. It seems more likely that the end was chiefly caused by a widespread extrinsic shortage of food that was caused by a return to cyclonic weather (sec. 14.211).

## 14.14   Conclusion

The ecology of *Choristoneura* is painted boldly on a broad canvas. Areas and numbers are immense; so are the odds. But the explanation is simple: in the envirogram there are only a few important components; all are found on one pathway that converges on "food" (fig. 2.02). Although intrinsic shortages of food develop on a large scale as the outbreak approaches its peak, they have no influence on its inception and very little on its course or its ending. They are a consequence rather than a cause of the major events. The inception of the outbreak, its extension, and its ending seem to be adequately explained as responses to the alleviation and subsequent reimposition of extrinsic shortages of food.

The operation is best seen as a gigantic gamble. The stakes are large: to win is to gain access to thousands of square kilometers of good food for the present and future generations; to lose is to die. The budworm has evolved to take such chances; 90% or more of first-instar larvae were recorded as dispersing on one occasion; and the other two dispersive stages seem equally eager to disperse. In New Brunswick, on the average, a run of bad luck might last through thirty-two generations. During this time the budworms become very few and far between, but they do not die out. If they did they might easily be replaced by immigrants blown in from elsewhere in the vast Canadian forests. The gamble is conducted by populations that are scattered over enormous distances and areas, and not many of them would be in step with one another. In New Brunswick the budworms might expect, about three times in a hundred years, to experience a run of good luck that is likely to last through seven or eight generations; then they make up the ground that was lost during the previous twenty years.

To complete the story the important questions are: What causes an outbreak to begin? What causes an outbreak to end? Why is *Choristoneura* so rare between outbreaks? Why does *Choristoneura* remain extant? Such questions can be met with plausible explanations or hypotheses within the theory of environment especially extrinsic shortage of food and spreading the risk (chap. 9). It is difficult to see the relevance of competition theory to what is known about the distribution and abundance of the spruce budworm.

## 14.2   The Queensland Fruit Fly *Dacus tryoni*

A brief account of the environment of *Dacus tryoni* is given in section 2.233, together with the envirogram in figure 2.06. In most parts of Australia, with the exception of the central and western deserts and southern Tasmania, the climate is hospitable for *D. tryoni*. However, its distribution is more restricted than the area that is climatically hospitable (fig. 14.05). The most important requirement of *D. tryoni* which accounts for this restricted distribution is a succession of fruits for each generation of larvae throughout the year when temperatures are suitable for breeding. Before fruit trees were cultivated in Australia, the distribution of *D. tryoni* may have been confined to tropical and subtropical forests where many trees have fleshy fruits that are suitable food for the larvae of the fruit fly. The cultivation of fruit trees would then have made a southward expansion of *D. tryoni* possible in the past hundred years (Lewontin and Birch 1966). These trees provided a succession of fleshy fruits in many places where they did not exist before. The absence of *D. tryoni* from the northwestern corner of Australia seems anomalous, for many native plants there have fleshy fruits, though there is little in the way of cultivated fruit trees. The insert map in figure 14.05 shows that this area is occupied by a sibling species, *D. aquilonis*, which is slightly different in color and in the composition of the pheromones of the male (T. E. Ballas, pers. comm.). *D. aquilonis* presumably is different in other respects as well, for it is not recorded as a pest in such cultivated fruits as are grown in that part of Australia. The insert map in figure 14.05 also shows the distribution of a second sibling species, *D. neohumeralis*. Its distribution is completely contained within that of *D. tryoni* but is more restricted. Its larvae occur in the same fruits as those of *D. tryoni* (Gibbs 1967). Its taxonomic characters integrate completely with those of *D. tryoni* (Vogt and McPherson 1972; Vogt 1977). Isozyme analysis showed that the frequency of alleles is the same at fifteen loci but different at four others (McKechnie 1974, 1975). Fletcher and Bellas (1979) were unable to find any significant differences in the composition of the pheromones of the males of the two species. However, whereas *D. tryoni* mates at dusk, *D. neohumeralis* mates in bright illumination at noon (Gibbs 1967 and sec. 4.211). *D. neohumeralis* has little influence on the distribution and abundance of *D. tryoni* (sec. 7.14) except in one possible way. There is some evidence, which is not altogether conclusive, that introgressive

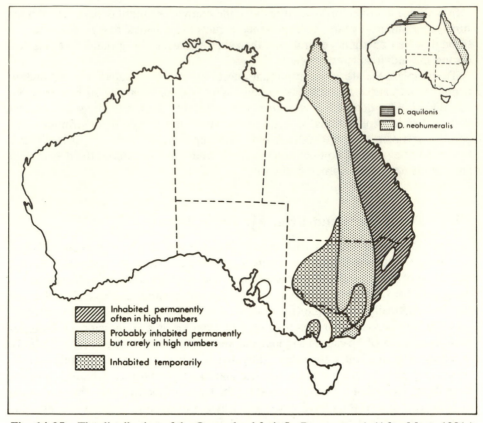

**Fig. 14.05**   The distribution of the Queensland fruit fly *Dacus tryoni*. (After Meats 1981.) Inset map: the distribution of two sibling species, *D. aquilonis* (G. Fitt, pers. comm.) and *D. neohumeralis* (Drew, Hooper, and Bateman 1978).

hybridization occurs in nature; such hybridization could have facilitated the extension of *D. tryoni* into colder southern areas through natural selection (Lewontin and Birch 1966; Birch and Vogt 1970). In the tropical part of its distribution *D. tryoni* shares host fruits with other species, the most common being *D. jarvisi* and *D. kraussi* (sec. 8.414).

The distribution of *D. tryoni* shown in figure 14.05 is divided into three areas: permanently inhabited and fruit flies abundant (near the coast); permanently inhabited but fruit flies not abundant (in a strip west of the coastal distribution); and temporarily inhabited from time to time. They correspond to areas of increasing dryness from the coast inland and decreasing quantities of heat from north to south. The influence of dryness on distribution and abundance is twofold: dryness reduces the quantity of fleshy fruit available except in irrigated areas, and there is an additional influence of rainfall, perhaps on the fecundity of adults and the survival of pupae, which is not yet well understood (Bateman 1968). The influence of heat on distribution and abundance has been analyzed in detail by Meats (1981). He made maps of the "bioclimatic potential" of *D. tryoni* in Australia based on the influence of heat on the number of generations that could be completed in a year and the rate of increase possible during the favorable period, which is mainly in summer, when fruit is more abundant than at other times.

Within areas that have a high "bioclimatic potential" the distribution and abundance of *D. tryoni* depend largely on the availability of a succession of fruits for larvae. A typical locality occupied by *D. tryoni* consists of a mosaic of places where fruit-bearing trees are interspersed with places without fruit. This is obvious in the case of cultivated orchards, which may be separated from one another by forests or cultivated land. But it is also true for endemic forests in the tropics and subtropics. Here fruiting trees, at any one time, are interspersed among many trees without fruits. And there is unlikely to be a succession of fruits in the same place in a forest over any length of time. Moreover, it is rare to see large quantities of fruit on individual forest trees at any one time. It seems that many fruits are eaten by mammals such as possums (R. A. I. Drew, pers. comm.) and birds such as fruit-eating pigeons (Crome, 1975); others are knocked to the ground by these animals, and there they are unavailable for fruit flies to lay their eggs in. This adds further to the patchiness of distribution of fruit.

Both in the native habitat of forests and in orchard habitats, it is unlikely that larvae experience an intrinsic shortage of food. Larvae are rarely so crowded in individual fruits as to cause an intrinsic shortage of food, even in places where adults are very abundant (Bateman 1968). An average female lays only about four eggs into a single fruit at a time, irrespective of the size of the fruit. She then moves on to another fruit. Other females may lay in the same fruit after her. A relative extrinsic shortage of fruit for adult females about to lay eggs is more likely because of the patchy dispersal in space and time of trees bearing fruit. Furthermore, as fruits become sparse in a locality, adults fly away to other localities. For the most part there is neither an extrinsic nor an intrinsic shortage of food for larvae of *D. tryoni*.

For two years, 1963 and 1965, Bateman (1968) counted all the fruits available to *D. tryoni* and recorded the total number of pupae that emerged in an experimental orchard at Wilton, southwest of Sydney and several kilometers inland from the coast (fig. 14.06). The total number of pupae produced was largely independent of the quantities of fruit available. It was determined by other components of environment. In 1963 there was ample rainfall, but 1965 was a year of severe drought. The amount of fruit available was about the same in both years. In 1963 the orchard became heavily infested, and the weekly production of pupae rose to a peak of eighty thousand. In 1965 there were few larvae in the fruit and the weekly production reached only twelve thousand. In both years the rate of production of pupae was independent of the quantity of fruit available. Figure 14.06 also shows that a large proportion of the fruit that was available in every season was not used by *D. tryoni*. Even in 1963, a favorable year for *D. tryoni*, only a small proportion of fruit that was available before the beginning of February was used. The new season's population is very slow in starting to grow. The clue is to be found in figure 14.07, which shows the number of immigrants that initiated the new season's population at Wilton in 1968. Fruit was on the trees in early January. Only a few immigrants arrived in mid-January. By mid-March they had built up to a peak of several thousand females. These flies were the second or third generation from flies that had overwintered elsewhere and had reproduced probably closer to the coast, which is

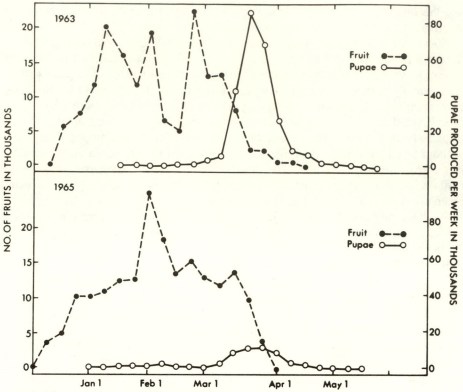

**Fig. 14.06**   The number of fruits and the number of pupae of *Dacus tryoni* in an orchard at Wilton (southwest of Sydney). The quantity of fruit is about the same in the two years, but the numbers of pupae are quite different. In 1963 the summer rainfall was ample; 1965 was a year of drought. (After Bateman 1968)

warmer. They had laid few if any eggs by the time they reached the orchard at Wilton. There they had an abundance of fruit in which to oviposit. Fletcher (1973) was able to distinguish between the flies that had migrated into the orchard and those that were probably the progeny of flies that had overwintered in the Wilton area and had found the orchard about the same time as the immigrant flies. Very few of them, as shown in figure 14.07, were resident gravid females.

Studies at Wilton over twenty years have shown that the size the population reaches each year is largely determined by the number of adults that migrate into the orchard during February and early March. These immigrants are responsible for most of the eggs laid into ripening fruit (Bateman 1968; Fletcher 1973). In localities that are warmer than Wilton in spring and early summer, such as suburbs of Sydney, the picture is probably different. There the overwintering adults are probably responsible for quite a large buildup in numbers much earlier in the season. They provide the immigrants for localities such as Wilton.

Figures 14.06 and 14.07 depict years in which there was a substantial crop of fruit at Wilton in early summer. In a couple of years in more than twenty years of study the fruit crop was drastically reduced by dry weather in spring and early summer, and good rains came late in the summer. The number of *D. tryoni*

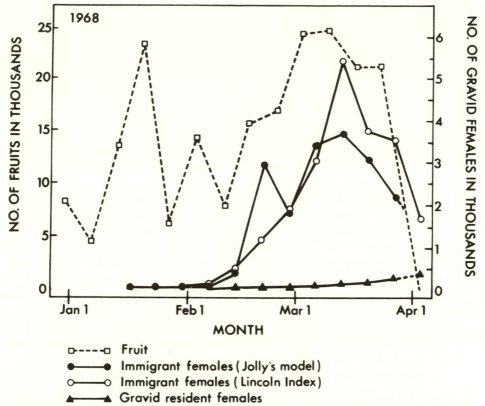

**Fig. 14.07** The number of female adults of *Dacus tryoni* that had immigrated into an orchard at Wilton (southwest of Sydney); estimates were based on both the Lincoln index and on Jolly's model applied to recaptured flies that had been previously captured and marked. Gravid resident females were offspring of flies that had overwintered in the neighborhood. The graph also shows that fruit is abundant long before the flies increase in numbers. (After Fletcher 1973)

remained low throughout the season despite the favorable weather from late summer onward. There was just not enough fruit about for the population to increase much (Bateman, pers. comm.). Unlike the years shown in figures 14.06 and 14.07, there was in these years with the dry spring and early summer, a correlation between the number of *D. tryoni* and the amount of fruit. The quantity of fruit set the limit to the number of *D. tryoni*.

So far we have emphasized the patchiness of the habitat for *D. tryoni* as a consequence of the patchy dispersion of fruits. But there is a second aspect in which the habitat is patchy in temperate Australia. To survive the winter, adults must overwinter in special places that are sheltered and contain evergreen trees and shrubs. These may be in valleys of creeks or in groups of evergreen trees in other sheltered areas (Fletcher 1975, 1979).

The two sorts of patchiness of the habitat make it necessary that the adults be good dispersers. The capacity for dispersal is a key to understanding the ecology of *D. tryoni*. Four phases of dispersal have been recognized as important for survival (Bateman and Sonleitner 1967; Bateman 1968; Fletcher 1973, 1974a).

When adults emerge from pupae in soil, they enter their first phase of dispersal. Instead of staying around where they emerged, even though there may be an abundance of fruit there, some 75% of them move away from the site of their emergence within the first week of their life. They spend at least the next two weeks becoming sexually mature. The shortest time for sexual maturation of 50% of females in population cages over trees in the field was fourteen days in midsummer (Pritchard 1970). The first phase of dispersal ends when the adults, many of which are now sexually mature, locate an area where there are ripening fruits suitable for egg laying. There they complete their sexual maturation and mate. Mating has its special requirements (sec. 4.211). Having mated, females lay their eggs into fruits, on average about four at a time, and continue to do so as they mature eggs. Most females mate only once in their lives and are able to fertilize further batches of eggs with sperm stored in their spermatheca. Adult flies may remain in the locality where they laid their first batch of eggs as long as fruits suitable for egg laying are plentiful there. When the supply of fruits starts to diminish, and usually before all the fruit has fallen from the trees, the adults fly away from the orchard or forest in search of places where there is fruit. This is the second phase of dispersal. How far they can fly is not known, but Fletcher (1974a) recaptured marked flies within three weeks at his most distant traps, which were 24 km from the site where the flies were marked and released. Later in the year, with the approach of winter, a third phase of dispersal begins. The last generation that emerged shortly before winter remain as immature adults without any chance of becoming sexually mature and laying eggs that season. If they are in a deciduous orchard they fly away from it, even though fruit may still be on the trees. They come to rest in whatever leafy trees and groves they encounter and overwinter there. If they were already in a nondeciduous grove such as a citrus grove or forest, they fly away from the periphery to overwinter in the more protected central trees. Adults that are sexually mature as winter approaches may stay on longer in the orchards if fruit is available. But they too are dependent upon finding leafy overwintering sites if they are to have any chance of surviving. After this third phase of dispersal adults spend the winter as nondiapausing adults. No other stage overwinters. Winter is a critical period for the adult population. We do not know what proportion of flies from the summer and autumn generations find suitable overwintering sites. Some flies may be months old and others only weeks old as they enter the overwintering phase. But the mature flies have been reduced by a 20% mortality each week during the late summer and autumn months (Fletcher 1974a). Those that survive to overwinter have a death rate of about 8% each week in overwintering sites in the vicinity of Wilton (Fletcher 1979). In a twelve-week winter this gives an accumulated winter mortality of 63%, which has to be added to the prewinter mortality. Winter mortality may rise to 100% if there is a spell of nightly minimum temperatures below $-3°C$. During one exceptionally cold winter spell in July, when the temperature dropped to $-10°C$, Fletcher (1979) recorded the complete extinction of the overwintering population in one locality. Farther south of Sydney the chance of extinction in winter becomes greater. During overwintering, ovaries are resorbed (Fletcher 1975). On windy days flies may be blown out of their overwintering sites, especially on cold days when they are torpid (fig. 2.06). It is therefore to their advantage not to become torpid. Meats (1976a,b,c) has found that the temperature threshold for torpor in adults is lowered by acclimitization to cold.

The fourth type of dispersal begins as spring approaches. Adults that have survived the winter fly away from their overwintering sites and scatter throughout the surrounding areas, where they search for fruit. The survival of the population now depends upon females' finding mates and then fruit in which to lay their eggs.

The capacity for dispersal of *D. tryoni* is critical in enabling adults to find fruit that is patchy in its dispersion in the habitat and to find protective refuges during winter. On the other hand, the tendency to disperse, especially among juvenile adults, must reduce the chance that a small number of colonists will successfully establish a population in a new locality (Bateman 1977). A few fertilized females may fly to a new place and lay eggs into fruit. Since most of their offspring fly away as juvenile adults, the few that remain may find their numbers below the density necessary if females are to find a mate. Establishing a new colony may require immigration into that locality by a substantial number of adults. What that number is we do not know.

*D. tryoni* is like *Choristoneura* in its adaptation for coping with widely dispersed food, which in the case of *Dacus* is also patchy in its dispersion. Both insects have a high capacity for dispersal. *Choristoneura* differs from *Dacus* in that three stages of its life cycle are dispersers instead of only the adult as in *Dacus*. The pathway in the envirogram that leads to food of larvae in *Dacus* is a simple one (fig. 2.06), that of *Choristoneura* is complex (fig. 2.02). In spring and summer there is usually an abundance of food for larvae of *Dacus* within its distribution, but this is not the case with *Choristoneura*. Outbreaks of *Choristoneura* are dependent upon a particular succession of events in the pathway leading to food that results in an abundance of suitable flower buds, leaf buds, and young foliage. The necessary sequence of events occurs only rarely. Weather is critical for *D. tryoni*, but within the distribution of *D. tryoni* its influence is often largely independent of food. There can be an abundance of food but few *D. tryoni*. Unlike the situation with *Choristoneura*, the sequence of hospitable weather necessary for an outbreak of *D. tryoni* is a common event (annual in some places). For both *Dacus* and *Choristoneura* extrinsic shortages of food are important. For *Dacus* the shortage is usually extrinsic, whether it relates to the local or the natural population, as in figure 9.05. The same holds for *Choristoneura* between outbreaks. But the shortages that occur when the outbreak is at its peak or in decline are too complex to be represented by either diagram in figure 9.05. At the height of the outbreak there may be intrinsic shortages at the level of the local population, and there may be an important element of intrinsic shortage at the level of the natural population as well. A combination of figure 9.05 with one built on 9.03 might serve as an approximate pictorial representation of the complexity that arises after the outbreak reaches and passes its peak. But this could be pushing these diagrams beyond the simple purpose for which they were intended, which was to help the reader to visualize the multipartite structure of a natural population.

## 14.3  The Contrast between *Dacus tryoni* and *Dacus oleae*

Food is widely and patchily dispersed for both *Choristoneura* and *D. tryoni*. Adaptations for dispersal are prominent among the niches of both species; and the success that comes from actively dispersing is prominent in both their ecologies. Not so with *Dacus oleae*, the olive fly, which presents some striking contrasts in its ecology. The larvae of *D. oleae* live only in the fruits of the olive tree *Oleae europeae* and a few closely related species. The distribution of *D. oleae*, although extensive and covering a wide range of climatic zones, is restricted to places where olive trees are indigenous or have been planted by man. *D. oleae* is commonest in countries surrounding the Mediterranean Sea because of both the abundance of olive trees and the favorable climate.

Like *D. tryoni*, in the warmer parts of its range *D. oleae* breeds throughout most of the year despite having virtually only a single host plant. It can do this because olive fruits mature slowly and remain on the trees for a long period which, depending on variety, extends from late June or July to May or June of the following year.

The following account of the ecology of *D. oleae* is based largely on the studies that Ethemios Kapatos and Brian Fletcher (pers. comm.) carried out on the island of Corfu. There *D. oleae* has four generations a year. Breeding is restricted in winter by low temperatures (most individuals overwinter as nondiapausing pupae in the soil), and in summer by lack of olive fruits and temperatures in excess of 26°C, which inhibit sexual maturation. In areas that are colder or much hotter than Corfu there may be fewer generations in a year (McFadden et al., 1977).

In the spring adults (which have emerged from pupae in the soil or have overwintered successfully) lay their eggs into the ripe olive fruits that remain on the trees from the previous fruiting season. On Corfu there is usually a relatively large amount of fruit available in the spring because it is not harvested in autumn as occurs in most other olive-growing areas but is allowed to fall naturally. Like *D. tryoni*, the ovaries of *D. oleae* are immature at the time of emergence, and maturation rates are determined by a number of interacting components of environment, the most important of which is temperature. However, the presence of ripening or ripe fruits of its host seems to play a much more important part in controlling the maturation process in *D. oleae* than in *D. tryoni*. Immature females of *D. oleae* will puncture the fruit with the ovipositor and suck up some of the juice exuded from the puncture. The juice seems to act as a stimulus for sexual and ovarian maturation of the females. If females do not have access to fruits, maturation is much slower (Fletcher, Pappas, and Kapatos 1978).

When sexually mature, females produce a sex pheromone which attracts males (Haniotakis 1977), whereas in *D. tryoni* it is the males that produce a sex pheromone to attract females. As in *D. tryoni*, however, mating occurs in the late afternoon, and most females mate only once. Ovipositing females normally lay only 1 egg per fruit, and during the process they also feed on the olive juice. After depositing the egg beneath the skin, the female marks the surface of the fruit with some of the juice, which is regurgitated from the proboscis. This excretion inhibits other females from ovipositing in the same fruit. The repellent agent is the phenolic compounds in the water-soluble fraction of the juice (Vita et al. 1977). The olive juice sucked up by an ovipositing female serves two purposes: the part that is ingested promotes egg production and the part that is deposited on the surface of the fruit repels other females. Once the larvae hatch, they are dependent upon symbiotic organisms (*Pseudomonas* sp.) that are transferred from the reproductive system of the female to the fruit along with the eggs. These microorganisms digest the oily fruit, and the breakdown products are food for the larvae.

The adults of the spring generation which emerge in May and June find themselves in an inhospitable environment. The old fruit has virtually disappeared, the new crop of olives is not yet ripe enough for attack, and the weather is hot and dry. During the first five to six weeks of their lives they remain sexually immature. At the same time the ovaries of any surviving older females revert to an immature condition. Once the fruit has become sufficiently mature, however, the females are attracted to it. This normally happens in early July when mean fruit weight is about 0.3–0.4 g. Contact with the fruit and ingestion of the juice stimulates ovarian development and synchronizes maturation so that the females have mated and are ready to lay eggs by the time mean fruit weight has reached about 0.6 g. At this stage the larvae are able to survive and develop in the ripening fruit, which eventually may reach a weight of 1.3–1.5 g.

Studies on the relation between fruit size and onset of maturation during the summer have played an important part in developing an effective control program against the olive fly on Corfu. Briefly, this depends upon the application of an aerial bait spray at a critical time during the summer when the adults are most responsive to the protein bait (which coincides with the start of ovarian maturation) but have not yet started to lay their eggs. Analysis of data on rates of increase of fruit weight and the relation between fruit weight and the ovarian maturation of females indicated that the critical period for spraying a particular area could be predicted accurately by determining the mean fruit weight at the end of June and then calculating the date when it would reach the weight at which the females would be nearing sexual maturity.

The number of adults that emerge each generation is mainly determined by the number of eggs laid, and this is largely determined by the number of fruits on the trees. There is seldom, if ever, any intrinsic shortage of food for larvae. The females ensure this by laying only one egg into an olive fruit and by marking the fruit to stop other females from laying into it (sec. 3.121). The developing larva consumes only a small proportion of the fruit (5–20% depending on the variety). On the other hand, there may be an extrinsic shortage of fruits. Olive trees have a two-year cycle of high and low fruit production. Within a habitat the cycles of individual trees are often synchronized, so the crop of fruit may be very small one year and very large the next. Other predators of olives may cause an extrinsic shortage of food for *Dacus*; caterpillars of the moth *Prays oleae* eat both flowers and developing fruits; outbreaks of the scale insect, *Saissetia oleae* inhibit the growth of foliage and suppress fruiting; attack by a number of species of fungi can accelerate fruit fall, as do strong winds and storms. Also, in most areas, though not on Corfu, olives are harvested in November, resulting in an extrinsic shortage of fruit in the spring. This is a critical period because it is mainly the adults from the spring generation (which pass through the summer period in an immature state and then start to oviposit in July and August) that give rise to the first generation of flies in the new season's fruit crop and thus determine the size of the spring generation. The subsequent size of the adult population in early summer is thus seen to be largely dependent upon the number of fruits available in spring.

During four years of study of *D. oleae* on Corfu, the mortality of *D. oleae* from egg to adult during the first three generations was remarkably constant, about 80%. Deaths were due to high temperature in late summer, which was a malentity for eggs and first-stage larvae. Later in the season these stages usually survive the weather, but by then (i.e., between August and October) the third-instar larvae may be heavily parasitized by chalcid wasps. After October, mortality from parasitism is negligible because the wasps switch to alternative hosts for the remainder of the year. When the generation that matures in August and September pupates, birds and insects (ants and wasps) prey upon the pupae in the fruits. Only a small proportion of this generation pupate in the soil, but most of them die from the high temperature and dryness of the soil. During the winter months when a high proportion of the population pupates in the soil, low temperatures, flooding of the soil, and predation are the main causes of mortality. Overall mortality from egg to adult of the spring generation is less than for the other generations and is usually about 50–60%, mainly owing to the death of pupae in the soil.

Although some individual adults of *D. oleae* have been known to fly several kilometers, this is unusual. In contrast, most adults of *D. tryoni* probably fly several kilometers or more. Wide dispersal is not normally a prerequisite for survival for *D. oleae* as it is for *D. tryoni*. When an adult of *D. oleae* emerges, the chances are there will be fruit in the immediate vicinity. Indeed, several generations can succeed one another in the fruits of a single tree. This is not at all likely to happen with *D. tryoni*, which spends a lot of time moving from locality to locality in search of fruits. Nor do adults of *D. oleae* need to leave their host trees to search for overwintering sites. They either pass the winter as pupae in the soil beneath the trees or overwinter as adults in the foliage of the olive trees. Adults of *D. oleae* may leave a locality if there are no fruits on the trees, but they are likely to find fruits close by. Dispersal is most likely to occur in the spring when the flies emerge after winter, which is toward the end of the fruiting season and the time when fruit is least abundant. Because of the biennial fruiting cycle, adults that emerge in May and June may also have to disperse to find trees with a new season's crop. The contrast in the ecologies of *D. tryoni* and *D. oleae* is due to their adaptation to different environments. All the resources that *D. oleae* needs can be found in a grove of olive trees or even a single tree, whereas the resources of *D. tryoni* have to be sought in many localities that may be far apart.

# 15

# The Ecology of Man

## 15.0  Introduction

The environment of modern man differs immensely from what we imagine for our primitive ancestors. Most of the changes have been "invented" and engineered by humans themselves. Most animals, if they have the potential to become abundant, also have the potential to change their environment, by virtue of their numbers. Man is not qualitatively different from other animals in this regard, but the quantitative difference between man and all other animals is immense. The difference lies partly in man's extreme dispersiveness and relentless multiplication, but, more essentially, it lies in the fact, as Waddington (1960, chap. 10) and others have pointed out, that man proceeds up the ladder of "cultural evolution" far faster and further than any other animal. Cultural evolution is distinguished from genetic evolution, the changes that come from accumulating genetic differences. It is important to recognize, however, that cultural evolution itself causes changes in the environment that may lead to further genetic evolution (e.g., see Birch and Cobb 1981, chap. 2). Even in this man is not qualitatively unique. Other sorts of animals also accumulate traditions: birds learn their songs, and cats learn to hunt from their parents and associates (see Birch and Cobb 1981, chap. 2). But the quantitative difference is immense. Nevertheless, because the ecology of man is not qualitatively different from that of other animals, it is practicable and may be fruitful to fill in an envirogram for man just as we would for any other sort of animal. We also find that man's interaction with renewable resources can be represented by the sorts of models that we have discussed in chapter 9. We need at least two models to allow for the changes that have come about through cultural evolution. During the hunting/food-gathering phase, shortages of food and other renewable resources were mostly extrinsic, as in figure 9.05. During the modern industrial, technological phase, intrinsic shortages of such resources have become more prominent and seem likely to become more severe before long, in which case figure 9.03 (line III) will become more relevant. We do not have a model for intrinsic shortages of nonrenewable resources because chapter 9 was developed for nonhuman animals and it is only in the ecology of *Homo* that intrinsic shortages of nonrenewable resources are likely to become important. We do not have space or skill to discuss man's nonrenewable resources, but we think the discussion should embrace the full ecology of man. And we think such a discussion might be best conducted within the framework of our theory of environment and of the distribution and abundance of animals.

Two points deserve emphasis. Humans, especially since they have become superabundant, have begun to feature strongly in the web of their own envirogram as a modifier that, in various ways, spoils or destroys their own habitat. Also, as they have become so numerous they have come increasingly to feature as a modifier in the webs of other animals, spoiling or destroying their habitats. In the ensuing parts of this section we try to keep these perspectives as we write about the ecology of *Homo sapiens*. We start with a brief review of what might more properly be called the paleoecology of *Homo* spp. because it is based on the evidence from fossils and artifacts (sec. 15.1). Then, in section 15.2 we look at the three phases in human ecology as revealed by archeology and history and at the inferences that can be drawn from studying present-day communities that seem not yet to have moved out of one of the earlier phases.

## 15.1   Fragments from the Paleoecology of *Homo sapiens*

According to Tobias (1981), there are records of the family Hominidae in Africa going back 3.8 million years and of the genus *Homo* going back 2 million years. According to Cronin et al. (1981), the origin of "modern" *Homo sapiens* is still obscure but may date from about a quarter of a million years ago. What have been classified as "ancient forms" of *Homo sapiens* first appeared in the record about 0.6 million years ago. Earlier fossils have been referred to the species *Homo erectus*. Washburn (1978) called *H. erectus* "the first true man" and said that the species first appeared in Africa about 1.5 million years ago. Curtis (1981) dated the first *H. erectus* at 1.6 million years ago.

By about 100,000 to 400,000 years ago *H. erectus* had spread right across the world, from China and Java in the east to Britain in the west and to southern Africa. The total population might have been about one million (table 15.01). *H. sapiens* is thought to have evolved from *H. erectus* in Africa. *H. erectus* became extinct and the new species spread first to Europe, then to Asia, reaching Sarawak at least 40,000 years ago, Australia at least 40,000 years ago, and North and South America more than 20,000 years ago (Davis 1974; table 15.01). Man, and species dependent upon man, has achieved a wider distribution than any other terrestrial animal.

## 15.2   Three Phases in Human Ecology

The rate of growth of the "human" population (*Homo* spp.) during the first million years, that is, for about the first 99% of its history, must have been about fifteen per million per year (0.0015%/year) (Coale 1974). During all this time *Homo erectus* and later *Homo sapiens* were hunters and food gatherers. They were efficient at making tools for the hunt as well as for other purposes. Practicing this form of culture, the human population grew to about eight million by some ten thousand years ago. This phase of human population growth is represented by the first curve on the log-log scale in figure 15.01. (On an arithmetic graph it is not possible to show at all effectively the huge change in numbers over such a long

**Table 15.01** Distribution and Abundance of Man through Different Stages of Cultural Evolution

| Years Ago | Year | Cultural Stage | Continents Populated | Total Populations (millions) |
|---|---|---|---|---|
| 1,000,000 | | Lower Paleolithic | Africa | 0.125 |
| 300,000 | | Middle Paleolithic | Africa, Asia, Europe | 1 |
| 40,000 | | Upper Paleolithic | Australia | 3 |
| 10,000 | 8000 B.C. | Mesolithic; beginning of agriculture | All continents (excluding Antarctica) | 8 |
| 6,000 | 4000 B.C. | Village farming and early urban | All continents (excluding Antarctica) | 150 |
| 2,000 | 20 B.C. | Village farming and urban | All continents (excluding Antarctica) | 300 |
| 300 | A.D. 1680 | Farming and early industry | All continents (excluding Antarctica) | 600 |
| 200 | A.D. 1780 | Farming and industrial | All continents (excluding Antarctica) | 800 |
| 130 | A.D. 1850 | Farming and industrial | All continents (excluding Antarctica) | 1,000 |
| 50 | A.D. 1930 | Farming and industrial | All continents (excluding Antarctica) | 2,000 |
| 10 | A.D. 1970 | Farming and industrial | All continents (excluding Antarctica) | 3,500 |
| | A.D. 2000 | Farming and industrial | All continents | 6,000+ |

*Source:* Modified from Deevey (1960) and Freedman and Berelson (1974).

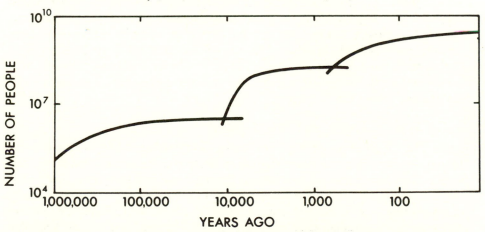

**Fig. 15.01** The growth of the human population plotted on a log-log scale showing three surges in population growth associated with the hunter/gatherer culture, the agricultural revolution, and the industrial revolution. (After Deevey 1960)

time. The log-log scale is an unfamiliar one, and for our purposes it is used solely to indicate the three phases of growth of the human population).

A second phase of population growth began about ten thousand years ago with the agricultural revolution (second curve, figure 15.01). Within some nine thousand years the population increased about eightyfold. The rate of growth was still low compared with what was to come, but it was about a hundred times higher than in the preceding phase of human history, being about 0.1% per year.

About three hundred years ago a third phase of population growth began with the commercial revolution, which led into the industrial revolution. During this three hundred years the world population multiplied about sixfold, reaching its maximum growth rate for all time (of about 1.9% per annum) about 1970. This last stage is usually referred to as the modern population explosion.

To summarize: During the hunting, food-gathering phase the human population increased at almost zero growth rate, averaging about 0.0015% per year. In the phase set off by the agricultural revolution the human population grew almost one hundred times as fast at about 0.1% per year. In the last industrial phase the population reached its maximum rate of growth of about 1.9% per year. Man's chance to survive and reproduce changed dramatically during these three phases of his history.

The changes, as we shall show, were associated with the niches through which man chose to exploit his environment. As a result of the new methods of exploitation the environment was changed dramatically. It was virtually all done by cultural evolution, although of course conventional genetic evolution had paved the way.

The growth of the human population can be appreciated in another way (table 15.01). The numbers of mankind (using that word to include *Homo sapiens* and an unknown number of paleospecies that may or may not have been ancestral to *H. sapiens*) reached about one million approximately two million years after the first species of *Homo* appeared. About half a million years later (A.D. 1850), with only *H. sapiens* still extant, the numbers had reached one billion (one thousand million); a second billion was added in eight years (1850–1930). The third billion was added in only thirty years (1931–60). The fourth billion took only fifteen years (1961–75). If this rate of growth continues the fifth billion will be added in slightly more than a decade, by about 1987 (Ehrlich, Ehrlich, and Holdren 1977, 183).

What caused these enormous changes in man's chance to survive and reproduce throughout human history? To answer that we need to consider the envirogram of man and see how it has changed.

### 15.21  Man the Hunter and Food-gatherer

Figure 15.02 is an envirogram of man during the first phase of population growth. Hunting and food-gathering were the two niches that characterized man during this period. We can reconstruct this part of human ecology partly on archeological evidence and partly from recent anthropological evidence derived from hunting, food-gathering men on the earth today, such as Eskimos, the Bushmen of southern Africa, and the Australian aborigines. Braidwood and Reed (1957) considered that the density of the population (in populated regions) during the hunting and food-

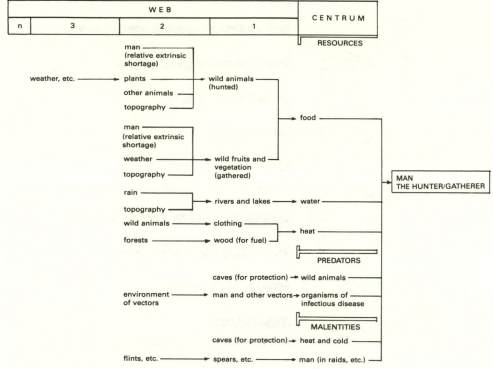

**Fig. 15.02** The envirogram of man the hunter/gatherer.

gathering phase was about 2 per 100 km². Deevey (1956) says that the aboriginal American Indian averaged about 4 per 100 km² and the Australian aborigine about 3 per 100 km². Both Braidwood and Reed (1957) and Deevey (1956) have made estimates of the amount of food available per square kilometer and both conclude that hunting, food-gathering men were limited not by the amount of food in their habitat but by their ability to get it during the lean season. For example, Deevey (1956) estimated that the production of catchable animals in American Indian terrain would have supported as many as 1,450 Indians per 100 km². That is about three hundred times the actual density of the Indian population. In our terminology (sec. 3.131), the hunter, food-gatherer experiences a relative extrinsic shortage of food. No doubt the shortage would have become less severe as they improved their means of hunting and food gathering, as indicated by the top two sections of figure 15.03. The quantity of food (expressed as grams of carbon) caught and transformed into man is considerably less for the lake spear-fisherman than for the marine net-fisherman or the rabbit-hunter with bow and arrow. The amount of food available for the hunter, food-gatherer is limited by his inability to find and catch it. Braidwood and Reed (1957) considered that some hunting, food-gathering men may have experienced hunger and starvation from time to time, but this is not common in existing hunting, food-gathering societies. Moreover, when not in close contact with urban society, their state of health is relatively good. Survival of more than half the population to age 15 is generally to be expected (Dumond 1975).

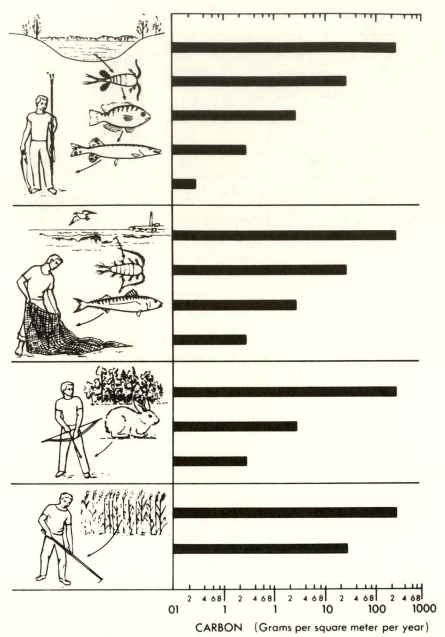

CARBON (Grams per square meter per year)

**Fig. 15.03** The efficiency of four human economies in providing food. The log scale at the bottom indicates the production of living matter in terms of carbon content. The top chart represents a primitive lake-fishing culture. The plants produce about three hundred grams of plant carbon per square meter per year (*top bar*). The next trophic level utilizes about 10% of the plants, that is, thirty grams of carbon. At the next trophic level, the fish utilize about 10% of the plant eaters, and the fish in the trophic level below that utilize about 10% of the prey. The fisherman utilizes about 10% of the carbon in the large predatory fish, that is, 0.03 grams of carbon. The chart second from the top represents a more advanced culture that is more effective in providing food for man. Of about the same efficiency is the land hunting

Carefully collected demographic information from a population of !Kung[1] Bushmen in Botswana shows that the expectancy of life at birth was of the order of 32.5 years, with 60% of all born surviving to age 15. Although low by modern standards, these rates of survival are such that, were they accompanied by birthrates as high as those in many developing countries, there would be a !Kung population explosion. But there is no evidence of this. Instead, the mean number of children born to women who survived to the end of the childbearing period of 45–50 years was no more than five, a figure just sufficient to produce a stationary or very slowly growing population. Studies of other hunting food-gathering societies provide further evidence of relatively low birthrates (Dumond 1975). Figure 15.04 provides further evidence of relatively low death rates in hunting, food-gathering societies. These survivorship curves are based on fossil remains and in the later societies, on skeletons. These and other data suggested to Dumond (1975) that there has been little change in human survivorship through hunting/gathering societies and primitive agricultural societies and perhaps for millennia afterwards. It seems likely that the life expectancy of 25 years plus or minus 5 years characterized these societies during the whole of this period. There is evidence, however, that survivorship increased at least from the eighteenth century onward (fig. 15.09). Curve B in figure 15.04 might be regarded as representative of a hunting, food-gathering society. Dumond (1975) has calculated that with this survivorship the population could have maintained its numbers or increased very slowly (net reproduction rate of 1.02), with five births per female distributed over ages similar to the spacing of births in the !Kung Bushmen of today.

What determines the low birthrate of the hunting, food-gathering society? Dumond (1975) argues that they had much potential fecundity that could have been actualized in a favorable environment, but that spacing of births reduced the birthrate to this low number. Such spacing was necessary if the mother was to perform her tasks of raising each child, including transporting it over long distances. To look after more than one at a time would have been difficult. Dumond (1975) points out that female mountain gorillas and chimpanzees transport their young and look after only one at a time. Births are spaced three to five years apart. This is thought to be under hormonal control. Dumond (1975) proposed that spacing of birth in the hunter, food-gatherer is under voluntary control. However, more recent studies (Short, 1976) provide evidence that overall fecundity in contemporary hunting, food-gathering societies is influenced by a combination of later age at the onset of fertility (compared with industrial man) and long intervals between births caused by suppression of ovulation while breast-feeding (amenorrhea). The evidence for later age of fertility derives from the following two sources. In hunting, food-gathering societies today sexual intercourse begins at

1. The exclamation mark denotes a click made with the tongue against the roof of the mouth.

culture represented in the chart second from the bottom. The bottom chart shows an agricultural society. The plants are utilized directly by man, so the amount of food available to man is one thousand times as much as in the primitive lake fishery. (From "The Human Crop," by Edward S. Deevey. Copyright © 1956 by Scientific American, Inc. All rights reserved)

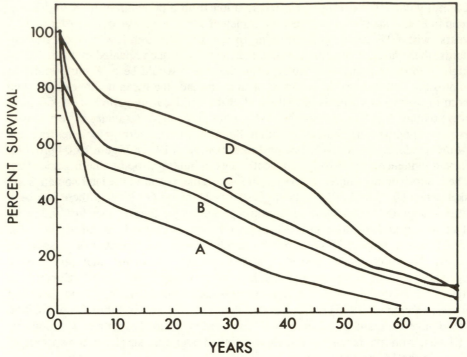

**Fig. 15.04**  Survivorship of early man for both sexes, derived from skeletal remains. (A) *Homo erectus* about 100,000 to 500,000 years ago. Life expectancy of thirteen years. (B) *Homo sapiens* about 20,000 years ago. Life expectancy of twenty-one years. (C) *Homo sapiens*. Agriculturalists of A.D. 100 to 400. Life expectancy of twenth-five years. (D) *Homo sapiens*. Agriculturalists of the late Roman era. Life expectancy of thirty-five years. (After Dumond 1975)

puberty, but the first pregnancy does not occur for several years thereafter, even though contraception is not practiced. In the !Kung hunting, food-gathering society the onset of menstruation (menarche) and marriage both occur at 15.5 years. The average age for birth of the first child is 19.5. Second, in developed countries there is some evidence that puberty has been accelerated in the past one hundred years. The main evidence, provided by Tanner (1962), is shown in figure 15.05. These data have been criticized on the basis of the small sample in the Scandinavian figures for the nineteenth century. With larger samples from a wider range of countries, Bullough (1981) finds that the decline in age of menarche from 1840 to the 1980s is from between 14 and 15 years to between 12 and 13 years. No one really knows the cause of this decline, but it is usually assumed that improved nutrition has been responsible.

Evidence for the spacing of births by lactation comes from the !Kung community (fig. 15.06). The people live as nomads, and the average interval between births is four years. Each child is breast-fed for three to four years. In recent years many of the !Kung have forsaken their nomadic life and settled in agricultural villages. Here the babies are weaned early and given a supplementary diet of grain meal and cow's milk. The interval between successive births in these village dwellers has dropped

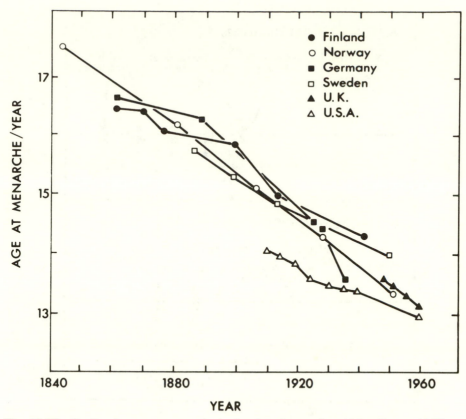

**Fig. 15.05** Decline in age at menarche in developed countries. (After Tanner 1962)

by 30% and the rate of increase of the population has increased accordingly. Many studies on other populations such as Eskimos and South American Indians show the influence of lactation in inhibiting ovulation. Indeed, Short (1976, 17) claims that "throughout the world as a whole, more births are prevented by lactation than all other forms of contraception put together." Short estimated that the women of the hunting, food-gathering society experience fifteen years of lactational amenorrhea and just under four years of pregnancy and menstrual cycles.

### 15.22 The Agricultural Revolution

When the last ice age was over, about 8000 B.C., the inhabitants of the hills around the "fertile crescent" of Mesopotamia, which stretched from the Mediterranean Sea through what is now Turkey and down through Iran and Iraq to the Persian Gulf, had begun to domesticate plants and animals that they had previously merely collected and hunted. At a slightly later time similar events happened in Central America and perhaps in the Andes, in Southeast Asia, and in China (Braidwood 1960). From these centers the new way of life, the farm, spread to the rest of the world. By about 4000 B.C. the people of southern Mesopotamia had achieved such increases in productivity that their farms were beginning to support an urban population. Some people were released from preoccupation with finding food to

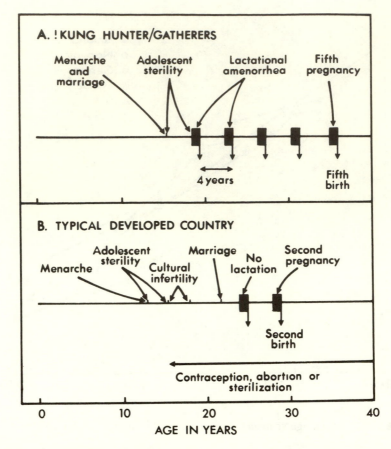

**Fig. 15.06** Fertility pattern of a hunter/gatherer community compared with that of people in developed countries today. (A) In !Kung hunter/gatherers menarche is late, at fifteen and a half years of age, and coincides with marriage. Because of adolescent sterility, the birth of the first child is deferred to nineteen and a half years. Lactational amenorrhea causes births to be spaced four years apart. Five children are born, of which three probably survive to reproductive age. (B) In a typical developed country menarche occurs at thirteen years of age. Adolescent sterility lasts until about fourteen years of age. Cultural infertility extends into the late teens. From then onward spacing of births is dependent upon contraception. Lactation is so short that it no longer induces amenorrhea. (After Short 1976)

other activities and innovations such as the discovery of the mechanical principles of weaving, the plow, the wheel, and metallurgy (Braidwood 1960). In the first 3,000–4,000 years after the cultivation of plants and the domestication of farm animals, the life of man changed more radically than in the preceding 250,000 years.

From an ecological point of view, farming was a new niche for man. It greatly reduced the chance of extrinsic shortages of food, and, because it enabled many people to do things other than produce food, it made possible the rise of cities. The first of them were in Mesopotamia. The development of cities depended upon there being supplies of food beyond the needs of the producers.

By 5500 B.C. or earlier, village farming was well developed in an area that extended from Mesopotamia into southern Asia. It began in the hills of Mesopotamia where there were the appropriate plants and animals, then moved down into the valleys where water was abundant and where irrigation was possible. In the following millennia the growth of cities in the Middle East culminated about 2500 B.C. in the Sumerian city-state with tens of thousands of inhabitants (Adams 1960).

The increase in the human population that followed the agricultural revolution is shown in figure 15.01 and table 15.01.

The envirogram of agricultural man in figure 15.07 summarizes some of the more obvious ways that the farmer's environment differed from that of his hunting/food-gathering predecessors (fig. 15.02). Important new pathways in the web converge on food and water. The protection provided by houses in villages generates new pathways in the web that converge on malentities and predators. The gathering together of people in towns and villages as a consequence of the development of agriculture might have led to a deterioration of hygiene and an increase in disease (Short 1976). The organization of man into large and wealthy communities may have increased the scope of aggression and led to war as a major new malentity. There are some who argue that war arose after the agricultural revolution and as one of its consequences (Alland 1972; Leakey and Lewin 1977; Shepard 1973, 1978). The capture of slaves and the conquest of territory were meaningless during the millions of years of hunting and food gathering.

What then was the influence of the agricultural environment on man's chance to survive and reproduce? The obvious answer is not necessarily the correct one. There is, for example, no clear evidence that increase in food resulted in a lower death rate. We have already indicated that disease might have increased. Indeed, some authors (Dumond 1975) argue that there was little change in the death rate of man from Neanderthal man until the 1800s A.D., though this seems to ignore the data presented in figure 15.08. What evidently did change was the birthrate. The change from a nomadic life to a relatively sedentary one on the farm removed the need for the mother to devote her whole attention to rearing a single child. For one thing, she no longer needed to carry it with her on long treks. In agricultural communities additional children are considered to be an economic benefit, and the average interval between births in such societies is two years (Dumond 1975). Urbanization also may have this influence. Short (1976) refers to studies on women in rural and urban areas in Rwanda. In rural areas where the women were malnourished and the baby was constantly carried on its mother's back, some twenty-

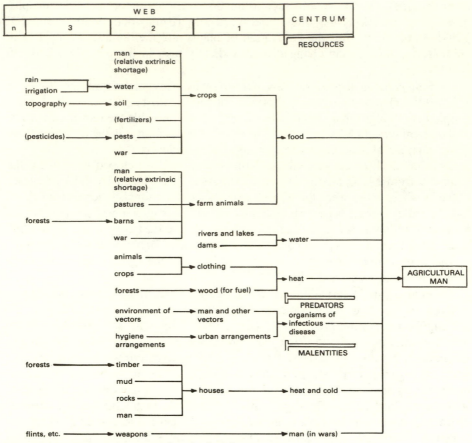

**Fig. 15.07**  The envirogram of agricultural man.

three months elapsed following birth before 50% of the women had conceived again. In urban areas, where mothers were better fed, 50% had conceived within nine months of the birth of a child. .

## 15.23  The Industrial Revolution

The third major phase of growth of the human population coincides with the industrial revolution—that is, from about 1850 (fig. 15.01, table 15.01). However, it was preceded by a preindustrial rise in numbers from 1650 to 1850, when the population of the world doubled. From 1650 to 1750 the rate of growth was 0.3% per year; from 1750 to 1850 it was 0.5% per year (Ehrlich, Ehrlich, and Holdren 1977, 190). The causes of this huge increase in the seventeenth and eighteenth centuries were numerous (Ehrlich, Ehrlich and Holdren 1977, 189). The peace that followed the Thirty Years War in Europe led to renewed economic development in a postfeudal environment. A commercial revolution and an agricultural revolution went hand in hand in Europe. The growth of cities created greater demands for agricultural goods, and we see the development of huge farms, improvement in

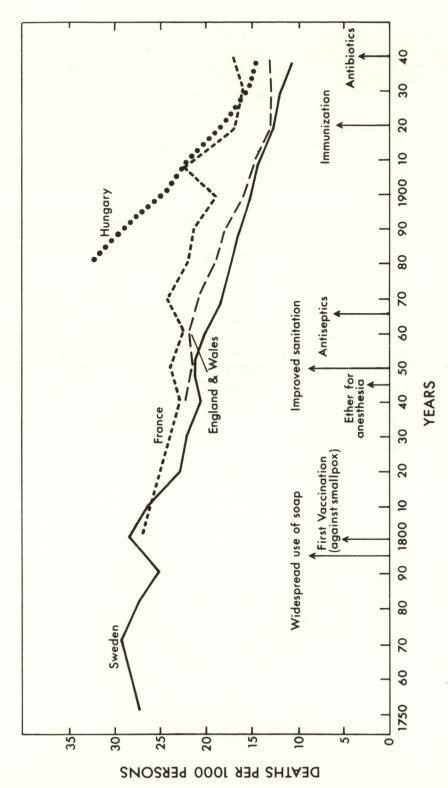

**Fig. 15.08** The annual death rates in England and Wales, Sweden, France, and Hungary from the time continuous records were kept. (Modified from McKeown, Brown, and Record 1972)

farming techniques, and improved crops. The Black Death had largely disappeared from Europe late in the seventeenth century, probably owing to the building of houses less suited for rats (sec. 5.222), though, as Langer (1972) points out, a severe outbreak of plague as late as 1720 left forty thousand dead of the ninety thousand citizens of Marseilles. In addition to these changes in Europe, the New World was opened up. In 1600 the population of North America was about one million, and the population of Mexico, Central and South America, and the Caribbean islands was no more than twelve to fourteen million (Miles 1971). In two hundred years the population of the Americas nearly doubled. In the one hundred years after this it multiplied six times.

In North America especially, the new inhabitants who were mostly farmers, had large families. The addition of two new crops from the new world, maize and the potato, added to the crops of peoples all over the world. The population of Asia also nearly doubled between 1650 and 1850, most of the increase probably occurring in China. This growth is more difficult to understand (Ehrlich, Ehrlich, and Holdren 1977, 189).

Probably the most important cause of increased growth in the human population from 1650 to 1850 was reduction in the death rate in pre-industrialized Europe (fig. 15.08). The incidence of infectious diseases such as tuberculosis and smallpox decreased. In the latter part of the eighteenth century 10% of all deaths were ascribed to smallpox (Langer 1972). We tend to assume that the disappearance of infectious diseases was due to modern medicine, but the sole therapeutic innovation up to 1798 was vaccination against smallpox. It was first used in 1798 and was made compulsory in England in 1853 (fig. 15.08) in an act "to extend and make compulsory the practice of vaccination" (Dixon 1962). It seems likely that the decrease of infectious disease in pre-industrial Europe was primarily due to better nutrition following improvements in agriculture and to the adoption of better personal hygiene (Ehrlich, Ehrlich, and Holdren 1977, 190). Except for the inhabitants of North America, there is no good evidence of an increase in the birthrate in pre-industrial peoples. The modern phase of population growth dates from the industrial revolution, about 1850, but cause and effect are by no means clear. The industrial revolution had been preceded by the renaissance of science some one hundred years earlier. From 1850 to 1950 the world's population increased from 1 billion to 2.5 billion at a growth rate of 0.8% per year. Despite the appalling conditions in mines and factories in England and Europe in the nineteenth century, the overall health had improved. More people were employed and were earning the sort of income that enabled them to feed themselves adequately. Hygiene had also improved. It was not until 1850 that sewage and drinking water were separated in the great cities of the world (fig. 15.08). Statistical demonstration of the connection between disease and the environment, such as the relation of poverty, particular occupations, and water supplies to disease, led to agitation by the medical profession and lay groups. They helped to spread knowledge about personal hygiene, and as a result of political action water supply and sewage disposal were better controlled (McCall 1970).

Most of the improvements in public health were achieved by changing components of man's environment to make it more healthful. The effectiveness of such hygiene as was practiced was given its rational explanation later on by the science of bacteriology. The discovery of the anesthetic properties of nitrous oxide in 1800 and the practical demonstration of the use of ether for anesthesia in 1846 greatly extended the scope of surgery but did not decrease the death rate from surgery. It was not until the introduction of antiseptic procedures in 1867 that surgery became relatively safe (McKeown and Brown 1955). Immunization against tuberculosis did not commence until 1921. Immunization against diphtheria was introduced in 1940 (though the antitoxin was available in 1895). The sulfonamides first became generally available in 1939 and other antibiotics came into use in the 1940s (fig. 15.08) (McCall 1970).

Starting in the 1800s death rates began to fall in countries undergoing industrialization (figs. 15.08, 15.09). European death rates in 1850 were twenty-two to twenty-four per thousand persons. By 1900 the death rates had dropped to eighteen

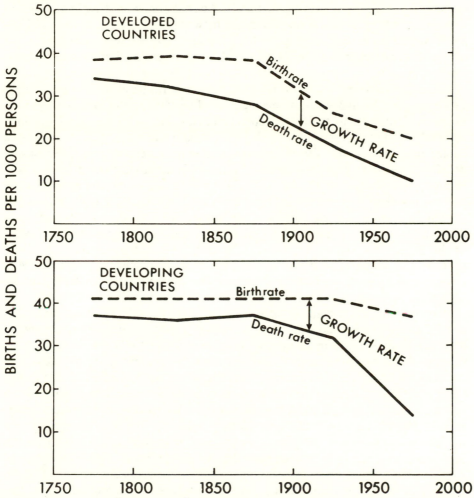

**Fig. 15.09** Annual death rates in developed and developing countries from 1750 to illustrate the "demographic transition" that began in developing countries in the latter part of the nineteenth century. (After Miles 1971)

to twenty per one thousand persons and to as low as sixteen per thousand in Sweden, Denmark, and Norway (Ehrlich, Ehrlich, and Holdren 1977, 192).

Before the end of the nineteenth century another trend began in industrialized countries. Birthrates began to fall (fig. 15.09). This was the start of the so-called "demographic transition", the fall in birthrates following the fall in death rates in Europe and North America associated with industrialization. The cause is not known for certain. In general it has been assumed that the changed conditions of life following industrialization led to a change in attitudes toward family size and to consequent limitation of births. But as Ehrlich, Ehrlich, and Holdren 1977, 193) point out, this cannot have been the only factor even if it were the main one.

The demographic transition continued into the twentieth century, yet the populations of many of these countries were still growing rapidly. By the 1930s the decrease in birthrate in some European countries had outpaced the decline in death rates, and population growth began to fall. After World War II birthrates rose in the 1940s. European growth rates averaged between 0.5% and 1.0% from 1945 to 1970. Since then decline in birthrates has continued, and in a few countries the rate of increase had declined to zero by the mid-1970s (Ehrlich, Ehrlich, and Holdren 1977, 196–197).

In most developing countries the death rate began a rapid decline from about 1940 (fig. 15.09). In Mexico it began earlier. This decline was largely due to the use of modern drugs to combat infectious diseases, the use of insecticides against malarial mosquitoes, and the use of public health measures. The result was a control of such infectious diseases as malaria, smallpox, yellow fever, tuberculosis, and dysentery, to an extent not previously known in developing countries (Ehrlich, Ehrlich, and Holdren 1977, 197). This decline in the death rate produced the postwar population explosion and resulted in the highest rate of world population growth for all time—about 1.9% per year in about 1970 (Brown 1979). After this there has been a slow decline to 1.8% in 1983 (World Population Data Sheet, 1983, of the Population Reference Bureau). The decline is due to China's successful birth-control program, decline in births in the developed countries, the decrease in birthrates of some Asian countries, and the rise in death rates in others (Brown 1979).

How has the envirogram of man changed with the advent of the industrial revolution and its extension into modern times? Perhaps this can be appreciated by referring to the envirogram of agricultural man in figure 15.07 and visualizing the changes that have to be made to that diagram. It becomes quite complex. The components of the centrum remain much the same, but the influence of most of them is different. Because of the large numbers of people, there is not only an extrinsic shortage of food and water when seasons are bad but an intrinsic shortage in local populations, especially in developing countries. The total quantity of food that the world produces increases year by year, but more people experience an intrinsic shortage of food because there are so many of them. This is likely to get worse unless population growth can be curtailed, because there is a limit to the quantity of food and water that the earth can produce. There are indications that some limits are now being rapidly approached. Man is utterly dependent on the products of the earth's pastures, crops, fisheries, and forests. The first three produce all our food; all four produce much of the raw material that is used by secondary industry. There is some evidence that the global production per person of each of these primary (food-producing) industries has peaked and is now declining (Brown 1979; Ridker and Cecelski 1979). There are many reasons: erosion, pollution, overgrazing, and overfishing are among the prominent ones. Underlying all of them is the basic problem of too many people. The pathways that lead to food and water in the envirogram of the industrial state incorporate huge secondary industries that provide farm machinery, fertilizers, and pesticides for producing food and another set of industries for distributing and marketing it. Much the same applies to the water supply, which needs to be dammed and reticulated from great distances to where people need it. All these industries require minerals and energy in dimensions that grow exponentially while the proportion of people actually producing food grows smaller.

As we have already indicated, the role of predators in the environment of man has changed in the industrial state. Infectious diseases have virtually disappeared from a large part of the earth and are no longer the scourge they were a few generations ago. In time infectious disease may disappear altogether from man's envirogram, thanks first to hygiene and then to modern therapeutic practices. And extremes of weather are no longer serious malentities for industrialized man because tremendous quantities of energy, timber, and minerals are used in cooling, warming, and otherwise protecting man in his industrial cities from heat and cold and wind and storm.

But a new set of malentities have arisen. Man has generated his own malentities in a way virtually unknown in the rest of the animal kingdom. So we need to add a rider to the second part of the general definition of a malentity at the beginning of chapter 6, namely, "The activity of the malentity remains unchanged (or decreases) as the primary animal becomes more numerous except insofar as the primary animal (e.g. man) manufactures malentities against itself". Because these malentities are important we indicate below some segments of this part of man's envirogram.

Insofar as the gun or the bomb of the terrorist may be counted as an extension of the animal, the terrorist per.se. may be counted as an aggressive malentity (sec. 6.2). On the other hand, the smog left behind by industrialists or by commuters, in their cars, is the malentity; and the humans who made it have a place in the web as modifiers of a malentity. The analogy is with the bullock that leaves deep footprints in its path (sec. 6.1). In either case the conclusion seems clear. The environment of *Homo* is likely to deteriorate on a worldwide scale as the population grows.

Some forms of pollution have a direct impact on man's well-being, notably atmospheric pollution from motor vehicles and industries and chemicals such as mercury and other heavy metals that pollute water and food.

Factories ⎤
            ⎥→Pollutants in air and water causing disease
Motor vehicles ⎦

Pollution control usually involves a shifting of impact from one place to another. Smokestacks are made taller in England, and acid rain falls in Scandinavia as a consequence. The ultimate solution to pollution is negative growth in pollution until the nuisance is alleviated, even if this entails drastic modification or even negative growth in certain industries (Birch 1976). As yet the modern world has hardly faced up to this prospect.

In the modern industrial world through political warfare, which is entirely manipulated by himself, man, together with the weapons that he wields, becomes a malentity of gargantuan proportions in the web of his own environment. As weapons become more deadly and more accurate in reaching their targets, the threat to human life becomes enormous. The pathway of the envirogram that leads to man as a malentity in warfare is complex, and we merely indicate a few of its components.

Nation-states ⎤
and their demands ⎟
⎟
Resources (especially ⎬→soldiers, terrorists, gangsters,
their unequal ⎟    and their organizers
distribution) ⎟
⎟
Political oppression ⎦

We refrain from discussing the "resources" of industrial man because the subject is too big for us, but we note that, insofar as modern war requires huge quantities of minerals and energy and the work of great numbers of people other than soldiers, war appears in the envirogram as a modifier in the pathway that leads to resources.

Accidents account for an increasing proportion of deaths and maiming of humans in the modern industrial state. Table 15.02 shows the relative proportions of the major causes of death through the whole life-span of Australians in 1972. Accident was the most frequent cause of death in every age-class from 5–9 to 33–39. In the age-group 1–4, accident was the most frequent cause of death along with respiratory and intestinal infections. It is an even more striking cause of death in the age-group 1—19 years. The same applies to the age-group 20–29 years in both sexes. In the 30–39 age-group accidents are still the most important cause of death in man. Automobile crashes account for more than half the deaths due to accidents in Australia (Hetzel 1980, 118).

Infectious disease has so decreased in importance in the industrial society that it no longer features as a major cause of death in Australia (table 15.02). Two other groups of diseases, neoplasms (cancer) and cardiovascular disease (heart failure and high blood pressure), have risen to take their place, especially for those over 40. These diseases, especially those in the cardiovascular group, are linked with what has been called a "stressful life-style". The signs of stress are manifold: they include excessive consumption of tobacco and other drugs and of food, and inadequate recreation and exercise. In addition to being signs of stress, these behaviors may also contribute to the consequences of stress. But the fundamental cause of stress is failure to control emotions that are aroused by the individual's maladjustment to industrial society, either in the workplace or in the home, including in each instance the other people associated with the individual. It is now widely accepted that to prevent stress changes have to be made in "environment" and life-style (Hetzel 1980, 165).

**Table 15.02**   Major Causes of Death, Australia, 1972

| Cause of Death | Age in Years | | | | | | | | |
|---|---|---|---|---|---|---|---|---|---|
| | 1–4 | 5–9 | 10–14 | 15–19 | 20–24 | 25–29 | 30–34 | 35–39 | 40–44 |
| Neoplasms | 11 | 22 | 16 | 7 | 7 | 15 | 17 | 20 | 22 |
| Accidents | 40 | 43 | 45 | 67 | 60 | 42 | 33 | 19 | 13 |
| Suicide | — | — | 2 | 6 | 12 | 14 | 13 | 11 | 8 |
| Cerebrovascular disease | — | — | — | 2 | 1 | 1 | 5 | 7 | 8 |
| Ischemic heart disease | — | — | — | — | 0 | 3 | 7 | 13 | 22 |
| Other | 49 | 35 | 37 | 18 | 20 | 25 | 25 | 30 | 27 |

| | 45–49 | 50–54 | 55–59 | 60–64 | 65–69 | 70–74 | 75–79 | 80–84 | 85+ |
|---|---|---|---|---|---|---|---|---|---|
| Neoplasms | 25 | 27 | 25 | 24 | 23 | 19 | 16 | 13 | 11 |
| Accidents | 9 | 6 | 4 | 3 | 2 | 2 | 2 | 3 | 3 |
| Suicide | 5 | 3 | 2 | 1 | 1 | 0.4 | 0.4 | 0.2 | 0.1 |
| Cerebrovascular disease | 8 | 9 | 9 | 10 | 13 | 15 | 18 | 21 | 22 |
| Ischemic heart disease | 27 | 32 | 36 | 38 | 38 | 37 | 40 | 31 | 29 |
| Other | 26 | 23 | 24 | 24 | 23 | 27 | 24 | 32 | 35 |

*Source:* After Hetzel (1980, 114).

*Note:* Expressed as percentage of all causes for specific age-groups.

To fit this phenomenon into an envirogram we must remember that we are discussing malentities. By definition no external organism is involved as a cause of these stress-induced cardiovascular diseases or neoplasms. (If a virus or a microble were implicated it would appear in a different pathway, probably one leading through predators in the centrum). The diseases themselves, and the physiology and the behavior that are associated with stress, all arise as internal conditions of the patient (the primary animal). These conditions all arise, in the first instance, in response to exposure to unpleasant work, incompatible bosses or workmates, domineering relatives or friends, unattainable goals, and so on and on. The "things" that produce these dangerous stimuli must be tracked down and labeled "malentity" if we are to make a useful envirogram. And, of course, the food and drugs that the stressed person consumes to excess must be added to the list.

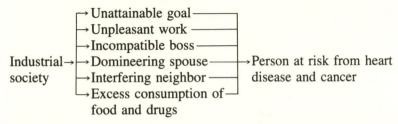

Although the causes of neoplasms (cancer) are only partly known, there is mounting evidence that causes are to be found in the environment in the form of malentities such as tobacco (causing lung cancer) and other pollutants and possibly things in the environment that lead to emotional stress (Hetzel 1980 170 et seq.). However, the most urgent questions about the future of man as a species are ecological. In modern times the impact of industrial man on the environment that determines the distribution and abundance of *Homo* has been so great that it may well be said that humans have themselves created the web that now characterizes their envirogram. Such is the extent to which invention and industry have given man "control over nature". The creation has been largely inadvertent, with little regard for, or knowledge of man's ecology. Doubtless some sort of interaction will continue. But if "progress" continues to be inadvertent, what is the chance that the habitat will remain competent to sustain dense populations of man even at the level of comfort that prevails today? On the other hand, what is the chance of inventing a "web" that would sustain an optimal number of people (doubtless many fewer than in 1984) in optimal comfort (sec. 15.3)? Even a scientific study of the optimal number for comfort might be a step in the right direction. As prerequisites it might be necessary to invent an economic system that could offer prosperity without growth, or to conceive a rational concept for money that saw money not as a good with its own price but merely as a device to facilitate the exchange of goods.

The envirograms for man (figs. 15.02, 15.07) omit some items that are essential to the ongoing life of man. This is because at present, as far as we know, their existing variation does not influence man's chance to survive and multiply; that is, their supply is reliable (see sec. 1.31). Oxygen is abundant in the atmosphere, and nothing that is happening to the atmosphere significantly affects the proportion of oxygen there. However, the proportion of carbon dioxide in the atmosphere has increased since the industrial revolution. From 1880 to 1975 the concentration of carbon dioxide increased by more than 10%, from about 290 parts per million to 320 (Ehrlich, Ehrlich, and Holdren 1977, 683). As yet, the influence of this increase on the temperature of the earth is uncertain. But if the earth's temperature does increase significantly by the year 2000 as a consequence of the increase in concentration of carbon dioxide, as some predict, then carbon dioxide would have to be included in the envirogram of man in a number of pathways where weather is already written in as a component. The same applies to particles in the atmosphere derived from industry (Ehrlich, Ehrlich, and Holdren 1977, 683). Likewise, ozone in the stratosphere between 15 and 40 km altitude plays a critical part in shielding the earth from ultraviolet solar radiation in the range of wavelengths that damage plants and animals. Destruction of the ozone is accelerated by certain catalysts including nitric oxide and chlorine, both of which are reaching the stratosphere in increasing proportions as a result of human activity on earth. For example, chlorine gets into the stratosphere from chlorofluorocarbons, widely used as refrigerants and as propellants in some aerosol spray cans. It is possible, but not yet confirmed, that the use of these gases on earth has already reduced the thickness of the ozone layer (Ehrlich, Ehrlich, and Holdren 1977, 673–74). If this is the case, then these pollutants have a place in a number of pathways in the envirogram.

Many microorganisms play an essential part in decomposing plant and animal remains and in converting nitrogen in the air to forms that plants can use. Others break down poisons such as carbon monoxide. These organisms are part of the "life-support system" of earth and provide essential "services" to the ongoing life of man and other animals. We have not included them in the envirogram, for as yet we are aware of no evidence that existing variation in their abundance influences man's chance to survive and multiply. But again the situation may change. If widespread persistent pollutants happened to be highly toxic to nitrogen-fixing organisms, their indirect influence on crop yields could be enormous. We would then have to put them in their appropriate pathways in the envirogram. This brief discussion should make it clear that the envirogram is not a diagram of everything that contributes to the life-style of man but includes only those components whose variation influences the chance of man to survive and reproduce and therefore his distribution and abundance. It also is a way of indicating the need to monitor the environment of man in all sorts of ways so that we identify new components of the envirogram before it is too late to attempt to reverse unfavorable trends. It is, of course, important that people realize the extent to which we are dependent on so many organisms and so many chemicals that we take for granted. They might then be less careless in their behavior in disposing of wastes and adding pollutants to the habitat. These aspects of human ecology are well covered by Ehrlich, Ehrlich, and Holdren (1977).

## 15.3   The Ecological Web of Man in the Future

In the context of our theory of environment the problem of the future of man centers on two questions:

1. Can resources (especially food and water) be made available in sufficient quantity and in the right places so that six billion people and more already destined to inhabit the earth will experience neither extrinsic nor intrinsic shortages of a major nature?
2. Can malentities be controlled so that their influence is not devastating to human life? As we pointed out above, these malentities include pollutants, soldiers, terrorists, and all the elements of unsatisfactory life-styles.

The answers to both questions depend upon quantities in the detailed pathways of the ecological web that lead to resources, malentities, and predators (figs 15.02, 15.07). The pathways include the number of people, components of environmental deterioration such as pollutants and soil erosion, and what are commonly called resources, such as soil, energy, minerals, and timber.

A human global society set on a path that leads to a "no" answer to these questions can be called an unsustainable society—that is, one that cannot continue to exist indefinitely at an acceptable standard of living. A human society set on a path that leads to the answer "yes" can be called sustainable: it can exist at an acceptable standard of living into the indefinite future (Birch and Cobb 1981, chap. 8). It is a matter of judgment what is and what is not a sustainable global society.

Ehrlich, Ehrlich, and Holdren (1977, 954–5) described as cornucopians those who consider the earth capable of sustaining a very much higher number of people than at present, all of them at a high standard of living, using large amounts of energy and materials per person. Cornucopians have faith in technology, mostly not invented yet, to solve problems of the finite supply of minerals and fossil fuels and problems of environmental deterioration. On the other hand, the neo-Malthusians believe there are much closer limits to the growth of the human population and its use of the earth, its minerals, its land surface, its energy, and so on. They therefore see a limit to economic growth that depends upon an indefinite increase in use of these things. Their concern is that continued population growth combined with continued economic growth will lead to an increase in the death rate or in the incidence of poverty.

The question of how many people the earth can support at a given standard of living is primarily one not of economics but of ecology. It is a matter of assessing the effects, negative and positive, of the arrows in the envirogram that converge on man. Ehrlich, Ehrlich, and Holdren (1977) have written a large book on the subject. Though they do not use our theory of the human environment (they did not know about it), we think their book is persuasive in showing that the way of the cornucopians is unsustainable. Only a change in the management of the human ecological web—that is, in our management of the earth—can lead to a sustainable future for man. This involves changes in traditional forms of agriculture, economics, politics, population policies, urban development, attitudes toward war, individual life-styles, and attitudes toward other inhabitants of earth besides ourselves (Birch and Cobb 1981, chaps. 5, 9, and 11). We pursue these subjects no further except for the last one, which we discuss in the section that follows.

## 15.4   Man in the Web of Nonhuman Life

The envirogram of an organism is one way of appreciating the part other organisms play in its chance to survive and reproduce. Many plants and animals and microorganisms contribute to man's chance to survive and reproduce, so there are sound pragmatic reasons for caring for our environment. Indeed, most arguments for conservation have emphasized the instrumental value of plants and animals to man. All the materials needed by industrialized societies come from living organisms except for water and minerals.

The part microorganisms play in the cycle of the elements, such as in the nitrogen cycle, renders essential services to man as well as to other living organisms. When we set aside national parks for our recreation and pleasure and when we preserve landscapes because of their beauty, we do so because of their instrumental value to us. And when pleas are made for the preservation of fast-disappearing rain forests on the grounds that they might contain species that could be useful to us in the future, this too appeals to the instrumental value of living organisms to man. Ehrlich and Ehrlich (1981) argue strongly for the preservation of all remaining species on the earth on the grounds that we do not yet know which ones are essential to the life-support systems of earth. Ehrlich and Ehrlich's (1981) book gives a vivid account of the influence of man on the extinction of species. In some cases this has been brought about by man deliberately acting as a malentity or a predator through hunting. But more usually extinctions come about through change in habitat as man develops land for agriculture, industry, and urban use (sec. 8.43). In these cases man becomes a part of the web modifying some essential component of the centrum such as food, shelter water or tokens (e.g., warrens of rabbits). Organic wastes and human excrement are normally broken down by microorganisms, and the breakdown products supply nutrients for plants. But when the quantities that are delivered into rivers and lakes become very great, as happens in some cities, eutrophication, as it is called, may deprive the water of virtually all its oxygen. Without oxygen, plants and animals die and organic residues accumulate (Ehrlich, Ehrlich, and Holdren 1977, 664 et seq.). In this case the envirogram of animals in the lake would show oxygen as a resource, with organic wastes entering the pathway that leads to it through plants.

Man enters the web of resources in other ways. The insecticide DDT, which man intended to be used as a malentity against insect pests, became a component of the web of many other organisms including man, entering their food chains through food and drinking water. The deleterious effect of this intrusion into the web of many organisms is now well known and needs no elaboration here (Ehrlich, Ehrlich, and Holdren 1977, 634 et seq.).

Man intrudes into another part of the environment by introducing new predators. The introduction of the *Myxoma* virus into the environment of rabbits in Australia is an example (fig. 2.01). Or species may be introduced in ignorance of their effects, as when the rabbit was brought into Australia in the first place. The rabbit eventually became an important modifier of food in the environment of sheep, to the detriment of sheep.

Man enters the web of many living organisms that have no instrumental value to humans at all or whose instrumental value is judged to be outweighed by some other value. Elephants in Africa have an instrumental value as tourist attractions, but this may be outweighed by the need to use the land for crops, as has happened in Rwanda. If plants and animals are valued only for their instrumental value to man, the time may come when the needs of an exploding human population will result in the annihilation of countless species that are in man's way. One response could be to try and rearrange man's ecological relations so this does not happen, as has already been suggested in section 15.3. Birch and Cobb (1981, chap. 5) suggest that this would almost certainly involve a new concept of what we mean by development, tying it less to the notion of economic growth in material goods. They suggest that we are unlikely to seek such alternatives unless we attribute a value to plants and animals beyond any instrumental value they may or may not have. They argue for the intrinsic value of the creature—value attributed to life just because it is life with its own experience. Life is presumably a value to the creature that experiences it, if not to others. Such an attitude toward value leads to the need for an ethic of instrinsic value that includes all life, not just human life. This might be called an ecological ethic (Birch and Cobb 1981, chap. 5).

If we value plants and animals for what they are, in addition to what use they may have for us, we shall find ourselves exploring more and more avenues to give life a chance on earth. This will mean we shall become more familiar with the way humans enter the web of life of other creatures as well as themselves. Conservation might then have a sounder ethical and scientific base. Conservation does not mean leaving living things alone. It involves management so that inevitable human intrusion is least disruptive. We agree with Charles Elton, who said conservation means "the co-existence between man and nature, even if it has to be a modified kind of man and a modified kind of nature" (Elton 1958, 145).

# Appendix: The Formal Definition of the Environment of an Animal

## B. S. Niven

The object of this appendix is to give precise, formal definitions of the notions of "centrum" and "web," which are then used to define the "environment" of an animal. Precision is achieved by using the ordinary symbols of modern logic, supplemented by certain other symbols introduced for the express purpose of formalizing the concept of the environment of an animal.

The variable $a$ designates an arbitrary animal, and $E_t a$ the environment of $a$ at time t.

The animal is conceived to be surrounded by objects (things), some of which are animate. Objects are assumed to change in time—thus, for example, a particular apple at time t may differ from the same apple at time $(t + \delta t)$ where the change in time, $\delta t$, is of magnitude relevant to the ecological situation under study.

By $\xi_t xy$ I mean that object x is brought into close physical proximity with object y at time t, evoking immediately in y some physical, physiological, or behavioral response or a change of position in space, and that no other object reduces or enhances this effect of x on y—that is, that x affects y directly.

For an arbitrary $a$, $H_t(a)$ is a positive real number that is some function of:
(i) the expectation of life of $a$ at birth or on entering its present stage of the life cycle and
(ii) the probability that $a$ will have an offspring.

An(x) is used to mean "x is an animal."

By x $\mathrm{Off}_t$ y·I mean "x is an offspring of y at time t."

By Prob $\{X\}$ I mean "the probability that the event X occurs."

I distinguish certain classes of objects surrounding $a$ at time t, which I call $R_{1t}, R_{2t} \ldots R_{Kt}, M_t, P_t,$ and $C_t$. $R_{it}$ is the class of all potential resources of $a$ of a particular kind (the ith kind), $M_t$ the class of all potential mates, $P_t$ of predators, and $C_t$ of malentities, all at time t. The elements belonging to these classes are defined below.

The environment of $a$ at time t is to be defined as the union of the centrum of $a$ at t and the web of $a$ at t:

$$E_t a =_{\mathrm{Df}} (c_t a \cup w_t a),$$

where $c_t a$, the centrum of $a$ at time t, is a set containing $(K + 3)$ elements, as follows:

$$C_t a =_{\mathrm{Df}} \{R_{1t}, \ldots, R_{Kt}, M_t, P_t, C_t\}.$$

Write

$r_i$ Res$_t a$, for "$r_i$ is a resource of $a$ at time t,"

m Mat$_t a$ for "m is a mate of $a$ at time t,"

p Pred$_t a$ for "p is a predator of $a$ at time t,"

c Mal$_t a$ for "c is a malentity of $a$ at time t." Then

$r_i$ Res$_t a$ $=_{\text{Df}} \xi_t r_i a \supset [\{H_t(a) > H_{t-\alpha}(a)\}$ & $\{\text{An}(r_i) \supset_{r_i} H_t(r_i) \leq H_{t-\gamma}(r_i)\}]$,

m Mat$_t a$ $=_{\text{Df}} \xi_t ma \supset \text{Prob}\{(\exists x)(x \text{ Off}_{t+\beta} a$ & x Off$_{t+\beta}$m$)\} > 0$,

p Pred$_t a$ $=_{\text{Df}} \xi_t pa \supset [\{H_t(a) < H_{t-\alpha}(a)\}$ & $\{H_t(p) > H_{t-\gamma}(p)\}]$,

c Mal$_t a$ $=_{\text{Df}} \xi_t ca \supset [\{H_t(a) < H_{t-\alpha}(a)\}$ & $\{\text{An}(c) \supset_c H_t(c) \leq H_{t-\gamma}(c)\}]$.

($\alpha$, $\beta$, and $\gamma$ are positive constants chosen to suit the particular situation; $\beta$ is related to the gestation time of the animal.) By these definitions the animal $a$ may not be a resource or predator or malentity of itself. It may, however, be a mate of itself; this is the case of parthenogenesis.

$w_t a$, the web of $a$ at time t, is the set of modifiers (also called Maelzer modifiers after D. A. Maelzer) and is defined in terms of the centrum.

The modifiers are objects in $E_t a$ that are linked to $a$ only through some other object or objects. A first-order modifier w is "nearer" $a$ than modifiers of higher order in the sense that only one object intervenes between w and $a$.

Write x Cent$_t a$ for "object x is in the centrum of animal $a$ at time t" and w Mod$_t^1 a$ for "w is a first-order (Maelzer) modifier in the environment of $a$ at time t." Then

$$w \text{ Mod}_t^1 a =_{\text{Df}} (\exists x)\Big[\{[(\xi_t wx \vee \xi_t wa) \supset x\text{Cent}_t a] \&$$

$$[(\sim\xi_t wx \vee \sim\xi_t wa) \supset \sim x\text{Cent}_t a]\}$$

$$\vee \{[(\sim\xi_t wx \vee \sim\xi_t wa) \supset x\text{Cent}_t a] \&$$

$$[(\xi_t wx \vee \xi_t wa) \supset \sim x\text{Cent}_t a]\}\Big].$$

The modifiers of higher order are successively "further" from $a$ (not necessarily spatially but in an ecological sense). Write w Mod$_t^n a$ for "w is a modifier of order n in the environment of $a$ at time t." Then

$$w \text{ Mod}_t^{n+1} a =_{\text{Df}} (\exists x)\Big[\{[(\xi_t wx \vee \xi_t wy) \supset x \text{ Mod}_t^n a] \&$$

$$[(\sim\xi_t wx \vee \sim\xi_t wy) \supset \sim x \text{ Mod}_t^n a]\}$$

$$\vee \{[(\sim\xi_t wx \vee \sim\xi_t wy) \supset x \text{ Mod}_t^n a] \&$$

$$[(\xi_t wx \vee \xi_t wy) \supset \sim x \text{ Mod}_t^n a]\}\Big],$$

n = 1, 2, 3,. . . ; y is either $a$ or a modifier of order $\leq(n - 1)$ for n > 1.

So to classify an object w as a first-order modifier we must be able to find an object x that is, or is not, an element of the centrum, depending on the presence or absence of w (or vice versa). Similarly, a second-order modifier modifies a first-order modifier, and so on. By the definitions every animal is a modifier in its own environment. This is a limiting case; in general we expect two similar animals to be modifiers in each other's environments.

$=_{Df}$   Definition. Read x $=_{Df}$ y as "x is interchangeable with y" or "x is equal by definition to y."

$\sim$   Negation. Read $\sim$A as "not A" or "A does not occur."

$\cup$   Union. A $\cup$ B is the set of all objects belonging to either A or B (or both).

&   Conjunction. Read A & B as "A and B."

$\vee$   Disjunction. Read A $\vee$ B as "either A or B (or both)."

$\supset$   The conditional. Read A $\supset$ B as "if A then B."

$\supset_x$   Read A $\supset_x$ B as "for all values of x, if A then B."

$\exists$   The existential quantifier. Read ($\exists$x) as "there exists an x (such that) . . . ," or "There is at least one x (such that). . . ."

# References and Author Index

The page citations in boldface following the bibliographical entries give the locations of references to the given title in the text and replace the customary author index.

Adams, R. M. 1960. The origin of cities. *Sci. Amer.* 203(3):153–68. **P. 445.**

Ahmad, T. 1936. The influence of ecological factors on the Mediterranean flour moth *Ephestia kühniella* and its parasite *Nemeritus canescens*. *J. Anim. Ecol* 5:67–83. **P. 108.**

Alland, A., Jr. 1972. *The human imperative*. New York: Columbia University Press. **P. 445.**

Anderson, P. K. 1970. Ecological structure and gene flow in small mammals. *Symp. Zool. Soc. Lond.* 26:299–325. **P. 262.**

Anderson, W.; Dobzhansky, Th.; Pavlovsky, O.; Powell, J.; and Yardley, D. 1975. Genetics of natural populations. 42. Three decades of genetic change in *Drosophila pseudoobscura*. *Evolution* 29:24–36. **P. 274.**

Andrewartha, H. G. 1933. The bionomics of *Otiorrhynchus cribricollis* Gyll. *Bull. Ent. Res.* 24:373–84. **P. 59.**

_____. 1934. Thrips investigation. 5. On the effect of soil moisture on the viability of the pupal stages of *Thrips imaginis* Bagnall. *J. Coun. Sci. Ind. Res. Aust.* 7:239–44. **P. 220.**

_____. 1940. The environment of the Australian plague locust (*Chortoicetes terminifera* Walk.) in South Australia. *Trans. Roy. Soc. S. Aust.* 64:76–94. **Pp. 166, 168, 169, 338, 346.**

_____. 1943. The significance of grasshoppers in some aspects of soil conservation in South Australia, and Western Australia. *J. Dept. Agric. S. Aust.* 46:314–22. **P. 340.**

_____. 1944. The distribution of plagues of *Austroicetes cruciata* Sauss (Acrididae) in Australia in relation to climate, vegetation and soil. *Trans. Roy. Soc. S. Aust.* 68:315–26. **P. 164.**

_____. 1952. Diapause in relation to the ecology of insects. *Biol. Rev.* 27:50–107. **P. 58.**

_____. 1961. *Introduction to the study of animal populations*. London: Methuen. **P. 54.**

_____. 1964. How animals can live in dry places. *Proc. Linn. Soc. N.S.W.* 89:287–94. **Pp. 57, 59.**

_____. 1970. *Introduction to the study of animal populations*. 2d. ed. London: Chapman and Hall. **Pp. 194, 200, 245, 364.**

Andrewartha, H. G., and Birch, L. C. 1954. *The distribution and abundance of animals*. Chicago: University of Chicago Press. **Passim.**

_____. 1982. *Selections from the distribution and abundance of animals*. Chicago: University of Chicago Press. **P.xiv.**

Andrewartha, H. G., and Browning, T. O. 1961. An analysis of the idea of "resources" in animal ecology. *J. Theor. Biol.* 1:83–97. **Pp. 51, 55, 248.**

Andrewartha, H. G.; Davidson, J.; and Swan, D. C. 1938. Vegetation types associated with plague grasshoppers in South Australia. *Bull. Dept. Agric. S. Aust.*, no. 333. **Pp. 166, 169, 338, 385.**

Andrewartha, H. G., and Kilpatrick, D. T. 1951. The apple thrips. *J. Agric. S. Aust.*, July, 1–8. **Pp. 216, 219.**

Andrewartha, H. G.; Miethke, P. M.; and Wells, A. 1974. Induction of diapause in the pupa of *Phalaenoides glycinae* by a hormone from the suboesophageal ganglion. *J. Insect Physiol.* 20:679–701. **P. 64.**

Anonymous. 1944. British Ecological Society: Symposium on the ecology of closely allied species. *J. Anim. Ecol.* 13:176. **P. 223.**

———. 1975. *Research on the anchovy—models and reality*. Lima: Pub. Instituto del Mar Peru (IMARPE). **P. 403.**

Aragão, H. de B. 1920. Transmissão do virus do myxoma dos coelhos pelas pulgas. *Brazil-Med.* 34:753. **P. 358.**

Ayala, F. J. 1965a. Evolution of fitness in experimental populations of *Drosophila serrata*. *Science* 150:903–5. **Pp. 270, 271.**

———. 1965b. Relative fitness of populations of *Drosophila serrata* and *Drosophila birchii*. *Genetics* 51:527–44. **P. 270.**

———. 1966. Evolution of fitness. 1. Improvement in the productivity and size of irradiated populations of *Drosophila serrata* and *Drosophila birchii*. *Genetics* 53:883–95. **P. 271.**

———. 1968. Evolution of fitness. 2. Correlated effects of natural selection on the productivity and size of experimental populations of *Drosophila serrata*. *Evolution* 22:55–65. **Pp. 270, 301.**

———. 1969. Evolution of fitness. 5. Rate of evolution in irradiated populations of *Drosophila*. *Proc. Nat. Acad. Sci. USA* 63(3):790–93. **Pp. 271, 272.**

———. 1970. Speciation in an Australian group of sibling species of *Drosophila*. *Simposio Internacional de Zoofilogenia* (Facultad de Ciencias, Universidad de Salamanca) 1:429–41. **P. 300.**

———. 1972. Frequency-dependent mating advantage in *Drosophila*. *Behav. Genet.* 2(1):85–91. **P. 297**

Ayala, F. J., and Campbell, C. A. 1974. Frequency-dependent selection. *Ann. Rev. Ecol. Syst.* 5:115–38. **Pp. 297, 298.**

Baars, M. A. 1979. Catches in pitfall traps in relation to mean densities of carabid beetles. *Oecologia* (Berlin) 41:25–46. **Pp. 177, 201.**

Babers, F. H. 1949. Development of insect resistance to insecticides. 1. *Bull. U.S. Bur. Ent.*, no. E-776, 1–31. **P. 295.**

Baker, A. E. M. 1981. Gene flow in house mice: Introduction of a new allele into free-living populations. *Evolution* 35:243–58. **P. 262.**

Baker, J. R., and Ransom, R. M. 1932. Factors affecting the breeding of the field mouse (*Microtus agrestis*). 1. Light. *Proc. Roy. Soc. Lond.*, ser. B., 110:313–21. **P. 65.**

Balch, R. E. 1946. The spruce budworm and forest management in the Maritime Provinces. *Can. Dept. Agric. Entomol. Div. Proc.*, no. 60. **P. 410.**

Ballantyne, G. H., and Harrison, R. A. 1967. Genetic and biochemical comparisons of organophosphate resistance between strains of spider mites (*Tetranychus* species: Acari). *Ent. Exp. Appl.* 10:231–39. **P. 295.**

Barker, J. S. F. 1963. The estimation of relative fitness of *Drosophila* populations. 3. The fitness of certain strains of *Drosophila melanogaster*. *Evolution* 17:138–46. **P. 298.**

Barlow, N. D., and Dixon, A. F. G. 1980. *Simulation of lime aphid population dynamics*. Wageningen: Centre for Agricultural Publishing and Documentation. **Pp. 37, 184.**

Bateman, M. A. 1955. The effect of light and temperature on the rhythm of pupal ecdysis in the Queensland fruit fly *Dacus (Strumeta) tryoni* (Frogg.). *Aust. J. Zool.* 3:22–33. **P. 68.**

———. 1967. Adaptations to temperature in geographic races of the Queensland fruit fly, *Dacus (Strumeta) tryoni*. *Aust. J. Zool.* 15:1141–61. **Pp. 34, 268, 300, 302.**

_____. 1968. Determinants of abundance in a population of the Queensland fruit fly. In *Insect abundance*, ed. T. R. E. Southwood. *Symp. Roy. Entomol. Soc. Lond.* 4:119–31. **Pp. 33, 426, 427, 428, 429.**

_____. 1977. Dispersal and species interaction as factors in the establishment and success of tropical fruit flies in new areas. *Proc. Ecol. Soc. Aust.* 10:106–12. **Pp. 75, 431.**

Bateman, M.A.; Friend, A. H.; and Hampshire, F. 1966. Population suppression in the Queensland fruit fly, *Dacus (Strumeta) tryoni.* 2. Experiments on isolated populations in western New South Wales. *Aust. J. Agric. Res.* 17:699–718. **Pp. 77, 78.**

Bateman, M. A., and Sonleitner, F. J. 1967. The ecology of a natural population of the Queensland fruit fly, *Dacus tryoni.* 1. The parameters of the pupal and adult populations during a single season. *Aust. J. Zool.* 15:303–35. **P. 429.**

Bell, R. H. V. 1971. A grazing ecosystem in the Serengeti. *Sci. Amer.* 225(1):86–93. **P. 156.**

Bennett, F. D., and Hughes, I. W. 1959. Biological control of insect pests in Bermuda. *Bull. Ent. Res.* 50:423–36. **P. 238.**

Berenbaum, M. R. 1981. Patterns of furanocoumarin production and insect herbivory in a population of wild parsnip (*Pastinacea sativa* L.). *Oecologia* (Berlin) 49:236–44. **P. 93.**

Berger, E. M. 1971. A temporal survey of allelic variation in natural and laboratory populations of *Drosophila melanogaster. Genetics* 67:121–36. **P. 282.**

Bernard, C. 1957. *An introduction to the study of experimental medicine.* New York: Dover. **P. 214.**

Bernays, E. A.; Chamberlain, D. J.; and Leather, E. M. 1981. Tolerance of acridids to ingested condensed tannin. *J. Chem. Ecol.* 7:247–56. **P. 93.**

Bess, H. A.; Van den Bosch, R.; and Haramoto, F. H. 1961. Fruit fly parasites and their activities in Hawaii. *Proc. Hawaii Ent. Soc.* 17:367–78. **Pp. 147, 148, 149.**

Birch, L. C. 1955. Selection in *Drosophila pseudoobscura* in relation to crowding. *Evolution* 9:389–99. **Pp. 276, 277.**

_____. 1961. Natural selection between two species of tephritid fruit fly of the genus *Dacus. Evolution* 15:360–74. **P. 151.**

_____. 1971. The role of environmental heterogeneity and genetical heterogeneity in determining distribution and abundance. In *Dynamics of populations,* ed. P. J. den Boer and G. R. Grandwell, 109–28. Wageningen: Centre for Agricultural Publishing and Documentation. **P. 167.**

_____. 1976. *Confronting the future—Australia and the world: The next hundred years.* Ringwood, Victoria: Penguin Books. **P. 452.**

_____. 1979. The effect of species of animals which share common resources on one another's distribution and abundance. *Fortschr. Zool.* 25:197–221. **P. 128.**

Birch, L. C., and Andrewartha, H. G. 1942. The influence of moisture on the eggs of *Austroicetes cruciata* Sauss (Orthoptera) with reference to their ability to survive desiccation. *Aust. J. Exp. Biol. Med. Sci.* 20:1–8. **P. 59.**

Birch, L. C., and Battaglia, B. 1957. Selection of *Drosophila willistoni* in relation to food. *Evolution* 11:94–105. **P. 286.**

Birch, L. C., and Cobb, J. B. 1981. *The liberation of life: From the cell to the community.* Cambridge: Cambridge University Press. **Pp. 12, 436, 455, 456, 457.**

Birch, L. C.; Dobzhansky, Th.; Elliott, P. O.; and Lewontin, R. C. 1963. Relative fitness of geographic races of *Drosophila serrata. Evolution* 17:72–83. **Pp. 299, 300, 301, 304.**

Birch, L. C., and Vogt, W. G. 1970. Plasticity of taxonomic characters of the Queensland fruit flies *Dacus tryoni* and *Dacus neohumeralis* (Tephritidae). *Evolution* 24:320–43. **Pp. 269, 426.**

Bishop, J. A. 1972. An experimental study of the cline of industrial melanism in *Biston betularia* (L.) (Lepidoptera) between urban Liverpool and rural North Wales. *J. Anim. Ecol.* 41:209–43. **P. 289.**

Bishop, J. A., and Cook, L. M.; 1975. Moths, melanism and clean air. *Sci. Amer.* 232(1):90–99. **Pp. 289, 290.**

———. 1980. Industrial melanism and the urban environment. *Adv. Ecol. Res.* 11:373–404. **Pp. 288, 289.**

Bishop, J. A.; Cook, L. M.; and Muggleton, J. 1976. Variation in some moths from the industrial north-west of England. *Zool. J. Linn. Soc.* 58:273–96. **P. 291.**

———. 1978. The response of two species of moths to industrialization in north-west England. 1. Polymorphisms for melanism. *Phil. Trans. Roy. Soc. Lond.*, ser. B, 281:489 514. **Pp. 287, 289.**

Bishop, J. A.; Cook, L. M.; Muggleton, J.; and Seaward, M. R. D. 1975. Moths, lichens and air pollution along a transect from Manchester to North Wales. *J. Appl. Ecol.* 12:83–98. **P. 289.**

Bissonette, T. H. 1935. Modification of mammalian sexual cycles. 3. Reversal of the cycle in male ferrets (*Putorius vulgaris*) by increasing periods of exposure to light between October second and March thirtieth. *J. Exp. Zool.* 71:341–67. **P. 65.**

———. 1941. Experimental modification of breeding cycles in goats. *Physiol. Zool.* 14:379–83. **P. 66.**

Blair-West, J. R.; Coghlan, J. P.; Denton, D. A.; Nelson, J. F.; Orchard, E.; Scoggins, B. A.; Wright, R. D.; Myers, K.; and Junqueira, C. L. 1968. Physiological morphological and behavioural adaptation to a sodium deficient environment by wild native Australian and introduced species of animals. *Nature* 217:922–28. **Pp. 350, 351.**

Blakley, N. R., and Dingle, H. 1978. Competition: Butterflies eliminate milkweed bugs from a Caribbean island. *Oecologia* (Berlin) 37:133–36. **P. 182.**

Blower, J. G.; Cook, M. L.; and Bishop, J. A. 1981. *Estimating the size of animal populations.* London: George Allen and Unwin. **P. 162.**

Bonnell, M. L., and Selander, R. K. 1974. Elephant seals: Genetic variation and near extinction. *Science* 184:908–9. **Pp. 265, 268.**

Bower, C. C. 1977. Inhibition of larval growth of the Queensland fruit fly, *Dacus tryoni* (Diptera: Tephritidae) in apples. *Ann. Entomol. Soc. Amer.* 70:97–100. **P. 34.**

Bowers, W. S.; Ohta, T.; Cleere, J. S.; and Marsella, P. A. 1976. Discovery of insect anti-juvenile hormones in plants. *Science* 193:542–47. **P. 94.**

Braidwood, R. J. 1960. The agricultural revolution. Sci. Amer. 203(3):130–48. **Pp. 443, 445.**

Braidwood, R. J., and Reed, C. A. 1957. The achievement and early consequences of food-production: A consideration of the archaeological and natural-historical evidence. *Cold Spring Harbor Symp. Quant. Biol.* 22:19–31. **Pp. 438, 439.**

Braithwaite, L. W. 1975. Managing waterfowl in Australia. *Proc. Ecol. Soc. Aust.* 8:107–28. **Pp. 388, 392, 393.**

Braithwaite, L. W., and Frith, H. J. 1969a. Waterfowl in an inland swamp in New South Wales. 1. Habitat. *CSIRO Wildl. Res.* 14:1–16. **P. 388.**

———. 1969b. Waterfowl in an inland swamp in New South Wales. 3. Breeding. *CSIRO Wildl. Res.* 14:65–109. **Pp. 388, 393.**

Branch, G. M. 1976. Interspecific competition experienced by South African *Patella* species. *J. Anim. Ecol.* 45:507–29. **P. 135.**

Breen, P. A., and Mann, K. H. 1976. Changing lobster abundance and the destruction of kelp beds by sea urchins. *Mar. Biol.* 34:137–42. **P. 105.**

Breitenbrecher, J. K. 1918. The relation of water to the behaviour of the potato beetle in a desert. *Pub. Carnegie Inst. Washington* 263:343–84. **P. 59.**

Brewer, R. H. 1971. The influence of the parasite *Comperiella bifasciata* How. on the populations of two species of armoured scale insects, *Aonidiella aurantii* (Mask.) and *A. citrina* (Coq.) in South Australia. *Aust. J. Zool.* 19:53–63. **P. 229.**

Brian, M. V. 1956. Segregation of species of the ant genus *Myrmica*. *J. Anim. Ecol.*

25:319–37. **Pp. 124, 125.**

———. 1965. *Social insect populations*. London: Academic Press. **Pp. 124, 125.**

Broadhead, E., and Wapshere, A. J. 1966. *Mesopsocus* populations on larch in England: The distribution and dynamics of two closely-related coexisting species of Psocoptera sharing the same food resource. *Ecol. Monogr.* 36:327–88. **P. 134.**

Bronowski, J., and Mazlish, B. 1960. *The Western intellectual tradition*. London: Hutchinson. **P. 188.**

Brown, I. L., and Ehrlich, P. R. 1980. Population biology of the checkerspot butterfly, *Euphydryas chalcedona*. Structure of the Jasper Ridge colony. *Oecologia* (Berlin) 47:239–51. **P. 162.**

Brown, J. H. 1971. Mechanisms of competitive exclusion between two species of chipmunks. *Ecology* 52:305–11. **P. 129.**

Brown, L. R. 1979. *Resource trends and population policy: A time for reassessment*. Worldwatch Paper 29. Washington, D.C.: Worldwatch Institute. **P. 450.**

Browning, T. O. 1954. Water balance in the tick *Ornithodorus moubata* Murray, with particular reference to the influence of carbon dioxide on the uptake and loss of water. *J. Exp. Biol.* 31:331–40. **P. 451.**

———. 1962. The environments of animals and plants. *J. Theor. Biol.* 2:63–68. **Pp. 7, 10, 14, 120.**

———. 1963. *Animal populations*. London: Hutchinson. **P. 7.**

Bull, L. B., and Mules, M. W. 1944. An investigation of *Myxomatosis cuniculi* with special reference to the possible use of the disease to control rabbit populations in Australia. *J. Coun. Scient. Ind. Res. Aust.* 17:79–93. **P. 357.**

Bullough, V. L. 1981. Age of menarche: A misunderstanding. *Science* 213:365–66. **P. 442.**

Bullough, W. S. 1951. *Vertebrate sexual cycles*. London: Methuen. **P. 65.**

Burnett, T. 1949. The effect of temperature on an insect host-parasite population. *Ecology* 30:113–34. **Pp. 56, 106, 107.**

Bush, G. L. 1969. Sympatric host race formation and speciation in frugivorous flies of the genus *Rhagoletis* (Diptera, Tephritidae). *Evolution* 23:237–51. **P. 284.**

———. 1974. The mechanism of sympatric host race formation in the true fruit flies (Tephritidae). In *Genetic mechanisms of speciation in insects,* ed. M. J. D. White, 3–23. Sydney: Australia and New Zealand Book Company. **Pp. 282, 284.**

———. 1975. Sympatric speciation in phytophagous parasitic insects. In *Evolutionary strategies of parasitic insects and mites,* ed. P. W. Price, 187–206. New York: Plenum Press. **Pp. 282, 283.**

Bush, G. L.; Neck, R. W.; and Kitto, G. B. 1976. Screwworm eradication: Inadvertent selection for noncompetitive ecotypes during mass rearing. *Science* 193:491–93. **P. 282.**

Busvine, J. R.; Bell, J. D.; and Guneidy, A. M. 1963. Toxicology and genetics of two types of insecticide resistance in *Chrysomyia putoria* (Wied). *Bull. Ent. Res.* 54:589–600. **P. 295.**

Buxton, P. A. 1955. *The natural history of tsetse flies*. London: H. K. Lewis. **Pp. 54, 207.**

Cain, A. J., and Currey, J. D. 1963. Area effects in *Cepeae*. *Phil. Trans. Roy. Soc.,* ser. B., 246:1–81. **P. 290.**

Cain, A. J., and Sheppard, P. M. 1954. Natural selection in *Cepaea*. *Genetics* 39:89–116. **P. 292.**

Calaby, J. H. 1951. Notes on the little eagle, with particular reference to rabbit predation. *Emu* 51:33–57. **P. 363.**

Caldwell, L. D. 1964. An investigation of competition in natural populations of mice. *J. Mamm.* 45:12–30. **P. 153.**

Caldwell, L. D., and Gentry, J. B. 1965. Interactions of *Peromyscus* and *Mus* in a one-acre field enclosure. *Ecology* 46:189–92. **P. 153.**

Calvert, W. H.; Hedrick, L. E.; and Brower, L. P. 1979. Mortality of the monarch butterfly

(*Danaus plexippus* L.): Avian predation at five overwintering sites in Mexico. *Science* 204:847–50. **P. 101.**

Carrick, R. 1963. Ecological significance of territory in the Australian magpie, *Gymnorhina tibicen. Proc. 13th Int. Ornithol. Congr.*, 740–53. **Pp. 47, 66.**

———. 1972. Population ecology of the Australian black-backed magpie, royal penguin and silver gull. U.S. Dept. of Interior, *Wildlife Research Report* 2:41–99. **Pp. 40, 66, 254, 372, 375, 377, 380, 381, 382, 383, 384, 385, 386, 392.**

Carroll, C. R., and Hoffman, C. A. 1980. Chemical feeding deterrent mobilized in response to insect herbivory and counter adaptation by *Epilachna tredecimnotata. Science* 209:414–16. **P. 95.**

Carter, M. A. 1968. Studies on *Cepaea*. 2. Area effects and visual selection in *Cepaea nemoralis* (L.) and *Cepaea hortensis. Phil. Trans. Roy. Soc. Lond.*, ser. B., 253:397–446. **Pp. 290, 293.**

Cartier, J. J.; Isaak, A.; Painter, R. H.; and Sorensen, E. L. 1965. Biotypes of pea aphid *Acyrthosiphon pisum* (Harris) in relation to alfalfa clones. *Can. Ent.* 97:754–60. **P. 294.**

Case, T. J., and Sidell, R. 1983. Pattern and chance in the structure of model and natural communities. *Evolution* 37:832–49. **P. 141.**

Caswell, H. 1978. Predator-mediated coexistence: A nonequilibrium model. *Amer. Nat.* 112:127–54. **P. 210.**

Caughley, G. 1970. Eruption of ungulate populations with emphasis on Himalayan thar in New Zealand. *Ecology* 51:53–72. **Pp. 55, 56, 105.**

———. 1976. Wildlife management and the dynamics of ungulate populations. In *Applied biology*, ed. T. H. Coaker, 1:183–246. London: Academic Press. **Pp. 142, 145.**

———. 1977. *Analysis of vertebrate populations*. London: John Wiley. **Pp. 88, 143, 144, 145, 256.**

Chan, B. G.; Waiss, A. C.; and Lukefahr, M. 1978. Condensed tannin, an antibiotic chemical from *Gossypium hirsutum. J. Insect Physiol.* 24:113–18. **P. 93.**

Chappell, M. A. 1978. Behavioral factors in the altitudinal zonation of chipmunks (*Eutamias*). *Ecology* 59:565–79. **P. 129.**

Chetverikov, S. S. 1926. On certain aspects of the evolutionary process from the standpoint of genetics. *Zh. Exp. Biol.* 1:3–54 (in Russian); English translation in *Proc. Amer. Phil. Soc.* 105(1959):167–95. **P. 267.**

Chitty, D. 1967. The natural selection of self-regulatory behaviour in animal populations. *Proc. Ecol. Soc. Aust.* 2:51–78. **P. 194.**

Chitty, H. 1950. Canadian Arctic wild life enquiry, 1943–49: With a summary of results since 1933. *J. Anim. Ecol.* 19:180–93. **P. 246.**

Clark, L. R. 1947. An ecological study of the Australian plague locust, *Chortoicetes terminifera* Walk. in the Bogan-Macquarie outbreak area in N.S.W. *Bull Count. Sci. Indu. Res. Aust.*, no. 226. **Pp. 167, 168.**

———. 1964. Predation by birds in relation to the population density of *Cardiaspina albitextura* (Psyllidae). *Aust. J. Zool.* 12:349–61. **P. 101.**

Clark, W. C.; Jones, D. D.; and Holling, C. S. 1978. *Patches, movements and population dynamics in ecological systems: A terrestrial perspective*. Publication R-13-B. Vancouver: Institute of Resource Ecology, University of British Columbia. **P. 187.**

Clarke, B. C., and Murray, J. J. 1962. Changes in gene frequency of *Cepaea nemoralis* (L). *Heredity* 17:445–65. **P. 293.**

Clarke, C. A., and Sheppard, P. M. 1966. A local survey of the distribution of industrial melanic forms in the moth *Biston betularia* and estimates of the selective values of these in an industrial environment. *Proc. Roy. Soc. Lond.*, ser. B, 165:424–39. **P. 287.**

Clausen, C. P.; Clancy, D. W.; and Chock, Q. C. 1965. Biological control of the oriental fruit fly (*Dacus dorsalis* Handel) and other fruit flies in Hawaii. *U.S. Dept. Agr. Tech. Bull.*, no. 1322. **P. 148.**

Coale, A. J. 1974. The history of the human population. *Sci. Amer.* 231(3):41–51. **P. 436.**

Cody, M. L. 1968. Interspecific territoriality among hummingbird species. *Condor* 70:270–71. **P. 128.**

_____. 1974. *Competition and the structure of bird communities.* Princeton: Princeton University Press. **P. 128.**

Cole, L. C. 1960a. Competitive exclusion. *Science* 132:348–49. **P. 223.**

_____. 1960b. Further competitive exclusion. *Science* 132:1675–76. **P. 223.**

Colinvaux, P. A. 1973. *Introduction to ecology.* New York: John Wiley. **Pp. 223, 244, 245, 247.**

Conant, J. B. 1951. *Science and common sense.* New Haven: Yale University Press. **P. 214.**

Connell, J. H. 1961. The influence of interspecific competition and other factors on the distribution of the barnacle *Chthamalus stellatus. Ecology* 42:710–23. **Pp. 122, 123.**

_____. 1970. A predator-prey system in the marine intertidal region. 1. *Balanus glandula* and several predatory species of *Thais. Ecol. Monogr.* 40:49–78. **Pp. 109, 132.**

_____. 1972. Community interactions on marine rocky intertidal shores. *Ann. Rev. Ecol. Syst.* 3:169–92. **Pp. 124, 132.**

_____. 1975. Some mechanisms producing structure in natural communities: A model and evidence from field experiments. In *Ecology and evolution and evolution of communities,* ed. M. L. Cody and J. M. Diamond, 460–90. Cambridge: Harvard University Press. **Pp. 97, 147.**

_____. 1980. Diversity and the coevolution of competitors; or, The ghost of competition past. *Oikos* 35:131–38. **P. 210.**

Cooke, B. D. 1974. Food and other resources of the wild rabbit *Oryctolagus cuniculus* (L). Ph.D. thesis, University of Adelaide. **Pp. 325, 326, 327, 328, 329, 343, 344, 348, 349.**

Cornish, E. A. 1936. On the secular variation of the rainfall at Adelaide, South Australia. *Quart. J. Roy. Met. Soc.* 62:481–92. **P. 346.**

Craighead, J. J., and Craighead, F. C. 1969. *Hawks, owls and wildlife.* New York: Dover. **P. 110.**

Crampton, E. W., and Lloyd, L. E. 1959. *Fundamentals of nutrition.* San Francisco: W. H. Freeman. **P. 326.**

Creese, R. G. 1982. The distribution and abundance of the limpet, *Patelloida latistrigata,* and its interaction with barnacles. *Oecologia* (Berlin) 52:85–96. **Pp. 28, 30.**

Creese, R. G., and Underwood, A. J. 1982. Analysis of inter- and intraspecific competition amongst intertidal limpets with different methods of feeding. *Oecologia* (Berlin) 53:337–46. **P. 151.**

Crissey, W. F., and Darrow, R. W. 1949. *A study of predator control on Valcour Island.* Research Series no. 1. Albany: New York State Conservation Department, Division of Fish and Game. **P. 112.**

Croghan, P. C. 1958. The osmotic and ionic regulation in *Artemia salina* (L.). *J. Exp. Biol.* 35:219-33. **Pp. 59, 62.**

Crome, F. H. J. 1975. The ecology of fruit pigeons in tropical northern Queensland. *Aust. Wildl. Res.* 2:155–85. **P. 427.**

Cronin, J. E.; Boaz, N. T.; Stringer, C. B.; and Rak, Y. 1981. Tempo and mode in hominid evolution. *Nature* 292:113–22. **P. 436.**

Crow, J. F. 1957. Genetics of insect resistance to chemicals. *Ann. Rev. Entomol.* 2:227–46. **Pp. 295, 296.**

Curtis, G. H. 1981. Man's immediate forerunners. *Phil. Trans. Roy. Soc. Lond.,* ser. B., 292:7–20. **P. 436.**

Cushing, D. H. 1975. *Marine ecology of fisheries.* Cambridge: Cambridge University Press. **Pp. 404, 405, 406.**

da Cunha, A. B. 1951. Modification of the adaptive values of chromosomal types in *Drosophila pseudoobscura* by nutritional variables. *Evolution* 5:395–404. **Pp. 285, 286.**

Daly, J. C. 1981. Effects of social organization and environmental diversity on determining the genetic structure of a population of the wild rabbit, *Oryctolagus cuniculus. Evolution* 35:689–706. **P. 262.**

Dantzig, T. 1947. *Number: The language of science.* London: George Allen and Unwin. **Pp. 7, 8, 159.**

Darwin, Charles. 1859. *On the origin of species by means of natural selection; or, The preservation of favoured races in the struggle for life.* 1st ed. London: John Murray. Reprinted in facsimile with an introduction by Ernst Mayr, Cambridge: Harvard University Press, 1964. **Pp. 5, 6, 70, 141, 297.**

Davidson, J. 1936. Climate in relation to insect ecology in Australia. 3. Bioclimatic zones in Australia. *Trans. Roy. Soc. S. Aust.* 60:88–92. **Pp. 168, 337, 338.**

Davidson, J., and Andrewartha, H. G. 1948a. Annual trends in a natural population of *Thrips imaginis* (Thysanoptera). *J. Anim. Ecol.* 17:193–99. **Pp. 12, 208, 215, 216.**

———. 1948b. The influence of rainfall, evaporation and atmospheric temperature on fluctuations in the size of a natural population of *Thrips imaginis* (Thysanoptera). *J. Anim. Ecol.* 17:200–222. **Pp. 12, 208, 215, 217, 221.**

Davis, J. 1973. Habitat preferences and competition of wintering juncos and golden-crowned sparrows. *Ecology* 54:174–80. **P. 128.**

Davis, K. 1974. The migrations of human populations. *Sci. Amer.* 231(3):93–105. **P. 436.**

Dayton, P. K. 1971. Competition, disturbance, and community organization: The provision and subsequent utilization of space in a rocky intertidal community. *Ecol. Monogr.* 41:351–89. **Pp. 103, 117, 121, 124, 151, 154, 155, 210.**

———. 1975. Experimental studies of algal canopy interactions in a sea otter-dominated kelp community at Amchitka Island, Alaska. *Fish. Bull.* 73:230–37. **P. 105.**

De Bach, P. 1946. An insecticidal check method for measuring the efficacy of entomophagous insects. *J. Econ. Ent.* 39:695–97. **P. 114.**

———. 1949. Population studies of the long-tailed mealybug and its natural enemies on citrus trees in southern California, 1946. *Ecology* 30:14–25. **P. 104.**

———. 1969. Biological control of diaspine scale insects on citrus in California. In *Proc. First Int. Citrus Symp.* (Riverside, Calif., 1968, 2:801–15. Riverside Calif.: Riverside Color Press. **P. 56.**

De Bach, P.; Hendrickson, R. M., Jr.; and Rose, M. 1978. Competitive displacement: Extinction of the yellow scale, *Aonidella citrina* (Coq.) (Homoptera: Diaspididae), by its ecological homologue, the California red scale *Aonidiella aurantii* (Mask.) in southern California. *Hilgardia* 46:1–35. **Pp. 226, 227, 229, 234.**

De Bach, P., and Huffaker, C. B. 1971. Experimental techniques for evaluation of the effectiveness of natural enemies. In *Biological control,* ed. C. B. Huffaker, 113–40. New York: Plenum. **Pp. 115, 116.**

De Bach, P., and Sundby, R. A. 1963. Competitive displacement between ecological homologues. *Hilgardia* 34:105–66. **P. 237.**

Deevey, E. S. 1956. The human crop. *Sci. Amer.* 194(4):105–12. **Pp. 438, 439, 440.**

———. 1960. The human population. *Sci. Amer.* 203(3):194–204. **P. 437.**

DeLong, K. T. 1966. Population ecology of feral house mice: Interference by *Microtus. Ecology* 47:481–84. **P. 129.**

den Boer, P. J. 1968. Spreading of risk and stabilization of animal numbers. *Acta Biotheor.* 18:165–94. **Pp. 175, 180, 181, 184, 199, 260, 298.**

———. 1977. Dispersal power and survival: Carabids in a cultivated countryside. *Landbouwhogeschool Wageningen, Misc. Papers* 14:1–190. **Pp. 162, 176, 177, 178, 179.**

———. 1979. The significance of dispersal power for the survival of species, with special reference to the carabid beetles in a cultivated countryside. *Fortschr. Zool.* 25:79–94. **Pp. 162, 180.**

———. 1980. Exclusion or coexistence and the taxonomic or ecological relationship between species. *Neth. J. Zool.* 30:278–306. **Pp. 137, 138, 139, 153, 210.**

_____. 1981. On the survival of populations in a heterogeneous and variable environment. *Oecologia* (Berlin) 50:39–53. **Pp. 202, 205.**

Denley, E. J. 1981. The ecology of the intertidal barnacle *Tesseropora rosea*. Ph.D. thesis, University of Sydney. **P. 28.**

Denley, E. J., and Underwood, A. J. 1979. Experiments on factors influencing settlement, survival, and growth of two species of barnacles in New South Wales. *J. Exp. Mar. Biol. Ecol.* 36:269–93. **P. 28.**

de Wit, C. T. 1971. On the modelling of competitive phenomena. In *Dydnamics of populations,* ed. P. J. den Boer and G. R. Gradwell, 269–81. Wageningen: Centre for Agricultural Publishing and Documentation. **P. 241.**

Diamond, J. M. 1978. Niche shifts and the rediscovery of interspecific competition. *Am. Sci.* 66:322–31. **P. 152.**

_____. 1979. Population dynamics and interspecific competition in bird communities. *Fortschr. Zool.* 25:389–402. **Pp. 128, 152.**

Diamond, J. M., and Veitch, C. R. 1981. Extinctions and introductions in the New Zealand avifauna: Cause and effect? *Science* 211:499–501. **P. 183.**

Dickson, R. C. 1949. Factors governing the induction of diapause in the oriental fruit moth. *Ann. Entomol. Soc. Amer.* 42:511–37. **P. 59.**

_____. 1960. Development of the spotted alfalfa aphid population in North America. *Int. Cong. Entomol.,* 11, 2:26–28. **P. 266.**

Dingle, H., ed. 1978. *Evolution of insect migration and diapause.* Berlin: Springer-Verlag. **P. 307.**

Dixon, C. W. 1962. *Smallpox.* London: J. A. Churchill. **P. 448.**

Dobzhanksy, Th. 1943. Genetics of natural populations. 9. Temporal changes in the composition of populations of *Drosophila pseudoobscura. Genetics* 28:162–86. **Pp. 273, 274, 275.**

_____. 1947. Genetics of natural populations. 14. A response of certain gene arrangements in the third chromosome of *Drosophila pseudoobscura* to natural selection. *Genetics* 32:142–60. **Pp. 276, 278.**

_____. 1948. Genetics of natural populations. 16. Altitudinal and seasonal changes produced by natural selection in certain populations of *Drosophila pseudoobscura* and *Drosophila persimilis. Genetics* 33:158–76. **Pp. 254, 280.**

_____. 1951. *Genetics and the origin of species.* 3d. ed. rev. New York: Columbia University Press. **Pp. 276, 277.**

_____. 1970. *Genetics of the evolutionary process.* New York: Columbia University Press. **P. 273.**

Dobzhansky, Th., and Epling, C. 1948. The suppression of crossing over in inversion heterozygotes of *Drosophila pseudoobscura. Proc. Nat. Acad. Sci. USA* 34:137–41. **P. 274.**

Dobzhansky, Th.; Lewontin, R. C.; and Pavlovsky, O. 1963. The capacity for increase in chromosomally polymorphic and monomorphic populations of *Drosophila pseudoobscura. Heredity* 19:597–614. **P. 301.**

Dobzhansky, Th., and Pavlovsky, O. 1957. An experimental study of interaction between genetic drift and natural selection. *Evolution* 11:311–19. **P. 265.**

Dobzhansky, Th., and Spassky, B. 1953. Genetics of natural populations. 21. Concealed variability in two sympatric species of *Drosophila. Genetics* 38:471–84. **Pp. 267, 272.**

_____. 1963. Genetics of natural populations. 34. Adaptive norm, genetic load and genetic elite in *D. pseudoobscura. Genetics* 48:1467–85. **P. 267.**

Dodd, A. P. 1936. The control and eradication of prickly pear in Australia. *Bull. Ent. Res.* 27:503–17. **P. 55.**

_____. 1940. *The biological campaign against prickly pear.* Brisbane: Government Printer. **P. 53.**

Dolinger, P. M.; Ehrlich, P. R.; Fitch, W. L.; and Breedlove, D. E. 1973. Alkaloid and predation patterns in Colorado lupine populations. *Oecologia* (Berlin) 13:191–204. **Pp. 93, 94.**

Dowdeswell, W. H.; Fisher, R. A.; and Ford, E. B. 1940. The quantitative study of populations in the Lepidoptera. I. *Polyommatus icarus*. Ann. Eugen. 10:123–36. **P. 162.**

Doyle, R. W. 1975. Settlement of planktonic larvae: A theory of habitat selection in varying environments. *Amer. Nat.* 109:113–26. **P. 175.**

Drew, R. A. I.; Hooper, G. H. S.; and Bateman, M. A. 1978. *Economic fruit flies of the South Pacific region*. Queensland: Department of Primary Industries. **P. 426.**

Duggins, D. O. 1980. Kelp beds and sea otters: An experimental approach. *Ecology* 61:447–53. **P. 105.**

Dumond, D. E. 1975. The limitation of human population: A natural history. *Science* 187:713–21. **Pp. 439, 441, 442, 445.**

Dunsmore, J. D. 1966a. Nematode parasites of free-living rabbits, *Oryctolagus cuniculus* (L.), in eastern Australia. 1. Variations in the numbers of *Trichostrongylus retortae formis* (Zeder). *Aust. J. Zool.* 14:185–99. **P. 350.**

————. 1966b. Nematode parasites of free-living rabbits, *Oryctolagus cuniculus* (L.) in eastern Australia. 2. Variations in the numbers of *Graphidium strigosum* (Dujardin), Railliet and Henry. *Aust. J. Zool.* 14:625–34. **P. 350.**

————. 1966c. Nematode parasites of free-living rabbits, *Oryctolagus cuniculus* (L.) in eastern Australia. 3. Variations in the numbers of *Passalurus ambiguus* (Rudolphi). *Aust. J. Zool.* 14:635–45. **P. 350.**

————. 1971. A study of the biology of the wild rabbit in climatically different regions in eastern Australia. 4. The rabbit in the south coastal region of New South Wales, an area in which parasites appear to exert a population-regulating effect. *Aust. J. Zool.* 19:355–70. **P. 364.**

————. 1974. The rabbit in subalpine south-eastern Australia. 1. Population structure and productivity. *Aust. Wildl. Res.* 1:1–16. **Pp. 350, 364.**

Dunsmore, J. D., and Dudzinski, M. L. 1968. Relationship of numbers of nematode parasites in wild rabbits, *Orcyctolagus cuniculus* (L.), to host sex, age and season. *J. Parasitol.* 54:462–74. **Pp. 350, 364.**

Dunsmore, J. D.; Williams, R. T.; and Price, W. J. 1971. A winter epizootic of myxomatosis in subalpine south-eastern Australia. *Aust. J. Zool.* 19:275–86. **P. 356.**

Ebersole, J. P. 1977. The adaptive significance of interspecific territoriality in the reef fish *Eupomacentrus leucostictus*. *Ecology* 58:914–20. **P. 127.**

Edminster, F. C. 1939. The effect of predator control on ruffed grouse populations in New York. *J. Wildl. Manage.* 3:345–52. **P. 112.**

Edmunds, G. F., and Alstad, D. N. 1978. Coevolution in insect herbivores and conifers. *Science* 199:941–45. **P. 261.**

————. 1981. Responses of black pineleaf scales to host plant variability. In *Insect life history patterns: Habitat and geographic variation,* ed. R. F. Denno and H. Dingle, 29–38. New York: Springer-Verlag. **P. 261.**

Edney, E. B. 1947. Laboratory studies on the bionomics of the rat fleas *Xenopsylla brasiliensis* Baker and X. cheopis, Roths. 2. Water relations during the cocoon period. *Bull. Ent. Res.* 38:263–80. **P. 62.**

Ehrlich, P. R. 1979. The butterflies of Jasper Ridge. *CoEvol. Quart.,* Summer, 50–55. **P. 170.**

Ehrlich, P. R., and Birch, L. C. 1967. The "balance of nature" and "population control." *Amer. Nat.* 101:97–107. **P. 172.**

Ehrlich, P. R.; Breedlove, D. E.; Brussard, P. F.; and Sharp, M. A. 1972. Weather and the "regulation" of subalpine populations. *Ecology* 53:243–47. **P. 180.**

Ehrlich, P., and Ehrlich, A. H. 1981. *Extinction: The causes and consequences of the disappearance of species*. New York: Random House. **Pp. 119, 183, 456.**

Ehrlich, P. R.; Ehrlich, A. H.; and Holdren, J. P. 1977. *Ecoscience: Population, resources, environment.* San Francisco: W. H. Freeman. **Pp. 438, 446, 448, 450, 454, 455, 457.**

Ehrlich, P. R., and Murphy, D. D. 1981. The population biology of checkerspot butterflies. (*Euphydryas*). *Biol. Zbl.* 100:613–29. **Pp. 162, 170, 172.**

Ehrlich, P. R., and Raven, P. H. 1965. Butterflies and plants: A study in coevolution. *Evolution* 18:586–608. **Pp. 92, 93.**

Ehrlich, P. R.; White, R. R.; Singer, M. C.; McKechnie, S. W.; and Gilbert, L. E. 1975. Checkerspot butterflies: A historical perspective. *Science* 188:221–28. **Pp. 170, 171, 172.**

Eisner, T. E. 1970. Chemical defense against predation in arthropods. In *Chemical ecology,* ed. E. Sondheimer and J. B. Simeone, 157–217. New York: Academic Press. **P. 96.**

Elson, P. F. 1962. Predator-prey relationships between fish-eating birds and Atlantic salmon. *Bull. Fish. Res. Bd. Can.,* no. 133. **P. 117.**

El-Tabey, A. M.; Shihata, A.; and Mrak, E. M. 1952. Intestinal yeast floras of successive populations of *Drosophila. Evolution* 6:325–32. **P. 286.**

Elton, C. 1927. *Animal ecology.* London: Sidgwick and Jackson. **Pp. xiii, 224.**

———. 1942. *Voles, mice and lemmings.* Oxford: Oxford University Press. **P. 245.**

———. 1946. Competition and the structure of ecological communities. *J. Anim. Ecol.* 15:54–68. **P. 223.**

———. l949. Population interspersion: An essay on animal community patterns. *J. Ecol.* 37:1–23. **Pp. xiii, 44, 223.**

———. 1958. *The ecology of invasions by animals and plants.* London: Methuen. **P. 457.**

Elton, C. S., and Miller, R. S. 1954. The ecological survey of animal communities: With a practical system of classifying habitats by structural characters. *J. Ecol.* 42:460–96. **Pp. xiii, 17, 223.**

Errington, P. L. 1943. An analysis of mink predation upon muskrats in north-central United States. *Res. Bull Iowa Agr. Exp. Sta.* 320:797–924. **P. 111.**

———. 1945. Some contributions of a fifteen-year local study of the northern bobwhite to a knowledge of population phenomena. *Ecol. Monogr.* 15:1–34. **P. 111.**

———. 1946. Predation and vertebrate populations. *Quart. Rev. Biol.* 21:145–77, 221–45. **P. 118.**

———. 1963. *Muskrat populations.* Ames: Iowa State University Press. **P. 68.**

———. 1967. *Of predation and life.* Ames: Iowa State University Press. **Pp. 110, 112.**

Evans, J. W. 1932. The bionomics and economic importance of *Thrips imaginis* Bagnall. *CSIR Australia,* pamphlet no. 30,1–48. **P. 42.**

———. 1933. Thrips investigation. 1. The seasonal fluctuations in numbers of *Thrips imaginis* Bagnall and associated blossom thrips. *J. Coun. Sci. Ind. Res. Aust.* 6:145–59. **P. 217.**

———. 1934. Thrips investigation. 2. Some factors that regulate the abundance of *Thrips imaginis* Bagnall. *J. Coun. Sci. Ind. Res. Aust.* 7:61–69. **Pp. 217, 220.**

———. 1935. Thrips investigation. 6. Further observations on the seasonal fluctuations in numbers of *Thrips imaginis* Bagnall and associated blossom thrips. *J. Coun. Sci. Ind. Res. Aust.* 8:86–92. **Pp. 217, 220.**

Ewer, R. F. 1968. *The ethology of mammals.* London: Logos Press. **P. 73.**

Fairweather, P. G., and Underwood, A. J. 1983. The apparent diet of predators and biases due to different handling times of their prey. Oecologia (Berlin). 56:169–79. **P. 29.**

FAO. 1981. *Review of the state of world fishery resources.* Fisheries Circular 710 (revision 2), Rome: Food and Agriculture Organization of the United Nations. **P. 407.**

Faure, J. C. 1932. The phases of locusts in South Africa. *Bull. Ent. Res.* 23:293–424. **P. 194.**

Feder, H. M. 1963. Gastropod defensive responses and their effectiveness in reducing predation by starfishes. *Ecology* 44:505–12. **P. 97.**

Feeny, P. P. 1968. Effect of oak leaf tannins on larval growth of the winter moth *Operophtera brumata*. *J. Insect Physiol.* 14:805–17. **P. 93.**

_____. 1970. Seasonal changes in oak leaf tannins and nutrients as a cause of spring feeding by winter moth caterpillars. *Ecology* 51:565–81. **P. 93.**

_____. 1976. Plant apparency and chemical defense. In *Biochemical interaction between plants and insects*, ed. G. W. Wallace and R. L. Mansell, 1–39. New York: Plenum Press. **P. 94.**

Feinsinger, P. 1976. Organization of a tropical guild of nectarivorous birds. *Ecol. Monogr.* 46:257–91. **P. 128.**

Fenner, F. 1953. Changes in the mortality-rate due to myxomatosis in the Australian wild rabbit. *Nature* 172:228–30. **P. 355.**

_____. 1965. *Myxoma* virus and *Oryctolagus cuniculus:* Two colonizing species. In *The genetics of colonizing species*, ed. H. G. Baker and G. L. Stebbins, 485–501. New York: Academic Press. **P. 294.**

Fenner, F., and Ratcliffe, F. N. 1965. *Myxomatosis.* Cambridge: Cambridge University Press. **Pp. 313, 353, 354, 355, 357, 358, 359, 361.**

Ferrari, J. A., and Georghiou, G. P. 1981. Effects of insecticidal selection and treatment on reproductive potential of resistant, susceptible, and heterozygous strains of the southern house mosquito. *J. Econ. Entomol.* 74:323–27. **P. 296.**

Fisher, R. A. 1958a. The nature of probability. *Cent. Rev.* 2:261–74. **P. 189.**

_____. 1958b. *The genetical theory of natural selection*. Rev. ed. New York: Dover Publications. **Pp. 266, 306.**

_____. 1959. *Smoking: The cancer controversy*. Edinburgh: Oliver and Boyd. **P. 189.**

Fisher, R. A., and Ford, E. B. 1947. The spread of a gene in natural conditions in a colony of the moth *Panaxia dominula* (L.). *Heredity* 1:143–74. **P. 164.**

Flanders, S. E. 1947. Elements of host discovery exemplified by parasitic Hymenoptera. *Ecology* 28:299–309. **P. 53.**

_____. 1948. A host-parasite community to demonstrate balance. *Ecology* 29:123. **P. 108.**

Fletcher, B. S. 1973. The ecology of a natural population of the Queensland fruit fly, *Dacus tryoni*. 4. The immigration and emigration of adults. *Aust. J. Zool.* 21:541–65. **Pp. 136, 428, 429.**

_____. 1974a The ecology of a natural population of the Queensland fruit fly, *Dacus tryoni*. 5. The dispersal of adults. *Aust. J. Zool.* 22:189–202. **Pp. 31, 430.**

_____. 1974b. The ecology of a natural population of the Queensland fruit fly, *Dacus tryoni*. 6. Seasonal changes in fruit fly numbers in the areas surrounding the orchard. *Aust. J. Zool.* 22:353–63. **Pp. 31, 32.**

_____. 1975. Temperature-regulated changes in the ovaries of overwintering females of the Queensland fruit fly, *Dacus tryoni*. *Aust. J. Zool.* 23:91–102. **Pp. 429, 430.**

_____. 1979. The overwintering survival of adults of the Queensland fruit fly, *Dacus tryoni*, under natural conditions. *Aust. J. Zool.* 27:403–11. **Pp. 34, 429, 430.**

Fletcher, B. S., and Bellas, T. E. 1979. Identification of the major components in the secretion from the rectal pheromone glands of the Queensland fruit flies *Dacus tryoni* and *Dacus neohumeralis* (Diptera: Tephritidae). *J. Chem. Ecol.* 5:795–803. **P. 425.**

Fletcher, B. S., and Giannakakis, A. 1973. Factors limiting the response of females of the Queensland fruit fly, *Dacus tryoni*, to the sex pheromone of the male. *J. Insect Physiol.* 19:1147–55. **P. 74.**

Fletcher, B. S.; Pappas, S.; and Kapatos, E. 1978. Changes in the ovaries of olive flies (*Dacus oleae* [Gmelin]) during the summer, and their relationship to temperature, humidity and fruit availability. *Ecol. Entomol.* 3:99–107. **P. 432.**

Foerster, R. E., and Ricker, W. E. 1941. The effect of reduction of predaceous fish on survival of young sockeye salmon at Cultus Lake. *J. Fish. Res. Bd. Can.* 5:315–36. **P. 127.**

Fogleman, J. C.; Starmer, W. T.; and Heed, W. B. 1981. Larval selectivity for yeast species by *Drosophila mojavensis* in natural substrates. *Proc. Nat. Acad. Sci. USA* 78:4435–39. **P. 286.**

Ford, E. B. 1945. *Butterflies*. London: Collins. **P. 17.**

Foster, G. G.; Kitching, R. L.; Vogt, W. G.; and Whitten M. J. 1975. Sheep blowfly and its control in the pastoral ecosystem of Australia. *Proc. Ecol. Soc. Aust.* 9:213–29. **P. 72.**

Fox, L. R., and Morrow, P. A. 1981. Specialization: Species property or local phenomenon? *Science* 21:887–93. **P. 97.**

Fraser, D. F. 1976a. Coexistence of salamanders in the genus *Plethodon:* A variation of the Santa Rosalia theme. *Ecology* 57:238–51. **P. 141.**

———. 1976b. Empirical evaluation of the hypothesis of food competition in salamanders of the genus *Plethodon. Ecology* 57:459–71. **P. 141.**

Frazer, B. D., and Gilbert, N. 1976. Coccinellids and aphids: A quantitative study of the impact of adult ladybirds (Coleoptera: Coccinellidae) preying on field populations of pea aphids (Homoptera: Aphididae). *J. Entomol. Soc. Brit. Columb.* 73:33–56. **Pp. 89, 91, 186.**

Freedman, R., and Berelson, B. 1974. The human population. *Sci. Amer.* 231(3):31–51. **P. 437.**

Frith, H. J. 1957. Breeding and movements of wild ducks in inland New South Wales. *CSIRO Wildl. Res.* 2:19–31. **Pp. 67, 167, 388.**

———. 1959a. The ecology of wild ducks in inland New South Wales. 1. Waterfowl habitats. *CSIRO Wildl. Res.* 4:97–107. **Pp. 388, 390, 395.**

———. 1959b. The ecology of wild ducks in inland New South Wales. 2. Movements. *CSIRO Wildl. Res.* 4:108–30. **Pp. 388, 397, 398, 399.**

———. 1959c. The ecology of wild ducks in inland New South Wales. 3. Food habits. *CSIRO Wildl. Res.* 4:131–55. **Pp. 388, 396.**

———. 1959d. The ecology of wild ducks in inland New South Wales. 4. Breeding. *CSIRO Wildl. Res.* 4:156–81. **Pp. 388, 394.**

———. 1962. Movements of the grey teal, *Anas gibberifrons* Muller (Anatidae). *CSIRO Wildl. Res.* 7:50–70. **Pp. 388, 390, 397, 399, 400, 401.**

———. 1963. Movements and mortality rates of the black duck and the grey teal in southeastern Australia. *CSIRO Wildl. Res.* 8:119–31. **Pp. 388, 402.**

———. 1967. *Waterfowl in Australia*. Sydney: Angus and Robertson. **Pp. 388, 390, 402.**

———. 1973. *Wildlife conservation*. Sydney: Angus and Robertson. **Pp. 153, 154, 183.**

Frith, H. J.; Braithwaite, L. W.; and McKean, J. L. 1969. Waterfowl in an inland swamp in New South Wales. 2. Food. *CSIRO Wildl. Res.* 14:17–64. **P. 388.**

Fuller, M. E. 1934. The insect inhabitants of carrion: A study in animal ecology. *Bull. Coun. Sci. Ind. Res. Aust.*, no. 82. **P. 16.**

Futuyma, D. J., and Mayer, G. C. 1980. Non-allopatric speciation in animals. *Syst. Zool.* 29:254–71. **P. 283.**

Futuyma, D. J., and Wasserman, S. S. 1980. Resource concentration and herbivory in oak forests. *Science* 210:920–22. **P. 102.**

Garrod, D. J. 1977. The North Atlantic cod. In *Fish population dynamics*, ed. J. A. Gulland, 216–42. London: John Wiley. **P. 403.**

Gause, G. F. 1934. *The struggle for existence*. Baltimore: Williams and Wilkins. **P. 108.**

Gentry, J. B. 1966. Invasion of a one year abandoned field by *Peromyscus polionotus* and *Mus musculus. J. Mamm.* 47:431–39. **P. 153.**

Georghiou, G. P. 1972. The evolution of resistance to pesticides. *Ann. Rev. Ecol. Syst.* 3:133–68. **P. 295.**

Georghiou, G. P., and Mellon, R. B. 1982. Pest resistance in time and space. In *Pest resistance to pesticides: Challenges and prospects*, ed. G. P. Georghiou and T. Saito. New York: Plenum. **P. 295.**

Georghiou, G. P., and Taylor, C. E. 1976. Pesticide resistance as an evolutionary phenomenon. *Proc. 15th Int. Cong. Ent.*, 759–85. **P. 295.**

Gerking, S. D. 1959. The restricted movement of fish populations. *Biol. Rev.* 34:221–42. **P. 127.**

————, ed. 1978. *Ecology of freshwater fish production*. Oxford: Blackwell Scientific Publications. **P. 407.**

Gerold, J. L., and Laarman, J. J. 1967. Behavioural responses to contact with DDT in *Anopheles atroparvus*. *Nature* 215:518–20. **P. 295.**

Gibbs, G. W. 1967. The comparative ecology of two closely related, sympatric species of *Dacus* (Diptera) in Queensland. *Aust. J. Zool.* 15:1123–39. **Pp. 33, 425.**

Giesel, J. T. 1976. Reproductive strategies as adaptations to life in temporarily heterogeneous environments. *Ann. Rev. Ecol. Syst.* 7:57–79. **P. 264.**

Gilbert, L. E. 1971. Butterfly-plant coevolution: Has *Passiflora adenopoda* won the selection race with Heliconiine butterflies? *Science* 172:585–86. **P. 95.**

Gilbert, N., and Gutierrez, A. P. 1973. A plant-aphid parasite relationship. *J. Anim. Ecol.* 42:323–40. **P. 114.**

Gilbert, N.; Gutierrez, A. P.; Frazer, B. D.; and Jones, R. E. 1976. *Ecological relationships*. Reading and San Francisco: W. H. Freeman. **Pp. 89, 91, 186.**

Gilbert, N., and Hughes, R. D. 1971. A model of an aphid population: Three adventures. *J. Anim. Ecol.* 40:525–34. **P. 114.**

Glantz, M., and Thompson, J. D., eds. 1981. *Resource management and environmental uncertainty: Lessons from coastal upwelling fisheries*. New York: Wiley-Interscience. **P. 404.**

Glover, P. E.; Jackson, C. H. N.; Robertson, A. G.; and Thomson, W. E. F. 1955. The extermination of the tsetse fly, *Glossina morsitans* Westw., at Abercorn, Northern Rhodesia. *Bull. Ent. Res.* 46:57–67. **P. 74.**

Goodhart, C. B. 1962. Variation in a colony of the snail *Cepaea nemoralis* (L.). *J. Anim. Ecol.* 31:207–37. **Pp. 290, 291.**

Grant, P. R. 1972. Interspecific competition among rodents. *Ann. Rev. Ecol. Syst.* 3:79–106. **P. 147.**

Griffin, D. R. 1969. The physiology and geophysics of bird navigation. *Quart. Rev. Biol.* 44:255–76. **P. 69.**

Grinnell, J. 1904. The origin and distribution of the chestnut-backed chickadee. *Auk* 21:364. **P. 223.**

Gulland, J. A. 1976. Production and catches of fish in the sea. In *The ecology of the seas*, ed. D. H. Cushing and J. J. Walsh, 283–314. Oxford: Blackwell Scientific Publications. **P. 403.**

Gutierrez, A. P.; Morgan, D. J.; and Havenstein, D. E. 1971. The ecology of *Aphis craccivora* Koch and subterranean clover stunt virus. 1. The phenology of aphid populations and the epidemiology of virus in pastures in south-east Australia. *J. Appl. Ecol.* 8:699–721. **P. 113.**

Gutierrez, A. P.; Summers, C. G.; and Baumgaertner, J. 1980. The phenology and distribution of aphids in California alfalfa as modified by ladybird beetle predation (Coleoptera: Coccinellidae). *Can. Entomol.* 112:489–95. **P. 113.**

Gwynne, M. D., and Bell, R. H. V. 1968. Selection of vegetation components by grazing ungulates in the Serengeti National Park. *Nature* 220:390–93. **P. 156.**

Hagen, K. S. 1966. Dependence of the olive fly, *Dacus oleae*, larvae on symbiosis with *Pseudomonas savastanoi* for the utilization of olive . *Nature* 209:423–24. **P. 226.**

Hall, F. G. 1922. The vital limits of exsiccation of certain animals. *Biol. Bull.* 42:31–51. **P. 58.**

Haniotakis, G. E. 1977. Male olive fly attraction to virgin females in the field. *Ann. Zool. Ecol. Anim.* 9:273–76. **P. 432.**

Hardin, G. 1960. The competitive exclusion principle. *Science* 131:1292–97. **P. 223.**

———. 1961. *Biology: Its principles and implications*. San Francisco: W. H. Freeman. **P. 213.**

Hardy, A. C. 1956. *The open sea: Its natural history*. Part 1. *The world of plankton*. London: Collins. **P. 37.**

Hardy, A. C., and Milne, P. S. 1937. Insect drift over the North Sea. *Nature* 139:510–11. **P. 37.**

Harger, J. R. E., and Landenberger, D. E. 1971. The effect of storms as a density dependent mortality factor on populations of sea mussels. *Veliger* 14:195–201. **P. 121.**

Harker, J. E. 1956. Factors controlling the diurnal rhythm of activity of *Periplaneta americana* L. *J. Exp. Biol.* 33:224–34. **P. 68.**

Hassell, M. P. 1976. *The dynamics of competition and predation*. London: Edward Arnold. **P. 147.**

———. 1978. *The dynamics of arthropod predator-prey systems*. Princeton: Princeton University Press. **Pp. 98, 100.**

Hatchett, J. H., and Gallun, R. 1970 Genetics of the ability of the Hessian fly, *Mayetiola destructor* to survive on wheats having different genes for resistance. *Ann. Entomol. Soc. Amer.* 63:1400–1407. **Pp. 284, 294.**

Haukioja, E. 1980. On the role of plant defences in the fluctuation of herbivore populations. *Oikos* 35:202–13. **P. 95.**

Hayward, J. S. 1961. The ability of the wild rabbit to survive conditions of water restriction. *CSIRO Wildl. Res.* 6:160–75. **P. 330.**

Hearn, A. B.; Ives, P. M.; Room, P. M.; Thomson, N. J.; and Wilson, L. T. 1981. Computer-based cotton pest management in Australia. *Field Crops Res.* 4:321–22. **P. 187.**

Hefley, H. M. 1928. Differential effects of constant humidities on *Protoparce quinquemaculatus* Haworth and its parasite *Winthemia quadripustulata*. *J. Econ. Entomol.* 21:213–21. **P. 108.**

Hershberger, W. A., and Smith, M. P. 1967. Conditioning in *Drosophila melanogaster*. *Anim. Behav.* 15:259–62. **P. 284.**

Hetzel, B. S. 1980. *Health and Australian society*. Ringwood, Victoria: Penguin Books. **Pp. 452, 453.**

Hinton, H. E. 1960. Cryptobiosis in the larva of *Polypedilum vanderplanki* Hint. (Chironomidae). *J. Insect Physiol.* 5:286–300. **P. 58.**

Hochbaum, H. A. 1942. Sex and age determination of waterfowl by cloacal examination. *Trans. 7th North Amer. Wildl. Conf.*, 299–307. **P. 392.**

Höhn, E. O. 1947. Sexual behaviour and seasonal changes in the gonads and adrenals of the mallard. *Proc. Zool. Soc. Lond.* 117:281–304. **P. 392.**

Holdren, C. E., and Ehrlich, P. R. 1982. Ecological determinants of food plant choice in the checkerspot butterfly *Euphydryas editha* in Colorado. *Oecologia* (Berlin) 52:417–23. **P. 173.**

Hölldobler, B., and Machwitz, U. 1965. Der Hochzeitsschwarm der Rossameise, *Camponotus herculeanus* L. (Hym. Formicidae). *Z. Vergl. Physiol.* 50:551–68. **P. 76.**

Holling, C. S. 1959. The components of predation as revealed by a study of small mammal predation of the European pine sawfly. *Can. Entomol.* 91:293–320. **P. 99.**

———. 1965. The functional response of predators to prey density and its role in mimicry and population regulation. *Mem. Entomol. Soc. Can.*, no. 45. **Pp. 98, 100.**

———. 1966. The functional response of invertebrate predators to prey density. *Mem. Entomol. Soc. Can.*, no. 48. **P. 98.**

———. 1973. Resilience and stability of ecological systems. *Ann. Rev. Ecol. Syst.* 4:1–23. **P. 187.**

Holmes, J. C. 1961. Effects of concurrent infections on *Hymenolepis diminuta* (Cestoda) and *Moniliformis dubius* (Acanthocephala). 1. General effects and comparison with crowding. *J. Parasitol.* 47:209–16. **P. 151.**

————. 1962. Effects of concurrent infections on *Hymenolepsis diminuta* (Cestoda) and *Moniliformis dubius* (Acanthocephala). 2. Effects on growth. *J. Parasitol.* 48:87–95. **P. 152.**

Hoover, E. E., and Hubbard, H. E. 1937. Modification of the sexual cycle in trout by control of light. *Copeia.* 4:206–10. **P. 66.**

Horn, H. S., and May, R. M. 1977. Limits to similarity among coexisting competitors. *Nature* 270:660–61. **P. 147.**

Howard, L. O. 1931. *The insect menace.* New York: Century Press. **P. 54.**

Howard, L. O., and Fiske, W. F. 1911. The importation into the United States of the parasites of the gipsy moth and the brown-tail moth. *Bull. U.S. Bur. Ent.,* no. 91. **P. 54.**

Huettel, M. D., and Bush, G. L. 1972. The genetics of host selection and its bearing on sympatric speciation in *Procecidochares* (Diptera: Tephritidae). *Entomol. Exp. Appl.* 15:465–80. **P. 284.**

Huffaker, C. B. 1958. Experimental studies on predation: Dispersion factors and predator-prey oscillations. *Hilgardia* 27:343–83. **Pp. 119, 256.**

Huffaker, C. B., and Kennett, C. E. 1956. Experimental studies on predation: Predation and cyclamen-mite populations on strawberries in California. *Hilgardia* 26:191–222. **P. 108.**

Huffaker, C. B., and Messenger, P. S., eds. 1976. *Theory and practice of biological control.* New York: Academic Press. **Pp. 108, 114, 148.**

Huffaker, C. B.; Messenger, P. S.; and De Bach, P. 1971. The natural enemy component in natural control and the theory of biological control. In *Biological control,* ed. C. B. Huffaker, 16–67. New York: Plenum Press. **P. 208.**

Hughes, R. D. 1981. The Australian bushfly: A climate-dominated nuisance pest of man. In *The ecology of pests: Some Australian case histories,* ed. R. L. Kitching and R. E. Jones, 177–91. Melbourne: CSIRO. **P. 16.**

Hughes, R. L., and Myers, K. 1966. Behavioural cycles during pseudopregnancy in confined populations of domestic rabbits and their relation to the histology of the female reproductive tract. *Aust. J. Zool.* 14:173–83. **P. 316.**

Idyll, C. P. 1973. The anchovy crisis. *Sci. Amer.* 228(6):22–29. **P. 407.**

Iles, T. D., and Sinclair, M. 1982. Atlantic herring: Stock discreteness and abundance. *Science* 215:627–33. **P. 405.**

Immelmann, K. 1963. Tierische Jahresperiodik in ökologischer Sicht. *Zool. Jb. Syst.* 91:91–200. **P. 67.**

Ishigaki, K. 1966. The interspecific territorialism between the shrikes *Lanius bucephalus* and *L. cristatus* in their cohabiting area. *Jap. J. Ecol.* 16:87–93. **P. 127.**

Istock, C. A. 1981. Natural selection and life history variation: Theory plus lessons from a mosquito. In *Insect life history patterns: Habitat and geographic variation,* ed. R. F. Denno and H. Dingle, 113–27. New York: Springer-Verlag. **P. 307.**

Istock, C. A.; Zisfein, J.; and Vavra, K. J. 1976. Ecology and evolution of the pitcher-plant mosquito. 2. The substructure of fitness. *Evolution* 30:535–47. **P. 307.**

Jackson, C. H. N. 1936. Some new methods in the study of *Glossina morsitans. Proc. Zool. Soc. Lond.* 1936:811–96. **Pp. 48, 49, 54.**

————. 1939. The analysis of an animal population. *J. Anim. Ecol.* 8:238–46. **P. 162.**

Jackson, P. B. N. 1961. The impact of predation, especially by the tiger-fish (*Hydrocyon vittatus* Cast) on African freshwater fishes. *Proc. Zool. Soc. Lond.* 136:603–22. **Pp. 97, 110.**

Jaeger, R. G. 1970. Potential extinction through competition between two species of terrestrial salamanders. *Evolution* 24:632–42. **P. 152.**

————. 1971. Competitive exclusion as a factor influencing the distributions of two species of terrestrial salamanders. *Ecology* 52:632–37. **P. 152.**

Jaenike, J. 1982. Environmental modification of oviposition behavior in *Drosophila. Amer. Nat.* 119:784–802. **P. 285.**

Janke, R. A.; McKaig, D.; and Raymond, R. 1978. Comparison and presettlement and modern upland boreal forests on Isle Royale National Park. *Forest Sci.* 24:115–21. **P. 89.**

Janzen, D. H. 1966. Coevolution of mutualism between ants and acacias in Central America. *Evolution* 20:249–75. **P. 95.**

———. 1979. New horizons in the biology of plant defenses. In *Herbivores: Their interaction with secondary plant metabolites,* ed. G. A. Rosenthal and K. H. Janzen, 331–50. New York: Academic Press. **P. 96.**

Johnson, C. G. 1951. The study of wind-borne insect populations in relation to terrestrial ecology, flight periodicity and the estimation of aerial populations. *Sci. Prog.* 39:41–62.

———. 1957. The vertical distribution of aphids in the air and the temperature lapse rate. *Quart. J. Roy. Met. Soc.* 83:194–201. **P. 54.**

Johnson, C. G., and Taylor, L. R. 1957. Periodism and energy summation with special reference to flight rhythms in aphids. *J. Exp. Biol.* 34:209–21. **P. 54.**

Johnson, F. M., and Powell, A. 1974. The alcohol dehydrogenases of *Drosophila melanogaster:* Frequency changes associated with heat and cold shock. *Proc. Nat. Acad. Sci. USA* 71:1783–84. **P. 282.**

Johnson, F. M., and Schaffer, H. E. 1973. Isozyme variability in species of the genus *Drosophila.* 7. Genotype environment relationships in populations of *D. melanogaster* from the eastern United States. *Biochem. Genet.* 10:149–63. **P. 282.**

Johnson, L. K., and Hubbell, S. P. 1974. Aggression and competition among stingless bees: Field studies. *Ecology* 55:120–27. **P. 127.**

———. 1975. Contrasting foraging strategies and coexistence of two bee species on a single resource. *Ecology* 56:1398–1406. **P. 127.**

Johnson, M. S. 1971. Adaptive lactate dehydrogenase variation in the crested blenny, *Anoplorchus. Heredity* 27:205–26. **P. 281.**

Jones, J. S.; Leith, B. H.; and Rawlings, P. 1977. Polymorphism in *Cepaea:* A problem with too many solutions? *Ann. Rev. Ecol. Syst.* 8:109–43. **Pp. 290, 293.**

Jones, J. S., and Yamazaki, T. 1974. Genetic background and the fitness of allozymes. *Genetics* 78:1185–89. **P. 266.**

Jones, N. S., and Kain, J. M. 1967. Subtidal algal colonization following the removal of *Echinus. Helgolaender Wiss. Meeresuntersuch* 15:460–66. **P. 105.**

Jordan, P. A.; Shelton, P. C.; and Allen, D. L. 1967. Numbers, turnover and social structure of the Isle Royale wolf population. *Amer. Zool.* 7:233–52. **P. 86.**

Keast, J. A., and Marshall, A. J. 1954. The influence of drought and rainfall on reproduction in Australian desert birds. *Proc. Zool. Soc. Lond.* 124:493–99. **P. 67.**

Keiser, I.; Kobayashi, R. M.; Miyashita, D. H.; Harris, E. J.; Schneider, E. L.; and Chambers, D. L. 1974. Suppression of Mediterranean fruit flies by oriental fruit flies in mixed infestations in Guava. *J. Econ. Entomol.* 67:355–60. **Pp. 150, 224, 226.**

Keith, L. B. 1974. Some features of population dynamics in mammals. *Trans. Eleventh Cong. Int. Union Game Biol.* (Stockholm), 17–58. **P. 101.**

Keith, L. B., and Windberg, L. A. 1978. A demographic analysis of the snowshoe hare cycle. *Wildl. Monogr.,* no. 58. **P. 101.**

Kennedy, J. S. 1961. A turning point in the study of insect migration. *Nature* 189:785–91. **P. 54.**

Kettlewell, H. B. D. 1955. Selection experiments on industrial melanism in the Lepidoptera. *Heredity* 9:323–42. **P. 287.**

———. 1956. Further selection experiments on industrial melanism in the Lepidoptera. *Heredity* 10:287–301. **P. 287.**

———. 1973. *The evolution of melanism.* Oxford: Clarendon Press. **Pp. 287, 290.**

Key, K. H. L. 1945. The general ecological characteristics of the outbreak areas and the outbreak years of the Australian plague locust (*Chortoicetes terminifera*, Walk.). *Bull. Coun. Sci. Indust. Res. Aust.*, no. 186. **Pp. 167, 169.**

———. 1978. *The conservation status of Australia's insect fauna*. Occasional Paper no. 1. Canberra: Australian National Parks and Wildlife Service. **P. 119.**

King, J. L. 1972. Genetic polymorphisms and environment. *Science* 176:545. **P. 263.**

Kitching, J. A., and Ebling, F. J. 1961. The ecology of Lough Ine. 11. The control of algae by *Paracentrotus lividus* (Echinoidea). *J. Anim. Ecol.* 30:373–83. **Pp. 97, 105.**

Kitching, J. A.; Sloane, J. F.; and Ebling, F. J. 1959. The ecology of Lough Ine. 8. Mussels and their predators. *J. Anim. Ecol.* 28:331–41. **P. 97.**

Klein, D. R. 1968. The introduction, increase and crash of reindeer on St. Matthew Island. *J. Wildl. Manage.* 32:350–67. **Pp. 142, 143.**

Klomp, H. 1962. The influence of climate and weather on the mean density level, the fluctuations and the regulation of animal populations. *Archs. Néerl. Zool.* 15:68–109. **P. 221.**

Kluijver, H. N. 1951. The population ecology of the great tit *Parus major*. *Ardea* 39:1–135. **P. 66.**

Knerer, G., and Atwood, C. E. 1973. Diprionid sawflies: Polymorphism and speciation. *Science* 179:1090–99. **P. 284.**

Knipling, E. F. 1955. Possibilities of insect control or eradication through the use of sexually sterile males. *J. Econ. Entomol.* 48:459–62. **P. 79.**

———. 1960. The eradication of the screw-worm fly. *Sci. Amer.* 203(4):54–61. **P. 79.**

Kodric-Brown, A., and Brown, J. H. 1978. Influence of economics, interspecific competition, and sexual dimorphism on territoriality of migrant rufous hummingbirds. *Ecology* 59:285–96. **P. 128.**

Koehn, R. K. 1969. Esterase heterogeneity: Dynamics of a polymorphism. *Science* 163:943–44. **P. 281.**

Koehn, R. K., and Rasmussen, D. I. 1967. Polymorphic and monomorphic serum esterase heterogeneity in catostomid fish populations. *Biochem. Genet.* 1:131–44. **P. 281.**

Krasnianski, L. M., and Nessonowa, N. 1934. Bestimmung ver Erhaltungsfutternormen fur asgewachsene Kaninchen bei wirtschafthchwissen-shaftlichen Versuchen. *Kleint. Peltz.* 10:249–51. **P. 326.**

Krebs, C. J. 1964. *The lemming cycle at Baker Lake Northwest territories, during 1959–62.* Technical Paper no. 15. Calgary: Arctic Institute of North America. **P. 254.**

———. 1972. *Ecology: The experimental analysis of distribution and abundance.* New York: Harper and Row. **Pp. 54, 207.**

Krebs, J. R. 1970. Regulation of numbers in the great tit (Aves: Passeriformes). *J. Zool.* 162:317–33. **P. 207.**

Krogh, A. 1939. *Osmotic regulation in aquatic animals.* Cambridge: Cambridge University Press. **P. 56.**

Kuenen, D. J. 1958. Some sources of misunderstanding in the theories of regulation of animal numbers. *Archs. Néerl. Zool.* 13, 1, suppl. 335–41. **P. 221.**

Lack, D. 1954. *The natural regulation of animal numbers.* Oxford: Clarendon Press. **P. 207.**

———. 1966. *Population studies of birds.* Oxford: Clarendon Press. **P. 207.**

Langer, W. L. 1972. Checks on population growth: 1750–1850. *Sci. Amer.* 226(2):92–99. **P. 448.**

Laughlin, R. 1965. Capacity for increase: A useful population statistic. *J. Anim. Ecol.* 34:77–91. **P. 367.**

Lawton, J. H., and Strong, D. R., Jr. 1981. Community patterns and competition in folivorous insects. *Amer. Nat.* 118:317–38. **P. 210.**

Leakey, R. E., and Lewin, R. 1977. *Origins: What new discoveries reveal about the emergence of our species and its possible future*. New York: Dutton. **P. 445.**

Lees, A. D. 1947. Transpiration and the structure of the epicuticle in ticks. *J. Exp. Biol.* 23:379–410. **P. 62.**

———. 1967. The production of apterous and alate forms in the aphid *Megoura viciae* Buckton with special reference to the role of crowding. *J. Insect Physiol.* 13:289–318. **P. 194.**

Leopold, A. 1933. *Game management*. New York: Charles Scribner's Sons. **P. 97.**

Lerner, I. M., and Ho, F. K. 1961. Genotype and competitive ability of *Tribolium* species. *Amer. Nat.* 95:329–43. **P. 298.**

Levene, H.; Pavlovsky, O.; and Dobzhansky, Th. 1954. Interactions of the adaptive values in polymorphic experimental populations of *Drosophila pseudoobscura*. *Evolution* 8:335–49. **P. 297.**

Levin, D. A. 1973. The role of trichomes in plant defense. *Quart. Rev. Biol.* 48:3–15. **P. 95.**

Lewontin, R. C. 1962. Interdeme selection controlling a polymorphism in the house mouse. *Amer Nat.* 96:65–78. **P. 297.**

———. 1965. Selection for colonizing ability. In *The genetics of colonizing species,* ed. H. G. Baker and G. L. Stebbins, 77–91. New York: Academic Press. **Pp. 303, 304.**

———. 1974. *The genetic basis of evolutionary change*. New York: Columbia University Press. **P. 280.**

Lewontin, R. C., and Birch, L. C. 1966. Hybridization as a source of variation for adaptation to new environments. *Evolution* 20:315–36. **Pp. 136, 268, 269, 425, 426.**

Lewontin, R. C., and Hubby, J. L. 1966. A molecular approach to the study of genic heterozygosity in natural populations. 2. Amount of variation and degree of heterozygosity in natural populations of *Drosophila pseudoobscura*. *Genetics* 54:595–609. **P. 267.**

Lichtwardt, E. T. 1964. A mutant linked to the DDT-resistance of an Illinois strain of house flies. *Ent. Exp. Appl.* 7:296–309. **P. 295.**

Lindauer, M. 1961. *Communication among social bees*. Cambridge: Harvard University Press. **P. 69.**

Lock, M., and Reynoldson, T. B. 1976. The role of interspecific competition in the distribution of two stream dwelling triclads, *Crenobia albina* (Dana) and *Polycelis felina* (Dalyell), in North Wales. *J. Anim. Ecol.* 45:581–92. **P. 152.**

Lofts, B., and Murton, R. K. 1968. Photoperiodic and physiological adaptations regulating avian breeding cycles and their ecological significance. *J. Zool., Lond.* 155:327–94. **P. 65.**

Lorenz, K. 1941. Vergleichende Bewegungsstudien an Anatiden. *J. Ornithol.* 89 (suppl. vol. 3):194–293. **P. 392.**

Lotka, A. J. 1932. The growth of mixed populations: Two species competing for a common food supply. *J. Washington Acad. Sci.* 22:461–69. **P. 255.**

Low, R. M. 1971. Interspecific territoriality in a pomacentrid reef fish, *Pomacentrus flavicauda* Whitley. *Ecology* 52:648–54. **P. 127.**

Lyon, D. L. 1976. A montane hummingbird territorial system in Oaxaca, Mexico. *Wilson Bull.* 88:280–99. **P. 127.**

Macan, T. T., and Worthington, E. B. 1951. *Life in lakes and rivers*. London: Collins. **P. 174.**

MacArthur, R. H., and Wilson, E. O. 1967. *The theory of island biogeography*. Princeton: Princeton University Press. **P. 304.**

McCall, M. G. 1970. Man and medicine. In *Man and his environment,* ed. R. T. Appleyard, 21–47. Perth: University of Western Australia Press. **Pp. 448, 449.**

McDonald, D. J. 1963. Natural selection in experimental populations of *Tribolium*. 3.

Characteristics of *Tribolium confusum* populations in equilibrium. *Amer. Nat.* 97:383–96. **P. 186.**

MacDonald, G. 1957. *The epidemiology and control of malaria.* London: Oxford University Press. **P. 362.**

McDonald, J. F., and Ayala, F. J. 1974. Genetic response to environmental heterogeneity. *Nature* 250:572–74. **P. 263.**

McFadden, M. W.; Kapatos, E.; Pappas, S.; and Carvounis, G. 1977. Ecological studies on the olive fly *Dacus oleae* Gmel. in Corfu. 1. The yearly life cycle. *Portici* 34:43–50. **P. 432.**

McKechnie, S. W. 1974. Allozyme variation in the fruit flies *Dacus tryoni* and *D. neohumeralis* (Tephritidae). *Biochem. Genet.* 11:337–46. **P. 425.**

————. 1975. Enzyme polymorphism and species discrimination in fruit flies of the genus *Dacus* (Tephritidae). *Aust. J. Biol. Sci.* 28:405–11. **P. 425.**

McKenzie, H. L. 1937. Morphological differences distinguishing California red scale, yellow scale, and related species. *Univ. Calif. Pub. Entom.* 6:323–36. **P. 226.**

McKeown, T., and Brown, R. G. 1955. Medical evidence related to English population changes in the eighteenth century. *Pop. Studies* 9:119–41. **P. 449.**

McKeown, T.; Brown, R. G.; and Record, R. G., 1972. An interpretation of the modern rise of population in Europe. *Pop. Studies* 26, no. 3 (November). **P. 447.**

McKey, D. 1979. The distribution of secondary compounds within plants. In *Herbivores: Their interaction with secondary plant metabolites,* ed. G. A. Rosenthal and D. H. Janzen, 55–133. New York: Academic Press. **P. 94.**

McLeod, M. J.; Hornbach, D. J.; Guttman, S. I.; Way, C. M.; and Burky, A. J. 1981. Environmental heterogeneity, genetic polymorphism, and reproductive strategies. *Amer. Nat.* 118:129–34. **P. 264.**

MacMillen, R. E., and Lee, A. K. 1969. Water metabolism of Australian hopping mice. *Comp. Biochem. Physiol.* 38:493–514. **Pp. 57, 62.**

McNeil, S., and Southwood, T. R. E. 1978. The role of nitrogen in insect-plant relations. In *Biochemical aspects of plant and animal co-evolution,* ed. J. B. Harborne, 77–98. London: Academic Press. **P. 253.**

McNeill, W. H. 1976. *Plagues and peoples.* New York: Anchor Press/Doubleday. **P. 103.**

Maelzer, D. A. 1965. A discussion of components of environment in ecology. *J. Theor. Biol.* 8:141–62. **P. 7.**

Main, A. R.; Shield, J. W.; and Waring, H. 1959. Recent studies on marsupial ecology. *Monograph. Biol.* 8:315–31. **P. 50.**

Mann, K. H. 1973. Seaweeds: Their productivity and strategy for growth. *Science* 182:975–81. **P. 105.**

Manning, A. 1967. Pre-imaginal conditioning in *Drosophila. Nature* 216:338–40. **P. 285.**

Marsden, H. M., and Holler, N. R. 1964. Social behaviour in confined populations of the cottontail and the swamp rabbit. *Wildl. Monogr.,* no. 13. **P. 73.**

Marshall, F. H. A. 1942. Exteroceptive factors in sexual periodicity. *Biol. Rev.* 17:68–89. **P. 65.**

Marshall, I. D. 1959. The influence of ambient temperature on the course of myxomatosis in rabbits. *J. Hyg., Camb.* 57:484–97. **P. 356.**

Mathews, G. V. T. 1955. *Bird navigation.* Cambridge: Cambridge University Press. **P. 69.**

May, A. W. S. 1963. An investigation of fruit flies (Fam. Trypetidae) in Queensland. 1. Introduction, species, pest status and distribution. *Queensland J. Agric. Sci.* 20:1–82. **P. 33.**

May, R. M. 1973a. Time-delay versus stability in population models with two and three trophic levels. *Ecology* 54:315–25. **P. 89.**

_____. 1973b. *Stability and complexity in model ecosystems*. Princeton: Princeton University Press. **P. 187.**

_____, ed. 1976. *Theoretical ecology: Principles and applications*. Oxford: Blackwell Scientific Publications. **P. 187.**

May, R. M.; Beddington, J. R.; Clark, C. W.; Holt, S. J.; and Laws, R. M. 1979. Management of multispecies fisheries. *Science* 205:267–76. **P. 403.**

Meats, A. 1971. The relative importance to population increase of fluctuations in mortality, fecundity and the time variables of the reproductive schedule. *Oecologia* (Berlin) 6:223–37. **P. 304.**

_____. 1976a. Developmental and long-term acclimation to cold by the Queensland fruit fly (*Dacus tryoni*) at constant and fluctuating temperatures. *J. Insect Physiol.* 22:1013–19. **P. 430.**

_____. 1976b. Thresholds for cold-torpor and cold-survival in the Queensland fruit fly, and predictability of rates of change in survival threshold. *J. Insect Physiol.* 22:1505–9. **P. 430.**

_____. 1976c. Seasonal trends in acclimatization to cold in the Queensland fruit fly (*Dacus tryoni*, Diptera) and their prediction by means of a physiological model fed with climatological data. *Oecologia* (Berlin) 26:73–87. **P. 430.**

_____. 1981. The bioclimatic potential of the Queensland fruit fly, *Dacus tryoni*, in Australia. *Proc. Ecol. Soc. Aust.* 11:151–61. **Pp. 31, 426, 436.**

Mech, L. D. 1966. *The wolves of Isle Royale*. Fauna Series 7, Washington, D. C.: U.S. National Park Service. **Pp. 82, 84, 85.**

_____. 1970. *Wolf: The ecology and behavior of an endangered species*. Garden City, N.Y.: Natural History Press. **Pp. 97, 98.**

_____. 1977. Wolf-pack buffer zone as prey reservoirs. *Science* 198:320–21. **P. 118.**

Mech, L. D., and Karns, P. D. 1976. *Role of the wolf in a deer decline in the Superior National Forest*. Forest Service Research Paper NC-148. Washington, D.C.: U.S. Dept of Agriculture. **P. 118.**

Melander, A. L. 1914. Can insects become resistant to sprays? *J. Econ. Entomol.* 7:167–72. **P. 294.**

Menge, B. A. 1976. Organization of the New England rocky intertidal community: Role of predation, competition, and environmental heterogeneity. *Ecol. Monogr.* 46:355–93. **P. 133.**

_____. 1978a. Predation intensity in a rocky intertidal community: Effect of an algal canopy, wave action and desiccation on predator feeding rates. *Oecologia* (Berlin) 34:17–35. **P. 132.**

_____. 1978b. Predation intensity in a rocky intertidal community: Relation between predator foraging activity and environmental harshness. *Oecologia* (Berlin) 34:1–16. **P. 133.**

Merritt, R. B. 1972. Geographic distribution and enzymatic properties of lactate dehydrogenase allozymes in the fathead minnow, *Pimephales promelas*. *Amer. Nat.* 106:173–84. **P. 281.**

Messenger, P. S., and van den Bosch, R. 1971. The adaptability of introduced biological control agents. In *Biological control*, ed. C. B. Huffaker, 68–92. New York: Plenum Press. **P. 294.**

Metcalf, R. L.; Metcalf, R. A.; and Rhodes, A. M. 1980. Cucurbitacins as kairomones for diabroticite beetles. *Proc. Nat. Acad. Sci. USA* 77:3769–72. **P. 95.**

Miles, R. E. 1971. *Man's population predicament,* Bulletin 27(2). Washington, D.C.: Population Reference Bureau. **Pp. 448, 449.**

Miller, S.; Pearcy, R. W.; and Berger, E. 1975. Polymorphism at the $\alpha$-glycerophosphate dehydrogenase locus in *Drosophila melanogaster*. 1. Properties of adult allozymes. *Biochem. Genet.* 13:175–88. **P. 282.**

Milne, A. 1949. The ecology of the sheep tick, *Ixodes ricinus* (L.); Host relationships of the tick. 2. Observations on hill and moorland grazings in northern England. *Parasitology* 39:173–97. **P. 49.**

———. 1950. The ecology of the sheep tick, *Ixodes ricinus* (L.): Spatial distribution. *Parasitology* 40:35–45. **P. 71.**

———. 1951. The seasonal and diurnal activities of individual sheep ticks *Ixodes ricinus* (L.) *Parasitology* 41:189–208. **P. 49.**

———. 1957. Theories of natural control of insect populations. *Cold Spring Harbor Symp. Quant. Biol.* 22:253–66. **Pp. 192, 193.**

Mitton, J. B., and Koehn, R. K. 1975. Genetic organization and adaptive responses of allozymes to ecological variables in *Fundulus heteroclitus*. *Genetics* 79:97–111. **P. 280.**

Moeur, J. E., and Istock, C. A. 1980. Ecology and evolution of the pitcher-plant mosquito. 4. Larval influence over adult reproductive performance and longevity. *J. Anim. Ecol.* 49:775–92. **P. 307.**

Monastero, S. 1967. La prima grande applicazione de lotta biologica artificiale contro la mosca delle olive (*Dacus oleae* Gmel.). *Boll. 1st Entomol. Agr. Palermo* 7:63–101. **P. 204.**

Monro, J. 1966. Population flushing with sexually sterile insects. *Science* 151:1536–38. **P. 32.**

———. 1967. The exploitation and conservation of resources by populations of insects. *J. Anim. Ecol.* 36:531–47. **P. 53.**

Moore-Ede, M. C.; Sulzman, F. M.; and Fuller, C. A. 1982. *The clocks that time us: Physiology of the circadian timing system*. Cambridge: Harvard University Press. **P. 67.**

Moran, M. J. 1980. The ecology, and effects on prey, of the predatory intertidal gastropod, *Morula marginalba*. Ph.D. thesis, University of Sydney. **Pp. 28, 29, 94.**

Moran, N., and Hamilton, W. D. 1980 Low nutritive quality as defense against herbivores. *J. Theor. Biol.* 86:247–54. **P. 94.**

Morris, R. F., ed 1963. The dynamics of epidemic spruce budworm populations. *Mem. Entomol. Soc. Can.* 31:1–332. **Pp. 26, 41, 408, 410, 412, 413, 414, 415, 416, 418, 419, 420, 421, 423, 424.**

Morrow, P. A.; Bellas, T. E.; and Eisner, T. 1976. *Eucalyptus* oils in the defensive oral discharge of Australian sawfly larvae (Hymenoptera: Pergidae). *Oecologia* (Berlin) 24:193–206. **P. 96.**

Morse, D. E.; Hooker, N.; Duncan, H.; and Jensen, L. 1979. $\gamma$-Aminobutyric acid, a neurotransmitter, induces planktonic abalone larvae to settle and begin metamorphosis. *Science* 204:407–10. **P. 68.**

Morse, D. H. 1967. Foraging relationships of brown-headed nuthatches and pine warblers. *Ecology* 48:94–103. **P. 135.**

———. 1980. *Behavioral mechanisms in ecology*. Cambridge: Harvard University Press. **P. 96.**

Mourão, C. A.; Ayala, F. J.; and Anderson, W. W. 1972. Darwinian fitness and adaptedness in experimental populations of *Drosophila willistoni*. *Genetica* 43:552–74. **P. 267.**

Mozley, A. 1955. *Sites of infection*. London: H. K. Lewis. **Pp. 388, 389.**

Mullen, D. A. 1969. Reproduction in brown lemmings (*Lemus trimucronatus*) and its relevance to their cycle of abundance. *Univ. Calif. Pub. Zool.* 85:24. **P. 245.**

Murphy, D. D., and Ehrlich, P. R. 1980. Two California checkerspot butterfly subspecies: One new, one on the verge of extinction. *J. Lepidopt. Soc.* 34(3):316–20. **P. 180.**

Murray, B. G. 1971. The ecological consequences of interspecific territorial behavior in birds. *Ecology* 52:414–23. **P. 128.**

Myers, K. 1962. A survey of myxomatosis and rabbit infestation trends in the eastern Riverina, New South Wales, 1951–60. *CSIRO Wildl. Res.* 7:1–12. **Pp. 313, 331, 332, 353, 365.**

———. 1964. Influence of density on fecundity, growth-rates and mortality in the wild rabbit. *CSIRO Wildl. Res.* 9:134–37. **P. 364.**

———. 1971. The rabbit in Australia. In *Dynamics of populations,* ed. P. J. den Boer and G. R. Gradwell, 478–503. Wageningen: Centre for Agricultural Publishing and Documentation. **Pp. 350, 364, 366, 367, 368, 370.**

Myers, K., and Calaby, J. H. 1977. Rabbit. In *Australian encyclopaedia,* 1977 ed. Sydney: Grolier Society of Australia. **P. 355.**

Myers, K., and Gilbert, N. 1968. Determination of the age of wild rabbits in Australia. *J. Wildl. Manage.* 32:841–48. **P. 366.**

Myers, K., and Parker, B. S. 1965. A study of the biology of the wild rabbit in climatically different regions in eastern Australia. 1. Patterns of distribution. *CSIRO Wildl. Res.* 10:1–32. **Pp. 166, 330, 335, 336, 364.**

———. 1975a. A study of the biology of the wild rabbit in climatically different regions in eastern Australia. 6. Changes in numbers and distributions related to climate and land systems in semiarid north-western New South Wales. *Aust. Wildl. Res.* 2:11–32. **Pp. 346, 363.**

———. 1975b. Effect of severe drought on rabbit numbers and distribution in a refuge area in semiarid north-western New South Wales. *Aust. Wildl. Res.* 2:103–20. **P. 346.**

Myers, K.; Parker, B. S.; and Dunsmore, J. D. 1975. Changes in numbers of rabbits and their burrows in a sub-alpine environment in south-eastern New South Wales. *Aust. Wildl. Res.* 2:121–33. **P. 335.**

Myers, K., and Poole, W. E. 1959. A study of the biology of the wild rabbit *Oryctolagus cuniculus* (L.), in confined populations. 1. The effects of density on home range and the formation of breeding groups. *CSIRO Wildl. Res.* 4:14–26. **Pp. 315, 364.**

———. 1961. A study of the biology of the wild rabbit *Oryctolagus cuniculus* (L.) in confined populations. 2. The effects of season and population increase on behaviour. *CSIRO Wildl. Res.* 6:1–41. **Pp. 315, 319.**

———. 1962. A study of the biology of the wild rabbit, *Oryctolagus cuniculus* (L.) in confined populations. 3. Reproduction. *Aust. J. Zool.* 10:225–67. **P. 315.**

———. 1963a. A study of the biology of the wild rabbit *Oryctolagus cuniculus* (L.) in confined populations. 4. The effects of rabbit grazing on sown pastures. *J. Ecol.* 51:435–51. **P. 315.**

———. 1963b. A study of the biology of the wild rabbit *Oryctolagus cuniculus* (L.) in confined populations. 5. Population dynamics. *CSIRO Wildl. Res.* 8:166–203. **Pp. 315, 321, 323, 363.**

Myers, K., and Schneider, E. C. 1964. Observations on reproduction mortality and behaviour in a small, free-living population of wild rabbits. *CSIRO Wildl. Res.* 9:138–43. **Pp. 319, 363.**

Mykytowycz, R. 1953. An attenuated strain of the myxomatosis virus recovered from the field. *Nature* 172:448–49. **P. 355.**

———. 1958. Social behaviour of an experimental colony of wild rabbits, *Oryctolagus cuniculus* (L.). 1. Establishment of the colony. *CSIRO Wildl. Res.* 3:7–25. **P. 315.**

———. 1959. Social behaviour of an experimental colony of wild rabbits, *Oryctolagus cuniculus* (L.). 2. First breeding season. *CSIRO Wildl. Res.* 4:1–13. **Pp. 315, 363.**

———. 1960. Social behaviour of an experimental colony of wild rabbits, *Oryctolagus cuniculus* (L.). 3. Second breeding season. *CSIRO Wildl. Res.* 5:1–20. **Pp. 315, 335.**

———. 1961. Social behaviour of an experimental colony of wild rabbits, *Oryctolagus cuniculus* (L.). 4. Conclusion: outbreak of myxomatosis, third breeding season and starvation. *CSIRO Wildl. Res.* 6:142–55. **P. 315.**

————. 1968. Territorial marking by rabbits. *Sci. Amer.* 218(5):116–19. **P. 320.**

Mykytowycz, R., and Davies, D. W. 1959. *Pasteurella pseudotuberculosis* in the Australian black-backed magpie, *Gymnorhina tibicen* (Latham). *CSIRO Wildl. Res.* 4:61–68. **P. 385.**

Mykytowycz, R., and Gambale, S. 1965. A study of the inter-warren activities and dispersal of wild rabbits, *Oryctolagus cuniculus* (L.) in a 45-acre paddock. *CSIRO Wildl. Res.* 10:111–23. **P. 320.**

Nelson, M. E., and Mech, L. D. 1981. Deer social organization and wolf predation in northeastern Minnesota. *Wildl. Monogr.*, no. 77, suppl. to *J. Wildl. Manage.* vol. 45, no. 3. **Pp. 118, 119.**

Nevo, E. 1978. Genetic variation in natural populations: Patterns and theory. *Theor. Pop. Biol.* 13:121–77. **Pp. 267, 268.**

Newsome, A. E. 1965a. The abundance of red kangaroos, *Megaleia rufa* (Desmarest) in Central Australia. *Aust. J. Zool.* 13:269–87. **P. 167.**

————. 1965b. The distribution of red kangaroos, *Megaleia rufa* (Desmarest) about sources of persistent food and water in Central Australia. *Aust. J. Zool.* 13:389–99. **P. 167.**

————. 1966. Estimating severity of drought. *Nature* 209:904. **P. 347.**

————. 1967. A simple biological method of measuring the food-supply of house-mice. *J. Anim. Ecol.* 36:645–50. **P. 245.**

————. 1969a. A population study of house-mice temporarily inhabiting a South Australian wheatfield. *J. Anim. Ecol.* 38:341–59. **P. 167.**

————. 1969b. A population study of house-mice permanently inhabiting a reed bed in South Australia. *J. Anim. Ecol.* 38:361–77. **Pp. 166, 167.**

Newsome, A. E., and Corbett, L. K. 1975. Outbreaks of rodents in semi-arid and arid Australia: Causes, preventions, and evolutionary considerations. In *Rodents in desert environments,* ed. I. Prakash and P. K. Ghosh, 117–53. The Hague: W. Junk. **P. 101.**

Nicholson, A. J. 1947. Fluctuation of animal populations. *Rep. 26th Meet. ANZAAS, Perth,* 1–14. Perth: Government Printer. **P. 53.**

————. 1950. Population oscillations caused by competition for food. *Nature* 165:476–77. **P. 45.**

————. 1957. The self-adjustment of populations to change. *Cold Spring Harbor Symp. Quant. Biol.* 22:153–73. **P. 192.**

Niven, B. S. 1967. The stochastic simulation of *Tribolium* populations. *Physiol. Zool.* 40:67–82. **Pp. 185, 186.**

————. 1980. The formal definition of the environment of an animal. *Aust. J. Ecol.* 5:37–46. **Pp. xiv, 7.**

————. 1982. Formalization of the basic concepts of animal ecology. *Erkenntniss* 17:307–20. **P. xiv.**

Nordenskiold, E. 1928. *The history of biology: A survey.* New York: Tudor. **P. 212.**

Odum, E. P. 1959. *Fundamentals of ecology.* 2d ed. Philadelphia: Saunders. **P. 224.**

O'Loughlin, G. T. 1964. The Queensland fruit fly in Victoria. *J. Agr. Vic. Dep. Agric.* 62:391–402. **P. 33.**

Oppenoorth, F. J., and Welling, W. 1976. Biochemistry and physiology of resistance. In *Insecticide biochemistry and physiology,* ed. C. F. Wilkinson, 507–51. New York: Plenum. **P. 295.**

Orians, G. H., and Collier, G. 1963. Competition and blackbird social systems. *Evolution* 17:449–59. **P. 127.**

Orians, G. H., and Wilson, M. F. 1964. Interspecific territories of birds. *Ecology* 45:736–45. **P. 128.**

Osmond, C. B., and Monro, J. 1981. Prickly pear. In *Plants and man in Australia,* ed. D. J. and S. G. M. Garr, 194–222. New York: Academic Press. **Pp. 53, 258.**

Paine, R. T. 1966. Food web complexity and species diversity. *Amer. Nat.* 100:65–75. **Pp. 115, 117, 132.**

———. 1969. The *Pisaster-Tegula* interaction: Prey patches, predator food preference, and intertidal community structure. *Ecology* 50:950–61. **Pp. 103, 105, 106.**

———. 1971a. Energy flow in a natural population of the herbivorous gastropod *Tegula funebralis. Limnol. Oceanogr.* 16:86–98. **P. 106.**

———. 1971b. A short-term experimental investigation of resource partitioning in a New Zealand rocky intertidal habitat. *Ecology* 52:1096–1106. **P. 117.**

———. 1974. Intertidal community structure: Experimental studies on the relationship between a dominant competitor and its principal predator. *Oecologia* (Berlin) 15:93–120. **Pp. 110, 117.**

———. 1976. Size limited predation: An observational and experimental approach with the *Mytilus-Pisaster* interaction. *Ecology* 57:858–73. **Pp. 97, 116.**

———. 1977. Controlled manipulations in the marine intertidal zone, and their contributions to ecological theory. In The changing scenes in natural sciences, 1776–1976. Academy of Natural Sciences, Philadelphia, *Special Pub.* 12:245–70. **P. 117.**

Paine, R. T., and Vadas, R. L. 1969. The effects of grazing by sea urchins *Strongylocentrotus* spp. on benthic algal populations. *Limnol. Oceanogr.* 14:710–19. **P. 105.**

Parer, I. 1977. The population ecology of the wild rabbit, *Oryctolagus cuniculus* (L.) in a Mediterranean-type climate in New South Wales. *Aust. Wildl. Res.* 4:171–205. **Pp. 317, 330, 331, 333, 334, 335, 363, 364.**

Park, T. 1948. Experimental studies of interspecies competition. 1. Competition between populations of the flour beetles *Tribolium confusum* Duval and *Tribolium castaneum* Herbst. *Ecol. Monogr.* 18:265–308. **P. 340.**

Park, T.; Leslie, P. H.; and Mertz, D. B., 1964. Genetic strains and competition in populations of *Tribolium. Physiol. Zool.* 37:97–162. **Pp. 185, 186, 298.**

Paulik, G. J. 1971. Anchovies, birds and fishermen in the Peru current. In *Environment, resources, pollution and society,* ed. W. W. Murdoch, 156–85. Stamford, Conn.: Sinauer Associates. **Pp. 405, 407.**

Pearson, O. P. 1966. The prey of carnivores during one cycle of mouse abundance. *J. Anim. Ecol.* 35:217–33. **P. 101.**

———. 1971. Additional measurements of the impact of carnivores on California voles *(Microtus californicus). J. Mamm.* 52:41–49. **P. 101.**

Peterman, R. M.; Clark, W. C.; and Holling, C. S. 1979. The dynamics of resilience: Shifting stability domains in fish and insect systems. In *Population dynamics,* ed. R. M. Anderson, B. D. Turner, and L. R. Taylor, 321–41. Oxford: Blackwell Scientific Publications. **P. 187.**

Peterson, R. O. 1977. *Wolf ecology and prey relationships on Isle Royale.* Scientific Monograph Series, 11. Washington, D.C.: U.S. Department of Interior, National Park Service. **Pp. 82, 86, 87, 88.**

Phillips, D. W. 1976. The effect of a species-specific avoidance response to predatory starfish on the intertidal distribution of two gastropods. *Oecologia* (Berlin) 23:83–94. **P. 97.**

Phillips, P. A., and Barnes, M. M. 1975. Host race formation among sympatric apple, walnut and plum populations of the codling moth, *Laspeyresia pomonella. Ann. Entomol. Soc. Amer.* 68:1053–60. **Pp. 282, 283.**

Pickett, A. D. 1949. A critique on insect chemical control methods. *Can. Entomol.* 81:67–76. **P. 114.**

Pickett, A. D.; Patterson, N. A.; Stultz, H. T.; and Lord, F. T. 1946. The influence of spray programs on the fauna of apple orchards in Nova Scotia. 1. An appraisal of the problem and a method of approach. *Sci. Agr.* 26:590–600. **P. 114.**

Pimentel, D., and Stone, F. A. 1968. Evolution and population ecology of parasite-host systems. *Can. Entomol.* 100:655–62. **P. 294.**

Pimlott, D. H. 1975. Ecology of the wolf in North America. In *The wild canids,* ed. M. W. Fox, 280–85. New York: Van Nostrand Reinhold. **P. 86.**

Pitelka, F. A. 1951. Ecologic overlap and interspecific strife in breeding populations of Anna and Allen hummingbirds. *Ecology* 32:641–61. **P. 128.**

Pitelka, F. A.; Tomich, Q. Q.; and Treichel, G. W. 1955. Ecological relations of jaegers and owls as lemming predators near Barrow, Alaska. *Ecol. Monogr.* 25:85–117. **P. 101.**

Pomeroy, D. E. 1966. The ecology of *Helicella virgata* and related species in South Australia. Ph.D. thesis, University of Adelaide. **P. 61.**

Pooley, A. C., and Gans, C. 1976. The Nile crocodile. *Sci. Amer.* 234(4):114–24. **P. 102.**

Popper, K. 1959. *The logic of scientific discovery*. London: Hutchinson. **Pp. 189, 192.**

Powell, J. R. 1971. Genetic polymorphisms in varied environments. *Science* 174:1035–36. **Pp. 262, 263.**

Powell, J. R.; Levene, H.; and Dobzhansky, Th. 1973. Chromosomal polymorphism in *Drosophila pseudoobscura* used for diagnosis of geographic origin. *Evolution* 26:553–59. **P. 279.**

Powell, J. R., and Taylor, C. E. 1979. Genetic variation in ecologically diverse environments. *Amer. Sci.* 67:590–96. **Pp. 264, 280.**

Powell, J. R., and Wistrand, H. 1978. The effect of heterogeneous environments and a competitor on genetic variation in *Drosophila. Amer. Nat.* 112:935–47. **P. 264.**

Prakash, S., and Lewontin, R. C. 1968. A molecular approach to the study of genic heterozygosity in natural populations. 3. Direct evidence of coadaptation in gene arrangements of *Drosophila. Proc. Nat. Acad. Sci. USA* 59:398–405. **P. 273.**

Prebble, M. L., ed. 1975. *Aerial control of forest insects in Canada*. Ottawa: Department of Environment. **P. 408.**

Price, P. W. 1980. *Evolutionary biology of parasites*. Princeton: Princeton University Press. **P. 422.**

Pritchard, G. 1970 The ecology of a natural population of Queensland fruit fly, *Dacus tryoni*. 3. The maturation of female flies in relation to temperature. *Aust. J. Zool.* 18:77–89. **P. 430.**

Pulliam, H. R., and Enders, F. A. 1971. The feeding ecology of five sympatric finch species. *Ecology* 52:557–66. **P. 141.**

Quezada, J. R., and De Bach, P. 1973. Bioecological and population studies of the cottony cushion scale *Icerya purchasi* Mask., and its natural enemies, *Rodolia cardinalis* Mul. and *Cryptochaetum iceryae* Will. in southern California. *Hilgardia* 41:631–88. **Pp. 51, 52, 230, 231, 232.**

Ratcliffe, F. N. 1959. The rabbit in Australia. In *Biogeography and ecology in Australia,* ed. F. S. Bodenheimer, 545–64. Monographie Biologie, 8. The Hague: W. Junk. **P. 313.**

Reddingius, J. 1971. Gambling for existence: A discussion of some theoretical problems in animal population ecology. *Acta Biotheor.* 20 (suppl.):1–208. **P. 221.**

Reid, P. A. 1953. Some economic results of myxomatosis. *Quart. Rev. Agric. Econ.* 6:93–94. **P. 353.**

Reissig, W. H., and Smith, D. C. 1978. Bionomics of *Rhagoletis pomonella* in *Crataegus. Ann. Entomol. Soc. Amer.* 71:155–59. **P. 283.**

Rendel, J. M. 1971. Myxomatosis in the Australian rabbit population. *Search* 2:89–94. **P. 356.**

Reynoldson, T. B. 1957. Discussion in A. J. Nicholson, The self-adjustment of populations to change. *Cold Spring Harbor Symp. Quant. Biol.* 22:153–73. **P. 192.**

————. 1966. The distribution and abundance of lake-dwelling triclads: Towards a hypothesis. In *Advances in ecological research,* ed. J. B. Cragg, 3:1–71. London: Academic Press. **Pp. 238, 239, 240.**

————. 1975. Food overlap of lake-dwelling triclads in the field. *J. Anim. Ecol.* 44:245–50. **P. 238.**

Reynoldson, T. B., and Davies, W. R. 1970. Food niche and co-existence in lake-dwelling triclads. *J. Anim. Ecol..* 39:599–617. **Pp. 240, 241.**

Rhodes, D. F., and Cates, R. G. 1976. Toward a general theory of plant antiherbivore chemistry. In *Biochemical interaction between plants and insects,* ed. G. W. Wallace and R. L. Mansell, 168–213. New York: Plenum Press. **P. 94.**

Richardson, R. H.; Ellison, J. R.; and Averhoff, W. W. 1982. Autocidal control of screwworms in North America. *Science* 215:361–70. **P. 273.**

Ridker, R. G., and Cecelski, E. W. 1979., *Resources, environment, and population: The nature of future limits.* Population Bulletin 34, no. 3. Washington, D.C.: Population Reference Bureau. **P. 451.**

Riedl, H., and Croft, B. A. 1978. The effects of photoperiod and effective temperatures on the seasonal phenology of the codling moth (Lepidoptera: Tortricidae). *Can. Entomol.* 110:455–70. **P. 307.**

Riney, T. 1964. The impact of introductions of large herbivores on the tropical environment. *IUCN Publ.,* N.S., no. 4:261–73. **P. 256.**

Robinson, M. A. 1980. World fisheries to 2000: Supply, demand and management. *Marine Policy* 4(1):19–32. **P. 404.**

Root, R. B. 1973. Organization of a plant-arthropod association in simple and diverse habitats: The fauna of collards (*Brassica oleracea*). *Ecol. Monogr.* 43:95–124. **P. 102.**

Rosenthal, G. A., and Bell, E. A. 1979. Naturally occurring toxic nonprotein amino acids. In *Herbivores: Their interaction with secondary plant metabolites,* ed. G. A. Rosenthal and D. H. Janzen, 353–85. New York: Academic Press. **P. 94.**

Rosenzweig, M. L., and Sterner, P. W. 1970. Population ecology of desert rodent communities: Body size and seed-husking as bases for heteromyiid coexistence. *Ecology* 51:217–24. **P. 141.**

Rothschild, M. 1972. Some observations on the relationship between plants, toxic insects and birds. In *Phytochemical ecology,* ed. J. B. Harborne, 1–12. New York: Academic Press. **P. 96.**

Roughgarden, J. 1979. *Theory of population genetics and evolutionary ecology: An introduction.* New York: Macmillan. **P. 187.**

Rowley, I. 1975. *Bird life.* Sydney: Collins. **P. 67.**

Russell, B. 1946. *History of Western philosophy.* London: George Allen and Unwin. **P. 118.**

Ryan, C. A. 1979. Proteinase inhibitors. In *Herbivores: Their interaction with secondary plant metabolites,* ed. G. A. Rosenthal and D. H. Janzen, 599–618. New York: Academic Press. **P. 95.**

Sale, P. F. 1978. Coexistence of coral reef fishes: A lottery for living space. *Env. Biol. Fish.* 3:85–102. **P. 127.**

Sale, P. F., and Dybdahl, R. 1975. Determinants of community structure for coral reef fishes in an experimental habitat. *Ecology* 56:1343–55. **P. 127.**

Sammeta, K. P. V., and Levins, R. 1970. Genetics and ecology. *Ann. Rev. Genet.* 4:469–88. **P. 267.**

Sawicki, R. M., and Farnham, A. W. 1967. Genetics and resistance to insecticides of the

SKA strain of *Musca domestica*. 1. Location of the main factors responsible for the maintenance of high DDT-resistance in diazinon-selected SKA flies. *Ent. Exp. Appl.* 10:253–62. **P. 295.**

Schaffer, W. M. 1974. Optimal reproductive effort in fluctuating environments. *Amer. Nat.* 108:783–90. **P. 264.**

Schaffer, W. M., and Elson, P. F. 1975. The adaptive significance of variations in life history among local populations of Atlantic salmon in North America. *Ecology* 56:577–90. **P. 264.**

Schaller, G. B. 1972. *The Serengeti lion: A study of the predator-prey relations*. Chicago: University of Chicago Press. **P. 119.**

Scheffer, V. B. 1951. The rise and fall of a reindeer herd. *Sci. Month.* 75:356–62. **P. 142.**

Schmidt, P. 1918. Anabiosis of the earthworm. *J. Exp. Zool.* 27:57–72. **P. 58.**

Schmidt-Nielsen, B., and Schmidt-Nielsen, K. 1950. Pulmonary water loss in desert rodents. *Amer. J. Physiol.* 162:31–36. **Pp. 59, 60.**

Schmidt-Nielsen, K. 1964. *Desert animals: Physiological problems of heat and water*. London: Oxford University Press. **Pp. 56, 57, 58, 62.**

―――― . 1975. *Animal physiology, adaptation and environment*. Cambridge: Cambridge University Press. **Pp. 56, 60.**

Schroder, G. D., and Rosenzweig, M. L. 1975. Perturbation analysis of competition and overlap in habitat utilization between *Dipodomys ordii* and *Dipodomys merriami*. *Oecologia* (Berlin) 19:9–28. **P. 147.**

Scott, J. A. 1968. Hilltopping as a mating mechanism to aid the survival of low density species. *J. Res. Lepid.* 7:191–204. **P. 72.**

―――― . 1974. Mate-locating behavior of butterflies. *Amer. Mid. Nat.* 91:103–16. **P. 72.**

Selander, R. K. 1970. Behavior and genetic variation in natural populations. *Amer. Zool.* 10:53–66. **P. 262.**

Serventy, D. L. 1963. Egg-laying timetable of the slender billed shearwater, *Puffinis tenuirostris*. *Proc. 13th Int. Ornithol. Congr.*, 338–43. **P. 65.**

―――― . 1967. Aspects of the population ecology of the short-tailed shearwater, *Puffinis tenuirostris*. *Proc. 14th Int. Ornithol. Congr.*, 165–90. **P. 64.**

Service, M. W. 1964. Deildrin resistance in *Anopheles funestris* Giles from an unsprayed area in northern Nigeria. *J. Trop. Med. Hyg.* 67:190. **P. 296.**

Service, M. W., and Davidson, G. 1964. A high incidence of deildrin-resistance in *Anopheles gambiae* Giles from an unsprayed area in northern Nigeria. *Nature* 203:209–10. **P. 296.**

Shepard, P. S. 1973. *The tender carnivore and the sacred game*. New York: Charles Scribner's Sons. **P. 445.**

―――― . 1978. *Thinking animals: Animals and the development of human intelligence*. New York: Viking Press. **P. 445.**

Sheppard, P. M. 1951. Fluctuations in the selective value of certain phenotypes in the polymorphic land snail, *Cepaea nemoralis* (L.). *Heredity* 5:125–34. **Pp. 291, 293.**

Sherrington, C. 1955. *Man on his nature*. Harmondsworth, England: Penguin Books. **P. 188.**

Shields, O. 1967. Hilltopping. J. Res. Lepid. 6:69–178. **P. 72.**

Shorey, H. H. 1976. *Animal communication by pheromones*. New York: Academic Press. **Pp. 76, 77, 78.**

―――― . 1977. Manipulation of insect pests of agricultural crops. In *Chemical control of insect behavior: Theory and application*, ed. H. H. Shorey and J. J. McKelvey, 353–67. New York: John Wiley. **P. 78.**

Short, R. V. 1976. The evolution of human reproduction. *Proc. Roy. Soc. Lond.*, ser. B, 195:3–24. **Pp. 441, 443, 444, 445.**

Sigurjonsdottir, H., and Reynoldson, T. B. 1977. An experimental study of competition between triclad species (Turbellaria) using the de Wit model. *Acta Zool. Fennica* 154:89–104. **Pp. 238, 241.**

Simberloff, D. 1981. What makes a good island colonist? In *Insect life history patterns: Habitat and geographic variation,* ed. R. F. Denno and H. Dingle, 195–205. New York: Springer-Verlag. **P. 210.**

Sinclair, A. R. E. 1977. *The African buffalo: A study of resource limitation of populations.* Chicago: University of Chicago Press. **Pp. 26, 27.**

———. 1979. The eruption of the ruminants. In *Serengeti: Dynamics of an ecosystem,* ed. A. R. E. Sinclair and M. Norton-Griffiths, 82–103. Chicago: University of Chicago Press. **P. 146.**

Singer, M. C. 1971. Evolution of food-plant preference in the butterfly *Euphydryas editha. Evolution* 25:383–89. **P. 173.**

Slatyer, R. O. 1962. Climate of the Alice Springs area. In *Lands of the Alice Springs area, Northern Territory, 1956–7,* Melbourne: CSIRO. 109–28. Land Research Series, no. 6. **P. 347.**

Smith, F. E. 1961. Density-dependence in the Australian thrips. *Ecology* 42:403–7. **P. 215.**

Smith, H. S. 1935. The rôle of biotic factors in the determination of population densities. *J. Econ. Entomol.* 28:873–98. **P. 71.**

———. 1939. Insect populations in relation to biological control. *Ecol. Monogr.* 9:311–20. **P. 52.**

———. 1941. Racial segregation in insect populations and its significance in applied entomology. *J. Econ. Entomol.* 34:1–13. **Pp. 295, 296.**

Smith, P. H. 1979. Genetic manipulation of the circadian clock's timing of sexual behaviour in the Queensland fruit flies, *Dacus tryoni* and *Dacus neohumeralis. Physiol. Entomol.* 4:71–78. **Pp. 74, 75.**

Snell, T. W., and King, C. E. 1977. Lifespan and fecundity patterns in rotifers: The cost of reproduction. *Evolution* 31:882–90. **Pp. 305, 306.**

Snyder, J. D., and Janke, R. A. 1976. Impact of moose browsing on boreal-type forests of Isle Royale National Park. *Amer. Midl. Nat.* 95:79–92. **P. 89.**

Solomon, M. E. 1949. The natural control of animal populations. *J. Anim. Ecol.* 18:1–35. **P. 98.**

Southern, H. N. 1979. The stability and instability of small mammal populations. In *Ecology of small mammals,* ed. D. M. Stoddart, 103–34. London: Chapman and Hall. **P. 207.**

Southwood, T. R. E. 1966. *Ecological methods with particular reference to the study of insect populations.* London: Methuen. **P. 162.**

Southwood, T. R. E., and Comins, H. N. 1976. A synoptic population model. *J. Anim. Ecol.* 45:949–65. **P. 187.**

Sower, L. L.; Gaston, L. K.; and Shorey, H. H. 1971. Sex pheromones of noctuid moths. 26. Female release rate, male response threshold and communication distance for *Trichoplusia ni. Ann. Entomol. Soc. Amer.* 64:1448–56. **P. 76.**

Specht, R. L., and Rayson, P. 1957. Dark Island Heath (Ninety Mile Plain, South Australia). 1. Definition of ecosystem. *Aust. J. Bot.* 5:52–85. **P. 251.**

Spiess, E. B. 1950. Experimental populations of *Drosophila persimilis* from an altitudinal transect of the Sierra Nevada. *Evolution* 4:14–33. **P. 279.**

Spight, T. M. 1974. Sizes of populations of a marine snail. *Ecology* 55:712–29. **P. 174.**

Stalker, H. D. 1980. Chromosome studies in wild populations of *Drosophila melanogaster.* 2. Relationship of inversion frequencies to latitude, season, wing-loading and flight activity. *Genetics* 95:211–23. **P. 280.**

Stearns, S. C. 1976. Life-history tactics: A review of the ideas. *Quart. Rev. Biol.* 51:3–47. **P. 298.**

———. 1977. The evolution of life-history traits: A critique of the theory and a review of the data. *Ann. Rev. Ecol. Syst.* 8:145–71. **P. 298.**

————. 1980 A new view of life-history evolution. *Oikos* 35:266–81. **Pp. 298, 305.**

Stiles, F. G., and Wolf, L. L. 1970. Hummingbird territoriality at a tropical flowering tree. *Auk* 87:467–91. **P. 128.**

Stimson, J. 1970. Territorial behaviour of the owl limpet *Lottia gigantea*. *Ecology* 51:113–18. **P. 126.**

Stoddart, E., and Myers, K. 1966. The effects of different foods on confined populations of wild rabbits, *Oryctolagus cuniculus* (L.). *CSIRO Wildl. Res.* 11:111–24. **Pp. 320, 322, 323.**

Strong, D. R. 1981. The possibility of insect communities without competition: Hispine beetles on *Heliconia*. In *Insect life history patterns: Habitat and geographic variation,* ed. R. F. Denno and H. Dingle, 183–94. Berlin: Springer-Verlag. **Pp. 139, 210.**

Swaine, J. M., and Craighead, F. C. 1924. Studies on the spruce budworm. *Can. Dept. Agric. Tech. Bull.,* no. 37. **P. 410.**

Tabashnik, B. E.; Wheelock, H.; Rainbolt, J. D.; and Watt, W. B. 1981. Individual variation in oviposition preference in the butterfly, *Colias eurytheme*. *Oecologia* (Berlin) 50:225–30. **P. 285.**

Tahvanainen, J. O., and Root, R. B. 1972. The influence of vegetational diversity on the population ecology of a specialized herbivore, *Phyllotreta cruciferae* (Coleoptera: Chrysomelidae). *Oecologia* (Berlin) 10:321–46. **P. 102.**

Talbot, L. M., and Talbot, M. H. 1963. The wildebeest in western Masailand. *Wildl. Mongr.* 12:1–84. **P. 73.**

Tallamy, D. W., and Denno, R. F. 1981. Alternative life history patterns in risky environments: An example from lacebugs. In *Insect life history patterns: Habitat and geographic variation,* ed. R. F. Denno and H. Dingle, 129–47. Berlin: Springer-Verlag. **P. 305.**

Tanner, J. M. 1962. *Growth at adolescence.* 2d ed. Oxford: Blackwell Scientific Publications. **Pp. 442, 443.**

Taylor, C. E., and Mitton, J. B. 1974. Multivariate analysis of genetic variation. *Genetics* 76:575–85. **P. 281.**

Teas, H. J. 1967. Cycasin synthesis in *Seirarctica echo* (Lepidoptera) larvae fed methyl-azoxymethanol. *Biochem. Biophys. Res. Commun.* 26:686–90. **P. 93.**

Thompson, W. R. 1943. *A catalogue of the parasites and predators of insect pests.* London: Imperial Parasite Service. **P. 114.**

Thorpe, W. H. 1939. Further studies on pre-imaginal olfactory conditioning in insects. *Proc. Roy. Soc. Lond.,* ser. B, 127:424–33. **P. 285.**

Thresher, R. E. 1976. Field analysis of the territoriality of the threespot damselfish *Eupomacentrus planifrons* (Pomacentridae). *Copeia* 1976:266–76. **P. 127.**

Tinbergen, L. 1960. The natural control of insects in pinewoods. 1. Factors influencing the intensity of predation by songbirds. *Archs. Néerl. Zool.* 13:265–336. **P. 99.**

Tobias, P. V. 1981. The emergence of man in Africa and beyond. *Phil. Trans. Roy. Soc. Lond.,* ser. B, 292:43–56. **P. 436.**

Tothill, J. D.; Taylor, T. H. C.; and Payne, R. W. 1930. *The coconut moth in Fiji.* London: Imperial Bureau of Entomology. **P. 254.**

Tucker, H. A., and Ringer, R. K. 1982. Controlled photoperiodic environments for food animals. *Science* 216:1381–86. **P. 65.**

Tychsen, P. H. 1978. The effect of photoperiod on the circadian rhythm of mating responsiveness in the fruit fly, *Dacus tryoni*. *Physiol. Entomol.* 3:65–69. **Pp. 74, 75.**

Tychsen, P. H., and Fletcher, B. S. 1971. Studies on the rhythm of mating in the Queensland fruit fly *Dacus tryoni*. *J. Insect Physiol.* 17:2139–56. **P. 74.**

Ullyett, G. C. 1950. Competition for food and allied phenomena in sheep-blowfly populations. *Phil. Trans. Roy. Soc. Lond.,* ser. B, 234:77–174. **P. 50.**

Underwood, A. J. 1978. An experimental evaluation of competition between three species of intertidal prosobranch gastropods. *Oecologia* (Berlin) 33:185–202. **P. 135.**

———. 1980. The effects of grazing by gastropods and physical factors on the upper limits of distribution of intertidal macroalgae. *Oecologia* (Berlin) 46:201–13. **P. 28.**

———. 1981. Structure of a rocky intertidal community in New South Wales: Patterns of vertical distribution and seasonal changes. *J. Exp. Mar. Biol. Ecol.* 51:57–85. **P. 29.**

Underwood, A. J.; Denley, E. J.; and Moran, M. J. 1983. Experimental analyses of the structure and dynamics of mid-shore rocky intertidal communities in New South Wales. *Oecologia* (Berlin) 56:202–19. **P. 28.**

Underwood, A. J., and Jernakoff, P. 1981. Effects of interactions between algae and grazing gastropods on the structure of a low-shore intertidal algal community. *Oecologia* (Berlin) 48:221–33. **P. 29.**

Utida, S. 1957. Cyclic fluctuations of population density intrinsic to the host-parasite system. *Ecology* 38:442–49. **P. 256.**

Varley G. C.; Gradwell, G. R.; and Hassell, M. P. 1973. *Insect population ecology*. Oxford: Blackwell Scientific Publications. **P. 133.**

Virnstein, R. W. 1977. The importance of predation by crabs and fishes on benthic infauna in Chesapeake Bay. *Ecology* 58:1199–217. **P. 132.**

Vita, C.; Cirio, U.; Fedeli, E.; and Jacini, C. 1977. L'uso di sostanze naturali presenti nell'oliva come prospettiva di lotta contro il *Dacus oleae* (Gmel.). *Boll. Lab. Ent. Agr. Portici* 34:55–61. **P. 432.**

Vogt, W. G. 1977. A re-evaluation of introgression between *Dacus tryoni* and *Dacus neohumeralis* (Diptera: Tephritidae). *Aust. J. Zool.* 25:59–69. **P. 425.**

Vogt, W. G., and McPherson, D. G. 1972. The weighted separation index: A multivariate technique for separating members of closely-related species using qualitative differences. *Syst. Zool.* 21:187–98. **P. 425.**

Vogt, W. G., and Woodburn, T. L. 1979. Ecology distribution and importance of sheep myiasis flies in Australia. In *National Symposium on Sheep Blowfly and Flystrike in Sheep*, 23–32. Sydney: New South Wales Department of Agriculture. **P. 72.**

Volterra, V. 1931. Leçons sur la théorie mathématique de la lutte pour la vie. Cahiers Scientifiques, vol. 7 Paris: Gauthiers-Villars. **P. 225.**

Waddington, C. H. 1960. *The ethical animal*. London: Allen & Unwin. **P. 435.**

Waloff, Z. 1946. Seasonal breeding and migrations of the desert locust (*Schistocerca gregaria* Forsk.) in eastern Africa. *Mem. Anti-locust Res. Centre London*, vol. 1. **P. 73.**

Walsh, J. J. 1981. A carbon budget for overfishing off Peru. *Nature* 290:300–304. **P. 406.**

Walters, S. S. 1966. The conservation of water in starving *Tenebrio molitor*, L. Ph.D. thesis, University of Adelaide. **P. 61.**

Ward, B. L.; Starmer, W. T.; Russell, J. S.; and Heed, W. B. 1974. The correlation of climate and host plant morphology with a geographic gradient of an inversion polymorphism in *Drosophila pachea*. *Evolution*. 28:565–75. **P. 278.**

Washburn, S. L. 1978. The evolution of man. *Sci. Amer.* 239(3):146–54. **P. 436.**

Waterhouse, D. F. 1947. The relative importance of live sheep and of carrion as breeding grounds for the Australian sheep blowfly *Lucilia cuprina*. *Bull. Coun. Sci. Indust. Res. Aust.* no. 217. **P. 134.**

Way, M. J. 1953. The relationship between certain ant species with particular reference to biological control of the coreid, *Theraptus* sp. *Bul. Ent. Res.* 44:669–91. **P. 125.**

Wellington, W. G. 1949a. The effects of temperature and moisture upon the behaviour of the spruce budworm, *Choristoneura fumiferana* Clemens (Lepidoptera: Tortricidae). 1. The relative importance of graded temperatures and rates of evaporation in producing aggregations of larvae. *Sci. Agric.* 29:201–15. **P. 414.**

———. 1949b. The effects of temperature and moisture upon the behaviour of the spruce budworm, *Choristoneura fumiferana* Clemens (Lepidoptera: Tortricidae). 2. The responses of larvae to gradients of evaporation. *Sci. Agric.* 29:216–29. **P. 414.**

————. 1952. Air mass climatology of Ontario north of Lake Huron and Lake Superior before outbreaks of the spruce budworm, *Choristoneura fumiferana* Clemens (Lepidoptera: Tortricidae), and the forest tent caterpillar, *Malacosoma disstria* Hubn. *Can. J. Zool.* 30:114–27. **P. 411.**

Wellington, W. G.; Fettes, J. J.; Turner, K. B.; and Belyea, R. M. 1950. Physical and biological indicators of the development of outbreaks of the spruce budworm, *Choristoneura fumiferana* Clem. (Lepidoptera: Tortricidae). *Can. J. Res., ser. D,* 28:308–31. **Pp. 410, 414, 415, 416, 417.**

White, T. C. R. 1966. Food and outbreaks of phytophagous insects. Ph.D. thesis, University of Adelaide. **P. 50.**

————. 1969. An index to measure weather-induced stress of trees associated with outbreaks of psyllids in Australia. *Ecology* 50:905–9. **Pp. 50, 51, 248, 249, 252.**

————. 1970a. Some aspects of the life history, host selection, dispersal and oviposition of adult *Cardiaspina densitexta* (Homoptera: Psyllidae). *Aust. J. Zool.* 18:105–17. **P. 248.**

————. 1970b. The nymphal stage of *Cardiaspina densitexta* (Homoptera: Psyllidae) on leaves of *Eucalyptus fasciculosa.* *Aust. J. Zool.* 18:273–93. **P. 248.**

————. 1970c. The distribution and abundance of pink gum in Australia. *Aust. Forestry* 34:11–18. **Pp. 25, 250.**

————. 1971. Lerp insects (Homoptera: Psyllidae) on red gum (*E. camaldulensis*) in South Australia. *S. Aust. Nat.* 46:20–23. **P. 248.**

————. 1973. The establishment, spread and host range of *Paropsis charybdis* Stal. (Chrysomelidae) in New Zealand. *Pacific Insects* 15:59–66. **P. 248.**

————. 1974. A hypothesis to explain outbreaks of looper caterpillars, with special reference to populations of *Selidosema suavis* in a plantation of *Pinus radiata* in New Zealand. *Oecologia* (Berlin) 16:279–301. **Pp. 248, 254, 417, 422.**

————. 1976. Weather, food and plagues of locusts. *Oecologia* (Berlin) 22:119–34. **P. 248.**

————. 1978. The importance of a relative shortage of food in animal ecology. *Oecologia* (Berlin) 33:71–86. **Pp. 51, 248, 417, 418.**

Whitehead, A. N. 1926. *Science and the modern world.* New York: Macmillan. **P. 149.**

Whittaker, R. H., and Feeny, P. O. 1971. Allelochemics: Chemical interactions between species. *Science* 171:757–70. **P. 93.**

Wiens, J. A. 1977. On competition and variable environments. *Amer. Sci.* 65:590–97. **Pp. 141, 210.**

————. 1983a. On understanding a nonequilibrium world: Myth and reality in community patterns and processes. In *Ecological communities: Conceptual issues and the evidence,* ed. D. R. Strong, Jr., and D. S. Simberloff, and L. G. Abele. Princeton: Princeton University Press. Forthcoming. **Pp. 147, 210.**

————. 1983b. Avian community ecology: An iconoclastic view. In *Perspectives in ornithology,* ed. A. H. Brush and G. A. Clark, Jr. Cambridge: Cambridge University Press. Forthcoming. **Pp. 147, 210.**

Wiens, J. A., and Rotenberry, J. T. 1980. Patterns of morphology and ecology in grassland and shrubsteppe bird populations. *Ecol. Monogr.* 50:287–308. **P. 210.**

Williams, C. B. 1944. Some applications of the logarithmic series and the index of diversity to ecological problems. *J. Ecol.* 32:1–44. **P. 140.**

————. 1947. The generic relations of species in small ecological communities. *J. Anim. Ecol.* 16:11–18. **Pp. 139, 140, 223.**

————. 1951. Intra-generic competition as illustrated by Moreau's records of East African bird communities. *J. Anim. Ecol.* 20:246–53. **Pp. 139, 140.**

Williams, C. M. 1970 Hormonal interactions between plants and insects. In *Chemical ecology,* ed. E. Sondheimer and J. B. Simeone., 103–32. New York: Academic Press. **P. 94.**

Williams, R. T.; Fullagar, P. J.; Davey, C. C.; and Kogon, C. 1972. Factors affecting the survival time of rabbits in a winter epizootic of myxomatosis at Canberra. *J. Appl. Ecol.* 9:399–410. **Pp. 356, 362.**

Williams, R. T.; Fullagar, P. J.; Kogon, C.; and Davey, C. 1973. Observations on a naturally occurring winter epizootic of myxomatosis at Canberra, Australia, in the presence of rabbit fleas (*Spilopsyllus cuniculi* Dale) and virulent myxoma virus. *J. Appl. Ecol.* 10:417–27. **Pp. 356, 362.**

Williams, R. T., and Parer, I. 1972. The status of myxomatosis at Urana, New South Wales, from 1968 until 1971. *Aust. J. Zool.* 20:391–404. **Pp. 356, 362.**

Willis, E. O. 1966. Interspecific competition and the foraging behavior of plain-brown woodcreepers. *Ecology* 47:667–72. **P. 136.**

Wilson, E. O. 1951. Variation and adaptation in the imported fire ant. *Evolution* 5:68–79. **P. 125.**

———. 1971. *The insect societies.* Cambridge: Belknap Press of Harvard University Press. **P. 125.**

Wilson, F. 1961. Adult reproductive behaviour in *Asolcus basalis* (Hymenoptera: Scelionidae). *Aust. J. Zool.* 9:737–51. **P. 46.**

Wittenberger, J. F., and Tilson, R. L. 1980. The evolution of monogamy: Hypotheses and evidence. *Ann. Rev. Ecol. Syst.* 1:197–232. **P. 297.**

Wratten, S. D. 1973. The effectiveness of the coccinelid beetle, *Adalia bipunctata* (L.), as a predator of the lime aphid, *Eucallipterus tiliae* L. *J. Anim. Ecol.* 42:785–802. **Pp. 98, 114.**

Wright, S. 1977. *Evolution and the genetics of populations.* Vol. 3. *Experimental results and evolutionary deductions.* Chicago: University of Chicago Press. **P. 262.**

———. 1978. *Evolution and the genetics of populations.* Vol. 4. *Variability within and among natural populations.* Chicago: University of Chicago Press. **P. 262.**

Wright, S., and Dobzhansky, Th. 1946. Genetics of natural populations. 12. Experimental reproduction of some of the changes caused by natural selection in certain populations of *Drosophila pseudoobscura. Genetics* 31:125–56. **Pp. 273, 275, 276, 277.**

Yeates, N. T. M. 1949. The breeding season of the sheep with particular reference to its modification by artificial means using light. *J. Agr. Sci.* 39:1–43. **P. 65.**

# Subject Index